BIOLOGICAL SCIENCE

SIXTH EDITION

JAMES L. GOULD
WILLIAM T. KEETON
WITH CAROL GRANT GOULD

BIOLOGICAL SCIENCE

VOLUME ONE

SIXTH EDITION

W.W. NORTON & COMPANY NEW YORK LONDON

This book is composed in New Aster, Composition by New England Typographic
Service, Inc. Manufacturing by R. R. Donnelley & Sons, Company. Book design by
Martin Lubin. Cover illustration: Wendell Minor.

SIXTH EDITION

The Library of Congress has cataloged the one-volume edition as follows:

Gould, James L., 1945–
    Biological science / by James L. Gould, William T. Keeton with Carol Grant
Gould.—6th ed.
      p.   cm.
    Keeton's name appears first on the earlier edition.
    Includes bibliographical references and index.
    ISBN 0-393-96920-7
    1. Biology.   I. Keeton, William T.   II. Gould, Carol Grant.   III. Title.
  QH308.2.G69   1996
  574—dc20   95-44962

  ISBN 0-393-96948-7 (Vol 1, pbk.)

W. W. Norton & Company, Inc., 500 Fifth Avenue, New York, N.Y. 10110
W. W. Norton & Company, Ltd., 10 Coptic Street, London WC1A 1PU

1 2 3 4 5 6 7 8 9 0

■ ABOUT THE COVER

Cover design and painting by Wendell Minor, based on photographs by Peter T. Boag, Queens University

The painting on the cover illustrates a central theme of modern biology: the evolution of species by natural selection. The imaginary scene is based on the work of Peter and Rosemary Grant of Princeton University and their colleagues, whose long-term study of natural selection in Darwin's finches has been called "a modern classic in the study of evolution in action" (by Jonathan Weiner, author of *The Beak of the Finch*, winner of the 1995 Pulitzer Prize for nonfiction). The researchers, who have spent extended periods every year for more than two decades observing the finches in their native habitat in the Galapágos Islands, have noted that changes in the supply of seeds available as food can produce measurable changes in beak size (an inherited trait) from one generation to the next—a much shorter time than Darwin's original field observations had suggested. In one particularly striking case, represented here, the sequence of a typical wet year (left) followed by an unusually dry one (right) resulted in an evident change not only in the island's vegetation but also in the beaks of successive generations of the medium-sized ground finch *Geospiza fortis*. Normally, the finches prefer to eat small, soft seeds, altering the proportion of small and large seeds left on the ground. In the drought year, few or no new seeds were produced, leaving only the previous years' larger, harder seeds for food. Birds with more massive beaks, such as the one on the right, were better adapted to break these tougher seeds open, and hence were more likely to survive, while smaller-beaked birds such as the one on the left perished. The population evolved when the offspring of these survivors inherited the large-beak trait from their parents. The evolutionary significance of these findings is discussed further in the "How We Know" box on page 506.

CHAPTER 27

WHAT HETEROTROPHS NEED TO EAT

WHY ISN'T SUGAR ENOUGH?

WHAT VITAMINS DO

WHY MINERALS ARE NECESSARY

THE UNIQUE FEEDING STYLE OF FUNGI

INTERNALIZED DIGESTION: ANIMALS AND PROTOZOANS

HOW PROTOZOANS CAPTURE AND DIGEST FOOD

THE ANIMAL AS STOMACH: COELENTERATES

THE ALL-PERVADING STOMACH OF FLATWORMS

DISASSEMBLY LINES: THE COMPLETE DIGESTIVE TRACT

HOW THE VERTEBRATE DIGESTIVE SYSTEM WORKS

HOW FOOD IS DIGESTED

HOW CELLS HARVEST ENERGY ■ 173

now move through the cytochrome series (left) to cytochrome a_3. This last enzyme group uses the energy of the electrons to split molecular oxygen (O_2) and catalyze the following reaction:

$$1/2 O_2 + 4H^+ + 2e^- \xrightarrow{\text{cytochrome } a_3} H_2O + 2H^+$$

This reaction reduces by four the number of hydrogen ions in the inner compartment. Two of these hydrogen ions are exported to the outer compartment, while the other two are incorporated into the water molecule. About 100 electrons can pass through each transport chain each second.

Chemiosmotic synthesis (Stage V of aerobic respiration) All the energy stored in the NAD_{re} has now been used up without generating any new ATP. However, an electrostatic and osmotic concentration gradient has built up across the inner membrane: the inner compartment, having lost H^+, has become negatively charged, while the outer compartment—the space between the inner and outer membrane—is filled with H^+ ions and therefore is positively charged. The result is a molecular battery that uses the difference in charge across the inner mitochondrial membrane to make ATP from ADP. Most of the H^+ ions return from the outer to the inner compartment by way of the F_1 complex (Fig. 6.12), passing down a *chemiosmotic gradient*—the combined osmotic and electrostatic gradients—is harvested (Figure 6.13). The abundant energy generated by this elegant system is stored in the phosphate groups of ATP. The same H^+ gradient probably also powers the pump that exports ATP to the cytosol. Peter Mitchell won the Nobel Prize in 1978 for working out this indirect route by which mitochondria convert the energy of NAD_{re} into ATP.

It is a continuing challenge to explain how such efficient, highly refined examples of biological engineering as the electron-transport chain might have evolved. How do the first elements in such a complicated array, which seem useless without the rest, ever come into existence? To answer the question, biologists look at a wide variety of species for preadaptations—fragments of the system as we know it now that are employed in other ways and which might have been present in ancestral organisms and available to natural selection for "remodeling."

Consider the most intricate element of the mitochondrial system, the F_1 complex, which uses the H^+ gradient to make ATP. Relict species of anaerobic bacteria living today still lack an electron-transport chain but nevertheless have an F_1 assembly that works in reverse. These anaerobes burn ATP to pump H^+ out of their cells, eliminating the acidic by-products of fermentation. There is similar evidence to suggest that the cytochrome complexes evolved later in ancient bacteria, where they served similar detoxification functions. Natural selection must then have favored combining these various systems into a rudimentary electron-transport chain using sulfur, carbon, or nitrogen as the ultimate electron acceptor.

Summary of respiration energetics Now that we have looked at all five stages in the aerobic breakdown of glucose, we can summarize their combined energy yield. The overall flow of energy in aer-

6.13 Operation of the F1 complex
The F_1 complex is composed of at least 18 subunits. The core group carries the H^+ to one of three protein pairs of the head region where ATP is formed; each of the three pairs is undergoing one of the three steps in the cycle: ADP and phosphate binding, ATP formation, and ATP release. Many researchers believe that the head rotates during ATP synthesis.

506 ■ CHAPTER 18 SPECIATION AND PHYLOGENY

HOW WE KNOW:
THE BEAK OF THE FINCH

In order to trace evolution as it occurs, researchers must know the distribution of heritable phenotypes in a population, and then monitor any changes as they occur. In the case of the Galapagos finches, this means measuring every size parameter of every bird on the island under study (Fig. A), marking each bird individually, and then recapturing every bird annually both to monitor survivorship and to add each year's crop of youngsters to the inventory. The extraordinary difficulty of such an undertaking probably explains why few researchers have been able to track evolution generation by generation. Let's look at some of the data Peter Grant and his colleagues collected, and how they were thus able to document one bout of evolution.

Two basic requirements for evolution are in place: there is variation in beak parameters and it is heritable: the offspring of parents with large beaks tend to have large beaks themselves (Fig. B). The question Grant sought to answer was

whether beak size is subject to selection.

In 1976, before a drought, the beaks of the 741 resident medium ground finches on the island of Daphne Major had a mean depth of about 9.4 mm (Fig. C, blue curve). The range of depths reflects the range of seeds available as food. Thus birds with smaller beaks were better able to harvest smaller seeds, whereas larger or harder seeds were more readily opened by finches with larger beaks.

The drought led to a threefold reduction in seed number. The seeds that remained were 50% larger and harder than normal. Selection strongly favored larger beak size: only 9% of the birds with average beaks survived, whereas 20% of the individuals with beaks only 1 mm deeper lived through the crisis, as did 40% of the finches with beaks 2 mm deeper. The result was a upward shift in the mean beak depth to 10.2 mm (Fig. C, red curve); the only hatchling to survive had an unusually deep beak.

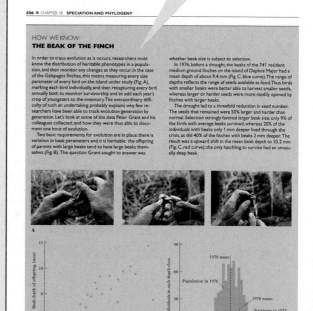

ANIMAL NUTRITION AND DIGESTION

Unlike plants and other autotrophs, heterotrophs cannot synthesize high-energy compounds from low-energy inorganic raw materials. To survive, they must feed on the high-energy molecules of other organisms. There are four main groups of heterotrophs: nonphotosynthetic bacteria, fungi, nonphotosynthetic protozoans, and animals. This chapter focuses on how animals obtain the nutrients they need.

Bacteria and fungi lack internal digestive systems and hence depend mainly on absorption for feeding. They are usually either *saprophytic* (living and feeding on dead organic matter) or *parasitic* (living on or in other organisms and feeding on them). By contrast, the principal mode of feeding for animals and protozoans is ingestion—taking in and digesting particulate or bulk food. Animals and protozoans may be *herbivores*, in which case they obtain high-energy compounds by eating plants or other photosynthesizers, or they may be *carnivores*, eating animals or protozoans that have fed on plants. Some animals, the *omnivores*, need both plant and animal material to survive. Much of the morphological diversity among living things reflects their different ways of employing the three major modes of nutrition—photosynthesis, absorption, and ingestion.

WHAT HETEROTROPHS NEED TO EAT

■ WHY ISN'T SUGAR ENOUGH?

Carbohydrates, fats, and proteins are the main energy sources for heterotrophs. Of these, carbohydrates alone would suffice if organic nutrients functioned only as an energy source. In fact, adult bees and hummingbirds can survive solely on carbohydrate-rich nectar until they attempt to reproduce (Fig. 27.1). Because all amino acids and nucleotides incorporate nitrogen, which carbohydrates lack, reproduction and growth require a source of fixed nitrogen. Many bacteria and fungi, as well as some protozoans, can flourish and reproduce on a diet consisting solely of carbohydrates, fixed nitrogen, and minerals; like plants, they can synthesize for themselves all the other classes of compounds needed for life.

Animals are especially deficient in synthetic ability; among their

27.1 A honey bee gathering pollen and nectar

Adult worker honey bees can survive on the sugars in the nectar they collect from plants; the pollen (which is packed into pollen baskets on the rear legs) is used as a protein source, and is fed to growing larvae and the egg-laying queen.

761

758 ■ CHAPTER 26 PLANT NUTRITION

long stiff teeth line the margin of each lobe. When an insect touches small sensitive hairs on the surface of the leaf, the lobes quickly change shape and come together with their teeth interlocked. When the trigger hairs are stimulated, they initiate a rapid pumping of H⁺ ions through the cell membrane, consuming about 30% of the cells' available ATP in 2 sec. The pH change in turn leads to a quick osmotic movement of water from the intercellular spaces into the cells at the base of the trap; their consequent rapid enlargement causes the leaf to close on its victim. The trapped animal is then digested by enzymes secreted from glands on the leaf surface, and the resulting amino acids are absorbed.

The leaves of sundews (Fig. 26.18) bear numerous hairlike tentacles, each with a gland at its tip. Small insects, attracted by the plant's odor, become trapped in the sticky fluid secreted by the glands. The stimulus from a trapped insect causes nearby tentacles to bend over toward the animal, further entangling it. The proteins of the insect are digested by enzymes, and the amino acids are absorbed.

26.18 Leaf of sundew (*Drosera intermedia*)

A damselfly is caught in the sticky fluid on the ends of the glandular hairs.

CHAPTER SUMMARY

AUTOTROPHS VERSUS HETEROTROPHS
Autotrophs can synthesize organic compounds from inorganic materials; heterotrophs require energy-rich organic molecules in their diet. (p. 743)

WHAT PLANTS ARE MADE OF
EARTH, AIR, OR WATER? The carbon and oxygen—together the elements that contribute most to a plant's mass of organic molecules—come from CO_2. The CO_2 comes from the air, most of the hydrogen comes from water, while the minerals used by plants mostly come from the earth. (p. 744)

WHY PLANTS NEED MINERALS Minerals are needed in amino acids and nucleic acids, at the active sites of many enzymes, and as components of coenzymes. (p. 745)

HOW PLANTS FEED
HOW ROOTS WORK Water containing minerals enters roots through root hairs or along cell walls; it crosses through endodermal cells, and then enters the xylem. The water usually enters because of a water-potential gradient generated by the evaporation of water from leaves. (p. 747)

WHERE PLANTS GET THEIR NITROGEN Nitrogen is obtained either as a dissolved soil compound (e.g., NH_3) or from symbiotic nitrogen-fixing bacteria, which convert N_2 into NH_3. (p. 755)

WHY DO SOME PLANTS EAT ANIMALS? Most often to obtain nitrogen that is lacking in the soil. (p. 756)

HOW THE LIGHT REACTIONS STORE LIGHT ENERGY ■ 185

A   **B**

7.7 How chlorophyll is mounted in the membrane
The antennas are organized as five to ten protein-bound units called light-harvesting complexes (LHC). Each LHC includes three subunits, viewed here from the stromal side of the membrane (A) and from the plane of the membrane (B). Each of these subunits, in turn, has approximately seven molecules of chlorophyll *a* (blue), five molecules of chlorophyll *b* (green), and several carotenoids (red); all are held in a protein matrix (tan). (The tails of the chlorophylls have been omitted for clarity.)

plants, *c* in brown algae, and *d* in red algae), and carotenoids. One group of four pigment molecules in each unit is distinct from all the rest; it is part of a complex that acts as a *reaction center* where energy conversion occurs. The other pigment molecules function somewhat like antennas for collecting light energy and transferring it to the reaction center.

When a photon strikes a chlorophyll (or a carotenoid) molecule and is absorbed, its energy is transferred to an electron of the pigment molecule; the excited electron moves up to a higher, relatively unstable energy level (Fig. 7.8). Not just any photon will do: because electrons occupy discrete energy levels, the photon must have a particular amount of energy to excite the pigment electron from its ground state to the higher level. This explains the well-

7.8 Effect of light on chlorophyll
When a photon is absorbed by a chlorophyll molecule, the photon's energy raises an electron to a higher energy level. This simplified representation shows a typical distribution of electrons at their lowest available energy levels, and the distribution as it is altered by absorption of a photon of red light, which raises one electron from its ground state in the first energy level to its first excited state in the second energy level. As we saw in Figure 7.6A, the absorption spectrum of a particular pigment can have several peaks. These exist because electrons can be excited from their ground state to their second excited state, two energy levels higher. In this illustration, another electron at the first level is shown being excited up to the third level—its second excited state—by a blue photon, which has enough more energy than a red photon that it can lift electrons two levels rather than one.

incoming photons
excited electrons
electrons in ground state
energy levels
nucleus

CONTENTS IN BRIEF

VOLUME ONE

VOLUME TWO

CONTENTS

The entire contents are presented here, but Volume One contains only pages 1–527 plus complete backmatter.

PART I

THE CHEMICAL AND CELLULAR BASIS OF LIFE

PART II
THE PERPETUATION OF LIFE

PART III
EVOLUTIONARY BIOLOGY

CHAPTER 17 ■ VARIATION, SELECTION, AND ADAPTATION 445

CHAPTER 18 ■ SPECIATION AND PHYLOGENY 487

PART IV

THE GENESIS AND DIVERSITY OF ORGANISMS

PART V

THE BIOLOGY OF ORGANISMS

PART VI
ECOLOGY

James Gould, who took over as sole author of *Biological Science* following William Keeton's untimely death in 1980, is professor of biology at Princeton University. A graduate of the California Institute of Technology, where he majored in molecular biology, Jim became interested in animal behavior as a graduate student, obtaining his Ph.D. in ethology from Rockefeller University in 1975. He has since gone on to become an internationally recognized expert on various aspects of animal behavior—communication, navigation, learning and cognition, and sexual selection—all research topics that demand a thorough mastery of physiology, ecology, and evolution.

In short, Jim's own scientific training and special areas of expertise encompass the core contents of all six parts of *Biological Science*.

An enthusiastic user of the classic Keeton text for his introductory biology course at Princeton even before his involvement in the task of updating and revising this book, Jim also teaches a popular undergraduate course in animal behavior there; his first major textbook for W. W. Norton, *Ethology*, was a natural outgrowth of this course. In addition, he is a frequent contributor to both *Science* and *Nature*, and his work has been featured in three BBC Horizon films, which he wrote and narrated: "What Makes an Animal Smart," "Making Sex Pay," and "In the Company of Ants and Bees."

Carol Grant Gould has contributed substantially to the clarity, grace, and educational value of the three editions of *Biological Science* she and Jim have worked on together, as she has on their many other joint projects, including three popular books for the Scientific American Library series (*The Honey Bee, Sexual Selection*, and *The Animal Mind*) as well as numerous coauthored articles for both professional and general-interest publications. Her own background includes a Ph.D. in English literature from New York University and wide-ranging experience as a science writer, editor, and teacher.

Jim and Carol live in Princeton, New Jersey with their two children, both of whom have been intimately involved in the preparation of the software ancillaries that accompany this book. Grant Gould, now an undergraduate at the Massachusetts Institute of Technology, was the coauthor (with his father) of *Norton Presentation Maker*, the versatile lecture creation and presentation tool that is being introduced with this edition. Clare Gould worked on the scanned images, diagrams, and other graphic material found on the CD-ROM that serves as a resource bank for users of *Biological Science*.

A final note: Since Jim Gould has now had primary responsibility for as many editions of this work as his predecessor, the sequence of names in the byline has been reversed with this edition. Henceforth all of us—authors, editors, and longtime users of the book included—will have to get used to the idea of referring to it familiarly not as "Keeton/Gould," but as "Gould/Keeton."

PREFACE

This newest edition of *Biological Science* continues Bill Keeton's vision of biology as an integrated science. In his first edition, published in 1967, Keeton revolutionized the teaching of introductory college biology by combining biochemistry, cell biology, microbiology, botany, zoology, and ecology into a single integrated text, with evolution as its central unifying theme. In time, most other texts followed his lead, putting all of biology between two covers. Even so, few caught the essential spirit of his vision, and to this day most pay little more than lip service to Keeton's theme of "unity in diversity." (Indeed, many continue to segregate plants so completely from animals that students might well be led to believe they have nothing whatever in common!) Outside of the chapters on evolution per se, most other texts simply pass over the many wonderful examples of how natural selection works at all levels to solve the universal problems of living faced by a diverse range of organisms.

In preparing the Sixth Edition we had three main objectives in mind: (1) to update every chapter, both to reflect new discoveries and to respond to the changing needs of students who plan to go on to upper-level courses; (2) to clarify and simplify the presentation wherever possible; (3) to revise existing material and add new examples that help illustrate the *process* of science. We have also made it easier to present the chapters in different sequences, and to teach plant biology separately (while continuing to make illuminating comparisons and contrasts between plant and animal solutions to analogous problems of living).

In the course of the revision, we have actually managed to shorten the text significantly, reversing the seemingly inexorable trend toward greater and greater verbosity in books for the introductory course. The space created by this efficient presentation has allowed us to enlarge the type size for better readability, without substantially increasing the total number of pages. We have also added a topics outline at the beginning of each chapter and a detailed summary at the end (replacing the Concepts for Review section) to help students with their review of the material, and to enable them to find their way around in the book more easily. To emphasize that science is a process and not just a collection of facts, there are many new How We Know boxes, in addition to the popular Exploring Further boxes, introduced in earlier editions. In the same spirit, most of the section headings have been recast to suggest questions and issues rather than conclusions; for easy orientation, the revised headings are repeated verbatim in both the chapter-opening topics outline and the concluding summary.

SOME SPECIFIC CHANGES

We have added much new material on a variety of important topics. These additions are far too numerous to list separately, but here are a few of them:

In Part I (The Chemical and Cellular Basis of Life) we have added discussions of atomic-force microscopy, caveolae, the internal structure of the nucleus, the structure of the F_1 complex in mitochondria, and the structure of the light-harvesting complexes of chloroplast antennas and the associated reaction centers.

In Part II (The Perpetuation of Life) we have included new data on the structure of DNA polymerase, the process of replication in *E. coli*, the tertiary structure of rRNA, the operation of self-splicing introns, the nature of cancer, the operation of the cell cycle, early steps in development, the role of DNA repair in longevity, the biology of AIDS, and the molecular basis of blood groups. In addition, the discussion of recombinant DNA has been greatly expanded. There are new figures and discussions illustrating recombinant transduction, recombinant screening, how recombinant fragments are employed to inactivate viruses, the use of anti-sense DNA, engineered disease resistance in plants, the DNA particle-gun and microinjection techniques, the results of microinjection of growth-hormone genes, and DNA fingerprinting.

In Part III (Evolutionary Biology) we've added material on frequency-dependent selection, directly observed natural selection in populations (the subject of the cover), the role of parasites and parasitoids, intriguing theoretical work on the evolution of the camera eye, as well as the consequences of minigenes and horizontal transmission. There is more on speciation via developmental changes, and field studies of selection and speciation. In particular, we have expanded and revised the section on phylogenetic methods, adding simple examples and figures illustrating and comparing modern phenetic and cladistic techniques.

In Part IV (The Genesis and Diversity of Organisms) we have revised the phylogenetic trees with the most recent and reliable data available. There is now more on prions; the presentation of protists has been reordered to correspond to phylogenetic reality; and the treatment of fungi has been expanded. Human evolution has been brought up to the minute with the latest results on Y-chromosome dating.

In Part V (The Biology of Organisms) we have added material on the role of mycorrhizae, the mechanisms of uptake in intestinal villi, and the effects of plant toxins on digestion. The discussion of the structure and mechanisms of exchange in alveoli has been expanded, as have the presentations on the evolution of lungs, the heart cycle, molecular movements underlying ion-regulation of the fluids of invertebrates, and the life histories of parasitic plants. On the human level, we have added recent work on the feedback control of testosterone levels and menstruation, and the molecular biology of RU 486.

In Part VI (Ecology) we have made numerous revisions, including a new section on the Law of Constant Yield. There is more on biological control of pests, the causes and effects in population growth and population cycles, the interaction between economics and birth rates, and more on the issue of diversity and stability. We have reordered Chapter 40 (Ecosystems) for a more effective presentation, and added discussions of the relationship between latitude and climate (rainfall and temperature) as well as the role of ocean currents, followed by a discussion of Lieth's and other models of world-wide terrestrial and aquatic primary productivity. There is now a more quantitative treatment of energy flow in ecosystems, more on the role of decomposers, more on the effects of human activity on cycles of materials (including the effects of deforestation on local climate and succession), and more on the consequences of polar melting, the effects of erosion, the changing carbon cycle, and the enormous impact of agriculture on the nitrogen cycle.

ANCILLARIES

The Sixth Edition continues to be available in a one-volume hardcover or two-volume paperback format. It is accompanied by a **CD-ROM** that contains not only all of the figures found in the book, but also a variety of animations, user-controlled simulations, and QuickTime movies. The two dozen simulations are particularly effective in illustrating how dynamic processes depend on relevant parameters. Two examples will illustrate the general strategy: One simulation graphs the oxygen-binding curve of hemoglobin for any combination of temperature, pH, and body size; the instructor can change any parameter and plot the new curve in one of several contrasting line styles. A more complex simulation allows the teacher to select any combination of population-growth parameter values (setting the life expectancy, age of first reproduction, number of offspring per female, birth interval, and offspring survival), adjust the distribution of individuals in various age classes in the starting population, and then see the population profile and the total population change as the years pass.

The simulations and animations will be useful in lecture, lab, and student review—indeed, I use the three population-growth simulations as the basis for one of the weekly labs in my course. To encourage the creation of additional simulations and animations, the publisher is sponsoring a competition. Entries (simulations in either HyperCard or ToolBook format; animations in QuickTime for Macintosh or Windows) will be judged twice a year (in January and July). Submissions by teachers and students will be evaluated separately by a panel that includes one of the authors, an editor from W. W. Norton, the developer of the BioXplorer software ancillaries, and another teacher who uses *Biological Science*. Winners will receive their choice of a modestly priced piece of computer hardware (a CD-ROM drive, for instance) or the equivalent value in software, computer supplies, or a generous number of Norton books. Details can be found in the Contest file on the CD-ROM, as well as on the publisher's World Wide Web home page (http://web.wwnorton.com). We will incorporate the best animations and simulations into future editions of the CD ancillary. (We will also create a version of each selected simulation in the other programming format.)

The increasing number of users with LCD-projection facilities will find the remarkable *Norton Presentation Maker* program on the CD of special value. This program was designed by Grant F. Gould and the authors specifically for college teaching. It allows the lecturer to assemble slides, animations, simulations, and movie clips from the CD, plus similar material from other sources, as well as sound files, text, and videodisk stills or clips. These items can be added to a presentation quickly and in any order. It then creates a lecture file capable of displaying these elements either in the original sequence, or in any other desired order. It has many additional useful features too numerous to detail here. It can also be used by instructors to create custom tutorials, incorporating the teacher's own notes and commentary in combination with the material from the CD or the instructor's own digital material.

ACKNOWLEDGMENTS

We are grateful to the many instructors and specialists who carefully and critically reviewed the Fifth Edition and the revised draft of this edition. Our special thanks go to Marvin R. Alvarez, University of South Florida; Ron Ash, Washburn University of Topeka; J. Barber, Imperial College of Science, Technology, and Medicine; Robert Beckmann, North Carolina State University; Annalisa Berta, San Diego State University; Domenic Castignetti, Loyola University; Thomas Cavalier-Smith, University of British Columbia; Edward C. Cox, Princeton University; Barbara Demmig-Adams, University of Colorado; Harold B. Dowse, University of Maine; Paul R. Elliott, Florida State University; Gary L. Galbreath, Northwestern University; Emily Giffin, Wellesley College; Lindsay Goodloe, Cornell University; Peter R. Grant, Princeton University; H. Ernest Hemphill, Syracuse University; Linda Kohn, Erindale College, University of Toronto; William Z. Lidicker, Jr., University of California, Berkeley; Axel Meyer, State University of New York, Stony Brook; Kevin Moses, University of Southern California; Stephen W. Pacala, Princeton University; David J. Patterson, University of Sydney; William Rumbach, Central Florida Community College; Fred Wilt, University of California, Berkeley; Charles Wyttenbach, University of Kansas.

We are also grateful to the many individuals at W. W. Norton who have contributed to this revision. In particular we thank our tireless editors Joe Wisnovsky and Jim Jordan, as well as Stephen King, who did much to facilitate the development of Norton Presentation Maker. We also thank Susan Middleton, our gifted copy editor; Mary T. Kelly, our able and attentive project editor; Rachel Warren, our ever-resourceful editorial assistant and photo researcher; Susan Crooks, our excellent layout artist; and Michael Reingold, Michael Goodman and John McAusland, whose remarkable artwork helps bring these pages to life.

Finally, we appreciate the many thoughtful and helpful comments and criticisms that users of the book provide. Although letters to the publisher and authors continue to be very welcome, we invite you to communicate by e-mail to the senior author (gould@princeton.edu) or the editor (jwisnovsky@wwnorton.com).

JLG
CGG

Princeton
August, 1995

CHAPTER I

SCIENCE AND EVOLUTION

Biology is the science of living things—what they are, how they work, how they interact, and how they change or evolve. Before we begin our exploration of biology, we must be clear about what we mean by life, science, and evolution.

WHAT IS LIFE?

Every day the sun radiates vast amounts of energy to the earth and our moon, and to all the other planets in our solar system. And every day the planets reradiate that energy into space. On earth there is a unique delay: for a brief moment, a minute portion of the sun's energy is trapped and stored (Fig. 1.1). This minor delay in the flow of cosmic energy powers life.

Plants capture sunlight and use its energy to build and maintain stems, leaves, and seeds, while animals secure the energy of sunlight by eating plants or by eating other animals that have eaten plants. Each stage in the many processes of living and dying produces waste heat, which joins the pool of energy being relayed back into space.

Biology is the study of this very special category of energy utilization—it is the science of living things. The world is teeming with life: millions of species of organisms of every description inhabit the earth, feeding directly or indirectly off lifeless sources of energy like the sun's light. What is there in this maze of diversity that unites all biology into one field? What do amoebae, redwoods, and people have in common that differentiates them from the nonliving? What, in short, is life?

Most dictionaries define *life* as the property that distinguishes the living from the dead, and define *dead* as being deprived of life. These singularly circular and unsatisfactory definitions give us no clue to what we have in common with protozoans and plants. The difficulty for the scientist as well as for the writer of dictionaries is that life is not a separable, definable entity or property; it cannot be isolated on a microscope slide or distilled into a test tube. To early mechanistic philosophers like Aristotle and Descartes, life was wholly explicable in terms of the natural laws of chemistry and physics. The vitalists, philosophers of the opposing school, were convinced there was a special property, a vital force, that was

1.1 Sunlight bursts through a forest canopy

A minute portion of the sun's energy powers all life's processes.

1

absent in inanimate objects and thus unique to life. Though some scientists continue to think in terms of an unnamed and intangible special property, vitalism has been discredited in biology for at least half a century. The more we learn about living things, the clearer it becomes that life's processes are based on the same chemical and physical laws we see at work in a stone or a glass of water.

If life is not a special property, what is it? One answer may be found by comparing living and nonliving things. Organisms from bacteria to humans have several attributes in common: all are chemically complex and highly organized; all use energy (metabolize), organize themselves (develop), and reproduce; all change (evolve) over generations. As far as we know, no nonliving thing possesses all these attributes. In addition, and perhaps most important, only living organisms have a set of instructions, or program, resident in the genes, that directs metabolism, organization, and reproduction, and is the raw material on which evolution acts.

HOW THE SCIENTIFIC METHOD WORKS

Science is based on disciplined curiosity. Children are notoriously curious and undisciplined. Professional scientists retain much of their youthful curiosity—springing in most cases from a profound respect for nature—but bring to their work a degree of discipline that can solve difficult problems, plus a basic belief that there is an underlying order to natural phenomena waiting to be uncovered.

The discipline that must accompany curosity in successful scientists is the rigorous and creative application of the *scientific method*. Like so many truly great ideas, the scientific method is a basically simple concept and is used to some extent by almost everyone every day. As the English biologist T. H. Huxley (1825–1895) put it, the scientific method is "nothing but trained and organized common sense." Its power in the hands of a scientist stems from the consistency and ingenuity of its application.

■ THE NEED FOR HYPOTHESES

Science is concerned with the material universe, seeking to discover facts about it and to fit those facts into conceptual schemes, called theories or laws, that will clarify the relations between them. Science must therefore begin with observations of objects or events in the physical universe. The objects or events may occur naturally, or they may be the products of planned experiments; the important point is that they must be *observed*, either directly or indirectly. Science cannot deal with anything that cannot be observed.

Science rests on the philosophical assumption (well justified by its past successes) that virtually all events of the universe can be described by physical theories and laws, and that we get the data with which to formulate those theories and laws through our senses. Needless to say, natural laws are descriptive rather than prescriptive; they do not say how things *should* be, but instead how things are and probably will be.

Scientists readily acknowledge the imperfection of human sensory perception: the major alterations our neural processing imposes upon our picture of the world around us are themselves a subject of scientific study. In addition, experience has shown that there is often an interaction between phenomenon and observer: however careful we may be, our preconceived notions and even our physical presence may affect our observations and experiments. But to recognize the imperfection of sensory perception and observation is not to suggest that we can get scientific information from any other source (spiritual revelation, for example). No other route is open to scientists.

The first step in the scientific method is to formulate the question to be asked. This is not as simple as it sounds: scientists must decide which of the endless series of questions our escalating knowledge inspires are important and worth asking; the question must then be honed into a form that will generate a clear answer. The next step is to make careful observations in an attempt to answer the question. The difficulties here are in deciding what to observe and (since measuring everything is impossible) what to ignore. The scientist must also decide how to make the measurements and how to record the data, so that any patterns will emerge and nothing will be lost that might answer future questions. This is no trivial matter: an oversight or a mistake can render years of work useless.

The data that emerge from a set of observations must be analyzed and fitted into some sort of coherent pattern or generalization. A formal generalization, or **hypothesis,** is a tentative causal explanation for a group of observations. The step from isolated bits of data to generalization can be taken with confidence only if enough observations have been made to give a firm basis for the generalization, and then only if the individual observations have been reliably made. But even when data have been carefully collected, a hypothesis does not automatically follow. Often data can be interpreted in several ways or may appear senseless.

■ TESTING HYPOTHESES

Coming up with a general statement, or hypothesis, is not the end of the process. Scientists must devise ways of testing their hypotheses by formulating predictions based on them and checking to see if these predictions are accurate. Again, this is often not easy: a hypothesis may supply the basis for many predictions, but probably only one can be tested at a time. A scientist must decide which predictions can be most readily tested, and of those, which one provides the toughest challenge for the hypothesis (Fig. 1.2).

Perhaps most difficult of all, when researchers find a discrepancy between a hypothesis and the results of their tests, they must be ready to change their generalization. If, however, all evidence continues to support the new idea, it may become widely accepted as probably true and be called a theory. It is important to realize that

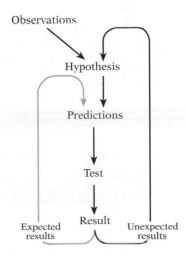

1.2 The scientific method

As this diagram suggests, experimental results are used, when necessary, to modify a hypothesis and to test predictions; the cycle of testing never really ends.

scientists do not use the term *theory* the way the general public does. To many people a theory is a highly tentative statement; when scientists dignify a statement by the name of theory, they imply that it has a very high degree of probability and that they have great confidence in it.

A *theory* is a hypothesis that has been repeatedly and rigorously tested. It is supported by all the data that have been gathered, and helps order and explain those data. Many scientific theories, like the cell theory, are so well supported by essentially all the known facts that they themselves are "facts" in the nonscientific application of that term. But the testing of a theory never stops. No theory in science is ever absolutely and finally proven. Good scientists must be ready to alter or even abandon their most cherished generalizations when new evidence contradicts them. They must remember that all their theories, including the physical laws, are dependent on observable phenomena, and not vice versa. Even incorrect theories, however, can be enormously valuable in science. We usually think of mistaken hypotheses as just so much intellectual rubbish to be cleared away before science can progress, but tightly drawn, explicitly testable hypotheses, whether right or wrong, catalyze progress by focusing thought and experimentation.

■ CONTROLLING EXPERIMENTS

In its simplest form the scientific method begins with careful observations, which must be shaped into a hypothesis. The hypothesis suggests predictions, which must be tested. The tests, furthermore, must be *controlled*, a matter of vital importance since only a controlled test has any hope of illuminating anything. Controlling a test, or experiment, means making sure that the effects observed result from the phenomenon being tested, not from some other source.

The most familiar way of controlling an experiment is to perform the same process again and again, varying only one discrete part of it each time, so that if a difference appears in the result, we can easily trace the cause. When Louis Pasteur, for instance, took his memorable stand against the prevailing theory of spontaneous generation (the idea that life can arise spontaneously from nutrients), he designed his experiment so that there could be no doubt about the outcome, whatever it might be (Fig. 1.3). He took identical flasks, filled them with identical nutrient solutions, subjected them to identical processes, but left some of them open to the air, while the others were effectively sealed. In time bacteria and mold developed in the open flasks, but the liquid in the sealed flasks remained clear. Since exposure to the outside air was the only variable in the procedure, it was obvious that the bacteria and the mold-producing organisms, rather than being spontaneously generated from the broth itself, must have come from the air.

■ THE ROLE OF INTUITION

In actual practice science is neither easy nor mechanical. Most scientists will readily admit that every stage in the scientific process requires not just careful thought but a large measure of intuition

1.3 Louis Pasteur

and good fortune. At the end of the last century, for instance, physicists saw many small problems, or anomalies, in Newtonian physics. Experience justifies living with small anomalies when a theory works well, and most minor problems prove in time to be irrelevant or mistaken—the result of observations that were faulty or interpretations that missed a point. Sometimes, though, anomalies in a theory signal a conceptual error.

Albert Einstein sensed that two of the many irritating anomalies in Newtonian physics were crucial difficulties, and rearranged the pieces of the puzzle to create a revolutionary new theory that accounted for those two anomalies. Like most new hypotheses, Einstein's theory of relativity did not win over the world of science at once. For one thing, it seemed far more complex than the Newtonian mechanics with which everyone was familiar, and for another, it left at least as many (though different) loose ends dangling as had the theory it sought to replace. The potential appeal of Einstein's theory, however, was great: it made some very unlikely but testable predictions—that the light from stars, for instance, should bend as it passed near the sun. When these were subsequently proven to be correct, the theory of relativity was accepted rather quickly, loose ends and all, where a far more plausible but less dramatic hypothesis might well have been ignored.

Scientific investigation, then, depends on a combination of subjective judgments and objective tests, a delicate mixture of intuition and logic. Done well, scientific research is truly an art: the ability to make insightful guesses and imagine clever and critical ways in which to test them is usually the distinguishing characteristic of great scientists. But in the final analysis, the basic rules are the same for all: observations must be accurate and hypotheses testable. And as testing proceeds, hypotheses must be altered when necessary to conform to the evidence.

THE ORIGINS OF MODERN BIOLOGY

■ WHEN DID BIOLOGICAL SCIENCE BEGIN?

Although naturalists and collectors had been at work for centuries observing living things, the application of the scientific method to biology is a relatively recent innovation. The earliest scientific biologist was Andreas Vesalius (1514–1564), who made the first serious studies of human anatomy by dissecting corpses. He discovered that the body is composed of numerous complex subsystems, each with its own function, and he pioneered the comparative approach, using other animals to work out the purpose and organization of these anatomical units. A typical (if grisly) example is his demonstration that the nerve from the brain to the throat, common to so many animals, is responsible for controlling vocalization. When he took a squealing pig, dissected out the nerve, and cut it, the animal instantly became mute even though its vocal apparatus remained intact (Fig. 1.4).

This powerful style of comparative and experimental study was carried forward by the English physician William Harvey

1.4   Cutting the vocal nerve
The experimenter pictured in the initial letter in Vesalius' *Fabric of the Human Body* (1542) is cutting the vocal nerve of a pig.

1.5   William Harvey

1.6 Antonie van Leeuwenhoek
The drawings are reproductions of his 1683
sketches of the intestinal protozoa of frogs.

(1578–1657), who showed conclusively that the heart pumps the blood and the blood circulates (Fig. 1.5). The heart, in short, is not in some metaphysical sense the seat of emotions but a mechanical device with a clear function. As a result of these studies and the anatomical work that followed, an increasingly mechanistic view of life began to develop.

The third of the pioneers was Antonie van Leeuwenhoek (1632–1723; Fig. 1.6). Leeuwenhoek, a cloth merchant, had the idea of using the powerful lens with which he inspected cloth to look at living things. The most important of his many discoveries were microorganisms (including bacteria), sperm and the eggs they fertilized, and the cells of which all living things seemed to him to be composed.

Centuries of painstaking observation were required to establish the science's fundamental generalizations. The cell theory, for instance, was not given its essentially modern form until 1858, and only about 130 years ago, in 1862, did Pasteur disprove the theory of spontaneous generation. With the realization that Leeuwenhoek's microorganisms might be responsible for disease, the English surgeon Joseph Lister proved the effectiveness of antiseptics (from *anti-*, "against," and *sepsis*, "decay") in 1865, and Pasteur greatly expanded the use of vaccination.

By far the most important figure in the history of biology, however, is Charles Darwin (1809–1882; Fig. 1.7). The publication in 1859 of his *Origin of Species*, presenting the theory of evolution by natural selection, suddenly provided a coherent, organizing framework for the whole of biology. His work sparked the explosive growth of biological knowledge that continues today. As the most important unifying principle in biology, the theory of evolution underlies the logic of every chapter of this book.

■ DID DARWIN DISCOVER EVOLUTION?

The theory of evolution by natural selection, as modified since Darwin, will be treated in detail in Chapter 17. Since we will need to use it in earlier chapters, however, we will examine the essence of the theory here. It consists of two major parts: the concept of evolutionary change, for which Darwin presented a great deal of evidence, and the quite independent concept of natural selection as the agent of that change.

Do species change?   Until only 200 years ago, it seemed self-evident that the world and the animals that fill it do not change: robins look like robins and mice like mice year after year, generation after generation, at least within the short period of written history. This commonsense view is very like our untutored impression that the earth stands still and is circled by the sun, moon, planets, and stars: it accords well with day-to-day experience, and until evidence to the contrary appeared, it provided a satisfying picture of the living world. The idea of an unchanging world also corresponded to a literal reading of the powerfully poetic opening of the Book of Genesis, in which God is said to have created each species independently, simultaneously, and relatively recently—a little over six thousand years ago by traditional scriptural reckonings.

But problems with the commonly held scriptural theory of creation arose from many sources. Scientists attempted first, quite naturally, to discount the evidence as ambiguous and then, when that proved impossible, to construct a new explanation. Let's look at the evidence for evolution that confronted Darwin and his contemporaries.

The most dramatic findings came from geology. By the eighteenth century, a picture of a changing earth had begun to emerge. Extinct volcanoes and their lava flows were discovered; most geological strata were found to represent sedimentary deposits, laid down layer upon layer a millimeter at a time in columns 3000 m (meters) or more deep; the gradual erosive action of wind and water were seen to have levelled entire mountains and carved out valleys, and unknown forces had caused mountains to rise where ocean floors had once been. This last fact in particular was impressed upon Darwin when he discovered fossilized seashells high in the Andes. Each of these phenomena implied continuous change during vast periods of time.

Another problem for the static view of life was presented by the New World fossils themselves. Many represented plants and animals wholly unknown in Europe, and though theologians had argued that the organisms these fossils represented were alive in the New World, increasingly intensive exploration of the Americas indicated that the hundreds of species of dinosaurs, for instance, were certainly extinct. In addition, many previously unknown and often bizarre animals inhabited the Americas (Fig. 1.8).

1.7 Charles Darwin

A

B

1.8 Challenges to traditional ideas of the origin of species

The discovery of fossils of now-extinct species brought into question the static view of life. Shown here are the remains of a baby mammoth (A) that had been preserved in the permafrost in Siberia. The discovery in the New World of organisms unfamiliar to Europeans, such as the anteater (B), also required a reinterpretation of traditional ideas of the origin of species.

1.9   Jean-Baptiste de Lamarck

The realizations grew that the number of animal species alive ran at least into the hundreds of thousands, and that the extinct species greatly outnumbered the living. New constellations of species had come and gone several times in the past. Moreover, the lowest, oldest rocks contained only the most primitive fossils—seashells, for instance—and these were followed in order by the more modern forms: fish appeared later, for example, reptiles still later, and then birds and mammals. The hypothesis of a young earth populated almost overnight by a single bout of creation began to seem very unlikely.

Jean-Baptiste de Lamarck (1744–1829; Fig. 1.9) was the first to offer the major alternative explanation of the fossil record: evolution. Lamarck had arranged fossils of various marine molluscs in order of increasing age; he saw clearly that certain species had changed slowly into others, and concluded that this process of slow change had continued right to the present day. As Lamarck put it in 1809, "it is no longer possible to doubt that nature has done everything little by little and successively," over a nearly infinite period of time. In Lamarck's view, the living world had begun with simple organisms in the sea, which eventually moved onto the land, and evolution had culminated in the appearance of our species, the inevitable result of the gradual trend toward change and "increasing perfection."

Lamarck was basically on the right track, but his mechanism for evolutionary change was incorrect. He was ignored for the very understandable reason that he could not offer sufficient evidence for the *fact* of evolution. Darwin, only 50 years later, was in a far better position: there was much more evidence of the sort Lamarck had pointed to, and Darwin, a respected geologist, was well acquainted with it. Furthermore, he had the ability to spot important data in the midst of apparent chaos.

One of the most important lines of evidence put forward by Darwin was the existence of morphological resemblances among living species (the findings of what we today call comparative anatomy). If we look at the forelimbs of a variety of different mammals, we see essentially the same bones arranged in the same order (Fig. 1.10). The basic bone structure of a human arm, a cat's front leg, and a seal's flipper is identical; the same bones are present even in a bird's wing. True, the size and shape of the individual bones vary from species to species, and some bones may be missing entirely in one species or another, but the basic construction is unmistakably the same. To Darwin the resemblance suggested that each of these species was descended from a common ancestor from which each had inherited the basic plan of its forelimb, modified to suit its present function.

The observation that structures with important functions in some species appear in vestigial, nonfunctional form in others further convinced Darwin of the reality of evolutionary change. Why otherwise would pigs, which walk on only two toes per foot, have two other toes that dangle uselessly well above the ground? Why would certain snakes, such as the boa constrictor, and many species of aquatic mammals, such as whales, have pelvic bones and small, internal hind-limb bones (Fig. 1.11)? Why would flightless birds such as penguins, ostriches, kiwis, and the cormorants of

1.10 A comparison of the bones in some vertebrate forelimbs

The labeled and color-coded bones of the human arm at left permit identification of the same bones in the other forelimbs depicted. In the bat the metacarpals (hand bones) and phalanges (finger bones) are elongated as supports for the membranous wing. In the seal the bones are shortened and thickened in the flipper. The cat walks on its phalanges (toes), the metacarpals having come to form a part of the leg. The sloth normally hangs upside down from tree limbs—hence its recurved claws. The horse walks on the tip of one toe, which is covered by a hoof (a specialized claw), while the sheep walks on the hoofed tips of two toes (and therefore cloven-hoofed, though only one hoof can be seen in this side view). The carpals (wrist bones) of both the horse and the sheep are elevated far off the ground, because the much-elongated metacarpals (hand bones) have become a section of the leg. Small splintlike bones that are vestiges of other ancestral metacarpals can be seen on the back of the upper portion of the functional metacarpals of both horse and sheep. All the animals mentioned so far—human, bat, seal, cat, sloth, horse, and sheep—are mammals, but the same bones can also be seen in the leg of a turtle and the wing of a bird. (All limbs are drawn to the same scale except that of the turtle, which is enlarged.)

1.11 Rudimentary hind limb of a whale

Whales lost their hind limbs long ago, when they returned to the sea from the land, but they retain rudimentary bones that correspond to the pelvic girdle and the thighbone (red).

A

1.12 Flightless male ostriches from the Namib desert in Africa

the Galápagos Islands still have rudimentary wings—or feathers, for that matter (Fig. 1.12)? Why would so many subterranean and cave-dwelling species have useless eyes buried under their skin?

Embryology—the study of how living things develop from eggs or seeds to their adult forms—also provided powerfully suggestive evidence. Darwin pointed out that in marine crustaceans as different as barnacles and lobsters the young larvae are virtually identical, implying a common origin. Telltale traces of their genealogy are obvious in vertebrates as well. Human embryos, for instance, have gill pouches and well-developed tails that disappear before the time of birth (Fig. 1.13). It seemed clear to Darwin that such inappropriate structures are inherited vestiges of structures that functioned in ancestral forms, and that may still function in other species descended from the same ancestors.

Another particularly convincing line of evidence offered by Darwin was the ability of breeders to produce dramatic changes in both plants and animals. Artificial selection can bring about vast changes given sufficient time. Great Danes, sheepdogs, Irish setters, Yorkshire terriers, poodles, bulldogs, and dachshunds, for in-

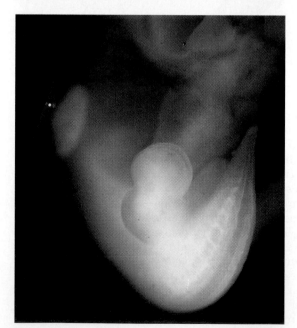

B

1.13 Embryological evidence of evolution

(A) Pharyngeal ("gill") pouches (arrows) in a 4-week human embryo.
(B) Tail in a 5-week human embryo.

stance, are all members of the same species, bred from tamed wolves to look like almost anything breeders have fancied.

Similarly, cabbage, brussels sprouts, cauliflower, broccoli, kohlrabi, rutabaga, collard greens, and savoy have all been bred from the same species, the wild form of which looks nothing like its domesticated progeny (Fig. 1.14). The many varieties of chickens, cattle, horses, flowers, grains, and so on have been bred to fill different needs over the years. Who could compare the colors and shapes of the wild rose or jonquil with the many colors and shapes of the far larger domesticated roses or daffodils and doubt that a species has the capacity to change enormously even in a mere century? In everything Darwin looked at—fossils, anatomy, embryology, and breeding—he saw the same message: species can and do change (Fig. 1.15).

What causes species to change? Lamarck's now-discredited hypothesis was one of the first attempts to explain why species evolve. Lamarck was impressed by how well suited each animal was to its particular position in the web of life, even though the environment had changed enormously again and again over countless millions of years. To account for this ability to adapt, he imagined God had given each species a tendency toward perfection that allowed for small alterations in morphology, physiology, and behavior to accommodate changes in the environment, and that these alterations, once made, could be inherited by the offspring.

Belief in Lamarck's idea of a natural tendency toward perfection and the inheritance of acquired characteristics required no more faith in his day than did belief in other invisible everyday forces, such as gravity and magnetism. But where in the vestigial legs of whales or the dangling toes of pigs was there evidence of perfection? Instead the clear mark of compromise was everywhere. Plants and animals were well adapted to their places in the environment, but they were by no means perfect.

Darwin proposed a different mechanism—***natural selection***—requiring no internal tendency other than the one toward variation so obvious in nature. Darwin had conceived the idea of natural selection two years after his return from his voyage to the Americas

1.14 Selective breeding of wild cabbage

The wild species *Brassica oleracea* (left) has been bred to create cauliflower (upper right), brussels sprouts (middle right), and cabbage (lower right). Each represents a selective exaggeration of one part of the wild plant—the flower heads for cauliflower, the side buds for brussels sprouts, and the leaves for cabbage. Despite the extreme morphological differences, however, the three domestic vegetables can be interbred.

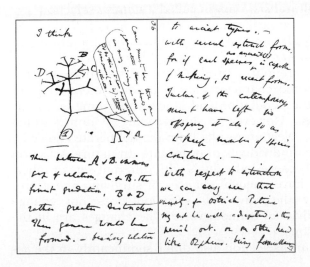

1.15 Pages from one of Darwin's notebooks

Scientists of the nineteenth century and earlier, among them Charles Darwin, often recorded in notebooks and journals their experiments, observations drawn from field trips, or simply their thoughts about hypotheses in the making. In addition to the many books he wrote, Darwin produced notebooks on a wide variety of subjects, including his trip to South America and the Galápagos (1831–1836), later described in the *Journal of Researches during the Voyage of the Beagle* (1840). The notebook from which this excerpt is taken was written in 1837, one year before Darwin saw that natural selection was the likely mechanism for evolution. These pages show Darwin's first recorded drawing of an evolutionary tree, a metaphor for the diversity and interrelatedness of species that has continued in the scientific literature to the present time. The trunk represents a common ancestor, the limbs major groups, and the twigs particular species, either extinct (indicated by a crossbar) or living.

1.16 Thomas Robert Malthus

on the *Beagle,* but was only goaded into publishing it 20 years later when in 1858 he received A. R. Wallace's manuscript proposing essentially the same theory. (We normally associate Darwin's name with the theory because of the impressive evidence he presented—he had been collecting it for two decades—and because of his thorough exploration of the theory's many ramifications.)

In essence, Darwin put together two ideas. The first was that numerous variations exist within species and that these variations are largely heritable. Immersed as he was in the Victorian preoccupation with plant and animal breeding, Darwin knew that, while cuttings produce plants identical to the parent, sexual reproduction produces individual offspring that differ both from their parents and from each other. Variation is a fact of life: breeders, as we know, are able to select for desirable traits and create new, morphologically distinct lines of plants and animals.

Darwin's second inspiration came when he reread the *Essay on the Principle of Population* by the economist Thomas Robert Malthus (Fig. 1.16). Malthus pointed out that humans produce far more offspring than can possibly survive; population growth always outruns any increase in the food supply and is held in check largely by war, disease, and famine. Vast numbers of people thus live perpetually on the edge of starvation.

Both Darwin and Wallace were struck by the consequence of applying the gloomy Malthusian logic to plants and animals: like humans, the creatures of each overpopulated generation must compete for the limited resources of their environment, and some—indeed most—must die. Each female frog, for example, produces thousands of eggs per year, and a fern produces tens of millions of spores, yet neither population is growing noticeably. Any organism with naturally occurring heritable variations that increase its chances in this life-or-death contest will be more likely than others to survive long enough to have offspring, some of which will inherit these variations. The offspring in turn will have an above-average chance to survive the struggle and so will form an increasingly large part of the population. As a result of this selection, the population as a whole will become better adapted—that is, it will evolve—and the never-ending struggle for existence will turn on the possession of still better adaptations. To distinguish this process from the sort of directed, artificial selection practiced by agriculturalists, Darwin called it *natural* selection.

The contrast between artificial and natural selection that served Darwin so well provides an instructive summary of the evolutionary process. In both, far more offspring are born than will reproduce; in both, differential reproduction, or selection, occurs, causing some inherited characteristics to become more frequent and prominent in the population and others to become less so as the generations pass. However, in the breeding of domesticated plants and animals, selection results from the deliberate choice by the breeder of which individuals to propagate. In nature, selection takes place simply because individuals with different sets of inherited characteristics have unequal chances of surviving and reproducing. Notice, by the way, that selection does not change individuals. An individual cannot evolve. The changes are in the makeup of populations.

Artificial and natural selection also differ significantly in the *de-*

1.17 Selective breeding of pigeons

By practicing rigorous selection, breeders can achieve major changes in relatively few generations. The ancestral rock dove is shown in the center. The domestic breeds (clockwise from upper left) are fantail, Silver Jacobin, Male Ice, Magpie Kormoner Tumbler, Norwich cropper, Voorburg Shield cropper, German Longface Tumbler.

gree of selection and its effect on the rate of change. Breeders can practice rigorous selection, eliminating all unwanted individuals in every generation and allowing only a few of the most desirable to reproduce. They can thus bring about very rapid change (Fig. 1.17). Natural selection, which involves a large measure of chance, is usually much less rigorous: some poorly adapted individuals in each generation will be lucky enough to survive and reproduce, while some well-adapted members of the population will not. Hence evolutionary change is usually slow; major changes may take thousands or even millions of years, depending on the degree of selection pressure imposed by the environment and by other species.

Darwin's evidence for evolutionary change and the common descent of at least the major groups of organisms was widely accepted in his time, but the idea of natural selection by small steps remained controversial until the 1930s. Some biologists had difficulty seeing how an elaborate and specialized structure like an eye, for instance, could evolve, since the first rudimentary but necessary steps might lack obvious survival value. An expanded understanding of the nature and organization of genes and their role in development has now made it clear that natural selection does explain most evolutionary change. Darwin extolled the beauty and simplicity of such a system in the final sentence of later editions of *The Origin of Species:* "There is grandeur in this view of life, with its several powers, having been originally breathed by the Creator into a few forms or into one; and that, whilst this planet has gone cycling on according to the fixed law of gravity, from so simple a beginning endless forms most beautiful and most wonderful have been, and are being evolved."

In summary, then, Darwin's explanation of evolution in terms of natural selection depends upon five basic assumptions:

1 Many more individuals are born in each generation than will survive and reproduce.

2 There is variation among individuals; they are not identical in all their characteristics.

3 Individuals with certain characteristics have a better chance of surviving and reproducing than individuals with other characteristics.

4 Some of the characteristics resulting in differential survival and reproduction are heritable.

5 Vast spans of time have been available for change.

All the known evidence supports the validity of these five assumptions.

■ THE TREE OF LIFE

Darwin's insights into evolution and its mechanisms, together with a variety of techniques (including, most recently, molecular analyses of genes themselves), have permitted a fairly reliable reconstruction of the course of evolution of life on earth and of the relationships between the species alive today. A diagram of the relationships among the eight major groupings (called kingdoms) of living species recognized at present is shown in Fig. 1.18. The organisms with the longest histories are the two kingdoms of bacteria, which are fundamentally different from the other groups because they lack nuclei; their chromosomes are mixed with the

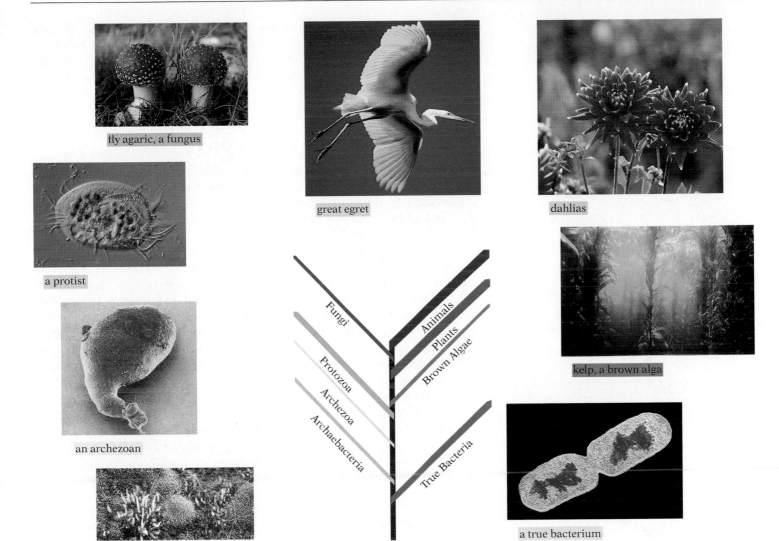

fly agaric, a fungus

great egret

dahlias

a protist

an archezoan

kelp, a brown alga

a true bacterium

archaebacteria

Fungi

Animals

Plants

Brown Algae

Protozoa

Archezoa

Archaebacteria

True Bacteria

1.18 Relationships among the eight kingdoms of life

In this representation, kingdoms are shown in color, and the width of the line indicates the relative number of species in each kingdom. The distance from the bottom of the figure at which a group's line departs from the central "trunk" of this evolutionary tree indicates the chronological order in which the kingdoms are thought to have arisen. Common names are used for most groups; technically, animals are in Kingdom Animalia, plants are in Plantae, brown algae are in Chromista, and "true" bacteria are in Eubacteria.

rest of the cellular contents instead of being segregated into a separate compartment. As a result, the bacteria are called ***procaryotes*** ("before nuclei"), while the other six kingdoms make up the ***eucaryotes*** (having "true nuclei").

The eucaryotes comprise the two kingdoms that now dominate the earth—the plants and the animals—as well as three smaller but significant kingdoms, the protozoans (most unicellular organisms), the fungi (mushrooms, for example), and the brown algae (like kelp); they also include one tiny, little-known group, the archezoans, which are by far the most primitive of the eucaryotes.

As we will see, the organisms in these kingdoms are marvelously diverse, having evolved to exploit a bewildering variety of habitats, and yet all share a common set of genetic and biochemical processes that unify the study of life.

CONTEMPORARY BIOLOGY

The work in the latter half of the 19th century by scientists like Darwin, Pasteur, Gregor Mendel (the monk who discovered the basic principles of inheritance), and a number of developmental biologists combined to set the stage for the emergence of the range of biological studies that go on today.

Although no one event formally marks the beginning of contemporary biology, the discovery of the structure of DNA in 1953 was an especially important foundation for later work. The double-helix model proposed by James D. Watson and Francis Crick (Fig. 1.19) provided a physical basis for Mendel's genes, which are the basic units of Darwinian selection. The structure immediately suggested how DNA could act as a template for its own reproduction and as a means for issuing instructions for building and operating cells.

Watson and Crick's model of DNA has helped fuel four decades of research on the messages encoded in this remarkable molecule; understanding the code is proving critical to every biological discipline. In relation to cell biology (our focus in Part I), research on DNA has led to a detailed appreciation of how cells are organized, both physically and chemically. One of the more impressive feats of investigators has been the deciphering of the chemical codes used as "mailing labels" on DNA products—codes that tell the elaborate transport system in the cell where to deliver each molecule.

In relation to genetics (Part II), we will see how research on DNA has been used to construct a nearly complete picture of how cells process information to regulate their metabolism, growth, and reproduction. This knowledge is critical in modern treatments of sickness and inborn defects, in the manufacture of chemicals of potentially enormous clinical and commercial importance like synthetic hormones and vaccines, and in the endowment of some plants with new genes that confer benefits such as disease resistance. An equally impressive by-product of DNA research is progress toward understanding the immune system—the body's main line of defense against pathogens—as well as the precisely orchestrated development of a complex, trillion-celled organism like a human being from a single, unspecialized cell. Another is the invention, since 1953, of many techniques for manipulating the genetic material and its protein products. Together the development of these techniques and the investigation of biological processes at the level of molecules are usually referred to as molecular biology.

The benefits of the new molecular biology for whole-organism studies are even more remarkable because they are less obvious. In Part III we will have a glimpse of the enormous impact molecular

1.19 Watson and Crick with their model of the DNA double helix

techniques are having on the study of evolution. Biologists now understand the chemical nature of the variation that is the basis of natural selection, and so are coming to grips with the question of how different species arise in nature and how they adapt to novel conditions. Small changes in genes controlling development, for example, can produce new species almost instantly.

We will see in Part IV that an understanding of how DNA and its chemical collaborators work is providing valuable insights into the origin and early evolution of life, as well as a reliable and often surprising reconstruction of the evolution of the major groups of organisms that populate the earth today. Molecular techniques now allow precise determination of species relationships; they reveal, for instance, that superficial similarities notwithstanding, pandas are not bears but overgrown raccoons.

In Part V, we will see how the molecular approach has exposed the workings of many physiological systems, including those responsible for circulation, gas exchange, movement, and internal coordination. For example, we can now analyze at the molecular level how neurons communicate and how hormones and other chemical signals (like insulin) regulate the activity of cells, tissues, and organs. So, too, the failures of organs to operate properly—breakdowns that lead to degenerative diseases—are beginning to be understood (and corrected) at the molecular level.

Finally, in Part VI, we will see how molecular techniques are aiding our understanding of ecological relationships. For example, the pesticide contaminant dioxin exerts its devastating ecological effects because the detoxification enzymes that animals use to dispose of unfamiliar (and thus potentially dangerous) chemicals alter it to an inappropriate form. And, of course, the potential of molecular engineering for solving or at least ameliorating environmental problems, like water pollution and even greenhouse-effect warming, is one avenue of hope for the future of our planet.

Now more than ever, as we stand able to read the genetic programs of organisms word for word, biology is the most exciting, intellectually stimulating, and promising discipline among the sciences.

CHAPTER SUMMARY

WHAT IS LIFE?

Living things have genetic instructions that direct their metabolism and reproduction, and they can evolve.

HOW THE SCIENTIFIC METHOD WORKS

THE NEED FOR HYPOTHESES The scientific method begins with formulating hypotheses that can account for observations. (p. 2)

TESTING HYPOTHESES Hypotheses are tested to see how well they predict events; failures lead us to revise our hypotheses. (p. 3)

CONTROLLING EXPERIMENTS A test usually compares two cases differing in only one variable; the hypothesis predicts the change that variable will cause compared with the "control" situation. (p. 4)

THE ROLE OF INTUITION Good judgment is required to select likely hypotheses for testing, and to design tests that are well controlled. (p. 4)

THE ORIGINS OF MODERN BIOLOGY

WHEN DID BIOLOGICAL SCIENCE BEGIN? The most important early biologists were Vesalius and Harvey, who studied functional anatomy, and Leeuwenhoek, who discovered microorganisms. Each helped make biology more experimental and mechanistic. (p. 5)

DID DARWIN DISCOVER EVOLUTION? Darwin collected much supporting data for evolution, but the idea was not his. Darwin discovered natural selection, which is the major mechanism of evolution: excess offspring are produced in each generation, and those with characteristics that are the most adaptive tend to live and reproduce, leading to an increase in the frequency of those characteristics in the next generation. (p. 6)

THE TREE OF LIFE The tree of life reconstructs the path evolution took to yield the diversity of organisms that inhabit the earth today. (p. 14)

CONTEMPORARY BIOLOGY

The discovery of DNA and the genetic code has led to a fuller understanding of the mechanistic bases of nearly every aspect of biology. (p. 16)

STUDY QUESTIONS

1 What roles do anomalies and intuition play in the scientific method? (pp. 3–5)

2 How might evolution occur in the absence of natural selection? (pp. 11–14)

3 What are three important requirements in a population for selection to take place? Can evolution occur in the absence of any one of these? (pp. 11–14)

4 How did Darwin's intimate knowledge of plant and animal breeding help him formulate his theory? (pp. 10–12)

SUGGESTED READING

COMROE, J. H., 1977. *Retrospectroscope*. Von Gehr Press, Menlo Park, Calif. *A fascinating study of how important scientific discoveries are made. The author concludes that great advances usually arise out of research directed at wholly unrelated problems.**

DARWIN, C., 1859. *The Origin of Species. Of the many reprints of this classic work, the edition by R. E. Leaky (Hill and Wang, New York, 1979) provides perhaps the best introduction and illustrations.**

* Available in paperback.

PART I

THE CHEMICAL AND CELLULAR BASIS OF LIFE

◄ **Powerhouse of the cell,** the elongated mitochondrion highlighted in yellow and red in this color-enhanced transmission electron micrograph (TEM) provides energy to the cell by completing the digestion of glucose, thereby releasing molecules of the cell's all-purpose energy currency: ATP (adenosine triphosphate). The black and blue membranous material surrounding the mitochondrion is mostly endoplasmic reticulum. The rounded, pinkish bodies are vesicles.

CHAPTER 2

SOME SIMPLE CHEMISTRY

All life processes obey the laws of chemistry and physics. In the behavior of molecules, atoms, and subatomic particles lies the key to such complex biological phenomena as the trapping and storing of solar energy by green plants, the extraction of usable energy from organic nutrients, the growth and development of organisms, the patterns of genetic inheritance, and the regulation of the activities of living cells. The study of biology, then, begins with the basic laws of chemistry and physics.

THE ELEMENTS ESSENTIAL TO LIFE

All the matter of the universe is composed of 92 basic substances called elements, some of which were formed in the original Big Bang that created the universe; others are released by the exploding stars called novas (Fig. 2.1). Each element is designated by one or two letters that stand for its English or Latin name. Thus H is the symbol for hydrogen, O for oxygen, C for carbon, Cl for chlorine, Mg for magnesium, K for potassium (Latin, *kalium*), Na for sodium (Latin, *natriu*), and so on. Only a few of the 92 elements are important to life (Table 2.1).

■ THE STRUCTURE OF ATOMS

Matter is composed of *atoms*. The atoms of a particular element differ in many measurable ways from the atoms of all other elements. Atoms are themselves composed of still smaller particles, three of which—the proton, the neutron, and the electron—play a central role in determining the activity of elements. In their interactions lie the power and the cohesion that make life possible.

Inside the nucleus All the positive charge and almost all the mass of an atom are concentrated in its nucleus, which contains two kinds of primary particles, the *proton* and the *neutron.* Each proton carries an electric charge of +1. The neutron has no charge. The proton and the neutron have roughly the same mass (Table 2.2).

The number of protons in the nucleus is unique for each element. This *atomic number* can be found in Table 2.2 for many ele-

2.1   The Supernova of 1987

This nova consists of enormous amounts of gas produced when an aging star exploded. The gas includes many heavy elements created inside the star out of lighter elements. When a new star like the sun condenses out of this gas, the surrounding planets that may form are enriched in the heavy elements, many of which are essential to life.

TABLE 2.1   Elements important to life

SYMBOL	ELEMENT	ATOMIC NUMBER/ TYPICAL MASS NUMBER	APPROXIMATE PERCENTAGE OF EARTH'S CRUST BY WEIGHT	APPROXIMATE PERCENTAGE OF HUMAN BODY BY WEIGHT
H	Hydrogen	1/1	0.14	9.5
B	Boron	5/11	trace	trace
C	Carbon	6/12	0.03	18.5
N	Nitrogen	7/14	trace	3.3
O	Oxygen	8/16	46.6	65.0
F	Fluorine	9/19	0.07	trace
Na	Sodium	11/23	2.8	0.2
Mg	Magnesium	12/24	2.1	0.1
Si	Silicon	14/28	27.7	trace
P	Phosphorus	15/31	0.07	1.0
S	Sulfur	16/32	0.03	0.3
Cl	Chlorine	17/35	0.01	0.2
K	Potassium	19/39	2.6	0.4
Ca	Calcium	20/40	3.6	1.5
V	Vanadium	23/51	0.01	trace
Cr	Chromium	24/52	0.01	trace
Mn	Manganese	25/55	0.1	trace
Fe	Iron	26/56	5.0	trace
Co	Cobalt	27/59	trace	trace
Ni	Nickel	28/59	trace	trace
Cu	Copper	29/64	0.01	trace
Zn	Zinc	30/65	trace	trace
Se	Selenium	34/79	trace	trace
Mo	Molybdenum	42/96	trace	trace
Sn	Tin	50/119	trace	trace
I	Iodine	53/127	trace	trace

TABLE 2.2   Fundamental particles

PARTICLE	MASS (DALTONS[a])	ELECTRIC CHARGE
Electron	0.001	−1
Proton	1.013	+1
Neutron	1.014	0

[a] One dalton equals 1.650×10^{-24} g, which is 1/16 of the mass of an oxygen atom. For definitions of units of measurement, see Glossary.

ments. The atomic number of hydrogen is 1; its nucleus contains only one proton. In contrast, each oxygen nucleus contains eight protons.

The total number of protons and neutrons in a nucleus is the **mass number;** it approximates the total mass (commonly called the **atomic weight**) of the nucleus. The mass number is usually written as a superscript immediately preceding the chemical symbol. For example, most atoms of oxygen contain eight protons and eight neutrons; the mass number is therefore 16, and the nucleus can be symbolized as ^{16}O.

Though the number of protons is the same for all atoms of the same element, the number of neutrons is not always the same, and neither, consequently, is the mass number. For example, most oxygen atoms contain eight protons and eight neutrons and have a

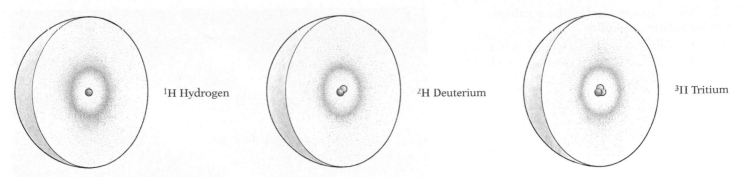

^{1}H Hydrogen ^{2}H Deuterium ^{3}H Tritium

2.2 Three principal isotopes of hydrogen

Each of the three isotopes has one proton (blue) in its nucleus and one electron orbiting the nucleus. The isotopes differ in that ordinary hydrogen (^{1}H) has no neutrons in its nucleus, deuterium (^{2}H) has one, and tritium (^{3}H), which is unstable, has two. The volume within which the single electron can be found 90% of the time (in this case, a sphere) is indicated by stippling; the denser the stippling in this cross-sectional view, the greater the likelihood that at any given moment the electron will be found in that portion of the sphere.

mass number of 16; some, however, contain nine neutrons and therefore have a mass number of 17 (symbolized as ^{17}O), and still others have ten neutrons and a mass number of 18 (symbolized as ^{18}O). Atoms of the same element that differ in mass, because they contain different numbers of neutrons, are called *isotopes;* ^{16}O, ^{17}O, and ^{18}O are three isotopes of oxygen.

Figure 2.2 illustrates three different isotopes of hydrogen: ^{1}H, which is the usual form of the element; ^{2}H, a stable isotope called deuterium; and ^{3}H, an unstable isotope called tritium. Isotopes of hydrogen and other elements are invaluable research tools for biologists (Fig. 2.3).

2.3 Magnetic Resonance Imaging technique

Isotopes whose atomic mass is an odd number (like ^{31}P) are weakly magnetic. When exposed to a strong magnetic field, a given isotope will absorb radio waves of a characteristic frequency and then reemit them when the signal is removed. In this image, the strong signal emitted by the relatively high concentration of organic compounds in a growing brain tumor stands out as a green area against the relatively watery cells of the healthy tissues. The tumor is growing in an auditory area, and is causing progressive deafness.

2.4 Atoms in a crystal of the semiconductor gallium arsenide

High-speed electrons create an image of hazy outer electron shells in this scanning tunneling micrograph. Color has been added to distinguish gallium atoms (blue) from arsenic atoms (red). (The operation of scanning tunneling microscopes, along with other sorts of microscopes, is described in Chapter 4.)

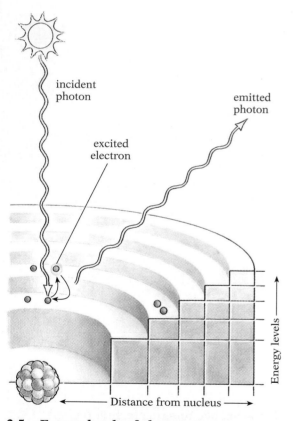

incident photon

excited electron

emitted photon

Energy levels →

← Distance from nucleus →

2.5 Energy levels of electrons

The electrons in an atom occupy discrete energy levels. If an electron absorbs the right amount of energy (shown as a photon, a discrete particle of light) and there is a vacancy in a higher energy level, the electron can move up the energy staircase to this level and so become excited. Usually an excited electron quickly reemits the absorbed energy (here again shown as a photon) and returns to its original energy level.

The electron cloud The portion of the atom outside the nucleus contains the third kind of primary particle—the *electron.* Though electrons have very little mass (see Table 2.2), their behavior is the single most crucial factor in the chemistry of life. Each electron carries a charge of −1, exactly the opposite of a proton's charge. In a neutral atom, the number of electrons around the nucleus is the same as the number of protons in the nucleus. The positive charges of the protons and the negative charges of the electrons cancel each other, making the total atom neutral.

The electrons are not in fixed positions outside the nucleus. Each is in constant motion, making 10^{15} to 10^{16} orbits of the nucleus each second. Hence, it is impossible to know exactly where any electron is; photographs made with the latest imaging technology record only a wispy shell boundary (Fig. 2.4). For this reason some illustrations of atoms, such as Fig. 2.2, do not show the electron itself but indicate the region where the electron is likely to be. All illustrations of atomic structure exaggerate the size of the nucleus. If a proton were the size indicated in Fig. 2.2, the outer edge of the electron cloud would extend 150 meters in all directions.

The average distance of an electron from the nucleus is a function of its energy; the higher its energy, the greater its probable distance from the nucleus. But in any particular atom, only certain discrete amounts, or levels, of energy—like steps on a staircase—are possible. To occupy a certain step, or energy level, an electron must possess a specific amount of energy. To achieve a higher energy level an electron must absorb additional energy from some outside source. Conversely, when an electron falls into the next lower level, it emits the same amount of energy it previously took to move up from that level (Fig. 2.5). We refer to an electron occupying the lowest step available to it in the atom as being in the ground state. Once it has absorbed enough energy to move up to the next energy level, it is said to be in an excited state.

An electron in the excited state has a strong tendency to return to its ground state by emitting, in some form, the additional energy just acquired. Most often the energy is released as light. It is the decay of excited electrons in the lining of fluorescent tubes, for instance, that produces light. As we will see, this fleeting moment of excitation, lasting 10^{-8} sec, is critical to living things: life is based on the ability of specialized molecules in photosynthetic organisms to capture the energy of excited electrons before they drop back down the energy staircase.

The volume within which an electron can be found 90% of the time is known as its ***orbital.*** In illustrations of atoms the orbital is often represented by a circle, with the electrons sometimes shown as round spots on the circle (Fig. 2.6).

The energy level of the hydrogen electron is the level nearest the nucleus (often referred to as the K level). The K level can contain only two electrons. In an atom such as oxygen, with eight electrons, only two can be accommodated at the K level; the other six occupy a higher energy level, farther from the nucleus, called the L level. A maximum of eight electrons can fit into the L level. Since the most stable configuration for an atom is one in which its electrons have minimum energy, the six electrons outside the K level in an oxygen atom are all at the L energy level. None are found in the next higher energy level—the M level. Only when elements have more than 10 electrons, or when one of the lower-level electrons has been excited beyond the L level, are electrons found at the M level.

The most likely distance from the nucleus of each of the six L

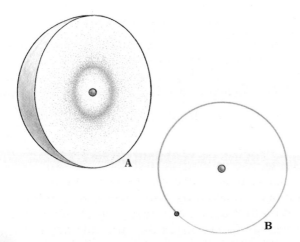

2.6   Two ways of representing the hydrogen atom

Since no one has ever seen the particles that make up an atom, all our knowledge of what atoms look like is indirect; we can only picture them as models that fit the data. (A) The nucleus is shown here as a central blue area, with the "cloud" around it in cross section representing the region where the electron is likely to be. The circle encloses the orbital of the electron—the spherical volume within which the electron will be found 90% of the time (see also Fig. 2.2). (B) Sometimes, for convenience, only the circle indicating the circumference of the orbital is shown; the electron may be represented by a small ball on the circle.

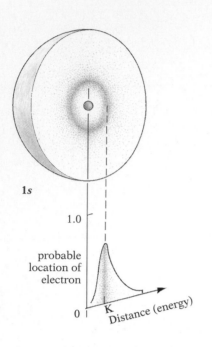

1s

1.0

probable location of electron

0 ┊ K
  Distance (energy)

2s

1.0

0 ┊ L

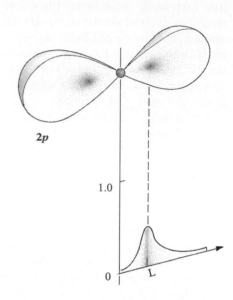

2p

1.0

0 ┊ L

2.7   Electron orbitals

Top: The orbitals of *s* electrons are approximately spherical, while those of *p* electrons are roughly dumbbell-shaped. The numerals before *s* and *p* indicate the energy level. Thus the 1*s* electron is at the first energy level (*K*), nearest the nucleus; the 2*s* and 2*p* electrons are at the *L* level, a higher energy level, and hence are at a greater average distance from the nucleus than the 1*s* electron. Note that, despite the very different shapes of their orbitals, the 2*s* and 2*p* electrons are at the same energy level—their most probable distances from the nucleus are the same, as shown in the graphs (bottom).

electrons of an oxygen atom is somewhat greater than the most likely distance of the *K* electrons (Fig. 2.7). While the orbital of two electrons at the *K* level is circular, the other three pairs occupy dumbbell-shaped orbitals oriented at right angles to one another (Fig. 2.8).

■ HOW ELECTRONS DETERMINE CHEMISTRY

When the elements are arranged in sequence according to their atomic numbers—beginning with hydrogen (atomic number 1) and proceeding to uranium (number 92)—elements with very similar properties are found at regular intervals in the list (Fig. 2.9). For example, fluorine (number 9) is more like chlorine (number 17), bromine (number 35), and iodine (number 53) than it is like oxygen (number 8) or neon (number 10), the two elements immediately adjacent to it in the list. This tendency for chemical properties to recur periodically throughout the sequence of elements is called the Periodic Law.

The explanation for this periodicity is that the reactivity and other chemical properties of elements are largely determined by the number of electrons in their outer shell (the highest energy level). If that shell is complete, as in helium (atomic number 2),

2.8   Three 2*p* electron orbitals

Each of the dumbbell-shaped orbitals is oriented in a different dimension of space, at right angles to the other two.

2.9   Partial periodic table with electron distributions

The first 20 elements are shown arranged according to their position in the periodic table. Elements in the same column share many chemical properties because they have the same number of electrons in their outer shell. (Helium is placed in column 8 even though it has only two outer electrons because, like neon, argon, and the other so-called noble gases, its outer shell is full, and its chemical properties are therefore those of a noble gas.)

neon (10), or argon (18), the element has very little tendency to react chemically with other atoms (Fig. 2.9). If the outermost shell has one electron fewer than the full complement, as in fluorine, chlorine, bromine, and iodine, the element has certain characteristic chemical properties; if, like oxygen, it lacks two electrons, the element has somewhat different properties.

A convenient way to represent the electron configuration of the outer shell is to symbolize each electron by a dot placed near the chemical symbol for the element under consideration. Thus fluorine and chlorine, which have seven electrons at their outer energy level, would have identical electron symbols:

$$:\!\ddot{F}\!\cdot \qquad\qquad :\!\ddot{Cl}\!\cdot$$

Similarly, hydrogen with one electron in its single shell, carbon with four in its outer shell, nitrogen with five, and oxygen with six are represented as follows:

$$\text{H}\!\cdot \qquad \cdot\!\dot{C}\!\cdot \qquad \cdot\!\ddot{N}\!\cdot \qquad \cdot\!\ddot{O}\!\!:$$

■ WHEN ATOMS DECAY

Though the various isotopes of an element carry different numbers of neutrons, their identical electron distributions give them the same chemical properties. Their physical properties, however, differ in two ways that are important to biological research and to human health.

Unusual isotopes are taken up by tissues just as well as the more common forms, but since isotopes differ significantly in atomic weight, they can be distinguished from each other by weight-sensitive techniques. For example, there was once controversy over whether the oxygen gas (O_2) released by plants comes from carbon dioxide (CO_2) or water (H_2O). The issue was settled by supplying plants with water containing a heavy isotope of oxygen (^{18}O rather than ^{16}O), while providing normal CO_2. The mass of the O_2 released by the plants was then compared with normal O_2 and found to be about 12% heavier, thus proving that water is the ultimate source for the oxygen in the air.

The other biologically significant physical property of some isotopes is their tendency to decay into a more stable form, giving off various particles in the process. The stability of an isotope is measured by the **half-life** of the isotope: the time it takes half the atoms in a sample to decay. Tritium (^3H), for instance, has a half-life of about 12 years; ^{32}P, roughly 14 days; ^{14}C, 5700 years; and so on.

Radioactive isotopes are extraordinarily useful in biology, since an isotope added to a sample emits radiation that acts as a label scientists can track (Fig. 2.10). With a labeled isotope of carbon in CO_2, for instance, we can trace how plants use carbon to build sugars. Because they are taken up by tissue as readily as their more stable counterparts, radioactive isotopes can also be used as tracers to help doctors locate circulatory blockages and pinpoint tumors. Naturally occurring isotopes make possible the dating of many rocks and fossils. The ratio of the radioactive isotope ^{14}C to the stable ^{12}C, for instance, is relatively constant in the CO_2 of the atmosphere, but once a plant has captured a CO_2 molecule and built it into a product like cellulose, the decay of ^{14}C atoms causes the ratio of ^{14}C to ^{12}C to decline steadily with time. Hence, the $^{14}C : ^{12}C$ ratio in a sample provides a moderately accurate measure of age.

Isotopes can be dangerous as well. A radioactive atom in a living cell poses two potential threats. First, it can, like uranium, decay into another element by losing protons, thereby altering the chemistry of its molecule completely. More often, radioactive isotopes produce highly reactive molecules that have too many or too few electrons to balance the electrical charge of their protons, and thus have a net charge. Since the behavior of electrons, as we will see, determines the chemistry of life, such unpredictable and uncontrolled movement of electrons can disrupt the precisely ordered and carefully regulated workings of the cell. For instance, a change in a critical part of a cell's DNA can trigger the complicated chain of events that leads to cancer. Changes of this sort arise in all of us every day from exposure to the sun's radiation and from the natural decay of radioactive elements in the earth's atmosphere and crust, and each cell has a battery of defense mechanisms to reduce the damage.

2.10   Autoradiography of a soybean plant

A radioactive isotope, ^{32}P, was taken up by the plant's roots and transported to sites of nucleic acid synthesis. To trace this movement, the plant was laid on a photographic emulsion, which recorded the sites of radioactive decay.

THE GLUE OF LIFE: CHEMICAL BONDS

The arrangement of electrons in the outer shell of the atoms of most elements gives those atoms an ability to bind to others. When two or more atoms are bound together in this fashion, the force of attraction that holds them together is called a chemical bond. The atoms of each particular element can form only a limited number of such bonds; the arrangement of its electrons gives each element its own characteristic bonding capacity.

■ BONDS BETWEEN IONS

Atoms are particularly stable when the outer electron shell is complete—that is, in most cases, when it contains eight electrons. There is consequently a general tendency for atoms to form complete outer shells by reacting with other atoms.

Consider, for example, an atom of sodium (atomic number 11). This atom has two electrons in its first shell, eight in the second, and only one in the third. One way sodium might gain a complete outer shell would be to acquire seven more electrons from some other atom or atoms. But the sodium atom would then have an enormous excess of negative charge, and since like charges repel each other, the electrons would tend to push each other away from the sodium nucleus. Instead it becomes an *electron donor,* giving up the lone electron in its third shell to some other atom (called the *electron acceptor),* which leaves the complete second shell as the new outer shell (Fig. 2.11A).

Next, consider an atom of chlorine (atomic number 17). This atom has two electrons in its first shell, eight in its second shell, and seven in its third shell. Its outer shell lacks only a single electron. It cannot lose the seven electrons in its outer shell because it would become too highly charged. However, by gaining an extra electron from some electron donor, chlorine acquires a complete outer shell.

If a strong electron donor like sodium (an atom with a strong tendency to lose an electron) and a strong electron acceptor like chlorine (an atom with a strong tendency to acquire an extra electron) come into contact, an electron may be completely transferred from the donor to the acceptor. The result, in the present example, is a sodium atom with one electron fewer than normal and a chlorine atom with one more electron than normal. Once it has lost an electron, the sodium is left with one more proton than it has electrons, and it therefore has a net charge of +1. Similarly, the chlorine atom that gained an electron has one more electron than it has protons and has a net charge of −1. Such charged atoms are called *ions,* and are symbolized by their chemical symbol with a superscript indicating the charge. Sodium and chlorine ions are written Na^+ and Cl^-.

A sodium ion with its positive charge and a chlorine ion (usually called a chloride ion) with its negative charge tend to attract each other, since opposite charges attract. This important kind of electrical interaction is known as *electrostatic attraction.* In this instance electrostatic attraction holds the two ions together to form the compound we know as table salt, or sodium chloride, NaCl (Fig. 2.11B). Such a bond, involving the complete transfer of an electron and the mutual electrostatic attraction of the two ions thus formed, is an *ionic bond.*

Ionic bonding may entail the transfer of more than one electron,

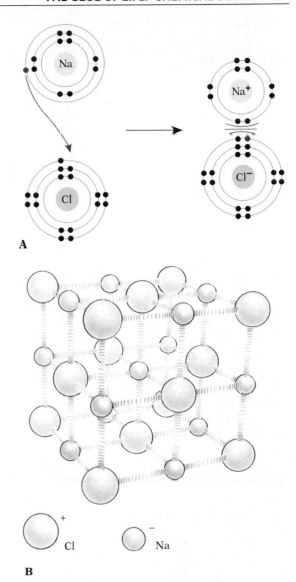

2.11 Ionic bonding of sodium and chlorine

Sodium has only one electron in its outer shell, while chlorine has seven (A). Sodium acts as an electron donor, giving up the one electron in its outer shell, whereupon the complete second shell functions as its new outer shell. Chlorine acts as an electron acceptor, picking up an additional electron to complete its outer shell. After donating an electron to chlorine, the sodium, left with one more proton than it has electrons, now has a positive charge. Conversely, the chlorine, with one more electron than it has protons, now has a negative charge. The two charged atoms, called ions, are attracted to each other electrostatically by their unlike charges. The result is sodium chloride (NaCl), or ordinary table salt, which in its crystalline state takes the form of a lattice-like structure (B).

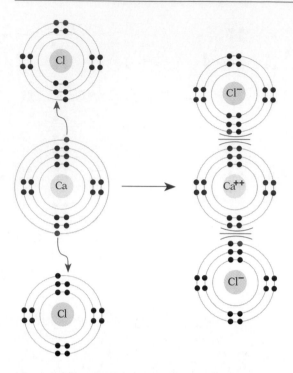

2.12 Ionic bonding of calcium and chlorine

Calcium has two electrons in its outer shell. It donates one to each of two chlorine atoms, and the two negatively charged chloride ions thus formed are attracted to the positively charged calcium ion to form calcium chloride ($CaCl_2$).

2.13 Ionization of sodium chloride

When in solution, the NaCl dissociates into separate Na^+ (yellow) and Cl^- (green) ions.

as in calcium chloride, another common salt. Calcium (atomic number, 20) has two electrons in its outermost shell, and it loses both to form the calcium ion, Ca^{++} (Fig. 2.12). Chlorine, however, requires only one electron to complete an octet in its outer shell. Hence it takes two chlorine atoms to act as acceptors for the two electrons from a single calcium atom; three ions bond together to form calcium chloride, symbolized as $CaCl_2$. Calcium has a bonding capacity, or *valence,* of +2, while sodium has a valence of +1 and chlorine a valence of −1.

Ionic bonding occurs between strong electron donors and strong electron acceptors. It is not common between configurations that have intermediate numbers of electrons in the outer shells, or between two strong electron donors or electron acceptors.

The binding of a calcium atom to two atoms of chlorine produces one molecule of calcium chloride. A *molecule* is generally defined as an electrically neutral aggregate of atoms bonded together strongly enough to be regarded as a single entity. In many instances, however, *ionization* (the transfer of one or more electrons from one atom to another to form ions) occurs without true molecular formation. Substances like sodium chloride (NaCl) and calcium chloride ($CaCl_2$), in which the bonds are ionic, tend to dissociate into separate ions when in solution. (We will look at why this happens shortly.) When they are ionized in solution, then, they exist as separate ions rather than molecules; NaCl, for instance, dissociates into a Na^+ ion and a Cl^- ion (Fig. 2.13).

Since ions are charged particles, they behave differently from neutral atoms or molecules, and substances wholly or partly ionized in water play many important roles in the functioning of biological systems. In later chapters we will see the effects of charge on the movements of materials through the membranes of cells.

■ ACIDS AND BASES

Cells are extemely sensitive to the chemistry of the fluids that bathe them. The delicate balance of acids and bases, in particular, is crucial to most tissues. An *acid* is a substance that increases the concentration of hydrogen ions (H^+) in water; a *base* is a substance that decreases the concentration of hydrogen ions, which in water is equivalent to increasing the concentration of hydroxyl ions (OH^-).

The degree of acidity or basicity (usually called alkalinity) of a solution is commonly measured in terms of *pH,* which reflects the concentration of hydrogen ions. The pH scale generally ranges from 0 on the acidic end to 14 on the alkaline end. A solution is neutral (containing equal concentrations of H^+ ions and OH^- ions) if its pH is exactly 7. Solutions with a pH of less than 7 are acidic (with a higher concentration of H^+ ions than of OH^- ions). Conversely, solutions with a pH higher than 7 are alkaline (with a higher concentration of OH^- ions than of H^+ ions). A change of one pH unit means a 10-fold change in the concentration of hydrogen ions. Thus the concentration of H^+ ions in the solution of a very strong acid (pH 0) is 100 trillion (10^{14}) times greater than in

2.14 The pH scale

The concentration of hydrogen ions in a solution is measured by pH. At pH 7, the concentration of hydrogen ions (H^+) exactly balances the concentration of hydroxyl ions (OH^-), and so the solution is neutral. At lower pH values (corresponding to higher H^+ concentrations) solutions are acidic; at higher pH values (corresponding to lower H^+ concentrations) solutions are alkaline, or basic. Notice that the pH number matches the concentration of H^+ in moles per liter—for example, pH 8 corresponds to a H^+ ion concentration of 10^{-8}. A mole of any compound contains the same number of atoms—Avogadro's number, or 6.02×10^{23}—as a mole of any other substance, and its weight in grams corresponds to the molecular weight (the summed atomic masses) of the substance.

pH is the negative logarithm of the hydrogen ion concentration:

$$pH = \log\left(\frac{1}{[H^+]}\right) = -\log[H^+]$$

	pH	Concentrations of ions (moles/liter)		
		H^+	OH^-	
Caustic soda (NaOH)	14	10^{-14}	10^0	ALKALINE
	13			ALKALINE
Detergent	12	10^{-12}	10^{-2}	ALKALINE
	11			ALKALINE
	10	10^{-10}	10^{-4}	ALKALINE
Baking soda	9			ALKALINE
Seawater	8	10^{-8}	10^{-6}	ALKALINE
Pure water	7	10^{-7}	10^{-7}	NEUTRAL
Saliva				
Unpolluted rainwater	6	10^{-6}	10^{-8}	ACIDIC
Coffee	5			ACIDIC
Typical acid rain / Beer	4	10^{-4}	10^{-10}	ACIDIC
Orange juice	3			ACIDIC
Carbonated soft drink	2	10^{-2}	10^{-12}	ACIDIC
Stomach acid	1			ACIDIC
Hydrochloric acid (HCl)	0	10^0	10^{-14}	ACIDIC

(Arrow: Increasing OH^- upward; Increasing H^+ downward)

the solution of a very strong base (pH 14). Figure 2.14 illustrates the range of pH values we normally encounter.

Except in parts of the animal digestive tract and a few other isolated areas, most cells function best when conditions are nearly neutral. Most of the interior material of living cells has a pH of about 6.8. The blood plasma and other fluids that bathe the cells in our own bodies have a pH of 7.2–7.3. Special mechanisms stabilize the pH of these fluids. In particular, chemicals known as buffers have the capacity to bond to H^+ ions, removing them from solution whenever their concentration begins to rise and, conversely, to release H^+ ions into solution whenever their concentration begins to fall. Buffers thus help minimize fluctuations in pH, which would otherwise be considerable since many of the biochemical reactions in cells either release or use up H^+ ions. The most important biological buffer in vertebrates is carbonic acid, which stabilizes the pH of the blood as it circulates through the tissues.

■ HOW STRONG BONDS ARE FORMED

Ionic bonds, as we have seen, involve the complete transfer of electrons from one atom to another. However, in most cases bonding occurs not by complete transfer, but by a sharing of electrons between the atoms involved. Bonds of this sort, based on shared electrons, are called *covalent bonds*. These may be nonpolar or polar.

Nonpolar bonds To see how covalent bonds are formed, consider an atom of hydrogen. A complete first shell for hydrogen would contain two electrons, one more than each atom has normally. Two atoms of hydrogen can bond to each other and form molecular hydrogen (H_2):

$$H\cdot + H\cdot \longrightarrow H\!:\!H$$

In this molecule, each atom shares its electron equally with the

2.15 Covalent bonding of two hydrogen atoms

The sharing of electrons may be indicated by overlapping electron clouds (A) or by interlocking orbital rings (B). The latter system of representation parallels the textual representation of H_2 as H : H.

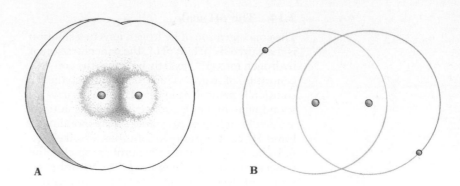

other atom, so that each hydrogen has, in a sense, two electrons (Fig. 2.15).

Covalent bonds are not limited to the sharing of one electron pair between two atoms. Sometimes two atoms share two or three electron pairs and form double or triple bonds. When two atoms of oxygen bond together, they form a double bond (since each oxygen atom needs two electrons to complete its outer shell), and when two atoms of nitrogen (atomic number 7) bond together, they form a triple bond, because each nitrogen atom needs three additional electrons to fill its outer shell:

$$\ddot{\text{O}}::\ddot{\text{O}} \qquad\qquad :\text{N}\vdots\text{N}:$$

A covalent bond may be represented simply by a line between two atoms, instead of a pair of dots; the other electrons in the outer shells are then ignored:

$$\text{H} - \text{H} \qquad \text{O} = \text{O} \qquad \text{N} \equiv \text{N}$$

The number of covalent bonds an atom can form is the same as its valence (ignoring the sign): hydrogen atoms tend to form only one bond, oxygen two bonds, and nitrogen three bonds. This covalent bonding capacity corresponds to the number of vacancies in the outer shell of an atom with one to three vacancies, and to the number of sharable electrons in an atom with one to three outer electrons. The maximum covalent bonding capacity is 4: in an atom with four outer electrons, this is the number both of sharable electrons *and* of vacancies. Carbon's bonding capacity of 4 confers on it the ability to form the maximum number of bonds.

Polar bonds Suppose that instead of being bonded to each other, two hydrogen atoms are covalently bonded to an oxygen atom, forming water (H_2O):

$$\text{H}\cdot + \text{H}\cdot + \cdot\ddot{\text{O}}: \longrightarrow \text{H}\overset{\cdots}{\underset{\text{H}}{:\!\ddot{\text{O}}:}}$$

Oxygen, with six electrons in its outer shell (Fig. 2.9), needs two more. By sharing electrons with two hydrogen atoms, the oxygen atom can obtain a full outer octet, while at the same time each hydrogen obtains a complete first shell of two electrons. A covalent bond between a hydrogen atom and an oxygen atom is somewhat different from one between two hydrogen atoms or between two oxygen atoms, however. No two elements have exactly the same

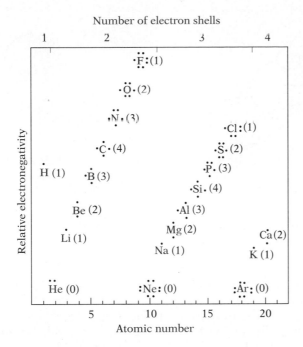

Number of electron shells

Relative electronegativity

Atomic number

2.16 Electronegativity

The relative tendency of an atom to attract electrons depends on the number of spaces in the outer shell left to be filled and on the distance of the outer shell from the nucleus. Hence, lithium (Li), with seven vacancies, is less attractive to electrons (that is, less electronegative) than carbon (C), which has four. Oxygen (O), with only two missing electrons, is yet more electronegative. This graph also helps explain why in methane (CH_4) the shared electrons will be nearer the carbon atom, while in carbon dioxide (CO_2) they will be nearer the oxygens: carbon is more electronegative than hydrogen but less electronegative than oxygen. The electronegativities of the noble gases (which have filled outer shells) are estimated. This knowledge of relative electronegativity permits us to make important predictions about many biochemical reactions. (The covalent bonding capacity of each atom is shown in parentheses.)

affinity for electrons. Consequently, when a covalent bond forms between two different elements, the shared electrons tend to be pulled closer to the more attractive element. Such a bond is called a *polar* covalent bond because the charge is distributed asymmetrically.

The formal measure of an atom's attraction for free electrons is its *electronegativity;* this depends on the number of vacancies in the outer shell—an atom like oxygen with only two electron openings is generally more electronegative than one with three or more—and on the distance of the outer shell from the nucleus (Fig. 2.16).

The phenomenon of polarity helps explain many of the properties of various molecules in living systems. Whole molecules can be polar as a result of the polarity of bonds within them. One example is the water molecule. We can imagine how, even though the two hydrogen-oxygen bonds are polar, the atoms in the water molecule might be aligned in a straight line so that the charge would be distributed symmetrically within the molecule, which would then be nonpolar:

$$\text{H} : \overset{..}{\underset{..}{\text{O}}} : \text{H}$$

However, this is not the actual arrangement. The covalent bonds formed by atoms are maximally separated from one another, and thus are oriented to the four corners of a tetrahedron (Fig. 2.17). The three atoms of water form a bent-chain, or V-shaped, structure, with the oxygen at the apex of the V and the two hydrogen atoms as the arms:

$$(+) \text{H} : \overset{(-)}{\underset{\overset{..}{\text{O}}}{\underset{\text{H}}{\overset{..}{\text{O}}}}}$$
$$(+)$$

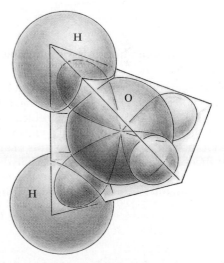

2.17 Structure of a water molecule

When oxygen bonds covalently with two hydrogen atoms, its second-level electrons become oriented to the four corners of a tetrahedron. As a result, the angle between the two hydrogens is neither 90° nor 180°, as might be expected from the perpendicular arrangement of the $2p$ orbitals, but rather is 104.5°.

2.18 Polarities of two biologically important molecules

(A) In carbon dioxide, each of the four electrons of the carbon's outer orbital is shared between the carbon (gray) and an oxygen (red). The result is a linear molecule with no electrical polarity. (B) In water, two hydrogen atoms (blue) are bonded covalently to one oxygen atom, but the shared electrons are pulled closer to the oxygen because of its higher electronegativity. Because the atoms in water are arranged at an angle of 104.5°, the charge distribution is asymmetrical, with negative charge concentrated at the oxygen end; as a result the molecule as a whole is polar.

CO_2

H_2O

A　　　　　　　　　　　　**B**

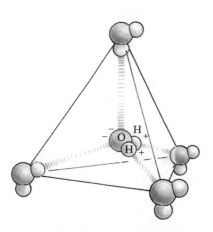

2.19 Hydrogen bonding between water molecules

Like the central H_2O molecule shown here, each water molecule can form hydrogen bonds (red bands) with four other water molecules. The array then assumes the shape of a tetrahedron. Water molecules near the edge of this imaginary tetrahedron can simultaneously form hydrogen bonds with two or three other water molecules, creating an interlocking array of tetrahedrons.

Since the electrons are drawn closer to the oxygen atom, there is a concentration of negative charge near the oxygen end of the molecule. Therefore, the molecule is polar (Fig. 2.18, right).

The carbon dioxide molecule, on the other hand, exhibits no polarity: its double bonds hold its atoms in rigid linear alignment (Fig. 2.19, left). Hence CO_2 is nonpolar.

As we will see shortly, the polarity of certain molecules has crucially important biological implications.

■ WHY ARE WEAK BONDS IMPORTANT?

Strong versus weak bonds To maintain internal stability, or **homeostasis,** living organisms must be able to adjust to the fluctuations of their environments. The changes all begin at the molecular level, and are powered by the liberated energy derived from strong, energy-rich covalent bonds. Covalent bonds are called strong because breaking them is hard, usually requiring between 50 and 110 kilocalories of energy per mole.[1] Bond breakage usually results from collisions of rapidly moving molecules. But since the energy from even the most rapidly moving molecules at physiological temperatures is almost never above 10 kcal/mole, covalent bonds are stable and show little tendency to rupture spontaneously.

However, life depends on a capacity for change, as well as on stability. The crucial sources of this ability to change are weak noncovalent bonds, which can readily be broken and re-formed. Ionic bonds in aqueous solutions, for instance, are relatively weak, averaging about 10 kcal/mole. The average duration of an ionic bond (the interval between formation of the bond and a collision with a

[1] A mole is the amount of a substance, in grams, that equals the combined atomic mass of all the atoms in a molecule of that substance; there are approximately 6×10^{23} molecules in a mole. A calorie (spelled with a small *c*) is defined as the quantity of energy, in the form of heat, required to raise the temperature of one gram of water from 14.5°C to 15.5°C. One **kilocalorie** (kcal) is 1000 calories. Nutritionists use a different scale to measure energy; their Calorie (spelled with a capital C) is equal to one kilocalorie on the standard scale.

BIOXPLORER PLUS

Student Software

DOWNLOAD a sample chapter from the World Wide Web
http://web.wwnorton.com/bioxplus.html

This enjoyable and easy-to-use program uses animation, simulated interactive experiments, quizzes, matching games, and an on-line glossary to help you understand and remember important concepts. **BioXplorer Plus** is carefully keyed to *Biological Science*, Sixth Edition, and includes page references to the text on every card.

REVIEW key text material, summarized on more than 250 brief notecards.

ANIMATIONS of complex biological processes, such as active transport across biological membranes, the mechanisms of photosynthesis and respiration, and recombinant DNA technology, make difficult concepts easier to grasp. Interactive **SIMULATIONS** allow you to manipulate variables within set experiments and see the results on-screen.

Test your understanding with the **CHAPTER QUIZ** feature. **BioXplorer Plus** includes more than 550 questions (with answers and feedback) covering every chapter of the text.

A **GLOSSARY**, complete with search function, provides handy definitions of key terms.

Detach and mail to:
W.W. Norton & Company
c/o National Book Company
800 Keystone Industrial Park
Scranton, PA 18512-4601
Attn: Mary C

System Requirements: The Macintosh version will run on any system that has sufficient memory to run Hypercard. The Windows version requires a 386 or higher processor operating under Windows 3.1 or higher and at least 8MB of memory.

YES, please send me BioXplorer Plus:

BioXplorer Plus	$17.95
Tax (in CA, NY, PA, IL)	_____
Total	_____

Choose one: ❑ Windows® (3.5")　　　❑ Macintosh®

❑ My check for $17.95 (plus tax, where applicable) is enclosed. Make checks payble to W. W. Norton & Company.

Please charge to my: ❑ Visa ❑ MasterCard ❑ American Express

Acct. # _____　　Exp. date _____　Telephone _____

Name _____　　Signature _____

Ship to: Address _____

City _____ State/Prov. _____ Zip/PC _____

Billing Address (if different): Address: _____

City: _____ State/Prov. _____ Zip/PC _____

molecule moving rapidly enough to break it) is quite short. As a result, several weak bonds must act in concert to produce molecular stability sufficient for most of life's metabolic processes. The weak bonds of biological significance include ionic bonds (discussed earlier), hydrogen bonds, and van der Waals interactions.

Hydrogen bonds The electrostatic attraction between oppositely charged portions of neighboring polar molecules creates *hydrogen bonds* (also called *polar bonds*). Water molecules provide an excellent example. The hydrogen atoms in each water molecule are covalently bonded to the oxygen atom, but because of the polarity of the bond—the electrons being closer to the oxygen end than to the hydrogen ends—each hydrogen has a net positive charge. The hydrogen atoms are therefore attracted to other oxygen atoms, with their net negative charge, in nearby water molecules. Specifically, each hydrogen can form a weak attachment with the oxygen of another water molecule, and the oxygen can form a weak attachment with two hydrogens in other water molecules. Thus each water molecule has the potential for being simultaneously linked by hydrogen bonds to four other water molecules (Fig. 2.19), which are themselves weakly bonded to yet other water molecules.

The distinction between hydrogen bonds and ionic bonds is clear: hydrogen bonds result from the electrostatic attraction between polar but electrically neutral molecules like water, while ionic bonds result from the electrostatic attraction between oppositely charged atoms (ions). There is also a hybrid of these two kinds of electrostatic attraction: ions can attract polar molecules, as we will see when we discuss, later in the chapter, the *hydration sphere*—the shell of polar water molecules drawn around an ion in solution. In aqueous solutions, hydrogen bonds usually have a bonding energy of about 4–5 kcal/mole; ionic bonds, about 10 kcal/mole; and hybrid polar-ionic bonds such as those of the hydration sphere, about 7–8 kcal/mole.

Van der Waals interactions Much weaker than ionic or hydrogen bonds are the linkages known as *van der Waals interactions*, which have bonding energies of only 1–2 kcal/mole. These linkages occur between electrically neutral molecules (or parts of molecules) when they are a precise distance from each other that allows the electrons in their outer orbitals to begin a synchronous, mutually avoiding motion. The normal repulsion between the two sets of outer electrons is lessened, and the atoms are able to bond weakly to each other. Van der Waals interactions play a crucial role in the enzymatic reactions that control virtually all the processes of life. They are also important in stabilizing aggregations of nonpolar molecules in cells.

The strength in weakness Weak bonds play a crucial role in stabilizing the shape of many of the large molecules—DNA and proteins in particular—and they often hold together groups of such molecules in orderly arrays. Like the minute hooks and eyes of a Velcro fastening, the bonds are individually quite weak (their average duration is only 10^{-11} sec), but they can be strong and stable

2.20   Stability from weak bonds

An array of many individually weak bonds—here represented as the many hooks and eyes of a Velcro fastener—can be surprisingly strong as a unit.

when many act together (Fig. 2.20). However, just as a Velcro patch is easy to unfasten from one end, a few "bonds" at a time, so too an array of molecules held together by weak bonds (as opposed to the more gripperlike covalent bonds) can be disassembled and rearranged with relative ease by forces that dislodge the fastening bonds one by one. An enormous number of life processes depend on just these sorts of changes.

WHY LIFE NEEDS INORGANIC MOLECULES

Chemists have traditionally referred to complex molecules containing carbon as *organic* compounds. All other compounds are called *inorganic.* Despite their name, many inorganic (non-carbon-based) substances are basic to the chemistry of life.

■ THE REMARKABLE CHEMISTRY OF WATER

Life on earth is totally dependent on water. Between 70% and 90% of all living tissue is water, and the chemical reactions of life all take place in an aqueous medium.

Water as a solvent   One of the main reasons water is so well adapted as the medium for life is that it is a superb solvent for many important classes of chemicals. This advantage arises from the marked polarity of water molecules and the corresponding propensity of water to form hydrogen bonds. Thanks to this polarity, both ionic and polar molecules are soluble in water (Fig. 2.21). Let us consider the effect of water on each.

2.21   Dead Sea

Because water is polar and NaCl is ionic in solution, relatively large amounts of salt can be dissolved in water. In the Dead Sea, nearly 30% of the fluid is dissolved salt, and the density is so great that humans float high in the water.

2.22   Hydration spheres of Na$^+$ and Cl$^-$

When dissolved in water, each Na$^+$ and Cl$^-$ ion is hydrated—that is, surrounded by water molecules electrostatically attracted to it. Note that the oxygen (red) of the water molecules is attracted to the positively charged Na$^+$, while the hydrogen (blue) of the water molecules is attracted to the negatively charged Cl$^-$. Water molecules in a hydration sphere are called bound water. This bonding between ion and polar molecules (red bands) makes evident the common electrostatic basis of ionic bonds and hydrogen (polar) bonds.

The ionic bonds linking the atoms of NaCl are relatively weak when the salt is in an aqueous medium, but within a dry crystal of the same salt the bonds are comparatively strong. Why the difference? When the crystal is put into water, the negatively charged oxygen ends of the water molecules are attracted to the positively charged sodium ions: similarly the positively charged hydrogen ends of the water molecules are attracted to the negatively charged chloride ions. By force of numbers these attractions are able to overcome the strong mutual attraction between the Na$^+$ and Cl$^-$ ions. In water, then, the ionic bonds are broken with extreme ease because of the competitive attraction of the water. The Na$^+$ and Cl$^-$ ions dissociate, and each of the ions is surrounded by a hydration sphere of regularly arranged water molecules electrostatically attracted to it—a process called *hydration* (Fig. 2.22). The ionic bonds between Na$^+$ and Cl$^-$ atoms are now weaker simply because the ions are kept far apart, and the strength of electrostatic attraction decreases exponentially with distance.

Water is also an excellent solvent for nonionic molecules if they are polar. Such molecules—ethyl alcohol, for instance—are called *hydrophilic* ("water-loving"). They dissolve in water because of the electrostatic attraction between the charged parts of the solute molecules and the oppositely charged parts of the water molecules. This occurs especially when the molecule has an oxygen with a hydrogen attached to it (−OH). As in water molecules, the hydrogen in such a group has a net positive charge and is therefore attracted by the negatively charged oxygen end of a nearby water molecule, with the result that a hydrogen bond is formed. The dissolved

In water

2.23   Polar basis of solubility

When a polar substance such as glucose, an energy-rich sugar (left), is placed in contact with
water, the water molecules are attracted to the
polar atoms of the sugar. (For clarity the polar
−OH groups—green and black spheres—are
shown for only one of the sugar molecules.) The
water forms hydrogen bonds with the substance,
surrounding it with water molecules, and so
dissolves it (right).

(solute) molecules and the water molecules thus become weakly
linked to each other (Fig. 2.23).

Substances that do not dissolve in water are electrically neutral
and nonpolar. They therefore show no tendency to interact electrostatically with water. A ***hydrophobic*** ("water-fearing") substance
like vegetable oil, when stirred into water, will soon begin to separate out, because the water molecules tend slowly to reestablish the
hydrogen bonds broken by the physical intrusion of the insoluble
material, and thus push it out. As a result, the nonpolar molecules
tend to coalesce to form droplets, which usually fuse and form a
separate layer outside the water (Fig. 2.24). As we will see, this tendency of hydrophobic molecules to be driven out of water is the
basis for the spontaneous formation of the cell membranes that
protect organisms as they grow and develop.

The polar nature of water also helps explain the mundane but
important workings of soaps and other detergents. Stains left by
water-soluble chemicals readily dissolve away when washed in
water, but grease and other hydrophobic materials leave stains that
cannot be lifted so easily. Detergents (Fig. 2.25) are able to remove
grease from a natural fabric because their molecules have ionic
heads and hydrophobic tails. The heads make them soluble in
water, while the tails tend to be driven into tight-packed company
with the grease molecules and dissolve them by incorporating
them into hydrophobic detergent droplets. Nonpolar fabrics (like
most synthetics) pose a special problem because water cannot easily wet them, which it must if the detergent is to be carried into the
fibers; even if the water does penetrate well, the water will often
"herd" the stain molecules into close association with the fabric

rather than the detergent. Molecules of fluids used by dry cleaners stand a better chance because they readily wet nonpolar materials and dissolve grease, yet because they are small enough to evaporate, will not simply remain with the fabric after displacing the stain. Covalently bonded stains like dried blood are notoriously difficult to remove; they require chemically active agents like bleach or digestive enzymes.

The stickiness of water Water molecules in the hydration spheres of ions are arranged in orderly arrays; such water, referred to as *bound water,* is essentially immobilized. The same is true of water molecules around polar groups of nonionic compounds. The orderly arrays of bound water are very different from those of pure water (see Fig. 2.19), and the physical properties of bound water are consequently different from those of free water; the greater the proportion of bound water in a given volume, the lower the freezing point and the higher the boiling point of that liquid. Since much of the water inside cells is bound water, the physical properties of the cell contents are very different from those of pure water.

The strong ordering of water molecules by hydrogen bonding has important implications for life. For example, water has a high *surface tension*: the surface of a volume of water is not easily broken. The effects of surface tension are evident when a water strider

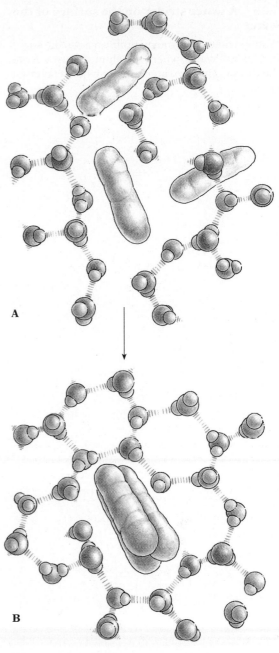

A

B

2.24 Water-induced clumping of hydrophobic molecules

Dispersed hydrophobic molecules disrupt the polar bonding pattern of pure water, so that few hydrogen bonds can form in the solution (A). As hydrophobic molecules (represented as brown ovals) encounter one another randomly in a solution of water, they tend to become trapped in clumps by polar bonding of water molecules to one another (B). Because there is more polar bonding when hydrophobic molecules are clumped, the solution becomes stabilized in this form.

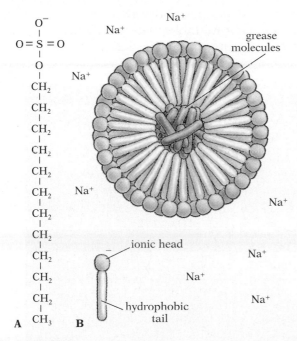

A

B

2.25 Detergents

Sodium dodecyl sulfate is a strong detergent often used to dissolve cell membranes and other hydrophobic molecules in experiments requiring the separation of these components for further analysis. It has a long, straight hydrophobic tail and, because it ionizes in water, a charged head (A). Water molecules dissolve the heads and drive the tails into tightly packed clumps that dissolve hydrophobic grease molecules (B). During washing, whether in a laboratory preparation or a home washing machine, the entire assembly is rinsed out.

2.26 A water strider on the surface of the water

Water striders can move rapidly across the surface of still water, where they hunt for prey. Note the dimples in the water surface where each foot rests.

or other insect walks on the surface of a pond without breaking the surface (Fig. 2.26) or when the water in a glass that is filled slightly above the rim does not spill. Water has a high surface tension because hydrogen bonds link the molecules at the surface to each other and to the molecules below them. Before the legs of the water strider (or any other object, for that matter) can penetrate the water's surface, they must break some of these hydrogen bonds and deform the orderly array of water molecules.

Just as water molecules are attracted electrostatically to areas of charge on dissolved molecules, so also are they attracted to the charged groups that characterize hydrophilic surfaces. Consequently such surfaces are **wettable**—that is, water spreads over them and binds loosely to them. By contrast, hydrophobic surfaces—those of most plastics and waxes, for example—lack surface charge and are not wettable; water on them will form isolated droplets, but will not spread out (Fig. 2.27).

2.27 Water beading

The polar bonding between water molecules causes droplets to form on hydrophobic surfaces like this feather.

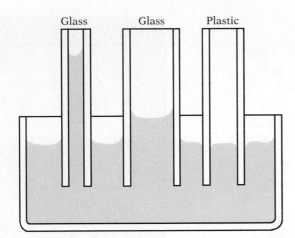

Glass   Glass   Plastic

2.28 Capillarity

Water rises higher in a glass tube of small bore (left) than in one of large bore (center) because in the smaller tube a higher percentage of the water molecules are in direct contact with the glass and can form hydrogen bonds with charged groups on the glass. By contrast, water cannot "stick" to the surface of a plastic tube (right) because plastic is uncharged.

The propensity of water to bind to hydrophilic surfaces explains the phenomenon of *capillarity*—the tendency of aqueous liquids to rise in narrow tubes. If the end of a narrow glass tube is inserted below the surface of a volume of water, water will rise in the tube to a level well above that of the water outside (Fig. 2.28). Because glass is hydrophilic, the water molecules are electrostatically attracted to the numerous charged groups on its surface and tend to creep up the inside of the tube. As the ring of water molecules in contact with the inner surface creeps upward, it pulls along other water molecules to which it is linked by hydrogen bonds. The water level stops rising when the pull of gravity just counteracts the electrostatic forces that contribute to capillarity. The larger the diameter of the tube, however, the smaller the percentage of water molecules in direct contact with the glass and, correspondingly, the smaller the rise in the water. Even though the relatively few molecules in contact with the glass have a tendency to creep upward, they are held back by their cohesion via the network of polar bonding with the rest of the water in the tube.

Capillarity is by no means restricted to glass tubes. Water will climb any charged surface. We are all familiar with the way it climbs up the fibers of paper towels and spreads through the fibers of many kinds of cloth.

Unlike most other substances, which become increasingly dense as the temperature falls, water first becomes denser as it is chilled to 4°C, and then begins to expand again below 4°C. This means that ice, being less dense than cold water, floats and—furthermore—that ponds and streams freeze from the top down rather than from the bottom up. The crust of ice that forms at the surface insulates the water below it from the cold air above and thereby often prevents the pond or stream from freezing solid, even in very cold weather. This special property of water makes life possible in the many ponds and streams that would otherwise freeze solid in the winter.

This curious behavior of cold water is a consequence of polar bonding. Though water molecules have the potential for forming hydrogen bonds with four other water molecules (see Fig. 2.19), this potential is not fully realized because molecular motion prevents stabilization. Hydrogen bonding reaches its full potential in

A

B

2.29 Molecular structure of ice

Because of the tetrahedral arrangement around each water molecule, the lattice is an open one, with considerable space between molecules (A). In liquid water the arrangement is not quite so rigid, and the packing of molecules is therefore slightly denser, but the general lattice arrangement is nonetheless largely preserved. (Planes have been added to help show the three-dimensional disposition of the molecules.) Note the hexagons created by the hydrogen bonding in this bit of ice. This conformation is the basis of the hexagonal shape of most snow crystals (B). Each snowflake contains about 10^{16} water molecules.

water that has frozen into ice. When all four possible bonds have formed, each is oriented in space at the greatest possible distance from the other three. Consequently the bonds are directed toward the four corners of the tetrahedron. The resulting three-dimensional lattice of water molecules in ice is an open one (Fig. 2.29); the rigidly tetrahedral structure maintains space between the molecules, so they can be only loosely packed compared to liquid water.

Water as a temperature regulator The hydrogen bonds in water give it a high internal cohesion, which enables it to absorb much heat without undergoing a large increase in temperature and to release much heat energy without undergoing a great drop in temperature. When most substances absorb heat, their molecules move more rapidly in relation to one another; temperature is an indication of the amount of such molecular motion. In water, by contrast, much of the absorbed heat energy is dissipated in increased vibration of the hydrogens, each of which is shared between the oxygen to which it is covalently bound and the oxygen of another water molecule, to which it is electrostatically bound. As a result, relatively little of the added heat energy is expressed as movement of whole water molecules, so the temperature increase is therefore modest. The high *heat capacity* of water (the amount of heat energy that must be added or subtracted to change the temperature by 1°C) and its high *heat of vaporization* (the amount of heat energy required for evaporation, or turning water from liquid to vapor) enables it to act as an effective buffer against extreme temperature fluctuations in the environment (Fig. 2.30). In this way water helps stabilize the earth's temperature within the range favorable to life.

■ CARBON DIOXIDE AND THE CYCLE OF LIFE

Carbon has four electrons in its outer electron shell and thus a covalent bonding capacity of 4. Carbon dioxide (CO_2) is formed when two atoms of oxygen bond to one atom of carbon. Though this substance contains carbon, it is generally considered inorganic because it is simpler than the compounds classified as organic.

Only a very small fraction of the atmosphere (0.033%) is CO_2; yet atmospheric CO_2 is the principal inorganic source of carbon, and carbon is the principal structural element of living tissue. Before CO_2 can take part in chemical reactions, it must usually first dissolve in water, which it does very readily on the thin aqueous films that coat most cells. Carbon dioxide then reacts with the water to form carbonic acid (H_2CO_3):

$$CO_2 + H_2O \longrightarrow H_2CO_3$$

This reaction involves so little change in energy that it is easily reversible, and CO_2 can readily be released from water solution when conditions are appropriate:

$$H_2CO_3 \longrightarrow CO_2 + H_2O$$

Carbon dioxide and water are the raw materials from which plants manufacture many complex organic compounds essential to life. When these complex compounds are metabolized, they are broken down again to carbon dioxide and water, and the carbon dioxide is eventually released into the atmosphere. Carbon dioxide, then, is the beginning and the end of the immensely complex carbon cycle in nature.

■ OXYGEN AND LIFE

Molecular oxygen (O_2) constitutes approximately 21% of the atmosphere. It is necessary for life in most organisms. It is utilized by both plants and animals in the process of extracting energy from nutrients. Its role, as we will see, is to serve as the ultimate acceptor of electrons. This is a crucial task: without oxygen to accept electrons, most cells can run at only 5% of their normal efficiency. Oxygen is not very soluble in water, but it dissolves enough to supply the needs of aquatic organisms provided (1) that the

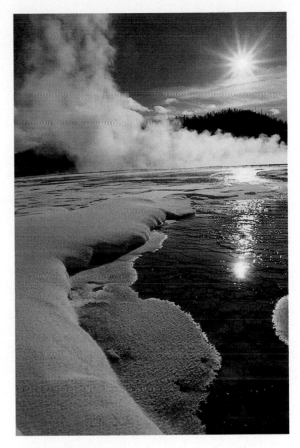

2.30 Water vapor
The three states of water are visible in this volcanic pool in Yellowstone Park during the winter.

water is not too hot and (2) that the water's surface is exposed to the air or, alternatively, plants are growing in it, releasing oxygen as they photosynthesize. Photosynthetic bacteria and plants are the source of virtually all atmospheric oxygen.

Though water, oxygen, and carbon dioxide are truly basic to life as we know it, still other compounds are used to capture, store, transport, and utilize the energy that fuels life. In the next chapter we will examine these complex, carbon-based compounds in an effort to understand the chemical reactions that make life possible.

CHAPTER SUMMARY

THE ELEMENTS ESSENTIAL TO LIFE

THE STRUCTURE OF ATOMS Negatively charged electrons occupy cloud-like regions around the positively charged protons of the nucleus. Electrons can have only certain energy levels, and they decay spontaneously to any vacancy in the lowest available level. (p. 21)

HOW ELECTRONS DETERMINE CHEMISTRY The chemistry of an element—particularly which other elements it will react with—is largely determined by the number of electrons in its outermost (highest) energy level. (p. 26)

WHEN ATOMS DECAY Atoms with unstable combinations of neutrons and protons tend to decay; the particles thrown off by such decay can be used to trace reactions and date fossils. (p. 27)

THE GLUE OF LIFE: CHEMICAL BONDS

BONDS BETWEEN IONS Atoms with very few outer electrons or very few outer vacancies tend to lose or gain electrons, respectively, becoming ions. Oppositely charged ions are attracted to one another electrostatically. (p. 29)

ACIDS AND BASES Acids increase the concentration of H^+ ions in a solution; bases increase the concentration of OH^- ions. Acidity is measured on a pH scale. (p. 30)

HOW STRONG BONDS ARE FORMED Atoms with complimentary numbers of outer electrons and outer-electron vacancies can share electrons, cooperating to complete each other's outer level and thus forming a strong covalent bond. When the shared electrons are attracted more strongly by one atom (the more electronegative atom), and the geometry of the resulting molecule does not cancel out this electron asymmetry, the distribution of charge in the molecule is unequal, generating a polar molecule. (p. 31)

WHY ARE WEAK BONDS IMPORTANT? Electrostatic bonds between polar molecules are weak, allowing flexibility in bonding but great strength when many of these weak hydrogen (polar) bonds cooperate. (p. 34)

WHY LIFE NEEDS INORGANIC MOLECULES

THE REMARKABLE CHEMISTRY OF WATER Water is polar and thus dissolves other polar molecules and ions while forcing nonpolar molecules into clumps. Hydrogen bonds between water molecules give water its surface tension, as well as its ability to climb polar surfaces and the capacity to retard temperature changes. (p. 36)

CARBON DIOXIDE AND THE CYCLE OF LIFE Carbon is the basis of all organic compounds; it comes initially from CO_2 captured by plants. (p. 43)

OXYGEN AND LIFE Oxygen is a by-product of photosynthesis and is essential to the efficient use of energy by both plants and animals. (p. 43)

STUDY QUESTIONS

1 Compare and contrast hydrogen (polar) bonds, ionic bonds, and covalent bonds. (pp. 29–34)

2 Why are electrons so important to the chemistry of life? (pp. 26–34)

3 What is electronegativity, and why is it important? (pp. 33–34)

4 What sorts of chemicals dissolve well in water? Explain their solubility. (pp. 36–39)

5 Do hydrogen bonds have to involve hydrogen? (p. 35)

SUGGESTED READING

ATKINS, P. W., 1987. *Molecules*. Scientific American Library, New York. *A beautifully produced "molecular glossary" illustrating the chemical formula, three-dimensional structure, and biological action of many common molecules.*

DICKERSON, R. E., AND I. GEIS, 1976. *Chemistry, Matter, and the Universe*. W. A. Benjamin, Menlo Park, Calif. *An excellent introduction to chemistry from a biological perspective; a real classic.*

CHAPTER 3

THE CHEMISTRY OF LIFE

Chemistry does what all good science does: it makes complex phenomena easier to predict. Chemistry explains why certain reactions among molecules will take place, and why particular molecular combinations will be stable. Biochemistry applies this understanding to organic compounds and the reactions critical to life. This chapter examines the molecular actors and how they interact in cells.

WHY CARBON?

The source of the vast chemical diversity in living things is the bonding capacity of just one element—carbon. Carbon's power lies in its versatile structure: four unpaired electrons in its outer shell, which allow it to form covalent bonds with up to four other atoms. These four bonding options make possible an almost endless variety of carbon-based, and thus organic, molecules. We will look first at the important kinds of organic compounds—carbohydrates, lipids, proteins, and nucleic acids—and then at how some of the crucial well-ordered chemical changes inside cells are orchestrated and controlled.

Though carbon can and does bond to a variety of elements, its four unpaired electrons are most commonly bonded to hydrogen, oxygen, nitrogen, or more carbon. Compounds containing only carbon and hydrogen, the *hydrocarbons*, are of central importance in organic chemistry; the number of different compounds of this kind is immense. The readiness with which carbon-carbon bonds can form and produce chains of varying lengths and shapes generates a great variety of hydrocarbons. Hydrocarbon chains may be simple, branched, or may form circles of varying numbers of carbons (Fig. 3.1). The more atoms a molecule contains, the more different arrangements of those atoms will be possible. Compounds with the same atomic content and molecular formula but different atomic arrangements are called *isomers* (Fig. 3.2). Very large organic molecules may have hundreds of isomers, with many differing physical properties.

Another source of variety in hydrocarbons is the capacity of ad-

straight chains

Propane

Butane

branched chains

Isobutane

Isopentane

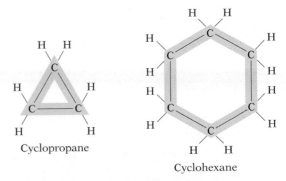

closed chains

Cyclopropane

Cyclohexane

3.1 Examples of hydrocarbons

The molecules appear flat in these conventionally drawn structural diagrams, though they are, in fact, three-dimensional. The bonds around a carbon atom that forms only single bonds are oriented toward the four corners of a tetrahedron.

Glucose   Galactose   Fructose

double bond   **triple bond**

Ethylene   Acetylene

3.3 Hydrocarbons with double and triple bonds

3.2 Three isomeric hexoses

Each of these six-carbon sugars has the same molecular formula, $C_6H_{12}O_6$; hence each is an isomer of the others.

jacent carbon atoms to form single, double, or triple bonds (Fig. 3.3). Substitution of other elements or groups of elements for hydrogen atoms in hydrocarbons makes possible an almost infinite number of **derivative hydrocarbons**. The total number of hydrocarbons and derivatives that form in nature exceeds half a million. This great capacity for diverse atomic organization makes hydrocarbons ideal for building chemicals with unique properties.

The four major classes of complex organic compounds are carbohydrates, lipids (both derivative hydrocarbons), proteins, and nucleic acids. Molecules in each of these classes are often identified on the basis of the subgroups they contain. Each subgroup, or **functional group**, has its own characteristic properties, which help determine solubility, reactivity, and other traits of the chemical "personality" of the whole molecule. We will refer to many of the groups listed in Table 3.1 in this and later chapters.

Table 3.1 Important functional groups

GROUP	NAME	PROPERTIES	IONIC FORM
—OH	**Hydroxyl**	Polar (soluble, because it is able to form hydrogen bonds)	
—C—OH (with H above and below)	**Alcohol**	Polar (soluble)	
—C(=O)—OH	**Carboxyl**	Polar (soluble); often loses its hydrogen, becoming negatively charged (an acid)	—C(=O)—O⁻
—N (with H above and below)	**Amino**	Polar (soluble); often gains a hydrogen, becoming positively charged (a base)	—N—H⁺ (with H above and below)
—C(=O)—H	**Aldehyde**	Polar (soluble)	
C=O	**Ketone**	Polar (soluble)	
—C—H (with H above and below)	**Methyl**	Hydrophobic (insoluble); least reactive of the side groups	
—P(=O)—OH (with OH below)	**Phosphate**	Polar (soluble); usually loses its hydrogens, becoming negatively charged (an acid)	—P(=O)—O⁻ (with O⁻ below)

■ THE ROLE OF CARBOHYDRATES

Carbohydrates are derivative hydrocarbons composed of carbon, hydrogen, and oxygen. In simple carbohydrates the hydrogen and oxygen are characteristically present in the same proportions as in water: there are two hydrogen atoms and one oxygen atom for each carbon atom. Here is its actual molecular arrangement:

$$H-\overset{\displaystyle |}{\underset{\displaystyle |}{C}}-OH$$

Some carbohydrates, such as starch and cellulose, are very large complex molecules. However, like most large organic molecules, they are composed of many simpler "building-block" compounds bonded together (Fig. 3.4). Understanding the constituent building blocks is the first step toward understanding more complex substances.

Ready energy: the simple sugars The basic carbohydrate molecules are simple sugars, or *monosaccharides*. All sugars when in straight-chain form contain a —C=O group (Fig. 3.2). If the double-bonded O is attached to a terminal C of a chain, the combination is called an aldehyde group; if it is attached to a nonterminal C, the combination is called a ketone group (Table 3.1). The —OH (hydroxyl) groups are polar. Hence sugars readily form hydrogen bonds with water and are soluble, unlike simple nonpolar hydrocarbon molecules, which tend to clump together in water.

3.4   A moderately complex organic compound

One of the largest of the moderately complex organic compounds is starch, the main energy-storage molecule in plants. Despite its apparent complexity, the starch molecule is actually composed of a repetitive string of glucose units, each represented here as a hexagon. Only a small part of one starch chain is shown. A branched molecule composed of units of several different compounds would be far more complex.

STRUCTURAL ISOMERS

Ethyl alcohol   Dimethyl ether

GEOMETRIC STEREOISOMERS

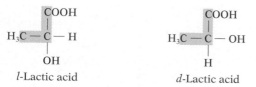

Maleic acid   Fumaric acid

OPTICAL STEREOISOMERS

l-Lactic acid   d-Lactic acid

3.5 Three types of isomerism

The two structural isomers differ in the basic grouping of their constituent atoms, one being an alcohol (characterized by an —OH group) and the other an ether (characterized by an oxygen bonded between two carbons). The two geometric stereoisomers are fixed in different spatial arrangements by their inability to rotate around the double bond between the middle two carbons. As Figure 3.6 shows, the two optical stereoisomers are asymmetric molecules that cannot be superimposed on each other.

3.6 Optical stereoisomers and chemical specificity

Of these two stereoisomers of lactic acid, only one (top) can fit all the holes in the hypothetical schematic molecule to which it binds. The other will not fit the holes no matter which way it is turned. Subtle distinctions of this kind are crucially important: they enable molecules to recognize and bind certain other particular molecules as part of the network of critical chemical reactions necessary for life.

The six-carbon sugars (hexoses) are the most important building blocks for more complex carbohydrates. There are many six-carbon sugars, **glucose** and fructose being two of the most common. All have the same molecular formula, $C_6H_{12}O_6$, and are therefore isomers. Glucose and fructose are structural isomers (see Fig. 3.2); the basic groupings of their constituent atoms are different, making one an aldehyde and the other a ketone sugar.

In addition to structural isomerism, there is another, more subtle, kind of isomerism called stereoisomerism. In a given pair of stereoisomers, identical groups are attached to the carbon atoms, but the spatial arrangements of the attached groups are different. The two middle compounds shown in Figure 3.5 are stereoisomers; if the carbon-carbon bonds in these two molecules were single, the two would in fact be the same compound, because free rotation is possible around a single bond. A double bond, however, holds its atoms in a rigid configuration.

Despite the way they are usually drawn, stereoisomers (Fig. 3.5, bottom) are not flat. In a carbon atom the four unpaired electrons form the corners of a tetrahedron; if in two molecules the groups attached to corresponding electrons are different, the resulting molecules will be different (Fig. 3.6). Glucose and galactose are stereoisomers because the position of one —OH group is different in each (see Fig. 3.2). The stereoisomers of a compound usually

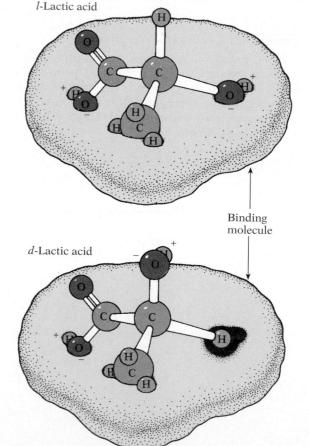

l-Lactic acid

Binding molecule

d-Lactic acid

3.7   Two forms of glucose

Glucose may exist in the straight-chain aldehyde form (left) or as a ring structure (center). The ring structure is the most common. (By convention, the unmarked corners of the hexagon signify carbon atoms.) Both representations fail to convey the true shape of the molecules, since the four bonds of each carbon atom are directed to the four corners of a tetrahedron. The illustration at right is a more realistic representation of the ring form, but such realism is possible for only the simplest organic molecules.

have quite different biological properties and behavior; subtle differences in shape may determine which isomer can bind to or react with a particular molecule and which cannot (Fig. 3.6). Hence, different isomers can play very different roles in the chemistry of cells.

Glucose, which usually exists in a ring form composed of five carbons and one oxygen (Fig. 3.7), plays a unique role in the chemistry of life. As the primary product of photosynthesis in plants, glucose becomes the ultimate source of all the carbon atoms in both plant and animal tissue. Moreover, the energy stored in its covalent bonds is usually, directly or indirectly, the source of the energy that powers cells. Other six-carbon monosaccharides, among them fructose and galactose, are constantly being converted into glucose or synthesized from glucose. Even fats and proteins can be converted into glucose or synthesized from glucose in cells.

In addition to ordinary monosaccharides composed only of carbon, oxygen, and hydrogen, there is a variety of derivative monosaccharides containing other elements. For example, some have a phosphate group attached to one of the carbons, and others an *amino* group: a nitrogen with two hydrogens, $-NH_2$ (Fig. 3.8).

Double sugars   The *disaccharides* are composed of two simple sugars bonded through reactions that remove a molecule of water. This kind of reaction series is consequently called a *condensation* or *dehydration reaction*. The disaccharide *maltose*, or *malt sugar*, for instance, is synthesized by a condensation reaction between two molecules of glucose. The overall reaction (omiting intermediate steps) is:

$$2C_6H_{12}O_6 \longrightarrow C_{12}H_{22}O_{11} + H_2O$$

The hydrogen atom from a hydroxyl group ($-OH$) of one glucose molecule combines with a complete hydroxyl group from another

Glucosamine   Glucose–6–phosphate

3.8   Two examples of derivative monosaccharides

Glucosamine, a chemical used in the synthesis of some protein building blocks as well as in insect exoskeletons, is a glucose molecule with an amino group ($-NH_2$) substituted for an $-OH$ group. Similarly, glucose-6-phosphate, which is used in an important step in the harvesting of energy from glucose, is glucose with a phosphate group added.

glucose molecule to form water (Fig. 3.9A). The oxygen valence −1 (created by the removal of hydrogen) and the carbon valence +1 (created by the removal of —OH) are neutralized by the bonding together of the oxygen of one glucose molecule with the carbon of the other glucose. As a result, the two glucose units are connected by an oxygen atom shared between them, producing the disaccharide maltose.

Sucrose—common table sugar—is also a disaccharide. It is synthesized by a condensation reaction between glucose and fructose. *Lactose*, or milk sugar, is a disaccharide composed of glucose and galactose joined by a condensation reaction (Fig. 3.9B). In fact, the synthesis of complex molecules from simpler units almost always produces water.

Disaccharides can be broken down to their constituent simple

A Glucose + Glucose = Maltose + Water

B Galactose + Glucose = Lactose + Water

C Sucrose + Water = Glucose + Fructose

3.9 Synthesis and digestion of disaccharides

Removal of a molecule of water (blue) between two molecules of sugar (a condensation reaction) results in the formation of a bond (red) between the two. In the first two examples shown here, intermediate steps are omitted. (A) A bottom-to-bottom (α) linkage forms between two glucose molecules, yielding maltose. (B) A top-to-bottom (β) linkage forms between galactose and glucose, yielding the milk sugar lactose; some adults suffer from milk intolerance because they cannot digest this linkage. (C) The hydrolysis reaction leading to the breakdown of sucrose involves adding back a water molecule.

3.10 Branched starch

Shown here is a small segment of a starch molecule. This starch is branched, but some forms are unbranched. Like the cellulose of plants, starch is a polymer of glucose, but in cellulose the glucose is connected by β linkages (see Fig. 3.9), whereas in starch the glucose is connected by α linkages.

sugars by the reverse process: adding back a water molecule. This reaction, called **hydrolysis**, splits a water molecule into a hydrogen atom and a hydroxyl group, and then adds them to the subunits through a series of steps. The summary reaction is:

$$C_{12}H_{22}O_{11} + H_2O \longrightarrow 2C_6H_{12}O_6$$

Hydrolysis reactions are particularly important in digestion, which breaks down complex molecules into simple building blocks, ready for subsequent use (Fig. 3.9C).

Storage and structure: the polysaccharides The prefix *poly-* means "many." Thus **polysaccharides** are complex carbohydrates composed of many simple-sugar building blocks bonded together in long chains (Fig. 3.10). They are synthesized by condensation reaction and can be broken down into their constituent sugars by hydrolysis.

A number of complex polysaccharides are of great importance in biology. **Starches** are the principal carbohydrate storage products of higher plants. They are composed of many hundreds of glucose units bonded together. **Glycogen** is the principal carbohydrate storage product in animals. Its molecules are much like those of starch; they have the same type of bond between adjacent glucose units. **Cellulose** is the most common carbohydrate on earth. It is a highly insoluble polysaccharide used by plants as their major supporting material. The bonds between its glucose units are β linkages rather than α linkages (Fig. 3.11); animals can digest the bonds of starch and glycogen, but most animals are unable to hydrolyze those of cellulose. **Chitin** (Fig. 3.11), the polysaccharide that serves as the major structural component of insect exoskeletons and fungal cell walls, is functionally equivalent to cellulose (Fig. 3.12).

All reactions in which small molecules (called **monomers**) bond together to form long chains are called polymerization reactions; polymerization of monosaccharides, for example, creates a polysaccharide. The products of polymerization are called **polymers**. Polymers play a critical role in biology, as we will see.

Cellulose

3.11 Cellulose and chitin

Cellulose is composed of long chains of β-linked glucoses. Chitin is composed of β-linked acetyl-glucosamines (glucoses with a combined acetyl-amino side group, —NHCOCH₃).

Chitin

B

3.12 (A) Mantis exoskeleton and (B) scanning electron micrograph (SEM) of paper fibers

The chitin of insect exoskeletons (visible here in the shed skin of a mantis) is chemically similar and functionally equivalent to the cellulose of plants (seen here as fibers in uncoated paper). Books like this one have a coating of clay over the fibers; this coating, which gives the paper a shine, prevents absorption and spreading of the ink, and thus permits the printing of high-resolution photographs.

A

■ THE ROLE OF LIPIDS

Like carbohydrates, a second major group of derivative hydrocarbons, the lipids, are composed principally of carbon, hydrogen, and oxygen, but they may also contain other elements, particularly phosphorus and nitrogen. In their simplest form, **lipids** are hydrocarbons with a carboxyl group (—COOH) at one end (Fig. 3.13). Such lipids are primarily nonpolar, by virtue of their long hydrocarbon tails. They are therefore relatively insoluble in water but will dissolve in organic solvents such as ether. Most lipids, as we will see, are more complex, having an ionic group attached to the carboxyl end; the long hydrophobic tails are universal features.

Storage and insulation: the fats and oils Among the best-known lipids are the neutral fats, which are called oils when liquid. Important as energy-storage molecules, the fats also provide insulation, cushioning, and protection for various parts of the body (Fig. 3.14). Each molecule of fat is composed of fatty acids joined together by glycerol. Glycerol (also called glycerin) has a backbone

3.13 A simple lipid

3.14 Fat-storage cells

Lipids are stored in spherical fat cells called adipocytes, seen here. Small blood vessels (capillaries) and support fibers (collagen) hold these cells in place. In the same manner as the styrofoam beads they resemble, these human adipocytes cushion and insulate underlying parts of the body. (From *Tissues and Organs: A Text-Atlas of Scanning Electron Microscopy* by Richard G. Kessel and Randy H. Kardon. Copyright 1979 W. H. Freeman and Company. Used with permission.)

of three carbon atoms, each carrying a hydroxyl —OH group (Fig. 3.15).

Fatty acids, like all organic acids, contain a carboxyl group. When both a double-bonded oxygen and an —OH group are attached to the same carbon atom, the double-bonded oxygen tends to cause the —OH part of this carboxyl to lose its hydrogen, making the group ionic and causing the compound to behave as an acid (see Table 3.1).

There are many different fatty acids, which vary in carbon-chain length, in the number of single or double carbon-carbon bonds, and in other characteristics. The fatty acids in edible fats and oils contain an even number of carbon atoms, and most of them have from 4 to 24 carbons in their backbones; three of the most common are stearic acid (18 carbons), palmitic acid (16 carbons), and linoleic acid (18 carbons; Fig. 3.16).

Organic acids and alcohols have a tendency to combine through condensation reactions. Since glycerol is an alcohol (by virtue of having a —CH$_2$OH group) and has three hydroxyls, it can combine with three molecules of fatty acid to form a molecule of fat (Fig. 3.15). Hence fats are sometimes also called triglycerides.

The various fats differ in the specific fatty acids, or types of fatty acids, composing them. You have doubtless read of the concern over saturated fats. *Saturated fats* are simply those incorporating fatty acids with the maximum possible number of hydrogen atoms attached to each carbon, and hence no carbon-carbon double bonds (Fig. 3.16). The fatty acids in *unsaturated fats* (actually oils, since they are usually liquid at room temperature) have at least one carbon-carbon double bond—that is, they are not completely saturated with hydrogen. The double bond induces a "kink" in the otherwise linear structure, which prevents solidification. There is now good evidence that an elevated intake of saturated fats is one of many factors predisposing humans to atherosclerosis—a disease of the arteries in which fatty deposits in the arterial walls cause partial obstruction of blood flow, which can lead to strokes and heart failure.

Since fats are synthesized by condensation reactions, they, like complex carbohydrates, can be broken down into their building-block compounds by hydrolysis, as happens in digestion. In addition, because fat, though slow to be metabolized, contains 2.5 times as much usable energy per gram as monosaccharides, it is a good substance for long-term energy storage. In fact, if all the energy the average individual stores in fat (about a month's supply) were to be maintained in the form of sugar, we would each be 25–30 kg heavier.

The key to membranes: phospholipids Various lipids contain a phosphate group at the carboxyl end of the chain. Among the most common of these phospholipids are those composed of one unit of glycerol, two units of fatty acid, and a phosphate group often linked with a nitrogen-containing group (Fig. 3.17). The phosphate group is bonded to the glycerol at the point where the third fatty acid would be in a fat. Because the phosphate group tends to lose a hydrogen ion, one of the oxygens usually becomes negatively charged; similarly, the nitrogen, being electronegative, tends to attract a hydrogen ion and thus to become positively charged. In

3.15 Synthesis of a fat

Removal of three molecules of water by condensation reactions results in the bonding of three molecules of fatty acid to a single molecule of glycerol. (Intermediate steps are omitted in this example.) Conversely, three molecules of water will be added by hydrolysis when this molecule of fat is digested. The carbon chains of the fatty acids are usually longer than shown here.

Glycerol   +   Fatty acids   =   Fat   +   Water

Palmitic acid

Linoleic acid

3.16 Examples of saturated and unsaturated fatty acids

Palmitic acid is saturated with hydrogen—that is, it contains the maximum number of hydrogens possible. By contrast, linoleic acid, with its two inflexible carbon-carbon double bonds, accommodates four fewer than the maximum number of hydrogens.

3.17 A phospholipid

The portion of the molecule with the phosphate and nitrogenous groups (brown) is soluble in water, whereas the two hydrocarbon chains are not. This particular phospholipid, ethanolamine phosphoglyceride, is one of the two most abundant in the cell membranes of higher plants and animals.

3.18 A steroid

All steroids have the same basic unit of four inter-locking rings, but they differ in their side groups. This particular steroid is cholesterol. (By convention, a hexagon signifies a six-carbon ring with its valences completed by hydrogens; see cyclohexane in Figure 3.1. A pentagon signifies a five-carbon ring, also with hydrogens attached to the carbons.)

short, the end of the phospholipid molecule with the phosphate and nitrogenous groups is strongly ionic and hence soluble in water, whereas the other end, composed of the two long hydrocarbon tails of the fatty acids, is nonpolar and insoluble. This curious property of solubility at one end but not at the other makes phospholipids especially well suited to function in cellular membranes, as we will see in Chapter 4.

Steroids   Though commonly classified as lipids because their solubility characteristics are similar to those of fats, oils, waxes, and phospholipids, the steroids differ markedly in structure from the other lipids we have discussed (Fig. 3.18). They are not based on fatty acid–alcohol bonds, but are complex molecules composed of four interlocking rings of carbon atoms, with various side groups attached to the rings. Steroids are very important biologically. Some vitamins and hormones are steroids, and steroids often occur as structural elements in living cells, particularly in cellular membranes.

■ THE ROLE OF PROTEINS

Far more complex than either carbohydrates or lipids, proteins are fundamental to both the structure and function of living material. They are directly responsible for controlling the delicate chemistry of the cell and exist in thousands of different forms. Like carbohydrates and lipids, proteins are composed of simple building-block compounds.

The building blocks and primary structure of proteins   All *proteins* contain four elements: carbon, hydrogen, oxygen, and nitrogen; most proteins also contain some sulfur. These elements are bonded together to form subunits called amino acids, which, being organic acids, contain the carboxyl (—COOH) group (Fig. 3.19). In addition, they each have an amino (—NH$_2$) group. Finally, each amino acid has a side chain, designated R:

NONPOLAR R GROUPS

Glycine (Gly)

Alanine (Ala)

Valine (Val)

Leucine (Leu)

Isoleucine (Ile)

Methionine (Met)

Phenylalanine (Phe)

Tryptophan (Trp)

Proline (Pro)

POLAR R GROUPS

Serine (Ser)

Threonine (Thr)

Cysteine (Cys)

Tyrosine (Tyr)

Asparagine (Asn)

Glutamine (Gln)

IONIC R GROUPS

Aspartic acid (Asp)

Glutamic acid (Glu)

Lysine (Lys)

Arginine (Arg)

Histidine (His)

3.19 The 20 amino acids common in proteins classified by R group

The amino acids are shown in their ionized form. All have the same arrangement of a carboxyl group and an amino group attached to the same carbon; they differ in their R groups (red or blue). Top: These nine amino acids have nonpolar R groups and are relatively insoluble in water. Glycine is an exception. Its R group, a single hydrogen atom, is nonpolar but is too small to outweigh the charge of the amino and carboxyl groups. The molecule therefore behaves more like a polar amino acid and is water-soluble. Proline is also unusual. It is technically not an amino acid because the nitrogen is bonded to part of the R group. However, it is included because it is regularly incorporated into proteins along with the true amino acids. Middle: These six amino acids have polar R groups and are soluble. Bottom: These five amino acids, with R groups ionized at intracellular pH levels, are electrically charged and thus water-soluble; the first two, being negatively charged, are acidic, whereas the last three, with a positive charge, are alkaline.

The side chains of the various amino acids can be very simple, as in glycine (only a hydrogen atom), or they may be very complex, as in tryptophan, which includes two ring structures. Twenty different amino acids are commonly found in proteins (Fig. 3.19). The various R groups give each of the amino acids different characteristics, which in turn greatly influence the properties of the proteins incorporating them. For example, some amino acids are relatively insoluble in water, owing to R groups that are nonpolar at a pH between 6.5 and 7 (Fig. 3.19, top), whereas other amino acids are water-soluble because their R groups are polar (middle) or ionic (that is, electrically charged; bottom row).

Proteins are long and complex polymers of the 20 common amino acids. The amino acid building blocks bond together by condensation reactions between the —COOH groups and the —NH₂ groups (Fig. 3.20). The covalent bonds between amino acids are called *peptide bonds*, and the chains they produce are called *polypeptide chains*. The amino acid units incorporated into a chain are called peptides. The number of peptides in a single polypeptide chain within a protein molecule is usually between 40 and 500; for any given protein, however, the chain length is constant. Proteins have a three-dimensional shape that is largely determined by the distribution of their polar, charged, and nonpolar R groups: the chain will tend to fold so that the nonpolar (hydrophobic) groups are inside the protein, while the hydrophilic groups are exposed at the surface, where they interact with nearby polar molecules, particularly water.

Protein molecules often consist of more than one polypeptide chain. The chains may be held together by numerous weak bonds, especially hydrogen bonds; for example, a single molecule of hemoglobin, the red oxygen-carrying protein in blood, has four polypeptide chains linked by hydrogen bonds. Insulin, an important hormone secreted by the pancreas in vertebrates, exemplifies a protein with polypeptide chains held together by both hydrogen and covalent bonds. The covalent bonds, called *disulfide bonds*,

3.20   Synthesis of a polypeptide chain

Condensation reactions between the —COOH and —NH₂ groups of adjacent amino acids result in peptide bonds (red) between the acids. Notice again that the process of combining units releases water.

3.21 Structural formula of a cystine bridge

A cystine bridge is formed when two cysteine peptides (orange) from different parts of a protein are linked by a disulfide bond.

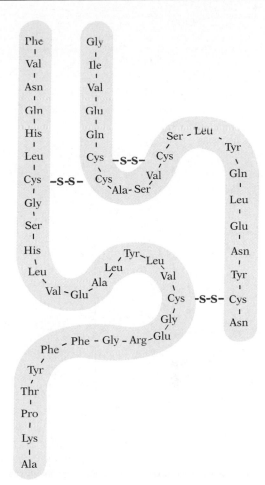

3.22 Structure of bovine insulin

The molecule consists of two polypeptide chains joined by two disulfide bonds. There is also one disulfide bond within the shorter chain (right). Hydrogen bonds (not shown) between the chains and between segments of the same chain are also present.

are between the sulfur atoms of two units of the amino acid cysteine (Fig. 3.19), which can react with each other to form a cystine bridge (Fig. 3.21). Disulfide bonds can also link two parts of a single polypeptide chain, maintaining it in a bent or folded shape (Fig. 3.22).

The number and sequence of amino acids is the ***primary structure*** of a protein molecule. The primary structure is different for every kind of protein; the potential number of different proteins is enormous. For a relatively short polypeptide chain of 100 amino acids, for instance, 20^{100} sequences are possible. Most organisms have between 1000 and 50,000 different proteins.

How proteins get their shape   Proteins are not laid out simply as straight chains of amino acids. Instead, they coil and fold into complex spatial conformations, which play a crucial role in determining the distinctive biological properties of each protein. Much of this three-dimensional character is a consequence of weak interactions between peptides in the protein. Certain precise degrees of coiling allow internal hydrogen bonds to form and stabilize what is called an ***alpha (α) helix*** (Fig. 3.23). A helix can be visualized as a ribbon wound around a regular cylinder (Fig. 3.23A). In a protein, each complete turn of the helix takes up approximately 3.6 amino acid units of the polypeptide chain (Fig. 3.23B). The chain is held

○ carbon ○ Nitrogen

○ Oxygen ○ Hydrogen

○ R group ┊ H bond

A

B

C

1 unit

1 unit

1 amino acid

1 amino acid

3.23 Alpha-helical secondary structure of some proteins

(A) The helix may be visualized as a ribbon wrapped around a regular cylinder. (B) The backbone of a polypeptide chain (the repeating sequence of N–C–C–N–C–C–N–C–C–) is shown coiled in a helix (with all other atoms and R groups omitted). It takes approximately 3.6 amino acid units (N–C–C) to form one complete turn of the helix. The hydrogen bonds shown extend between the amino group of one amino acid and the oxygen of the third amino acid beyond it along the polypeptide chain. (C) A ball-and-stick model of an α-helical section of a protein.

in this helical shape by hydrogen bonds formed between the amino group of one amino acid and the oxygen of the third amino acid beyond it along the polypeptide chain—which is the amino acid next to it in the axial direction of the helix (Fig. 3.23C). The *conformation*, or arrangement in space, of amino acids in a peptide chain is called *secondary structure*.

The helical pattern is seen at its simplest in some *fibrous proteins*. One category of these insoluble proteins includes the *keratins*, which provide the structural elements for many of the specialized derivatives of skin cells. Keratins with extensive α-helical secondary structure, such as nails, hooves, and horns, are hard and brittle. The hardness results from an extraordinarily large number of covalent cystine bridges: as many as 1 amino acid in 4 is cysteine. Others, such as hair and wool, are soft and flexible and can easily be stretched (especially when moistened and warmed). The stretching is possible because there are many fewer cystine bridges. The intrachain hydrogen bonds are easily broken, and the polypeptide chains can then be pulled out of their compact helical shapes into a more extended form. The chains tend to contract to their normal length, with re-formation of the hydrogen-bonded α helix, when the tension on them is released (or when they are dried and cooled).

Another stable arrangement of peptides within a polypeptide chain is the beta (β) structure, often called the *pleated sheet*. In this conformation, also seen at its simplest among some of the keratins, many side-by-side polypeptide chains are cross-linked by interchain hydrogen bonds (Fig. 3.24). The resulting arrangement is flexible and strong but resists stretching because the polypeptide chains are already almost fully extended. The best-studied β-keratin is silk; other examples include spider webs, feathers (Fig. 3.25), and the scales, claws, and beaks of reptiles and birds. A number of important variants of the β structure give other distinctive shapes to protein segments.

Another kind of fibrous protein, with its own distinctive secondary structure, is *collagen*, the most abundant protein in higher

3.24 Pleated-sheet secondary structure of some proteins

Diagrammatic representation of five parallel polypeptide chains in β conformation, with the imaginary pleated sheet shown for four of the chains.

3.25 The feather of a bird, a β-keratin

This SEM shows the base of a parakeet's tail feather.

3.26   Model of a portion of a molecule of collagen

Three polypeptide chains, each helically coiled, are wound around one another to form a triple helix. The "sheaths" here and in Figures 3.27 and 3.30 are intended as a reminder that each molecule consists not merely of a backbone, but also of R groups, which give it volume.

vertebrates. Collagen can constitute one-third or more of all the body protein and is especially abundant in skin, tendons, ligaments, bones, and the cornea of the eye. A molecule of collagen comprises three polypeptide chains, each first helically coiled and then wound around the other two to form a triple helix (Fig. 3.26). Collagen fibers are exceedingly strong and very resistant to stretching.

The polypeptide chains of globular proteins are folded into complicated spherical or globular shapes (Fig. 3.27). Because of charged and polar R groups on their exposed surfaces, globular proteins (which include organic catalysts known as enzymes), proteinaceous hormones, the antibodies of the immune system, and most blood proteins are usually water-soluble. Typically, they are made up of sections of α helix and β sheet interspersed with non-helical regions, strengthened by disulfide bonds (Fig. 3.28). The protein myoglobin, which is the oxygen-storage protein in muscles, provides a typical example of a helix-dominated protein. It consists of one polypeptide chain containing eight α helices connected by short regions of irregular (nonhelical) coiling. At each nonhelical region, the three-dimensional orientation of the polypeptide chain changes, giving rise to the protein's characteristic folding pattern. This three-dimensional folding pattern, which is superimposed on the secondary structure, is called *tertiary structure* (Fig. 3.27). In practice, tertiary structure is difficult to determine. The protein must first be crystallized. Then X rays are beamed through the crystals; deflected by the electrons of the thousands of atoms, they form a pattern that is then deciphered by a computer. This process is called X-ray crystallography (Fig. 3.29).

When a globular protein is composed of two or more independently folded polypeptide chains loosely held together (usually by weak bonds), the manner in which the already folded subunits fit together is called *quaternary structure* (Fig. 3.30).

Several aspects of a protein's primary structure (its amino acid sequence) contribute to producing its tertiary and quaternary

3.27   Tertiary structure of myoglobin

Myoglobin is a globular protein related to hemoglobin and, like hemoglobin, is characterized by a strong affinity for molecular oxygen. It has a single, complexly folded polypeptide chain of 151 amino acid units, and attached to the chain is a nonproteinaceous group called heme (represented by the disk). The polypeptide chain consists of eight sections of α helix (labeled A through H), with nonhelical regions between them. These nonhelical regions are a major factor in determining the tertiary structure of the molecule—that is, the way the helical sections are folded together. Section D is oriented perpendicular to the plane of the page.

3.28 Mixture of helices and sheets in an enzyme

The enzyme lactate dehydrogenase incorporates five α helices and six β sheets, indicated schematically here as coils (red) and ribbons (blue).

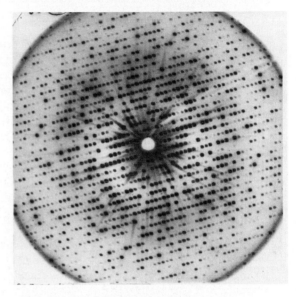

3.29 X-ray diffraction pattern of myoglobin

The intensity of each dot in this photograph of an X-ray diffraction pattern of sperm whale myoglobin provides information about the location of atoms in the molecule.

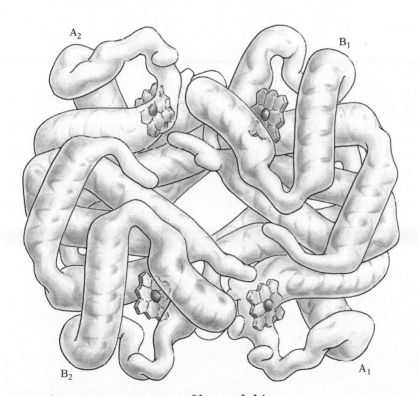

3.30 Quaternary structure of hemoglobin

A single molecule of hemoglobin is composed of four independent polypeptide chains, each of which has a globular conformation and its own prosthetic group. The spatial relationship between these four—the way they fit together—is called the quaternary structure of the protein. Chains A_1 and A_2 are identical, as are B_1 and B_2.

HOW WE KNOW:
CHROMATOGRAPHY

Chromatography is one of the most valuable techniques available for separating the substances found in blood or cells, or separating a single substance into its constituent parts. It is chromatography that first demonstrated that bacteria have more than 1000 different proteins, and allowed most of them to be individually isolated. After isolating a single protein using chromatography, we can hydrolyze it and then use chromatography again to separate the amino acids from one another. The many kinds of chromatography all share the same fundamental mechanism—the simultaneous exposure of the mixture being studied to two different substances, such as two solvents that will not mix, or a solvent and an adsorbent solid. (An adsorbent is so named because molecules of a gas, dissolved solute, or liquid adhere to its surface.) Each molecule of each solute in the mixture will diffuse back and forth between the two substances; the relative affinity of a solute molecule for those two solvents will determine how long, on average, it remains in each. For example, if solute A, a material highly soluble in water but only minimally soluble in phenol, is shaken in a jar with water and phenol (which do not mix), each molecule of A will divide its time between the two solvents in such a fashion that it will more often be in the water.

Paper chromatography is one of the simplest kinds of chromatography. Several drops of a mixture of unknown molecules are placed near one bottom corner of a piece of filter paper moistened with water (A, in the figure). The bottom edge of the paper is then dipped into a nonaqueous solvent such as phenol. As the solvent migrates up the paper by capillary action, those solutes in the mixture that have a much higher affinity for the solvent than for the water in the filter paper travel freely up the paper with the solvent. By contrast, those solutes that have a much higher affinity for the water do not travel far, but instead quickly transfer from the flowing solvent and bind to the stationary film of water on the filter paper. Materials with intermediate affinities for the solvent, compared with the water, travel intermediate distances. At the end of a measured time interval, the various solutes in the original mixture have come to rest at different places along the filter paper (B). Frequently, further separation is achieved by using a second solvent and allowing the solutes to travel across the paper in a new direction (C and D). The technique is then known as two-axis chromatography.

Another way to separate substances is electrophoresis. After chromatographic separation along one axis, an electric field is set up along the other; molecules with different ionic strengths and polarities migrate at different speeds toward one end or the other.

Column chromatography works like simple paper chromatography, except that instead of filter paper, a glass column packed with some hydrated adsorbent material, such as starch or silica gel, is used. The mixture to be tested, in a nonaqueous solvent, is poured into the top of the column and allowed to filter downward (or is pulled down by a pump). Each component in the mixture tends to move at its own rate, which depends on its relative affinity for the flowing solvent and the stationary beads of adsorbent material. When enough solvent is poured into the column to keep all the materials moving, each substance emerges from the bottom of the column at a different time, and they are collected in separate containers, ready for further analysis.

mixture of unknown molecules

molecules partly separated

molecules more completely separated

A   B   C   D

Solvent 1   Solvent 2

3.31 Folding of a globular protein in solution

As water molecules form hydrogen bonds with polar and ionic side groups, polypeptide chains can fold spontaneously so that the nonpolar groups are herded into the middle.

structure. If, for example, a polypeptide chain contains two cysteine units, the intrachain disulfide bond joining them may introduce a fold in the chain or stabilize one created in other ways (Fig. 3.22). The most common cause of folding is proline. Wherever there is a proline, a kink or bend occurs, because the structure of proline cannot conform to the geometry of an α helix. Four of the eight bends in globular myoglobin, in fact, result from the presence of prolines in the chain.

The distinctive properties of the various R groups of the amino acids also impose constraints on the shape of the protein. For example, hydrophobic groups tend to be close to each other in the interior of the folded chains—as far away as possible from the water that suffuses living tissue—whereas hydrophilic groups tend to be on the outside, in contact with the water (Fig. 3.31). In myoglobin, too, all the hydrophobic peptides are in the interior, and all but two of the hydrophilic peptides are on the outside. (The two exceptions, both ionic amino acids, hold the central, iron-containing heme group in place.) Thus the various kinds of weak bonds discussed earlier play crucially important roles in forming and stabilizing the tertiary structure of proteins.

Ultimately, the primary structure of a protein determines its spatial conformation. This is because there is one energetically most favorable, and therefore most stable, possible arrangement of the polypeptide chains. Further support for this comes from studies of **denatured** proteins—proteins that have lost most of their secondary, tertiary, and quaternary structure (and with it their normal biological activity) through exposure to high temperature or extreme pH. Under favorable test-tube conditions, some denatured proteins can spontaneously refold, regaining their native three-dimensional conformation, and thus recovering their normal biological activity. Since only the primary structure is available to dictate the folding pattern in such cases, it alone must be sufficient to determine all other aspects of protein structure (Fig. 3.32). In the crowded conditions within cells, however, many proteins need help from other molecules (called chaperones) to fold into a biologically active shape; most chaperones bind to the hydrophobic regions of unfolded protein chains and delay folding long enough to allow the protein to "discover" its most stable tertiary structure.

Conjugated proteins Attached to some proteins are nonproteinaceous groups called **prosthetic groups**; an example is the heme group of myoglobin, a disklike structure with an iron atom in its center (see Fig. 3.27). Prosthetic groups may be as simple as a single metal ion bonded to the polypeptide chain; they may be sugars or other carbohydrate entities, or they may be composed of lipids. The prosthetic group alters the properties of the protein in important ways. Without their heme groups, for example, myoglobin and

polar or ionic R groups

nonpolar R groups

unfolded polypeptide

folded conformation in solution

hydrophobic core region contains nonpolar R groups

hydrogen bonds form between water and both ionic and polar R groups on the outside of the molecule

A

B

3.32 Denaturation and renaturation of ribonuclease

When ribonuclease, a normally globular protein (A), is denatured, with both its weak bonds and its four intrachain disulfide bonds (green) broken, it unfolds into an irregularly coiled state (B). In this denatured condition the ribonuclease lacks its usual ability to digest RNA. When the denaturing agents are removed and favorable conditions restored, the protein spontaneously refolds into its native conformation (A) and regains its capacity for biological activity as an enzyme. Even the four disulfide bonds re-form correctly. Since there are 105 possible ways to join the cysteines but the enzyme folds only to bring the correct pairs together, the conformation does not result just from the positions of the cysteines; most disulfide bonds serve to stabilize, rather than to determine, structure.

3.33 Diagram of a nucleotide

A phosphate group and a nitrogenous base are attached to a five-carbon sugar. The sugar shown is a ribose, part of RNA; the sugar in DNA (deoxyribose) lacks the oxygen (red) at the bottom.

hemoglobin lose their high affinity for molecular oxygen. All proteins that contain nonproteinaceous substances are called **conjugated proteins**.

■ THE ROLE OF NUCLEIC ACIDS

Nucleic acids constitute a fourth major class of organic compounds crucial to all life. They are the materials of which genes, the units of heredity, are composed. They are also the messenger substances that convey information from the genes in the nucleus to the rest of the cell.

Nucleic acid molecules are long polymers of smaller building blocks called **nucleotides**, which are themselves composed of smaller units: a five-carbon sugar, a phosphate group, and an organic nitrogen-containing base. Both the phosphate group and the nitrogenous base are covalently bonded to the sugar (Fig. 3.33).

Information storage: deoxyribonucleic acid   This nucleic acid, commonly called **DNA**, is the one genes are made of. Four different kinds of nucleotide building blocks occur in DNA. All have deoxyribose as their sugar, but they differ in their nitrogenous bases. Two bases, **adenine** and **guanine**, are double-ring structures of a class known as purines; the other two, **cytosine** and **thymine**, are single-ring structures known as pyrimidines (Fig. 3.34).

The nucleotides within a DNA molecule are bonded so that the sugar of one nucleotide is attached to the phosphate group of the next nucleotide in the sequence (Fig. 3.35). Thus a long chain of alternating sugar and phosphate groups is established, with the nitrogenous bases oriented as side groups off this chain. The sequence in which the four different nucleotides occur is essentially constant in DNA molecules of the same species but differs between species. It is this sequence that determines the specificity of each type of DNA that encodes hereditary information, which is expressed largely through protein synthesis.

DNA molecules do not ordinarily exist in the single-chain form shown in Figure 3.35. Instead, two such chains, oriented in opposite directions, are arranged side by side like the uprights of a ladder, with their nitrogenous bases constituting the cross rungs of the ladder (Fig. 3.36). The two chains are held together by hydro-

3.34 Nitrogenous bases in DNA and RNA

The three single-ring bases—uracil, thymine, and cytosine—are pyrimidines; the two double-ring bases—adenine and guanine—are purines. Adenine, guanine, and cytosine are used in both DNA and RNA; thymine is restricted to DNA, while uracil takes its place in RNA.

PYRIMIDINES   PURINES

Uracil

Thymine   Adenine

Cytosine   Guanine

3.35 Portion of a single chain of DNA

Nucleotides are linked by bonds between the sugars and the phosphate groups. The nitrogenous bases (G, guanine; T, thymine; C, cytosine; A, adenine) are side groups.

3.36 Portion of a DNA molecule uncoiled

The molecule has a ladderlike structure, with the two uprights composed of alternating sugar and phosphate groups and the cross rungs composed of paired nitrogenous bases. Each cross rung has one purine base and one pyrimidine base. When the purine is guanine (G), the pyrimidine with which it is paired is always cytosine (C); when the purine is adenine (A), the pyrimidine is thymine (T). Adenine and thymine are linked by two hydrogen bonds (red bands), guanine and cytosine by three. Note that the two chains run in opposite directions—that is, the free phosphate is at the upper end of the left chain and at the lower end of the right chain.

3.37   A model of the DNA molecule

The double-chained structure is coiled in a helix. As shown in detail in the second segment, it consists of two polynucleotide chains held together by hydrogen bonds (red bands) between their adjacent bases.

gen bonds between adjacent bases. Finally, the entire double-chain molecule is coiled into a double helix (Fig. 3.37).

The regular helical coiling and the hydrogen bonding between bases impose two important constraints on how the cross rungs of the ladderlike DNA molecule can be constructed. First, each rung must be composed of a purine (double ring) and a pyrimidine (single ring); only in this way will all cross rungs be of the same length, allowing the formation of a regular helix. Second, if the purine is adenine, the pyrimidine must be thymine, and, similarly, if the purine is guanine, the pyrimidine must be cytosine; only these two pairs are capable of forming the required hydrogen bonds (Fig. 3.36). Since it does not matter in which order the members of a pair appear (A–T or T–A; G–C or C–G), the double-chain molecule can have four different kinds of cross rungs (Fig. 3.36). The biological significance of this arrangement is that the base sequence of one chain uniquely specifies the base sequence of the other, so that the two strands can be separated and exact copies made every time a cell divides.

Jack of all trades: ribonucleic acid   A second important category of nucleic acids comprises the ribonucleic acids, or **_RNA_**. There are several types of RNA, each with a different role in protein synthesis. Some act as messengers carrying instructions from the DNA to the sites of protein synthesis in the cell. Others are structural components of subcellular structures, called ribosomes, on which protein synthesis takes place. Still others transport amino acids to the ribosomes for use in synthesis. A few even direct chemical reactions. RNA differs from DNA in three principal ways: (1) The sugar in RNA is ribose, whereas that in DNA is deoxyribose (Fig. 3.33); (2) instead of the thymine found in DNA, RNA contains a very similar base called **_uracil_** (Fig. 3.34); (3) RNA is ordinarily single-stranded, whereas DNA is usually double-stranded.

The relative contributions of DNA, RNA, proteins, and lipids to the volume of a typical cell are compared in Figure 3.38.

3.38   Chemical composition of a typical mammalian cell

Values are expressed as a percent of weight.

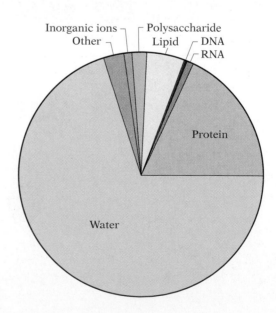

HOW CHEMICAL REACTIONS WORK

All the processes of life depend on the ordered flow of energy, and in particular on the behavior of electrons. Virtually all the energy for living things comes as light from the sun and is captured by electrons, which are thereby excited into higher-energy orbitals. The energy released by such electrons as they move to more highly electronegative atoms is harvested to fuel all the processes of life. To understand biology, then, it is essential to understand how the transfer of energy in chemical reactions takes place as one set of covalent bonds is replaced by another, an important subject in the field known as thermodynamics.

■ WHY FREE ENERGY ISN'T FREE

Suppose that two substances, A and B, can react with each other in solution to produce two new compounds, C and D:

$$A + B \longrightarrow C + D$$
reactants　　　products

What determines whether a reaction will take place spontaneously? The answer is that **energy**, defined as *the capacity to do work*, must be available. **Free energy** (as the term is used in chemistry and biology) is the energy in a system available for doing work under conditions of constant temperature and pressure. Where there is energy to be tapped—whether in the weight of the water stored behind a dam, in the covalent bonds of sugars like glucose, in an electron that has been excited into a higher orbital by sunlight, or in the tightly bound nuclei of the atoms in a nuclear reactor—the potential for work is present (Fig. 3.39).

The **First Law of Thermodynamics**—the Law of Conservation of Energy—tells us that *the total energy in the universe is constant*: if the energy needed to do work in a particular system—in a cell, for example—is not already available internally, it must be obtained from a source outside that system, which thereby loses a corresponding amount of energy. The **Second Law of Thermodynamics** states that *in the universe as a whole the total amount of free energy is declining*. This is because practically every energy transfer generates heat that is then no longer available for doing work. The magnitude of this waste is enormous, as the need for cooling towers in power plants and radiators in car engines to dissipate unused (and unusable) thermal energy makes evident. Even in our own bodies, waste heat can be a serious problem, and highly specialized mechanisms have evolved for releasing this waste energy to the environment.

The net change in free energy that accompanies a reaction determines whether or not it can proceed spontaneously. In other words, the course of the reaction depends on whether the total free energy of the covalent bonds in the reactants is greater or less than that of the covalent bonds in the products. We quantify any free-energy change as **ΔG** for a defined set of conditions: a temperature of 25°C, a pressure of 1 atm (atmosphere), pH 7.0, and with both

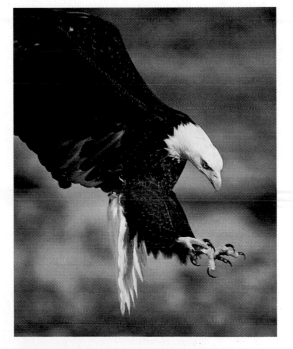

3.39　Bald eagle swooping down on prey
Many birds of prey use gravitational potential energy to power a rapid dive toward the animals they hunt. This eagle is braking from its dive, talons extended to grasp its prey.

3.40   Exergonic versus endergonic reactions

Reactions either liberate some of the free energy of the bonds within the reactants or consume free energy from outside the system. Those that release energy (in the amount represented by ΔG) and thus go downhill from the initial state to the final state are said to be exergonic (right), while those that require energy and go uphill are said to be endergonic (left). (As we will see, these diagrams describe ideal conditions and are therefore oversimplified.)

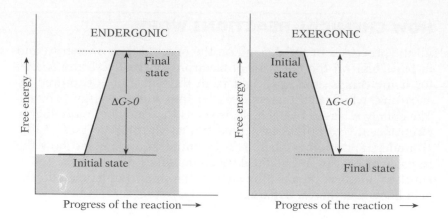

the reactants and products at a concentration of 1 mole/liter.[1] Any reaction from a thermodynamic perspective involves the free energy of the initial reactants (G_i), the free energy of the final products (G_f), and the change in free energy engendered by the reaction (ΔG):

$$\begin{array}{cc} \text{initial state} & \text{final state} \\ \text{(reactants)} \quad A + B \longrightarrow & C + D \quad \text{(products)} \end{array}$$

$$G_i \longrightarrow G_f \qquad (\Delta G = G_f - G_i)$$

The term ΔG (pronounced "delta G"), the change in free energy as a result of the reaction, is the crucial variable. *If the reaction results in products with less free energy in the covalent bonds than the reactants possessed, the reaction is "downhill" and can proceed spontaneously.* Since the free energy liberated from the covalent bonds of the reactants is usually released as heat, such reactions are said to be exothermic (heat-producing) or, more generally, *exergonic* (energy-releasing). If, on the other hand, the reaction requires an input of external energy, it cannot proceed spontaneously. An "uphill" reaction of this kind, in which the covalent bonds of the products have more free energy than those of the reactants, is said to be *endergonic*. These two alternatives are illustrated schematically in Figure 3.40.

All spontaneous reactions are downhill, or exergonic. This explains why all systems, living or inanimate, tend to lose free energy and ultimately reach a state in which their free energy is as low as possible, just as rocks on a hill tend to move down to the bottom rather than up to the top.[2] Throughout this book we will indicate the free-energy change, ΔG, resulting from biological reactions to

[1] In thermodynamics texts the formal symbol for changes in free energy under these conditions is $\Delta G'^\circ$. Since this is the only version of ΔG we will discuss, we can dispense with the superscripts.

[2] Physicists and chemists frequently relate free energy and order, which is defined as an arrangement of components in a system that is unlikely to occur by chance. Consider the unlikely circumstance in which all the molecules of air in a room are on one side, with a vacuum on the other. This orderly arrangement disintegrates rapidly as the molecules of air spread throughout the room, with a corresponding loss of free energy. The more orderly a system is, the more free energy it possesses; however, since increasing disorder is inevitable, so also is the spontaneous loss of free energy. The amount of disorder in a system—the energy unavailable for doing work—is called entropy.

the right of the reaction formula. Remember that when ΔG is negative (meaning that the covalent bonds of the reactants had more energy than those of the products), the reaction is exergonic; when ΔG is positive, the covalent bonds of the products have gained energy, so the reaction is endergonic. This relationship holds even when factors like temperature do not correspond to the standard conditions listed earlier: though the exact magnitude of ΔG will shift as conditions change, the *relative* magnitude of the ΔG values of different reactions in the same cell normally will not. Hence the ΔG values of different reactions under the same conditions can be directly compared, and many aspects of cell chemistry can then be predicted.

■ PREDICTING REACTIONS: THE EQUILIBRIUM CONSTANT

Our discussion so far has implied that two reactant molecules can combine spontaneously to create products only if free energy is liberated, and that if a reaction is exergonic, all the reactants in a mixture will ultimately be turned into products. However, we have ignored the possibility of a **back reaction**, with the products C and D combining to regenerate the reactants A and B:

$$C + D \longrightarrow A + B$$

Though this reaction is endergonic (uphill), it can take place—very slowly—because there is a source of energy within every system. This source, which we have ignored so far, is the energy of motion: the **kinetic** (or **thermal**) **energy** of the molecules. Every molecule has both a certain characteristic amount of stored energy in its bonds and energy of motion, which depends on its speed. In any solution, some molecules move very fast, others more slowly. In our example, when two fast-moving product molecules, C and D, collide, their energy of motion will sometimes be converted into covalent bond energy to produce two (slow-moving) reactants, A and B. Though this back reaction is rare, it is very important when the forward reaction is near completion. At this point the reactants A and B are so scarce that the rare, kinetic-energy-dependent back reaction may be just as likely as the forward reaction.

When the forward reaction, slowed in consequence of the increasing scarcity of reactants, is just counterbalanced by the rare back reaction, the two processes are in equilibrium and no further net change in the concentration of substances takes place. This ultimate stable ratio of products to reactants is the **equilibrium constant (K_{eq})**:

$$K_{eq} = \frac{\text{product concentration}}{\text{reactant concentration}}$$

This relationship succinctly summarizes the results of any reaction, and so complements ΔG in describing chemical reactions. For instance, an equilibrium constant of 10 means that when equilibrium has been reached under conditions of stable temperature and pressure, the products of the reaction outnumber the reactants by a factor of 10:

$$A + B \xrightarrow{K_{eq} = 10} C + D \qquad (\Delta G = -1.4 \text{ kcal/mole})$$

Table 3.2 Relationship between K_{eq} and ΔG at 25°C

ΔG (KCAL/MOLE)	K_{EQ}	
4.1	0.001	
2.7	0.01	Endergonic
1.4	0.1	reactions
0	1.0	
−1.4	10.0	
−2.7	100.0	Exergonic
−4.7	1000.0	reactions

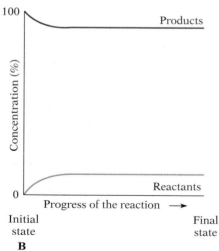

3.41 Relationship between the equilibrium constant and concentration

The equilibrium constant, specifying the final ratio of products to reactants, is independent of the starting concentration. In the example shown here, the equilibrium constant is 10, which means that products (red curves) will be 10 times more common than reactants (blue) when the reaction ends. This is the outcome whether we begin with pure reactants (A) or pure products (B).

Such a reaction is downhill: the products are more abundant than the reactants, and the reaction liberates energy. The predominance of the forward reaction is indicated by the longer arrow pointing from the reactants to the products. In a reaction with an equilibrium constant of 0.1, on the other hand, the reactants outnumber the products by 10 to 1:

$$E + F \underset{K_{eq} = 0.1}{\overset{}{\rightleftarrows}} G + H \qquad (\Delta G = +1.4 \text{ kcal/mole})$$

Such a reaction is uphill; it consumes energy, and so relatively few of the initial reactant molecules will be converted into products. And finally, an equilibrium constant of 1.0 indicates that the products and the reactants are equally abundant:

$$I + J \underset{K_{eq} = 1.0}{\overset{}{\rightleftharpoons}} K + L \qquad (\Delta G = 0)$$

As we would expect, there is no free-energy change in this reaction. The numerical relationship between K_{eq} and ΔG at 25°C is summarized in Table 3.2.

Keep in mind that the equilibrium constant of a reaction is a ratio—a simple empirical description of the outcome of the reaction—and depends, ultimately, on the ΔG of the reaction. Hence, the ultimate ratio of products to reactants in no way depends on the starting conditions. Whether we start with a great deal of reactant or only product, we wind up with the ratio specified by the equilibrium constant (Fig. 3.41).

■ HOW ACTIVATION ENERGY GUIDES REACTIONS

How do particular compounds come to follow particular reaction pathways? Suppose the cell manufactures compounds D and Z, as follows:

$$A + B \rightleftharpoons C + D \qquad W + X \rightleftharpoons Y + Z$$

What is to prevent B from reacting with W, X, Y, or Z instead of with A? Some of these combinations may well have negative ΔG values and so be energetically favorable. What prevents all but the "correct" reactants from combining?

For two molecules to combine, they must be brought unusually close to each other in a particular orientation, and frequently one or more preexisting bonds must be broken. This requires energy—specifically known as **activation energy (E_a)**—so even an exergonic reaction has an endergonic first step (Fig. 3.42). The only source of energy for this "priming" is the kinetic energy of collision.

Because covalent bonds are strong, substantial energy may be necessary to break the preexisting bonds of reactants—often far more than the amounts listed for the ΔG in Table 3.2. Consider this enormously exergonic reaction:

$$2H_2 + O_2 \longrightarrow 2H_2O$$

Even though this combination of oxygen and hydrogen can be explosive—for example, it provides much of the energy that pushes the space shuttle into orbit—the two reactants can coexist as a stable mixture indefinitely. A single spark, however, will initiate an explosive reaction. The same stability is a property of most reactants in living systems: the energy necessary to bring the reactants together and break their covalent bonds is far greater than the energy of all but the very few most rapidly moving molecules in a solution. The activation-energy barrier therefore prevents most reactions from taking place at a significant rate (Fig. 3.42). Without such a barrier the complex high-energy molecules on which life depends would be unstable and would break down.

Once a reaction does get started, the combination of one pair of reactants may release enough energy (usually in the form of heat) to activate the next pair, and so on, in a chain reaction. This is precisely what happens in a rocket engine and in the combustion of a dry piece of firewood. The wood, as we all know, can lie in a woodpile for years without bursting into flames spontaneously; once set on fire, it literally consumes itself as the free energy liberated by the combining of carbon and oxygen into CO_2 supplies enough activation energy to continue the burning. Put quite simply, heating a mixture will increase the *rate* of reaction (though it cannot affect the ultimate ratio of products to reactants, since that is a constant—K_{eq}). Cells, however, literally cook at temperatures much above the internal 37°C of most mammals and birds, so they cannot use added heat as a way of overcoming the activation-energy barrier. Moreover, cells must use some method that lowers the barrier selectively, so that some exergonic reactions run, while others do not. How is this crucial task managed?

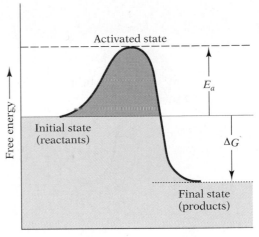

Free energy → / **Progress of the reaction** →

Activated state / E_a / Initial state (reactants) / ΔG / Final state (products)

3.42 Energy changes in an exergonic reaction

Though the reactants are at a higher energy level than the products, the reaction cannot begin until the reactants have been raised from their initial energy state to an activated state by the addition of activation energy (E_a). It is the need for activation energy that ordinarily prevents high-energy substances like glucose from breaking down, and hence makes them stable; the higher the activation-energy barrier, the slower the reaction and hence the more stable the substance. When activation energy is available, the reactants form a temporary and unstable activated complex, which breaks down to yield the end products of the reaction; in the process both activation energy and free energy (ΔG) are released.

■ **HOW CATALYSTS BREAK THE RULES**

Certain chemicals speed up reactions between other chemicals. As we have seen, a simple mixture of hydrogen and oxygen does not react. If we add a small quantity of platinum, however, the mixture will explode. After the reaction is over, the platinum will still be present, unchanged.

A substance that speeds up a reaction but is itself unchanged when the reaction is over is a **catalyst**. Catalysts affect only the *rate* of reaction; they simply speed up reactions that are thermodynamically possible to begin with. A catalyst cannot alter the direction of

3.43 Reduction of necessary activation energy by catalysts

The activation energy (E_a) necessary to initiate the reaction is much less in the presence of a catalyst than in its absence. It is this lowering of the activation-energy barrier by enzyme catalysts that makes possible most of the chemical reactions of life. Note that the amount of free energy liberated by the reaction (ΔG) is unchanged by the catalyst—it is the same for both the catalyzed and the uncatalyzed reaction—and that only the activation energy is changed.

a reaction, its final equilibrium, or the reaction energy involved.

Catalysts work by decreasing the activation energy needed for the reaction to take place (Fig. 3.43), thereby increasing the proportion of reactants energetic enough to react (Fig. 3.44). A catalyst binds reactants in an intermediate state in which they are correctly oriented and important internal bonds are weakened (Fig. 3.45). As a consequence of binding, then, conditions are highly favorable for the reaction.

An inorganic catalyst like platinum is relatively unselective about the reactants it "helps"; the spacing of its atoms on a crystal face happens to match the distance between pairs of small covalently bonded atoms, and its loosely bound outer electron readily interacts with the outer shells of other atoms. However, cells need catalysts that only promote specific reactions; other combinations of potential reactants are undesirable. Lacking the necessary activation energy, and without specific catalysts to lower this thermody-

3.44 Effect of a catalyst on the ability of kinetic energy to activate a reaction

As indicated by this bell curve, reactant molecules exhibit a wide range of kinetic (thermal) energy. Only a minute fraction (right) have enough energy to overcome the activation-energy barrier for most reactions. In the presence of a catalyst, such as an enzyme, the barrier is lower, so a much larger proportion of the reactants (middle) can combine to form products.

3.45 Action of platinum as a catalyst

Because of its loosely bound outer electron, platinum (Pt) is able to form weak temporary bonds with molecules of both hydrogen (blue) and oxygen (red). This binding draws the hydrogen and oxygen electrons away from their covalent positions, thus weakening the bonds within their respective molecules. In addition, the spacing of the platinum atoms tends to align the hydrogen and oxygen atoms in such a way that new bonds between hydrogen and oxygen can be more easily formed. Platinum is a catalyst in that it facilitates the reaction without being itself altered.

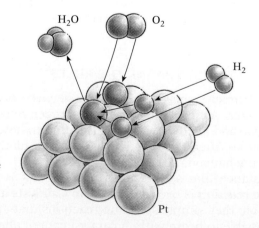

namic barrier, these reactants are unable to combine with one another. Cells have evolved an enormous variety of highly specialized organic catalysts, called **enzymes**, which direct cellular chemistry along useful pathways. Virtually all enzymes are globular proteins.

■ BIOLOGICAL CATALYSTS: ENZYMES

Enzymes make use of the basic facts of biological thermodynamics: (1) A chemical reaction can proceed spontaneously if it releases free energy. (2) Because the activation energy is relatively high, biological reactions occur only very slowly without catalysts. (3) Catalysts, including enzymes, alter neither the equilibrium constant of a reaction nor the net change in free energy—they cannot, by themselves, cause a reaction to run uphill, but they can make specific downhill (exergonic) reactions occur quickly.

Why enzymes are so specific   Enzymes are highly selective: a particular enzyme generally inteacts with only one type of reactant or pair of reactants, customarily called the **substrates**. The enzyme thrombin, for instance, acts only on certain proteins and only at very specific sites. It "recognizes" the bond between the amino acids arginine and glycine, which it then hydrolyzes (an important step in the formation of blood clots). Like all catalysts, enzymes lower the activation energy required (see Fig. 3.43), thereby increasing the proportion of substrate molecules energetic enough to react (see Fig. 3.44). As a result, enzymes vastly speed up the reactions they catalyze; a single molecule of enzyme may cause thousands or even hundreds of thousands of molecules of reactant to combine into product each second.

Because of their efficiency and specificity, enzymes both steer specific substrates into particular reaction pathways and block them from others, thus guiding the chemistry of life with great precision. This specificity depends on the enzyme's three-dimensional molecular conformation; a given enzyme interacts only with substrates whose molecular configurations (their conformation and location of charged groups) "fit" its surface. This explains why when proteins are denatured—when their three-dimensional conformation is disrupted—their enzymatic properties vanish (Fig. 3.46). It also explains why enzymes are highly sensitive to changes in pH (Fig. 3.47). A change in pH results in the breakage of many of the weak bonds that help stabilize the conformation of proteins, thus changing the shape of the protein.

Enzyme and substrate fit together like a lock and key, or like pieces of a puzzle. The reactive portion of the substrate molecule and the portion of the enzyme known as the **active site** must fit together in space intimately enough to become temporarily bonded, like the platinum with the hydrogen and oxygen in Figure 3.44. In this way they form a transient enzyme-substrate complex:

$$E + S \longrightarrow ES \longrightarrow E + P$$

(E stands for enzyme, S for substrate, and P for product.) But enzymes and their substrates do not always have to fit together exactly *before* the enzyme-substrate complex (ES) can form; many

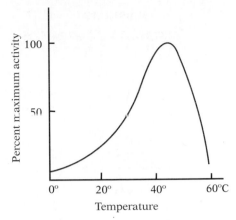

3.46   Enzyme activity as a function of temperature

Though temperature sensitivity varies somewhat from one enzyme to another, this curve applies to typical enzymes. Its activity rises steadily with temperature (approximately doubling for each 10°C increase) until thermal denaturation causes a sudden sharp decline, beginning between 40° and 45°. The enzyme becomes completely inactivated at temperatures above 60°, presumably because its three-dimensional conformation has been severely disrupted. Bacteria living in volcanic pools, by contrast, have well-stabilized enzymes with activity optima near 100°; below 60°, however, their enzymes have almost no activity.

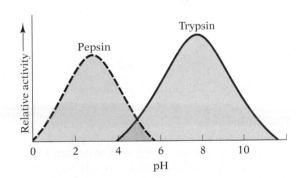

3.47   Enzyme activity as a function of pH

Most enzymes are very sensitive to pH, but they differ markedly in the pH value at which they are most active. Pepsin and trypsin are both enzymes that digest protein, but the pH ranges within which they are active overlap only slightly. Pepsin, a stomach enzyme, is most active under strongly acidic conditions, while trypsin, an enzyme secreted into the small intestine, is most active under neutral and slightly alkaline conditions.

SUBSTRATE

ENZYME

Enzyme-substrate complex

PRODUCT

Enzyme resumes
original configuration

3.48 Induced-fit model of enzyme-substrate interaction

The enzyme molecule has an active site onto which the substrate molecules can fit (top), forming an enzyme-substrate complex (middle). The binding of the substrate induces conformational changes in the enzyme that maximize the fit and force the complex into a more reactive state. The enzyme molecule reverts to its original conformation when the product is released (bottom).

(perhaps most) enzymes undergo conformational changes in the course of bonding, which create an ***induced fit*** (Fig. 3.48).

Spatial complementarity is only one of the prerequisites for enzyme-substrate interaction. Another is that enzyme and substrate be chemically compatible and capable of forming numerous and precise weak bonds with each other. The bonds between enzyme and substrate are usually the same types of weak bonds that stabilize protein conformation. These bonds can be made and broken rapidly in the collisions that result from random thermal motion at normal temperatures. The type of substrate to which a given enzyme molecule can become bonded depends on the exposed R groups of the amino acids at the active site and on their arrangement relative to one another. Suppose the active site of a particular enzyme is a curving groove and that most of the exposed R groups in this groove are electrically charged. The reactive portion of the substrate must have a complementary shape and charge or polarity; an electrically neutral nonpolar substrate molecule, or one with the same charge as the active site, could not bond to the active site of this enzyme. Conversely, only a hydrophobic substrate could interact with an active site made up largely of hydrophobic R groups.

Figure 3.49 shows what is currently known about the active site

3.49 Active site of an enzyme

Shown here in schematic form is the base of the cleft (white) where the active site of carboxypeptidase (green) is located. (The hydrophobic entrance to the cleft is out of the drawing, to the bottom. The entire enzyme, with the active site highlighted, is depicted in Figure 3.50). Part of a substrate molecule (red) is shown in the cleft, linked to the enzyme by five weak bonds (red bands). Seven of the amino acids of the active site are indicated by their abbreviated names; the numbers beside the names refer to the positions of the amino acids in the enzyme polypeptide chain. The function of this enzyme is to separate the terminal amino acid (top) from the rest of the substrate amino acid chain (extending down out of the figure) at the covalent bond indicated by the arrow. The highly electronegative zinc and the charged oxygen draw away the electrons of this bond and so initiate its rupture.

3.50 Location of the active site in an enzyme

The folding of the long chain of amino acids of carboxypeptidase brings together the zinc atom and three of the four amino acids that bind to the substrate at the active site, even though—as their numbers indicate—they are located at different places in the chain. The zinc atom itself is held in place by three different amino acids: nos. 69, 72, and 196 (not shown). The fifth part of the site folds into place (arrow) when the substrate binds to the enzyme. The region at the top center in dark color is the active site, while the area in light color is the hydrophobic entrance to the cleft. Most of the length of the polypeptide chain serves to (1) create an appropriately shaped cleft, (2) precisely position the R groups at the active site, (3) provide for the hingelike action of the induced-fit arm (containing amino acid 248), and (4) supply a second location for enzyme regulation (discussed later).

of the enzyme carboxypeptidase,[3] which catalyzes the removal of the terminal amino acid from one end of polypeptide chains during digestion. The enzyme's active site is a cleft into which the end of the substrate molecule can fit. The substrate forms several weak bonds with the R groups of amino acids at the active site. In addition, the substrate binds to a zinc held in place by the R groups of three other amino acids. Note that the critical amino acids in the active site (nos. 71, 196, 72, 69, 145, 248, and 270) are not adjacent to each other in the polypeptide chain, which means that the complex folding of the protein—its tertiary structure—has brought amino acids from several regions of the protein close together to form the active site (Fig. 3.50). Active sites nearly always include some nonadjacent amino acids. This helps explain why elevated temperature or a major pH change may greatly reduce an enzyme's activity: anything that changes the precise folding pattern of the polypeptide chain is likely to alter the critical arrangement of amino acids in the active site.

Carboxypeptidase is also typical in that the entrance to the cleft in which the active site is located is hydrophobic, so that the water molecules that surround the substrate are stripped away as it enters the groove. The binding energy of the weak bonds is just strong enough to stabilize the enzyme-substrate complex. The ability of carboxypeptidase to form these particular bonds accounts in

---

[3] The suffix -ase designates an enzyme, and is derived from the name diastase, coined in 1833 by two French chemists from the Greek diastasis, "to separate". The function of most enzymes can be guessed from their names. In the case of carboxypeptidase, -peptid- is from the Greek pepsis, "to digest," and carboxy- indicates that the digestion is from the carboxyl end of the protein.

ENZYME

Enzyme-substrate
complex

Competitive inhibitor
bound to enzyme

Noncompetitive inhibitor
bound to enzyme

3.51   Competitive and noncompetitive inhibition of an enzyme

Top: The substrate is bound to the active site of the enzyme. Middle: The binding of a competitive-inhibitor molecule to the catalytic site prevents the substrate from binding. Bottom: A noncompetitive inhibitor bound to a different site on the enzyme induces an allosteric change that prevents the active site from catalyzing reactions.

part for its specificity. The functioning of this enzyme also illustrates the induced-fit strategy: as the binding begins at the other four parts of the active site, the part of the chain containing tyrosine (no. 248) moves in from the periphery to trap the substrate (the polypeptide chain). The highly electronegative zinc atom and the charged oxygen of glutamate (no. 270) help activate the substrate by drawing away the electrons of the N—C bond that holds the terminal amino acid to the rest of the chain (Fig. 3.49). The resulting cascade of electron shifts between atoms of differing degrees of electronegativity makes the outcome—elimination of the terminal bond—inevitable. Finally, the two products, the terminal amino acid and the remainder of the chain, separate from the enzyme after the reaction because the new bonds in the products have redistributed their electrons. The tenuous array of hydrogen bonds and van der Waals interactions that depended on precise electron interactions with the enzyme at the active site has been disturbed, and the products drift free.

Figure 3.49 illustrates another important point. Many enzymes contain a prosthetic group essential to their activity. A metal atom is often part of the prosthetic group; in carboxypeptidase the metal is zinc. Most of the trace elements our bodies require (see Table 2.1) are needed for enzyme prosthetic groups.

Some enzymes that do not have a prosthetic group require a cofactor to which they bond briefly during the reactions they catalyze. The cofactors may be metal ions, or they may be nonproteinaceous organic molecules called *coenzymes*. Coenzyme molecules are much smaller and less complex than proteins. Like enzymes, they are not used up or permanently altered by the reactions in which they participate, and hence they can be used over and over again. Though only tiny amounts are needed, if the supply falls below normal, the health or even the life of the organism may be endangered. This is why vitamins, which act as parts of essential coenzymes, are so necessary in the diet.

How enzyme activity is controlled   A variety of mechanisms have evolved for controlling the activity of enzymes. These mechanisms often depend on chemical agents that mask, block, or alter the active sites of the enzymes they help regulate.

One common form of enzyme control, *competitive inhibition*, uses an inhibitor substance similar to the normal substrate of the enzyme. It binds reversibly to the enzyme's active site but is not chemically changed in the process; a bond normally broken in the substrate molecule, for instance, may in the inhibitor be too strong or too well "insulated" by nearby bonds to be severed. By binding to the active site, the inhibitor (I) masks the site and prevents the normal substrate molecules from gaining access to it (Fig. 3.51 middle). Thus the reaction

$$E + I \longrightarrow EI$$

competes with the reaction

$$E + S \longrightarrow ES \longrightarrow E + P$$

because both involve the same enzyme, which is present only in very small quantities.

Which of the two reactions will predominate depends on their relative energetics and, even more, on the relative concentrations of I and S. If there is much inhibitor and a low concentration of substrate molecules, a high percentage of the enzyme will be bound as EI and therefore unavailable. On the other hand, if there is much S and only a small concentration of I, then most of the enzyme molecules will be free to catalyze the reaction of substrate molecules to form the product.

Carbon monoxide poisoning is an extreme example of competitive inhibition. The carbon monoxide competes with oxygen for the active sites in hemoglobin (see Fig. 3.30), the molecule in the blood of vertebrates that carries oxygen to the body's cells. Carbon monoxide binds so strongly to the active sites that oxygen is excluded. Moreover, the inhibition is essentially irreversible: once bound, carbon monoxide remains in place. As a result of the oxygen deprivation that ensues, living tissue, particularly brain tissue, can be damaged or destroyed. Most competitive inhibitors, however, bind to their target only briefly; this reversible inhibition means that inhibitor and substrate molecules are always competing: a change in the concentration of either is quickly reflected by an alteration in enzyme activity.

A second category of reversible inhibition, **noncompetitive inhibition**, depends on two kinds of binding sites in the same enzyme molecules: the usual active sites to which substrate can bind, and other sites to which inhibitors can bind. Most noncompetitive inhibition is **allosteric inhibition**. An allosteric enzyme can exist in two distinct spatial conformations, which usually reflect alterations of tertiary structure. When the molecule is in one conformation, the enzyme is active; in the other conformation it is inactive (or less active) because the substrate-binding site is disrupted. The binding of inhibitor molecules—usually called negative **modulators**—stabilizes the enzyme in its inactive conformation (Fig. 3.51 bottom). Quite often the product itself (or the product of a later reaction in the same biochemical pathway) is the modulator: when present in such high concentrations that no more is needed, it turns off the process responsible for its own synthesis. This self-limiting strategy is known as **feedback inhibition**. Other allosteric enzymes have binding sites for positive modulators, which induce conformational changes enhancing enzyme reactivity.

Instead of having different kinds of binding sites, some allosteric enzymes have two or more sites of a single kind, and so can bind substrate at two or more locations simultaneously. The binding of substrate at one active site causes conformational changes that make the remaining sites more reactive. This phenomenon, called **cooperativity**, is exemplified in hemoglobin. A single molecule of hemoglobin is capable of carrying four oxygen molecules. The binding of the first oxygen molecule induces changes in the quaternary structure of the hemoglobin molecule that give the other three binding sites a higher affinity for oxygen. Thus in cooperativity, the first substrate molecule functions as the modulator. It stabilizes the allosteric protein in one of its possible conformations—in the case of hemoglobin, its most reactive conformation.

The techniques of molecular biology now permit selective substitution or deletion of individual peptides in enzymes. As a result,

LYS

ORN

3.52 A synthetic enzyme

This 73-peptide designer enzyme mimics the action of chymotrypsin, a protein-digesting enzyme 245 amino acids long that is secreted by the pancreas. The active site with the R groups shown in blue is just as specific as the natural enzyme, is blocked by the same competitive inhibitors, and increases the reaction rate 100,000 times over that of an enzyme-free solution. Nevertheless, this abbreviated version of chymotrypsin is 1000 times slower than its natural counterpart, indicating that the complex globular structure imparted by the other 172 amino acids probably serves to optimize catalytic activity. (The part of the substrate actually digested is shown in black.)

the action of any amino acid in a protein can be explored experimentally. One consequence of the knowledge this technique has brought is that biochemists can now plan and synthesize artificial enzymes (Fig. 3.52). These "designer enzymes" can be made with specificities and activities unknown in nature. Such enzymes could have nearly limitless potential in treating disease or even solving problems of environmental pollution.

CHAPTER SUMMARY

WHY CARBON?

Carbon is a versatile backbone for organic molecules because it can form four covalent bonds. Many properties of organic compounds are determined by the functional groups attached to carbon skeletons. (p. 47)

THE ROLE OF CARBOHYDRATES Carbohydrates include simple sugars (used for energy) and chains of sugars (polysaccharides) used for energy storage and structural strength. (p. 49)

THE ROLE OF LIPIDS Lipids include fats (used for insulation and energy storage), phospholipids (the major component of cell membranes), and steroids. (p. 55)

THE ROLE OF PROTEINS Proteins are composed of chains of amino acids. Amino acids have specific characteristics depending on their side group. Proteins fold into active shapes that depend on polar, ionic, and covalent interactions between the side groups of their amino acids, and between the side groups and water. The most common structures for parts of a protein are α helices and β pleated sheets. Proteins can act as structural elements or as enzymes. (p. 58)

THE ROLE OF NUCLEIC ACIDS DNA stores information in the sequence of its nucleic acids. RNA is used in a variety of ways; most is involved in carrying the information from genes to the sites of protein synthesis. (p. 68)

HOW CHEMICAL REACTIONS WORK

WHY FREE ENERGY ISN'T FREE Cells depend on free (biological potential) energy to fuel the processes of life. Energy-consuming endergonic reactions must be powered by energy-releasing exergonic ones. (p. 71)

PREDICTING REACTIONS: THE EQUILIBRIUM CONSTANT The ratio of products to reactants when a reaction reaches equilibrium—the equilibrium constant—is directly related to the free energy the reaction produces (or consumes). (p. 73)

HOW ACTIVATION ENERGY GUIDES REACTIONS Most energy-releasing reactions do not occur spontaneously. Weakening the bonds involved in the reaction requires more activation energy than is available from the thermal motion of the molecules involved. Selectively lowering the activation energy for a reaction causes that reaction to occur instead of other reactions the reactants could participate in. (p. 74)

HOW CATALYSTS BREAK THE RULES Catalysts lower activation energy by binding reactants in an intermediate state. (p. 75)

BIOLOGICAL CATALYSTS: ENZYMES Enzymes act as highly specific biological catalysts. Reactants bind at an active site, bonds are loosened, and the reaction occurs. Enzyme activity, and thus the reactions they catalyze, is regulated to match the needs of the cell. (p. 77)

STUDY QUESTIONS

1 How might a change in pH affect the "personality" of polar and ionic amino acids, and the activity or solubility of an enzyme? (pp. 77–78)

2 Why is an α-helix more elastic than a β-pleated sheet? (pp. 61–64)

3 Calculate the approximate bonding energy holding the two 1,000,000-nucleotide-long strands of *E. coli* DNA together. How do you suppose the strands are separated to allow duplication? (pp. 34–36, 68–70)

4 What effect does raising the temperature or increasing the concentration of the relevant enzyme have on an equilibrium constant? (pp. 73–75)

5 Why is activation energy a blessing for organisms? (pp. 74–75)

6 Why is it essential for a cell to be able to control enzyme activity? What are some ways this task is accomplished, and why are some more appropriate for certain sorts of chemical pathways? (pp. 80–82)

SUGGESTED READING

ATKINS, P. W., 1987. *Molecules*. Scientific American Library, New York. *A beautifully produced "molecular glossary" illustrating the chemical formula, three-dimensional structure, and biological action of many common or unusually interesting organic molecules.*

CRAIG, E. A., 1993. Chaperones: helpers along the pathway to protein folding. *Science* 260, 1902–1903. *A brief overview of chaperone-assisted folding.*

DOOLITTLE, R. F., 1985. Proteins, *Scientific American* 253 (4). *Reviews the properties of amino acids and the structure of proteins, and discusses the evolution of different modern proteins from common ancestral enzymes.*

DRESSLER, D., AND H. POTTER, 1991. *Discovering Enzymes*. W. H. Freeman, New York. *A well-written and illustrated history of the study of enzymes, with particular emphasis on how the digestive enzyme chymotrypsin works.*

STROUD, R. M., 1974. A family of protein-cutting proteins, *Scientific American* 231 (1). *A good discussion of how enzymes like chymotrypsin work.*

STRYER, L., 1988. *Biochemistry*, 3rd ed. W. H. Freeman, San Francisco. *A beautifully produced, clearly written, but highly technical exposition of biochemistry.*

CHAPTER 4

THE CELL MEMBRANE

Like the departments in a giant corporate conglomerate, the organic molecules that make up life depend on precise compartmentalization and organization. If we put a population of amoebae through a blender, the resulting stew of organic molecules will not reorganize itself spontaneously into living entities. A complex *plasma membrane* protects the interior of the cell—the *cytoplasm*—from assault by the external environment. The cytoplasm itself is intricately organized: it contains not just organic fluids (the *cytosol*), but many internal membranes and a variety of specialized, self-contained entities known as organelles, which separate themselves from the rest of the cell by means of their *own* membranes. In this way the cell's chemical processes are partitioned off—food, for example, is digested in membrane-enclosed compartments that prevent digestion of the rest of the cell. This chapter will examine the organization and operation of membranes, which are critical to life.

THE DISCOVERY OF CELLS

■ THE CELL THEORY

The discovery of cells and of their structure came with the development of magnifying lenses. In the 17th century Antonie van Leeuwenhoek and other interested amateurs constructed simple microscopes, and in 1665 Robert Hooke reported to the Royal Society of London "the first microscopical pores I ever saw, and perhaps, that were ever seen," in a piece of cork (Fig. 4.1). Hooke's pores were, in fact, cells.

The idea that all living things are composed of cells—the *cell theory*—was not proposed until about 1840. An important extension of the cell theory came 20 years later: all living cells arise from preexisting living cells. This theory of *biogenesis*, life coming from life, contradicted the prevailing belief in spontaneous generation. In 1862 Louis Pasteur proved that living cells cannot arise from nonliving matter (see *How We Know: Spontaneous Generation Disproved*, p. 87).

These two components of the cell theory—that all living things are composed of cells and that all cells arise from other cells—pro-

7.5 cm

A

B

4.1   Van Leeuwenhoek's microscope

(A) This magnifier was first developed to inspect samples of cloth but was soon adapted for viewing living things. (B) Drawings of cork that appeared in Robert Hooke's *Micrographia,* published in 1665.

vide a working definition of life: living things are chemical organizations composed of cells and are capable of reproducing themselves.

■ HOW CELLS ARE STUDIED

Much of our knowledge of subcellular organization has been made possible by the development of better and more powerful microscopes. In the detailed analysis of subcellular structure, three attributes of microscopes are of particular importance: magnification, resolution, and contrast. Magnification is a means of increasing the apparent size of an object. Resolution is the capacity to make adjacent objects distinct. Contrast is important in distinguishing one part of a cell from another.

In the ordinary compound light microscope, light passes through a specimen and is captured, bent, and brought into focus by lenses (Fig. 4.2A). Depending on the magnification and the size of the specimen, a whole cell or only a tiny part may be in the field of view at any one time.

A COMPOUND LIGHT MICROSCOPE

eyepiece

focus knob

objective

specimen

light source

B TRANSMISSION ELECTRON MICROSCOPE

source of electrons

specimen

C SCANNING ELECTRON MICROSCOPE

source of electrons

magnetic lenses

beam deflectors

magnetic lens

detector

specimen

image on phosphorescent screen

image on TV viewing screen

D SCANNING TUNNELING MICROSCOPE

specimen

electron = emitting probe

computer

probe movement controller

4.2 Microscopes

(A) In a compound light microscope, light passes through a specimen to the objective lens. Here the light is refracted and directed to the eyepiece, where it is focused for a camera or for the eye. (B) In a transmission EM, electrons provide the illumination, which passes through the specimen and is then focused by magnets to form an image for photographic film or on a phosphorescent screen. (C) In a scanning EM, a focused beam of electrons moves back and forth across the specimen, while a detector monitors the consequent emission of secondary electrons and reconstructs an image. (D) The scanning tunneling microscope (STM) moves an electron-emitting tip over the sample, constantly adjusting its height to maintain a fixed distance above the sample; these variations in tip elevation allow reconstruction of the specimen's contours. The minute movements necessary are accomplished by a mounting arm attached to piezoelectric crystals. Since the dimensions of such crystals change as different amounts of voltage are applied, it is possible to control the position of the tip exactly.

HOW WE KNOW:
SPONTANEOUS GENERATION DISPROVED

Pasteur's experiments on spontaneous generation are classically simple, well-controlled tests of a hypothesis. He placed various nutrient broths in long-necked flasks and then bent the necks into curves (see figure). Next he boiled the broths in the flasks to kill any microorganisms that might be in them. Then he left the flasks standing. Any microbe-laden dust particles in the air moving into the flasks were trapped in the films of moisture on the humid curves of the necks; the curved necks thus acted as filters. The broths in the swan-neck flasks remained unchanged indefinitely. On the other hand, identical broths boiled in flasks with straight necks—the control solutions—were soon teeming with life. Similarly, if the swan neck was broken off, the experimental broth rapidly developed colonies of molds and bacteria. Since the handling of the control solutions differed in only one respect from the experimental flasks—exposure to airborne microbes—the changed outcome had to be attributed to that difference. In the same way, Pasteur showed that the source of the microorganisms that ferment or putrefy milk, wine, and sugar-beet juice is the air; there was no spontaneous generation in the nutrient media.

PROLONGED HEATING

STERILE

germs from air

REINFECTED   STERILE   germs trapped here

Pasteur's experiment

Nutrient broths in two kinds of flasks, one with a straight neck, the other with a bent neck, were boiled to kill any microbes they might contain (top). The sterile broths were then allowed to sit in their open-mouthed containers for several weeks (middle). Microorganisms entering the straight-necked flask contaminated the broth, but those entering the bent neck of the other flask were trapped in films of moisture in the curves of the neck (bottom).

A | 0.1 mm

B | 0.1 mm

C | 0.1 mm

D | 0.1 mm

4.3 Views of the green alga *Scenedesmus* obtained with microscopes of various types

(A) Photograph of an unstained specimen as seen with a phase-contrast light microscope. (B) Photograph taken by the Nomarski process. (C) Transmission electron micrograph. Many of the membranous and particulate intracellular organelles show up much more clearly here than under the light microscope. They are made visible by a stain containing heavy-metal atoms, which combines differentially with various structures. (D) Scanning electron micrograph, providing a three-dimensional view of surface features. The scale bars below the images in this and later figures indicate relative dimensions.

We vary the magnification by using lenses with different shapes that accept light from larger or smaller sections of the specimen: the higher the magnification, the smaller the amount of light that reaches the eyepiece. The limit on useful magnification in light microscopes is not a matter of exhausting the illumination, however; but arises from the tendency of light to bend as it passes near an edge. This phenomenon, known as diffraction, spoils images by bending light out of the straight-line path as it moves from the source, through the specimen, and to the eyepiece, thus blurring the image. For ordinary light, useful magnification is limited to about 1000 times the actual size of the object in focus. Though magnification of 1000x is an enormous improvement over the unaided eye, it is still not enough to let us see many of the smaller subcellular structures.

Contrast is as important as magnification. Most cellular components are colorless and have essentially the same texture; they do not contrast with one another. However, different parts of the cell often differ in their affinities for various dyes, so when these areas are stained, they stand out from one another. Unfortunately, staining usually kills the cell and may thereby change its internal structure. Techniques such as phase-contrast and Nomarski optics, both of which depend on elaborate optical manipulation, greatly increase the value of the light microscope because they create contrast optically, without staining (Fig. 4.3A-B). Another technique uses small, nonlethal amounts of one or more fluorescent stains to create contrast inside cells. The cell being studied is illuminated with a highly focused laser beam of just the wavelength to cause fluorescence, and the focal point is systematically scanned back and forth through a specimen; a second highly focused optical system is used to pick up only the light emitted from the focal point of the scanning beam, and the resulting output is used to reconstruct a crisp cross-sectional image with a computer.

The electron microscope (EM) uses a beam of electrons instead of light as its source of illumination. Because resolution improves (that is, diffraction decreases) as the wavelength of the illumination becomes shorter, and because electron beams have much shorter wavelengths than visible light, electron microscopes can resolve objects about 10,000 times better than light microscopes (Fig. 4.3C-D).

In a transmission EM (Fig. 4.2B) the electrons, focused by magnets rather than by lenses, pass through the thinly sliced "sectioned" specimen and fall on a photographic plate or a phosphorescent screen, where they produce an image of the specimen. Because cells are essentially transparent to electrons, a specimen being prepared for the transmission EM must be stained with an electron-dense chemical that binds to specific cell structures. Many stains contain heavy-metal atoms, which bind to particular cellular structures, blocking the passage of electrons in these places (Fig. 4.3C). Another approach is to tilt the specimen to allow atoms of the electron-dense substance to fall onto it (Fig. 4.4). The shadows and highlights in the resulting EM picture create a three-dimensional effect—a kind of topographical map of the specimen that can reveal important surface details.

A scanning EM (SEM) can also produce a three-dimensional view. A specimen coated with atoms of metal is scanned from above by a moving beam of electrons (Fig. 4.2C). This focused probe does not penetrate the specimen but instead causes so-called secondary electrons to be emitted from the surface. The intensity of the emission of secondary electrons depends on the angle at which the probe beam strikes the surface, and therefore varies with the contours of the specimen. Hence a point-by-point recording of the emission produces a three-dimensional picture (Fig. 4.3D). Though the resolution of the scanning EM does not approach that of the transmission EM, its ability to create a better three-dimensional effect is an advantage for many applications. Since shadowing is not required, the same specimen can be turned repeatedly and observed from various perspectives.

A further elaboration of the electron microscope, the scanning tunneling microscope (STM), uses an ultrafine electron-emitting tip, which is moved back and forth over a metal-coated sample, scanning it systematically from top to bottom (Fig. 4.2D). The

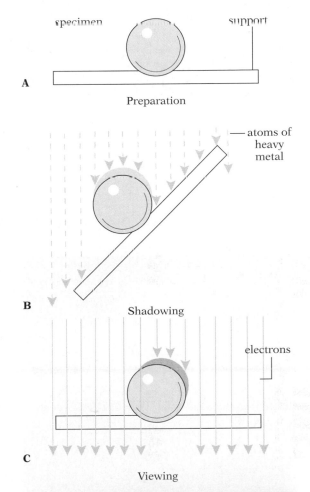

A — Preparation

B — Shadowing

atoms of heavy metal

C — Viewing

electrons

4.4 Shadow staining for the transmission EM

The specimen (A) is tilted and dusted, or "shadowed," with atoms of a heavy metal such as platinum (B). (Another technique is to "shoot" the metal atoms from an angled source onto a horizontal specimen.) When an electron beam is aimed at the coated specimen (C), the unevenly distributed metal plating prevents most of the electrons from penetrating the shadowed areas, and produces a three-dimensional view of the surface of the specimen, as in the electron micrograph of polio virus (D).

D

4.5   STM image of DNA

This computer-enhanced picture shows both ordinary DNA (thin coils) and DNA complexed with the protein SSB (discussed in detail in Chapter 8), which holds the double helix in a much more open configuration during the process of chromosomal duplication (thick coils).

amount of current that flows from the tip to the specimen depends on the gap between them; by recording the constant adjustments of tip height necessary to maintain a given current flow, a computer can reconstruct the topography of the sample and create a picture on a video monitor (Fig. 4.5). An analogous technique uses atomic forces to map surfaces (Fig. 4.6).

4.6   Atomic-force image of a moving cell

This living cell is in the process of flattening itself, part of its normal behavior in the nervous system where it envelopes neurons. The flattened outer edges of the membrane are searching for the molecules that identify the class of cells that it envelopes. The thickened center (N) contains the nucleus; the linear structures (F) lie just below the membrane and are part of the machinery that moves the cell.

HOW CELLS EXPLOIT DIFFUSION

At one time, the cell membrane was considered little more than a bag to hold in all the organic chemicals that somehow combine to produce life. In fact, the cell membrane is critical to cell movement, and to the cell's sampling of the extracellular environment; moreover, it bears the primary responsibility for regulating the chemical traffic between the precisely ordered interior of the cell and the essentially unfavorable and potentially disruptive outer environment. All substances moving into or out of a cell must pass through a membrane barrier, and the membrane of each cell can be quite specific about what is to pass, at what rate, and in which direction. The cell membrane exercises this control in two ways: by utilizing natural processes such as diffusion, and by transporting specific substances in and out.

■ DIFFUSION AS A CHEMICAL REACTION

As we have seen, temperature affects the rates of chemical reactions by increasing the kinetic energy (the average velocity) of the molecules involved. Imagine a box containing marbles in a cluster near one end (Fig. 4.7A). When we shake the box, the marbles disperse almost evenly over the bottom (Fig. 4.7B). This obvious result is worth a closer look, because the marbles can be thought of as molecules, and the shaking as adding kinetic, or thermal, energy to the system.

Of all the possible directions in which a given marble might move, more lead *away* from the center of the cluster than toward it. Hence *random movement will tend to disrupt the cluster rather than to maintain it.* In the absence of any counteracting external influence, the Second Law of Thermodynamics—in particular, the principle of **entropy**—dictates that a dynamic system will tend to move toward the more probable, disorderly state rather than toward the less likely, orderly state. This is what happens when a lump of sugar dissolves in a cup of warm coffee: the sugar molecules move from the region of high concentration (the sugar crystal) to regions of lower concentration; eventually the sugar molecules disperse throughout the liquid. The warmer the liquid, the more kinetic energy the molecules in solution will have on average, and the faster diffusion will take place.

We can now make an important generalization: all other factors being equal, *the net movement of the particles of a substance is from regions of higher concentration to regions of lower concentration of that substance.* Note that we speak of *net* movement. There will always be some particles moving in the opposite direction, but overall the movement will be away from the centers of concentration. An obvious result is that the particles of a given substance tend to become equidistant from one another within the available space. When this uniform density has been reached, the system is in equilibrium; the particles continue to move, but there is little net change in the system.

The random, unrestrained movement of molecules from one place to another is called **diffusion**. Diffusion is fastest in gases, where there is much space between the particles and hence relatively little chance of collisions that retard movement. Diffusion in liquids is much slower (Fig. 4.8); in the absence of convection currents a substance can take a very long time—years, in fact—to move in appreciable quantity only 1m through cold water. Diffusion in solids is, of course, much slower still: there is very little space between the molecules of a solid, and collisions occur almost before the molecules get going. In all these instances, however, regardless of the rate of diffusion, the net effect is movement away from regions of higher concentration, as long as all regions are at the same temperature and pressure. In living organisms, where molecules are generally in a warm aqueous solution and the distances involved are measured in fractions of a millimeter, diffusion is a highly significant process; an amino acid in an aqueous medium will typically diffuse about one cell diameter (10-50 μm; also called microns or micrometers) in less than 0.5 sec.

So far, we have discussed diffusion in terms of movement from a higher to a lower concentration along a gradient. In the living world, however, diffusion is rarely a function of concentration alone; differences in temperature and pressure, for instance, affect diffusion. It is more useful to look at diffusion in terms of the free

4.7 Mechanical model for diffusion

(A) All 17 marbles are placed in a cluster at one end of a rectangular box. (B) When the box is shaken to make the marbles move randomly, they become distributed throughout the box in nearly uniform density.

4.8 Diffusion in a liquid

Particles of solute are at the bottom of a flask of water (A). The particles slowly diffuse away from the cluster (B, C) until they are distributed with nearly uniform density through the water (D). If the water is cold and there are no convection currents to help move the particles, it may take a long time to reach uniform distribution.

energy of the particles involved. A local concentration of a substance is a relatively orderly and unlikely arrangement. We can see, for instance, that energy (among other things) would be needed to change a mixture of sugar molecules and coffee back into the original lump of sugar and unsweetened coffee. A random, disorderly arrangement of molecules has less potential for doing work than an orderly one, and accordingly it has less free energy. As we know from the Second Law of Thermodynamics, the amount of free (useful) energy in the universe is always decreasing; entropy is always increasing.

Diffusion, then, is spontaneous because orderly molecules, concentrated together, have greater free energy than dispersed molecules: it is a downhill reaction from order to disorder. The mixture (or product) has less free energy than the separate original substances (the reactants). If we begin with two pure substances, the rate of diffusion is fastest at the outset, and slows as the equilibrium point (complete mixture) is approached. If we could observe diffusion at the molecular level, we would see the sugar molecules speeding away from the lump during the early part of the reaction. Later, however, as the substances became more evenly mixed, the frequency of the back reaction returning sugar molecules to the location of the dissolving lump would rise until equilibrium was reached. In fact, diffusion is a chemical reaction with its own free energy, which depends on the characteristics of the substances involved:

$$\text{sugar + coffee} \rightleftharpoons \text{sweetened coffee} \qquad (\Delta G = -x)$$

Free energy tells us how a substance will diffuse; concentration gradients are only one component. Consider a situation in which there is a slight concentration gradient in one direction and a pronounced temperature gradient in the reverse direction (Fig. 4.9). The opposing effects of these two gradients produce a net movement of molecules that depends on the relative free energies of the two gradients; in this case it is from the region of higher temperature to the area of higher concentration.

The importance of diffusion and its basis in free energy with respect to cells is clear: the concentration of organic molecules and a select group of ions inside a cell is a very unlikely arrangement. Without the cell membrane, the free energy of the cellular chemistry would be lost as the contents diffused into the environment. There must be a barrier between the inside and the outside of the cell to maintain the integrity of the cellular chemistry. Moreover, the free-energy gradient across the cell membrane is available to do work.

4.9 Multiple gradients and free energy

This two-chambered vessel has one side (Y) with a slightly higher pressure than the other (Z), but the temperature in Z is much higher than that in Y. If concentration alone were important, the net diffusion would be from Y, the region of higher concentration, to Z, the region of lower concentration. But the higher the temperature, the greater the thermal motion of the particles in that system; and the greater the thermal motion, the greater the free-energy content. Because the difference in free energy associated with the temperature gradient from Z to Y outweighs the difference in free energy associated with the concentration gradient from Y to Z, net diffusion will be from Z to Y.

■ MANIPULATING DIFFUSION: OSMOSIS

To envision the way cell membranes can function, imagine a U-shaped chamber divided by a membrane (Fig. 4.10). Suppose that molecules of water can pass through the membrane while molecules of sugar cannot. How will this *differentially permeable* (or *selectively permeable*) membrane affect the diffusion of materials between the two halves of the chamber? Suppose that side A of the

4.10 U-tube divided by a selectively permeable membrane

The membrane at the base of the U-tube is permeable to water but not to sugar molecules (yellow balls). Left: Side A contains only water; side B contains a sugar solution. Initially the quantity of fluid in the two sides is the same. Center: A larger number of water molecules (blue balls) bump into the membrane per unit time on side A than on side B. Right: Because more water molecules move from A to B than from B to A, the level of fluid on side A falls while that on side B rises.

U-tube contains pure water and side B an equal quantity of sugar solution (sugar dissolved in water), and that both sides are at the same temperature and pressure. If the membrane is permeable to water but not to sugar, water molecules will be able to pass in both directions, from A to B and from B to A.

This movement of a solvent (usually water) through a selectively permeable membrane is called *osmosis*. Since water is already present on both sides of the membrane in the U-tube, it might seem that the movement of water molecules across the membrane should have no net effect. Consider the differences between the pure water and the sugar solution more carefully. We have seen that substances diffuse from regions of higher concentration to regions of lower concentration. Water, being more concentrated on side A, will tend to diffuse to side B; sugar molecules, trapped by the membrane on side B, will not cross.

We can also see why water molecules move from A to B by picturing to ourselves the events at the membrane itself. On side A, all the molecules that bump into the membrane during a given interval are water molecules, and because the membrane is permeable to water, many of these will pass through the membrane from A to B. By contrast, on side B, some of the molecules bumping into the membrane during the same interval will be water molecules, which may pass through, and some will be sugar molecules, which cannot pass through because the membrane is impermeable to them. Hence more water molecules will move across the membrane from side A to side B per unit time than in the opposite direction; the net osmosis will be from A to B.

We can also think of the matter more abstractly, in terms of entropy. The arrangement of water molecules in pure water is orderly, in that every molecular location is occupied by a water molecule, whereas the arrangement in the sugar solution is disorderly, in the sense that any given molecular location may be occupied by either a water molecule or a sugar molecule. Since an orderly system possesses more free energy than a disorderly one, it follows that the orderly water molecules in the pure water (side A) have more free energy than the disorderly water molecules in the sugar solution (side B). There is a free-energy gradient for water from side A to side B, so there will be a net movement of water down this gradient, from A to B. No matter how you choose to view it, remember that in the absence of temperature or pressure gradients, water moves from a region of high water concentration (that is, pure water) to one of lower water concentration.

We can make make another generalization about osmosis. *The free energy of water molecules is always decreased if osmotically active substances* (dissolved or colloidally suspended particles) *are present in the water.* (Colloidal particles are generally larger than the separated individual molecules of a dissolved substance, yet small enough so that—unlike the still larger particles of a true suspension—they do not settle out at an appreciable rate but remain dispersed within the fluid medium.) The **osmotic concentration** of a fluid—the number of osmotically active particles per unit volume—thus bears a direct relationship to that fluid's free energy. In the U-tube example, for instance, *the difference in the free energy of water molecules in a solution versus pure water is proportional to the osmotic concentration of the solution.* Since the osmotically active particles disrupt the orderly three-dimensional array of the water molecules, the fluid has less free energy.

Each solution, then, has a certain free energy, depending on its osmotic concentration. Under conditions of constant temperature and pressure, this free energy can be calculated; it is called **osmotic potential**. (Pure water is arbitrarily assigned an osmotic-potential value of zero. Since osmotic potential decreases as osmotic concentration increases, all solutions have values of less than zero.) If two different solutions are separated by a membrane permeable only to water, and temperature and pressure are constant, *the net movement of water will be from the solution with the lower osmotic concentration to the solution with the higher osmotic concentration.* The steeper the osmotic-concentration gradient, the more rapid the movement: the water flows from regions of higher osmotic potential to regions of lower potential at a rate proportional to the degree of difference in osmotic potential. These generalizations are summarized in Table 4.1.

If the net movement of water in the U-tube is from side A to side B, the volume of liquid will increase on side B and and decrease on side A. Does this process continue until side A is dry? Clearly, if the membrane is completely impermeable to sugar molecules, conditions on the two sides will never be equal, no matter how many water molecules move from A to B: the fluid in B will remain a sugar solution, though an increasingly weak one, and the fluid in A will remain pure water. Nevertheless, under normal conditions, the fluid level in B will rise to a certain point and then stop. Why? The column of fluid is, of course, being pulled downward by gravity. As the column rises, its weight exerts increasing downward hydrostatic pressure. As the pressure increases, the free energy of the water in the sugar solution rises, because pressure too is a form of free (useful) energy. Eventually the column of sugar solution becomes so high, and its pressure and free energy so great, that water molecules are pushed across the membrane from B to A as fast as they move into B from A.

When water is passing through the membrane in opposite direc-

Table 4.1 Some rules of osmotic diffusion

I. The free energy of water molecules is always decreased if osmotically active substances (dissolved or colloidally suspended particles) are present in the water.

II. The decrease in the free energy of water molecules is proportional to the osmotic concentration.

III. The net movement of water will be from the solution with the lower osmotic concentration to the solution with the higher osmotic concentration.

EXPLORING FURTHER

OSMOTIC POTENTIAL, OSMOTIC PRESSURE, AND WATER POTENTIAL

As we have seen, osmotic potential is useful for thinking about how two solutions with differing osmotic concentrations interact. It is easily related to the underlying difference between them in free energy, which results in the movement of the solvent. Many researchers, however, prefer to think in terms of the pressure that must be exerted on a solution to keep it in equilibrium with pure water when the two are separated by a selectively permeable membrane. In our U-tube example, this pressure corresponds to the hydrostatic pressure exerted by the sugar solution at equilibrium; it is known as **osmotic pressure**. Clearly, *the osmotic pressure of a solution is a measure of the tendency of water to move by osmosis into it.* The more dissolved particles in a solution, the greater the tendency of water to move into it and the higher the osmotic pressure of the solution. Thus, under constant temperature and pressure, water will move from the solution with the lower osmotic pressure to the solution with the higher osmotic pressure.

While the terms *osmotic pressure* and *osmotic potential* are regularly used by physiologists studying animals, plant physiologists more often refer to **water potential**, which is essentially the same as the free energy of water. At a pressure of 1 atm (atmosphere), pure water is assigned a water potential of zero. Since the water potential decreases as the osmotic concentration increases, all solutions have values of less than zero. In this sense, water potential is like osmotic potential. Unlike osmotic potential, which is a function of solute concentration alone, water potential (like free energy) is also a function of temperature and pressure. When two solutions are separated by a selectively permeable membrane, water will move from the solution with the higher water potential to the solution with the lower water potential. Throughout this book we will discuss fluid movement in plants in terms of water potential, while the analogous descriptions of fluids in animals will be in terms of osmotic potential.

A | B

Lower osmotic concentration | Higher osmotic concentration
Higher free energy of water | Lower free energy of water
Higher osmotic potential | Lower osmotic potential
Higher water potential | Lower water potential
Lower osmotic pressure | Higher osmotic pressure

tions at the same rate, the system is in dynamic equilibrium, with the free energy—the osmotic potential—of the pure water on one side of the membrane just matching the free energy—the osmotic potential and hydrostatic pressure—of the column of solution on the other side. The greater the concentration difference across the membrane, the higher the column of solution will rise before this osmotic difference is counterbalanced by the difference in hydrostatic pressure.

Osmotic concentration is not concentration by weight, it is molecular or ionic concentration—the total *number* of solute particles per unit volume. If there are several kinds of solutes in the same solution, then the osmotic concentration of that solution is determined by the total (per unit volume) of *all* the particles of all kinds. If a dissolved substance ionizes, then each ion functions osmotically as a separate particle: 1 mole of sodium chloride (NaCl) dissolved in water produces 2 moles of particles: 1 mole of Na^+ ions and 1 mole of Cl^- ions. Colloidal particles may also contribute to the total osmotic concentration.

■ BETWEEN SWELLING AND COLLAPSE

We have discussed diffusion and osmosis at such length because the cell membrane is selectively permeable, and the processes of diffusion and osmosis are fundamental to cell life. Though the membranes of different types of cells vary widely in their permeability characteristics—the membrane of a human red blood corpuscle,[1] for instance, is over 100 times more permeable to water than the membrane of *Amoeba*—we can make a few generalizations: Most cell membranes are relatively permeable to water and to certain simple sugars, amino acids, and lipid (fat and oil)-soluble substances. They are relatively impermeable to polysaccharides, proteins, and other very large molecules. In short, cell membranes pass the building blocks of complex organic compounds, not the compounds themselves. The permeability of cell membranes to small ions varies greatly, depending on the particular ion, but in general, negatively charged ions can cross more rapidly than positively charged ions, though neither can do so as readily as uncharged particles.[2]

What implications do these generalizations hold for life? On the one hand, selective permeability enables cells to retain the large organic molecules they synthesize. On the other, the tendency of water to pass through selectively permeable membranes into regions of higher osmotic concentration can be harmful or even fatal. When a cell is in a medium that is **hypertonic** relative to it (a medium to which it loses water by osmosis, usually because the medium contains a higher concentration of osmotically active particles), the cell tends to shrink (Fig. 4.11); if the process goes too far, it may die. Conversely, when a cell is in a medium **hypotonic** to it (one from which the cell gains water, usually because the medium contains a lower concentration of osmotically active particles), the cell tends to swell; unless it has special mechanisms for expelling excess water or special structures that prevent excessive swelling (as most plant cells do), it may burst. A cell in an **isotonic** medium (one with which the cell is in osmotic balance, usually because it contains the same concentration of osmotically active particles) neither loses nor gains much water by osmosis.

The osmotic relationship between the cell and the surrounding medium can be critical. Some cells are normally bathed by an isotonic fluid and therefore have no serious osmotic problems. Human red blood corpuscles are an example; they are normally

[1] A red blood corpuscle begins as a cell but (in mammals) loses its nucleus as it matures and becomes specialized to transport oxygen. However, it remains sufficiently like true cells to be used as an example of many cellular properties.

[2] The process in which, in addition to solvent, some solutes selectively cross the membrane is often called dialysis.

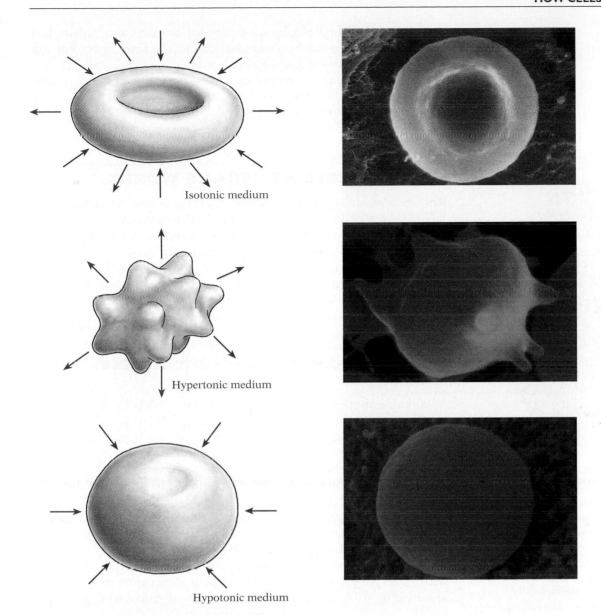

4.11 Osmotic relationships of a cell

In an isotonic medium, water gain and water loss are equal, and so the cell neither shrinks nor grows. In a hypertonic medium, there is a net loss of water from the cell, and the cell shrinks. In a hypotonic medium, the cell swells as water moves from the medium inside. The accompanying photographs are of human red blood cells.

bathed by blood plasma, with which they are in relatively close osmotic balance. The cells of most simple oceanic plants and animals have an osmotic concentration close to that of sea water. All cells, however, have a higher osmotic concentration than fresh water. Freshwater organisms therefore live in a hypotonic medium and have had to evolve ways either to avoid taking in excess water or to dispose of it when it too much does get in.

Controlling the flow of water is only one problem. Though the selective permeability of the membrane effectively traps large molecules inside, it does not provide any mechanism for concentrating the organic building blocks necessary for constructing substances like DNA, proteins, and polysaccharides in the first place. For nutrients to be captured and retained, wastes expelled, and cell volume controlled, the cell membrane must have the capacity to pass many chemicals in only one direction. The secret of this critical ability lies in the structure of the cell membrane itself.

HOW THE CELL MEMBRANE WORKS

Researchers struggled for decades to explain the remarkable behavior of the cell membrane. The key is the membrane's structure, which was inferred from its various properties. For instance, permeability studies showed that lipids and many substances soluble in lipids can move between the cell and the surrounding medium. Therefore the cell membrane must contain lipids to allow fat-soluble substances to cross. But how could a lipid membrane be stable?

■ LIPID BILAYERS: MEMBRANES HELD TOGETHER BY WATER

A workable model was proposed in the late 1930s by J. F. Danielli of Princeton University and H. Davson of University College, London. They suggested that the membrane might be composed of two layers of phospholipids oriented with their polar (hydrophilic) ends exposed at the two surfaces of the membrane and their nonpolar (hydrophobic) hydrocarbon chains buried in the interior, hidden from the surrounding water. A structure based on hydrophobic-hydrophilic interactions would be very stable because of the polar bonding of the hydrophilic ends with water. Indeed, as we now know, spheres—called *liposomes*—composed of phospholipid bilayers form spontaneously when phospholipids are mixed with water (Figs. 4.12, 4.13). Electron micrographs of cell membranes (Fig. 4.14) reveal a structure nearly identical to that of liposomes.

4.12 A liposome

Mixing phospholipids with water produces spherical phospholipid bilayers, each enclosing a droplet of water. The spontaneous formation of these spheres, called liposomes, is a result of the energetically favorable interaction of the hydrophilic ends of the phospholipids with the water molecules. Cell membranes are structured in exactly this way. Hence, they are basically stable, forming almost automatically and requiring no energy to maintain.

4.13 Single-layer liposomes

Liposomes can be manufactured to deliver concentrated dosages of drugs to sites of infection, inflammation, or cancerous growth. This electron micrograph shows a freeze-fracture of artificial liposomes in a cell membrane.

0.1 μm

4.14 Membrane of a red blood corpuscle

The cytoplasm of the corpuscle is in the dark grey area of this electron micrograph. The membrane consists of two dark lines (the phosphate-group "heads") separated by a lighter area (the hydrocarbon tails).

0.1μm

The phospholipid-bilayer model accounts for the stability, flexibility, and lipid-passing characteristics of the membrane, but it does not explain the selective permeability of the membrane to certain ions and chemicals. In 1972 S. J. Singer of the University of California at San Diego and G. L. Nicolson of the Salk Institute proposed the **fluid-mosaic model**, a hypothesis that is now almost universally accepted. The model incorporates the Davson-Danielli conception of a bilayer of phospholipids oriented with their hydrophobic tails directed toward the interior and their hydrophilic heads exposed to the aqueous environment on both surfaces. To explain the selective permeability, the fluid-mosaic model imagines various specialized proteins inserted in the membrane that act as transmembrane pores (Fig. 4.15).

4.15 The fluid-mosaic model of the cell membrane

A double layer of lipids forms the main continuous part of the membrane; the lipids are mostly phospholipids, but in plasma membranes of higher organisms cholesterol (brown) is also present. Proteins occur in various arrangements. Some, called peripheral proteins, lie entirely on the surface of the membrane, to which they are anchored by a covalent bond with a membrane lipid. Others, called integral proteins, are wholly or partly embedded in the lipid layers; some of these may penetrate all the way through the membrane. In the front right corner, note the three protein units, joined by covalent bonds (not shown) to form part of a single protein molecule bounding a membrane-spanning pore. Proteins make up about half of the weight of membranes. The hexagons represent carbohydrate groups.

FREEZE-FRACTURE AND FREEZE-ETCHING

Freeze-fracture etching, a technique for preparing specimens for electron microscopy, has become an indispensable tool for studying the details of membrane structure. The specimen is first rapidly frozen and then fractured along the plane of its bilipid membrane (A-B). Some of the ice is then removed from the specimen by sublimation (conversion directly to vapor), which exposes the inside surface of the membrane and gives the specimen an etched appearance (C). Carbon and a metal, usually platinum, are then applied at an angle to the specimen (D) so as to shadow any irregularities in the membrane. Next the original specimen is removed from the platinum cast or surface replica thus formed (E). The replica can now be examined by microscopy. EMs of freeze-etched cell membranes or other structures in cells usually have a three-dimensional appearance (Fig. 4.13).

Some membrane proteins (*peripheral proteins*) are confined to the surfaces; others, the *integral proteins*, are located wholly or largely within the lipid bilayer (Figs. 4.15, 4.16). Integral proteins may be buried within the bilayer, or have parts that project through the surface; some are confined to the outer half of the lipid core, and others to the inner half, while some extend entirely through the bilayer, projecting into the watery medium on both sides. Hydrophilic amino acids (those with polar or electrically charged R groups) predominate in the portions of the protein molecules that project out into the water, whereas hydrophobic (nonpolar) amino acids are abundant in the portions buried in the bilayer (Fig. 4.17). Indeed, the location of the hydrophilic and the hydrophobic amino acids in a membrane protein determines

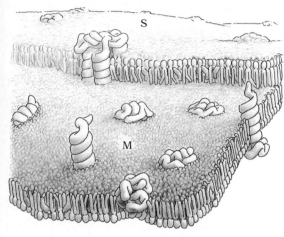

1μm

4.16 Plasma membrane of a red blood corpuscle

In this electron micrograph the plasma membrane has been frozen and then fractured along the plane between the two layers of lipids—that is, along the middle of the bimolecular lipid core (see sketch). The numerous spherical particles visible in the micrograph are proteins (see the gray coiled entities in sketch). S, outer surface of membrane; M, interior of fractured membrane.

4.17 Orientation of proteins within membranes

The parts of the polypeptide chain containing most of the hydrophilic amino acids (polar or charged R groups; blue) tend to project into the watery medium outside the lipid layers, whereas the parts of the chain with hydrophobic amino acids (brown) tend to be folded into the inner, lipid portion of the membrane. The relative diameter of the protein strands has been reduced for clarity.

which part of the protein will be anchored in the membrane and whether the protein will be integral or peripheral.

The structure of the membrane is not static. Most lipid molecules can move in the plane of the membrane, so a particular molecule found in one position at a given moment may be in an entirely different position only seconds later. Mobility of the lipids is greatest in membranes that are high in unsaturated phospholipids (Fig. 4.18) and contain no cholesterol. Speeds of $2\mu m$ (micrometers)/sec are possible—an astonishing mobility considering that many organisms (the bacterium *Escherichia coli*, for example) are only about $2\mu m$ long.

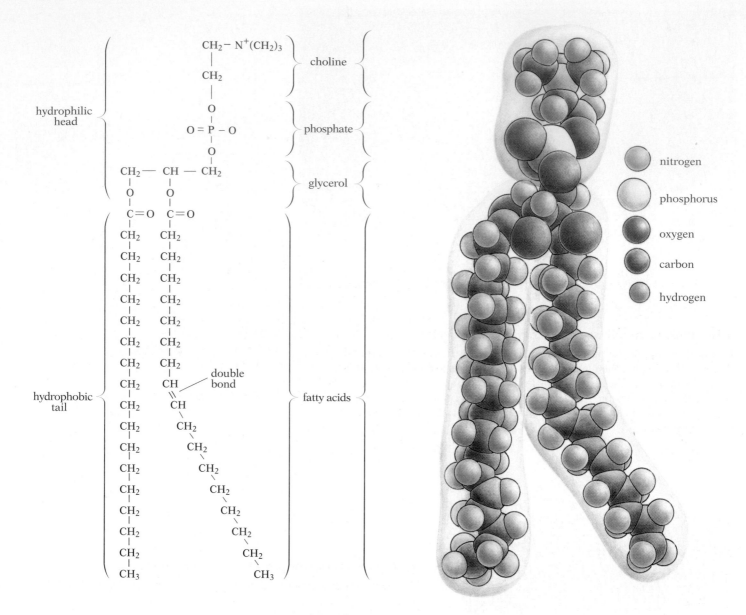

4.18 A phospholipid

The cell membrane is made up mostly of phospholipids. Phosphatidyl-choline, a common membrane phospholipid found almost exclusively in the outer half of bilayer membranes, consists of a polar head (a positively charged choline, a negatively charged phosphate, and an uncharged glycerol) joined to two hydrophobic fatty acid chains. The kink in the right tail is created by a double-carbon bond. Because this tail is unsaturated—that is, not every carbon has its full complement of hydrogen atoms—the phospholipid will be less tightly packed in the membrane and hence will be more mobile.

When cholesterol is present, it binds weakly to adjacent phospholipids and reduces the permeability of the membrane to small molecules. It can also have one of two other effects. If the membrane phospholipids are mostly saturated, cholesterol prevents them from packing so closely and regularly that they crystallize into a stiff and solid layer. When, on the other hand, the phospholipids are mostly unsaturated (so the "kinks" in their hydrophobic tails keep them loosely packed), cholesterol can fill these gaps, bind to adjacent phospholipids, and thus join them together (Fig. 4.19). Plant membranes lack cholesterol and instead depend on the cell wall for stability.

The cholesterol concentration and the degree of saturation vary enormously between species and even among tissues in the same organism, depending on the amount of flexibility needed. Though medical opinion is far from unanimous, there is great concern that too much cholesterol and saturated fat in the human diet can lead to hardening of cell membranes, particularly those lining the arteries. Hardening of the arteries (atherosclerosis) is a major cause of stroke and heart disease.

Some proteins in the cell membrane can also move laterally to some extent, but others are anchored in place, thus limiting the fluidity of the membrane.

■ HOLES IN THE MEMBRANE: CHANNELS AND PUMPS

The lipid bilayer forms spontaneously from phospholipids manufactured by the cell, creating a flexible but effective barrier between the inside of the cell and the world outside. However, the need to pass specific substances through the membrane—sometimes actively pumped against their osmotic-concentration gradients—requires specialized pores. The pores in the membrane are actually channels through one or (more often) a group of protein molecules (Fig. 4.15). The distinctive properties of the various R groups of the amino acids in the proteins give the pores some selectivity; not all ions or molecules small enough to fit in the pores can actually move through them. Since they enable specific chemicals to permeate the membrane, channels and pumps are known collectively as *permeases*.

How channels control membrane traffic The simplest of the permeases, the *membrane channels,* provide openings through which specific substances can diffuse across the membrane. These channels, though selective, are passive: they simply permit particular chemicals to move down their concentration gradients. This *facilitated diffusion* is the basis of the membrane's highly specific permeability. The protein channel for potassium ions (Fig. 4.20A) provides a good example. Cellular processes result in the accumulation of K^+ inside most cells. As a charged particle, K^+ is insoluble in the hydrophobic interior of the lipid membrane, but the channel specific for K^+ allows it to leak out slowly at a controlled rate. Without such leakage the internal K^+ concentration would become too high for the cell to function. The specificity of the K^+ channel results from its internal shape and its charge, though the details are not understood.

More complex channels can move two specific substances in concert. For example, ion-exchange channels, called *antiports*, work by passively trading two similarly charged ions. As one moves into the cell, the other exits, thereby maintaining an electrical balance. The many integral membrane proteins in Fig. 4.16 are channels that exchange Cl^- for HCO_3^- (dissolved carbon dioxide), as part of the red blood corpuscle's job of carrying waste CO_2 from cells to the lungs.

4.19 Cholesterol in the membrane

Cholesterol (brown) binds weakly but effectively to two adjacent phospholipids, thereby partially immobilizing them. The result is a less fluid and mechanically stronger membrane. The amount of cholesterol varies widely according to cell type; the membranes of some cells possess nearly as many cholesterol molecules as phospholipids, while others lack cholesterol entirely. For the structural formula of cholesterol, see Figure 3.18, p. 58.

4.20 Strategies of membrane transport

Many strategies for moving substances across the cell membrane are known. (A) In facilitated diffusion of the simplest kind, a protein channel, or pore, embedded in the membrane provides a direct path for the chemical it passes "down" its osmotic-concentration gradient. The channel's diameter and the chemical environment the channel creates (hydrophilic or hydrophobic, for example) serve to prevent all but the correct substance from passing. (B) Other channels pass two substances cooperatively, or exchange two substances. Illustrated here is a hypothetical mechanism for the channel that uses the highly favorable osmotic-concentration gradient for bringing sodium ions into the cell to overcome the osmotic-concentration gradient working against glucose (labeled G). When Na^+ binds to the channel (middle), it induces an allosteric change that enables glucose also to bind to the channel. This binding of glucose causes the channel to close to the outside and open to the inside. This change releases the glucose as well as the Na^+ into the interior (right). Having lost the Na^+ and glucose, the channel can reopen to the outside. (C) An allosteric interaction between a signal molecule (red) and the gated channel causes the gate to open, so that diffusion can take place down a favorable concentration gradient. Other molecular systems, not shown, then inactivate the signal molecule so that the channel can close again. (D) A mobile carrier would not provide transmembrane channels, but would itself migrate back and forth from one surface to the other. The existence of mobile carriers is controversial.

A FACILITATED-DIFFUSION CHANNEL

B COOPERATIVE ION CHANNEL (SYMPORT)

C GATED CHANNEL

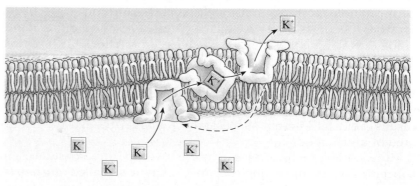

D MOBILE CARRIER

Other cooperative channels called *symports* move two substances in the same direction. Coordinated movement of this sort (called *cotransport*) moves glucose (the molecule most cells use as their major energy source) through the membrane in company with Na$^+$. Sodium ions are 11 times more concentrated outside the cell, and are therefore subject to a highly favorable osmotic gradient to the inside. Yet they must be accompanied by glucose to pass through the appropriate channel (Fig. 4.20B). Both Na$^+$ and glucose must bind to the outside of the channel before this special membrane pore will open. Thus the free energy of the osmotic-concentration gradient of Na$^+$ is exploited to overcome the smaller, unfavorable concentration gradient of glucose. In thermodynamic terms, the two diffusion "reactions" are linked: moving Na$^+$ "downhill" releases more free energy than is consumed in moving glucose "uphill."

The osmotic-concentration gradient of Na$^+$ accounts for only part of the energy used in moving glucose across the membrane. As you know, oppositely charged ions attract one another electrostatically, while ions with the same charge repel one another. As a result, if a cell has more negative than positive ions, positive ions will be attracted to it from the surrounding fluid. (Most cells actually have a net negative charge of about 70 mv (millivolts) relative to the fluids surrounding them.) The difference in charge across the membrane generates an *electrostatic gradient*, and when appropriate channels are open, positive ions tend to flow into the cell, while negative ions tend to flow out. For ions like Na$^+$, which are far more concentrated outside, the osmotic and electrostatic potentials combine to create a strong *electrochemical gradient* (Fig. 4.21). It is the free energy of these combined potentials that accounts for the particular effectiveness of Na$^+$ in moving glucose into cells.

Gates across membrane channels are another way to control movement across the membrane. One use is to convert molecular signals specialized for carrying information between cells into other signals more suitable for communicating inside the cell. When a molecular signal—a hormone or one of the transmitter substances that carry messages from one nerve to another—binds to an exposed part of a transmembrane *receptor*, an allosteric change (a change in shape) takes place. This conformational change allows the gate to open, and the second signal, usually an ion like Na$^+$ or Ca^{++}, then moves across, carrying the message into the cell (Fig. 4.20C). This *gated channel* strategy underlies the transmission of many chemical messages in both plants and animals, and of the nerve impulses by which animals sense the outside world and move their muscles. This strategy is also used to respond to internal signals: for instance, internal gating regulates flow through Cl$^-$ channels, which are important in regulating the secretion of fluids from cells. A defect in this channel can cause cystic fibrosis.

Another way molecules might get across the membrane would be for the permease to act as a *mobile carrier*, taking them through one by one (Fig. 4.20D). No definite example of such a permease is yet known, but valinomycin acts in just the way we would expect a mobile carrier to act. Valinomycin is a ring-shaped

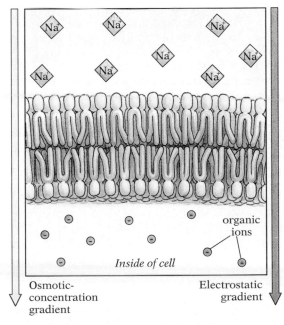

4.21 An electrochemical gradient

Cells have an electrical potential of about 70 mv negative with respect to the fluids that surround them. This gradient arises mainly because large numbers of negatively charged organic ions are trapped inside the cells, while a relatively high concentration of positively charged Na$^+$ exists outside. Sodium ions are therefore subject to both a strong osmotic-concentration gradient and a sizable electrostatic gradient. The effects of these two gradients combine to create an electrochemical gradient.

4.22 Valinomycin

This permease, which is synthesized by some bacteria, kills cells that take it up. It consists of three kinds of peptides: alanine, valine, and a chlorine-substituted valine. The potassium ion is held to the center by the oxygens while it is carried across the host membrane.

polymer with a hydrophobic exterior and a polar interior (Fig. 4.22). The polar pocket, lined with six oxygen atoms, can hold a single potassium ion. Apparently the complex bobs back and forth randomly, carrying K^+ ions in both directions. The net transfer of K^+ is a statistical consequence of the concentration gradient: valinomycin more often picks up K^+ inside the cell and releases it outside simply because there is more K^+ inside than out. Valinomycin is actually an antibiotic produced by certain bacteria to poison competing microorganisms by altering the selective permeability of their membranes. The evidence for mobile carriers in normal membranes is equivocal.

Moving molecules "uphill" Other permeases, known as *pumps*, do not depend on free-energy gradients. Instead, pumps use the cell's store of energy to move substances against their gradients. This ***active transport*** is important in ridding the cell of undesirable molecules. Pumps also transport many essential building blocks into the cell. The best-understood example of a membrane pump, however, is the ***sodium-potassium pump***, which maintains the electrochemical gradient across the membrane: (Fig. 4.23). The free-energy source for this pump, and for many cellular processes, is the cellular energy carrier ATP (adenosine triphosphate), which we will discuss in detail in later chapters. The pump utilizes this energy to exchange potassium and sodium ions across the membrane, thereby maintaining the electrochemical gradient.

The effects of this pump are far-reaching: it is responsible for the electrical activity of nerves and muscles; it indirectly supplies the free energy for many osmotic transport systems, including the one for glucose; it supplies the gradient that permits a Na^+-H^+ antiport

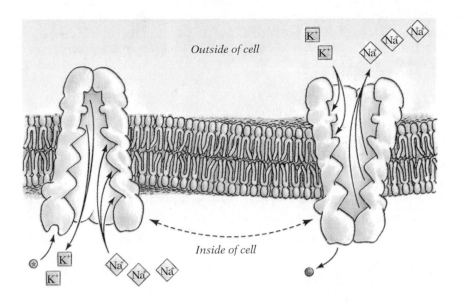

Outside of cell

Inside of cell

channel to control cellular pH; it facilitates the uptake of water by plant roots; and it helps regulate the volume of many cells by controlling osmotic potential. Indeed, when the pump is destroyed by the poison ouabain, cells swell uncontrollably with water until they burst.

In addition to the engineering feats of channels and pumps, cells control osmotic potentials across the membrane by chemical means. We saw how the Na$^+$ gradient can be used to "pull" glucose inside because the gradient against glucose is very small. Yet the world outside most cells has very little glucose to begin with, while most cells use it in large amounts. How can the concentration of glucose inside the cell be kept so low? The cell manages this trick by binding the glucose into another compound as soon as it is inside (Fig. 4.24). As a result, the free-glucose concentration of glucose remains artificially low inside, and the osmotic potential against glucose does not get out of hand.

■ LARGE-SCALE MOVEMENT: ENDOCYTOSIS AND EXOCYTOSIS

As we have seen, permeases are the means by which substances enter and leave cells through the cell membrane. Cells also have ways of admitting substances, usually in larger quantities, without having them pass through channels. Through *endocytosis*, a cell encloses the substance in a membrane-bounded vesicle that is pinched off from the cell membrane. There are three types of endocytosis, all of which depend on specialized membrane proteins:

4.24 Formation of a complex inside the cell

As fast as glucose molecules (yellow hexagons) enter the cell, they combine with acceptor molecules (brown) already in the cell to form a new substance. The concentration of free glucose inside the cell therefore remains low, so glucose continues to diffuse inward.

4.23 The sodium-potassium pump

In contrast to other methods of membrane transport, the pump strategy uses energy from the cell (rather than the free energy of a concentration gradient) and provides active transport of a substance against its gradient. In this case, three sodium ions are exchanged for two potassium ions; both kinds of ions are already more concentrated on the side to which they are being moved. In the model shown here, the release of K$^+$ ions brought in during the previous cycle is followed by the binding of three Na$^+$ ions and an energy source, ATP (circle), on the inside. The resulting conformational changes in the protein open it to the outside, reduce its affinity for Na$^+$, which is then released, and increase its affinity for K$^+$ ions. The binding of K$^+$ then causes the channel to open to the inside, increase its affinity for Na$^+$, and decrease its affinity for K$^+$, and the cycle begins again. The net ionic effect is to pump positive charges out of the cell, and the inside of the cell becomes negatively charged with respect to the outside. The electrical and osmotic potential created by the sodium-potassium pump ultimately makes possible the cooperative transport of glucose illustrated in Figure 4.21B.

A

1μm

B

4.25 Phagocytosis

(A) White blood cells, or leukocytes, in the bloodstream use phagocytosis to capture foreign organisms; here the leukocyte is engulfing a dividing bacterium. (B) In *Amoeba*, pseudopodia flow around the prey until it is entirely enclosed within a vacuole.

1 When the material engulfed is in the form of large particles, the process is called ***phagocytosis***, or "cellular eating" (Fig. 4.25). Usually, armlike ***pseudopodia*** ("false feet") flow around the material, enclosing it within a vesicle, which then detaches from the plasma membrane and migrates to the interior of the cell. Phagocytosis is triggered only when special membrane-mounted receptor proteins bind to a suitable target; this binding is analogous to enzyme-substrate binding. In vertebrates, phagocytosis is generally restricted to scavenger cells in the blood, which ingest debris and invading microorganisms (Fig. 4.25B).

2 When the engulfed material is liquid the process is called pinocytosis, or "cellular drinking." Pinocytosis is used to ingest extracellular fluid or to transport it across a cellular barrier (Fig. 4.26).

4.26 Pinocytosis

(A) Extracellular fluids are enclosed in vesicles on the cell surface through endocytosis. The vesicles may be used to transport the fluid across the cell, leading to exocytosis on the other side of the cell, or the fluid may be allowed to escape into the cell itself. (B) Complete transcellular movement can be seen in this electron micrograph of the cells lining a blood capillary (left) and a portion of the lung (right). The membrane facing the lung must be kept constantly moist for gases (O_2, CO_2) to dissolve and cross; since the water used constantly evaporates and is exhaled, pinocytotic vesicles are always in use moving water from the blood to the inner surface of the lung, across a basement membrane.

A

B

0.1μm

A

0.5μm

4.27 Receptor-mediated endocytosis

(A) Three stages of receptor-mediated endocytosis of a short polypeptide used in cell-to-cell communication are seen taking place in a cultured nerve cell. (B) Molecules bind to receptors on the cell surface, triggering vesicle formation and endocytosis.

3 When the material to be ingested is adsorbed on the cell membrane at selective binding sites, the process is termed *receptor-mediated endocytosis.* The loaded vesicles are formed and detach from the membrane at the cell surface (Fig. 4.27). In most cases, the site that collects a particular substance before trapping it in a vesicle appears as a coated pit, with receptor molecules clustered in one spot in the membrane (Fig. 4.28).

A

B

C

D

0.1 μm

4.28 Endocytosis by means of a coated pit

(A) Specialized receptors for lipoproteins have aggregated to form a coated pit in the membrane of an egg cell. (B–D) The pit is subsequently pinched off to form a vesicle. Most endocytotic vesicles are transported to lysosomes, organelles within the cell where the contents are enzymatically altered. The lipoprotein in the vesicle shown here will become part of the yolk.

A

0.2μm

4.29 Clathrin-coated pits versus caveolae

(A) Clathrin-coated pits, seen from the inside of a liver cell, are in the process of budding off from the membrane to form intracellular vesicles. The cablelike structures are part of the cytoskeleton, discussed in the next chapter. (B) Caveolae, seen from the inside of a fibroblast, lack the characteristic clathrin coating; one clathrin-coated pit is visible on the upper left-hand side.

B

These outward-facing receptors (and the substances they bind) produce what looks like a haze on the extracellular surface of the membrane (Fig. 4.28). A similar darkening appears on the intracellular surface of the membrane at the spot destined to form a vesicle. The best-understood component of this internal patch is the structural protein ***clathrin***, which begins assembling itself into the foundation for a membrane indentation, or pit, and then into the vesicle's coating. The clathrin scaffolding is clearly visible surrounding newly formed vesicles (Fig. 4.29A). Smaller invaginations, called caveolae, have a separate scaffolding protein and a different range of membrane-mounted receptors (Fig. 4.29B); they may form vesicles, or merely concentrate their target molecules and then pass them to the cytoplasm.

4.30 Endocytosis of cholesterol

(A) When a cell needs cholesterol, it synthesizes receptors for low-density lipoprotein (LDL) and incorporates them into the cell membrane, where they are free to migrate. (B) The receptors soon bind LDL, a carrier complex that transports cholesterol in the blood. The LDL complex includes some 2000 cholesterol molecules and an associated protein (called apoprotein) that binds to the LDL receptor. (C) Having bound the LDL, the receptors stop drifting in the membrane and stick to one another over a clathrin-rich patch of membrane. Even in the absence of bound LDL, many of the receptors eventually aggregate spontaneously. (D–E) Aggregations of LDL receptors trigger endocytosis, the first step of which is the formation of a coated pit. (F) The resulting vesicle is transported to the site of membrane synthesis. This method of cholesterol uptake may demonstrate the usual strategy by which cells obtain nutrients that cannot pass through the membrane directly.

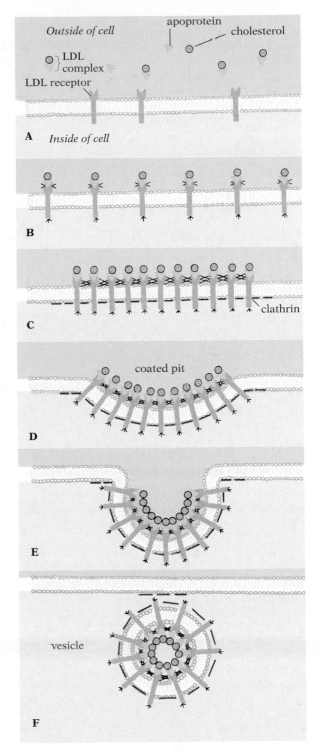

One example that illustrates the importance and specificity of receptor-mediated endocytosis involves cholesterol uptake by cells. Cholesterol is transported in the blood to cells by a carrier complex of low-density lipoprotein (LDL). When the cell needs cholesterol—usually for use in the manufacture of new membrane—LDL receptors are synthesized and incorporated into the cell membrane (Fig. 4.30). The LDL receptors aggregate spontaneously over membrane patches rich in clathrin, particularly once they have bound LDL. The cholesterol is then transported in endocytotic vesicles, known as **endosomes**, for use in the cell's vast complex of membranes. One cause of atherosclerosis involves failure of the receptors to bind LDL; another involves failure of the receptors, having bound LDL, to aggregate and to initiate endocytosis. Either defect can allow cholesterol-rich plaques to form on artery walls.

Material enclosed within endocytotic vesicles has not yet become part of the cytoplasm. It is still separated by a membrane; it must eventually cross that membrane (or the membrane must disintegrate) if it is to become incorporated into the cell. Normally the vesicle membrane forms from cell membrane. Meanwhile, vesicles full of digestive enzymes form within the cell. These **lysosomes** fuse with the endosomes to create miniature cellular stomachs, which break down the endosomal contents. After digestion, many of the products can cross the lysosome membrane into the cytoplasm, while others remain trapped inside. As a result the cell is able to ingest the substances it requires, while the unwanted parts of the meal, and the lysosome's destructive enzymes, remain segregated from the cell's delicate chemical interior. Endosomes and lysosomes will be described more fully in the next chapter.

Exocytosis is essentially the reverse of endocytosis. Materials contained in vesicles are conveyed to the periphery of the cell, where the vesicular membrane fuses with the cell membrane and then bursts, releasing the materials to the surrounding medium (Fig. 4.31). Many glandular secretions are released from cells in this way; the hormone insulin, for example, is released by exocytosis from the pancreatic cells that synthesize it. Exocytosis also exports waste products. The undigested remains of materials brought in by endocytotic vesicles, for example, are normally disposed of through exocytosis. In some cases a coupling of endocytosis and

A

B   1μm

C   1μm

4.31   Exocytosis

(A) A membranous vesicle moves to the periphery of the cell, where it bursts, releasing its contents to the exterior. (B) The final steps of exocytosis are seen here, as a vesicle containing tear fluid fuses with the plasma membrane and bursts.

exocytosis moves a substance entirely across a cellular barrier, such as the wall of a blood vessel. The substance is picked up by endocytosis on one side of the cell, and the vesicle moves through the cell to the other side, where the substance is released by exocytosis (see Fig. 30.16, p. 855).

Exocytosis is not simply a matter of a vesicle colliding with a membrane, fusing, and thus dumping its contents; if this were the case, vesicles containing toxic waste might deliver their contents to any number of membrane-enclosed intracellular organelles, with fatal results. Instead, vesicles have membrane molecules that allow them to attach only to their proper targets; even then, fusion proteins must be recruited and cooperate to form an initial pore that allows the vesicle membrane to fuse with the cellular membrane. The potent nerve toxins responsible for botulism and tetanus apparently exert their fatal effects by blocking the fusion of the specialized exocytotic vesicles of nerve cells responsible for communicating with other nerve and muscle cells.

OUTSIDE THE MEMBRANE: CELL WALLS AND COATS

Unlike animal cells, plant cells are encased in conspicuous cell walls. These walls, which are located outside the plasma membrane, are composed primarily of carbohydrates. Fungi and most bacteria also have strong, thick walls rich in carbohydrates. Animal cells, too, have carbohydrates on the outer surface of their membranes, but they do not form a wall; instead, they are attached as side groups to some lipids and membrane proteins. This cell "coat" plays an important role in determining certain properties of the cells.

■ HOW CELL WALLS ARE BUILT

The principal structural component of the cell wall that encloses plant cell membranes is the complex polysaccharide **cellulose**, which is generally present in the form of long threadlike structures called fibrils. The cellulose fibrils are cemented together by a matrix of other carbohydrate derivatives, including pectin and hemicellulose. The spaces between the fibrils are not entirely filled with matrix, however; they generally allow water, air, and dissolved materials to pass freely through the cell wall. The membrane located within the cell wall, rather than the wall itself, usually determines which materials can enter the cell.

The first portion of the cell wall laid down by a young growing cell is the **primary wall**. As long as the cell continues to grow, this somewhat elastic wall is the only one formed. Where the walls of two cells abut, a layer between them, known as the **middle lamella**, binds them together. **Pectin**, a complex polysaccharide generally present in the form of calcium pectate, is one of the principal constituents of the middle lamella. If the pectin is dissolved away, the cells become less tightly bound to one another. That is what happens, for example, when fruits ripen: the calcium pectate is partly converted into other, more soluble forms, the cells become looser, and the fruit becomes softer. Many of the bacteria and fungi that produce soft rots in the tissues of higher plants do so by first dissolving the pectin, reducing the tissue to a soft pulp that they can absorb. In cooking, extra pectin is added to thicken fruit preserves.

Cells of the soft tissues of the plant have only primary walls and intercellular middle lamellae. After ceasing to grow, the cells that eventually form the harder, more woody portions of the plant add further layers to the cell wall, forming the **secondary wall**. Since this wall, like the primary wall, is deposited by the cytoplasm of the cell, it is located inside the earlier-formed primary wall, lying between it and the membrane (Fig. 4.32). The secondary wall is

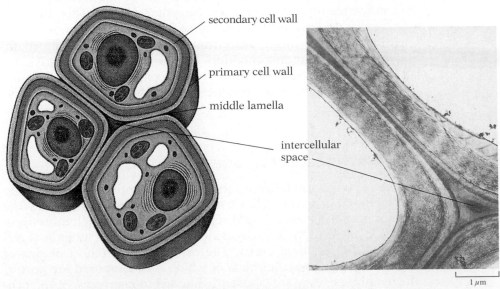

secondary cell wall

primary cell wall

middle lamella

intercellular space

1 μm

4.32   Cell walls and middle lamella of three adjacent plant cells

1 μm

4.33 Cellulose microfibrils from the cell wall of a green alga

The microfibrils in this electron micrograph are laid out in parallel lines in two directions; each is about 20 nm (nanometers) wide. Water and ions move freely through this meshwork.

often much thicker than the primary wall and is composed of a succession of compact layers, or lamellae. The cellulose fibrils in each lamella lie parallel to each other and are generally oriented at angles of 60-90° to the fibrils of the adjacent lamellae (Fig. 4.33). This arrangement gives added strength to the cell wall. In addition to cellulose, secondary walls usually contain other materials, such as *lignin*, which make them stiffer. Once deposition of the secondary wall is completed, many cells die, leaving the hard tube formed by their walls to function in mechanical support and internal transport for the body of the plant.

The cellulose of plant cell walls is commercially important as the main component of paper, cotton, flax, hemp, rayon, celluloid, and, obviously, wood itself. Lignin extracted from wood is sometimes used in the manufacture of synthetic rubber, adhesives, pigments, synthetic resins, and vanillin.

The cell walls of both fungi and bacteria differ from those of plant cells. In most fungi the main structural component of the wall is not cellulose but *chitin*, a polymer that is derived from the amino sugar glucosamine (see Fig. 3.8, p. 51); chitin is also the major component of insect exoskeletons. In bacteria the cell walls contain several kinds of organic substances, which vary from subgroup to subgroup. (One small group of bacteria lack cell walls altogether). The distinctive responses of these organic substances to diagnostic stains help identify bacteria in the laboratory. Structurally, however, the cell walls of all bacterial groups are alike in one respect. Part of each bacterial wall has a rigid framework of polysaccharide chains cross-linked covalently by short chains of amino acids.

The presence of cell walls means that the cells of plants, fungi, and bacteria can withstand exposure to fluids with low osmotic concentrations without bursting. In such media the cells are in a condition of *turgor* (distention). Water tends to move into them by osmosis, as a result of the high osmotic concentration of the cell contents. The cell swells, building up *turgor pressure* against the cell walls. Equilibrium is reached when the resistance of the wall is so great that no further increase in the size of the cell is possible and, consequently, no more water can enter the cell. Thus the cells of plants, fungi, and bacteria are not as sensitive as animal cells to the difference in osmotic concentration between the cellular material and the surrounding medium. Moreover, turgor pressure actually strengthens the mechanical structure of plants, just as inflating an initially limp balloon produces a much stiffer and stronger structure.

■ WHAT THE CELL COAT DOES

In plants, fungi, and bacteria, the cell wall is entirely separate from the membrane; if the cell shrinks in a hypertonic medium, the membrane shrinks with it and separates from the much more rigid wall (see Fig. 31.2, p. 881). By contrast, the "coat" of an animal cell is not an independent entity. The carbohydrates (short chains of sugars called oligosaccharides) of which it is composed are part of the membrane: they are covalently bonded to nearly 5% of the protein and lipid molecules in the plasma membrane (Fig. 4.34). The resulting complex molecules are termed glycoproteins and glyco-lipids, and the cell coat itself is often called the ***glycocalyx***.

The glycocalyx provides the recognition sites on the surface of the cell that enable it to interact with other cells. For example, if individual liver and kidney cells are mixed in a culture medium, the liver cells will recognize one another and reassociate; similarly, the kidney cells will seek out their own kind and reassociate. The nature of the carbohydrate markers varies consistently from tissue to tissue and from species to species. Cell recognition in the process of embryonic development must also depend, at least in part, on the glycocalyx, and the same is probably true for the control of cell growth. When normal cells grown in tissue culture touch each other, they cease moving, and their growth slows down or stops altogether. This phenomenon of ***contact inhibition*** is absent in most cancer cells, which continue growing without restraint because defective glycocalyces prevent them from interacting normally.

As a source of identity, the glycocalyx is also important in a variety of infectious diseases: malaria parasites, for instance, recognize their host (the red blood corpuscle) by a distinctive carbohydrate marker that the corpuscle manufactures for a completely different purpose. The recognition of host cells by invading viruses also often depends on the carbohydrate markers of the glycocalyx. The markers in the glycocalyces of foreign cells probably provide the cues that the immune system's antibody molecules use to recognize invaders.

HOW CELLS STICK TOGETHER

Having completed our examination of the structure and operation of individual cell membranes, we need to look briefly at how cells can bind to one another to create multicellular organisms like ourselves.

■ LIMITS TO CELL SIZE

There are enormous benefits to being multicellular. Larger size enhances an organism's ability to capture or harvest smaller organisms efficiently, to move farther and faster, and so on. However, large size cannot be achieved by simply increasing the size of a single cell indefinitely. A cell must take in its nutrients and oxygen across the membrane. As a cell triples in volume, so does its need for nutrients and oxygen, and yet its membrane surface area does not even double. Since metabolic needs increase faster than the surface area through which they are satisfied, a point arrives at which the membrane can no longer support its contents. The need for efficient diffusion, therefore, puts a strict limit on the surface-to-volume ratio of a cell and consequently limits cell size.

Many single-cell organisms are extraordinarily complex, for the one cell must do everything needed for survival. Although simple aggregations of identical cells do exist in nature, such as the 32-cell

A

B
0.2μm

4.34 Plasma membrane with glycocalyx

(A) The glycocalyx of an animal cell is composed of oligosaccharides (branching carbohydrates) attached to some of the protein and lipid molecules of the outer surface of the membrane. (B) In this electron micrograph the glycocalyx of a red blood corpuscle gives the outer surface of the membrane a fuzzy appearance.

disks of some green algae and the amoeboid clusters of slime molds, in most assemblies of cells more sophisticated specialization becomes possible. The course of evolution has demonstrated that arrangements in which certain cells concentrate on particular functions (propulsion, feeding, reproduction, and so on) can be far more effective than those in which each cell pursues a jack-of-all-trades strategy.

■ ALTERNATIVE WAYS TO CONNECT CELLS

There are two dramatically different strategies for giving form to multicellular aggregations, which would otherwise be amorphous lumps of cells. In certain animal tissues specialized cells called *fibroblasts* secrete fibrous proteins, among them collagen (see Fig. 3.27, p. 64) and elastin, which are components of an *intercellular matrix*. Cells are held in place by this structural network, and they grow and function there.

The other strategy for providing shape and strength is for cells to adhere to one another. The cells must specifically recognize which other cells are appropriate partners and then secure their membranes to them. This recognition, which is especially critical during embryological development, remains largely a mystery. At least certain types of cells have characteristic carbohydrate markers on the lipids and proteins of their outer membranes that are recognized by specialized receptors on other cells. Other cells have special glycoproteins, called cell-adhesion molecules, which bind directly to one another. A curious class of plant proteins, called lectins, recognize cells of specific tissue types on the basis of their carbohydrate identification tags. The role of lectins is unclear; apparently they help many plant cells adhere to one another, but there is also evidence that they glue together, thereby immobilizing the bacterial and fungal cells that otherwise cause plant diseases.

However cells locate each other, they frequently form strong junctions of several sorts. Most of the cellular attachment junctions utilized in multicellular organisms other than plants can be seen in the lining of the small intestine (Fig. 4.35). Intestinal cells are arranged with microvilli projecting into the intestine, where they absorb the nutrients released by digestion. They must adhere to one another not only to form the tubular channel of the gut but also to prevent the digestive enzymes (against which they are specifically armored) from leaking out and digesting the rest of the organism.

The general mechanical joining of two cells is accomplished by structures known as *spot desmosomes*. A spot desmosome consists of two cytoplasmic plaques, one just inside each of two adjacent cells (Fig. 4.35B). The outside faces of the plaques are joined by intercellular filaments, acting like rivets, while the inside faces are firmly attached to the cellular cytoskeleton, which will be described in the next chapter. There is a superficial resemblance between spot desmosomes and *belt desmosomes*, which are also composed of plaques and filaments (Fig. 4.35). Belt desmosomes, however, have no role in cell-to-cell adhesion. Instead, the circumferential plaque containing contractile fibers provides internal support for the cell.

4.35 Varieties of cellular junctions

The cells lining the mammalian small intestine are attached by several types of specialized junctions. An example of each is shown in enlarged detail. (A) A tight junction is composed of rows of transmembrane proteins in adjacent cells that bind to each other. (B) A spot desmosome consists of a pair of cytoplasmic plaques, each just inside the cell membrane and connected to the other across the intermembrane space by specialized intercellular filaments. Each plaque is also attached to fibers of the cytoskeleton within its cell. (C) A gap junction is formed by a pair of membrane channels aligned and bound together to create a specialized pathway between cells. The chemical structure of intermediate filaments and actin microfilaments is described in Chapter 5.

Cells in the intestine are also joined by *tight junctions*, in which specific transmembrane proteins attach directly to their counterparts in the adjacent cell (Fig. 4.35A). The cells are thus drawn together in such intimate association that there is no intercellular space and hence no possibility of leakage.

Finally, cells may be connected by *gap junctions*. A junction of this type is apparently formed by a pair of identical membrane channels in the two cells that line up with and bind to each other

(Fig. 4.35C). The result is both mechanical strength and the ability to share certain particular substances between the cells. Gap junctions are most common in developing tissue, and they may play a role in the arrangement and initial adhesion of cells.

The problems involved in achieving cellular adhesion and communication for most plant cells are very different from those for animal cells. In plants, rigid cell walls intervene between the plasma membranes of adjacent cells. Hence adhesion must be accomplished for the most part by relatively simple cross-linking between the polysaccharides of the cell walls. Effective adhesion and communication between plant cells is crucial, however, if water and inorganic nutrients are to be passed up from the roots, and the energy-rich products of photosynthesis are to be transmitted from the leaves to other parts of the plant. To accommodate this need, plant cell walls contain specialized openings—*plasmodesmata*—where the membranes of adjacent cells come into contact with each other. Some of these openings take the form of membrane-lined holes through which the cytoplasm of adjoining cells can mix directly. Others retain a double-membrane barrier; these play an important role in controlling the movement of both solutes and solvents between cells.

CHAPTER SUMMARY

THE DISCOVERY OF CELLS

THE CELL THEORY All cells arise from other cells; there is no spontaneous generation. (p. 85)

HOW CELLS ARE STUDIED Microscopes magnify objects and help resolve adjacent structures; contrast, often supplied by stains, helps distinguish objects from each other. (p. 86)

HOW CELLS EXPLOIT DIFFUSION

DIFFUSION AS A CHEMICAL REACTION Diffusion is the tendency of a substance to move from a region of high concentration to one of lower concentration; like ordinary reactions, it has an equilibrium and an associated free energy. (p. 90)

MANIPULATING DIFFUSION: OSMOSIS Osmosis occurs when a semipermeable membrane allows some but not all substances in a solution to diffuse across. (p. 92)

BETWEEN SWELLING AND COLLAPSE Cells in a hypotonic medium—one with less osmotically active solutes—tend to gain water and swell, while those in a hypertonic medium tend to lose water and collapse. (p. 96)

HOW THE CELL MEMBRANE WORKS

LIPID BILAYERS: MEMBRANES HELD TOGETHER BY WATER The phospholipids of the membrane form a bilayer with their hydrophilic ends exposed to the water of the cytoplasm and surrounding fluids, and their hydrophobic ends hidden together in the interior of the membrane. (p. 98)

HOLES IN THE MEMBRANE: CHANNELS AND PUMPS Channels are openings in the membrane that pass specific substances. Facilitated-diffusion channels simply pass their particular substance at a controlled rate; gated channels open in response to chemical or electrical stimuli. Antiports are channels that trade substances going in opposite directions; symports are channels that pass two substances the same direction cooperatively. Pumps use chemical energy to move substances against their osmotic gradient. (p. 103)

LARGE-SCALE MOVEMENT: ENDOCYTOSIS AND EXOCYTOSIS Large quantities of material can be brought into cells by trapping it endocytotically in vesicles; material can be expelled from cells through vesicle-mediated exocytosis. (p. 107)

OUTSIDE THE MEMBRANE: CELL WALLS AND COATS

HOW CELL WALLS ARE BUILT Cell walls, found in bacteria, some protists, fungi, and plants, can provide enough rigidity to keep cells from bursting in hypotonic media. (p. 113)

WHAT THE CELL COAT DOES The cell coat consists of membrane-mounted carbohydrates that help in cell adhesion and recognition. (p. 114)

HOW CELLS STICK TOGETHER

LIMITS TO CELL SIZE Cells cannot grow indefinitely since larger cells have proportionally less surface area, and thus have access to smaller relative quantities of nutrients and can dispose of proportionally less waste material. (p. 115)

ALTERNATIVE WAYS TO CONNECT CELLS Cells can be attached by a variety of junctions that hold them in place; some junctions permit adjacent cells to share contents or information. (p. 116)

STUDY QUESTIONS

1 How do water, saturated and unsaturated phospholipid tails, and cholesterol contribute to membrane stability and flexibility? (pp. 98–103)

2 Why does the sodium-potassium pump require chemical energy from the cell? Why does the sodium-glucose symport not require any? (pp. 103–107)

3 List five different cross-membrane transport strategies, and explain how each one works. (pp. 103–112)

4 What are the costs and benefits of the plant strategy of using cell walls to stabilize membranes? (pp. 113–114)

5 What changes in conditions would serve to increase the rate of diffusion of an ionic substance? (The list is not short.) (pp. 90–92)

6 Explain osmosis in terms of probability at the molecular level, and then in terms of thermodynamics. (pp. 92–96)

SUGGESTED READING

BRETSCHER, M. S., 1985. The molecules of the cell membrane, *Scientific American* 253 (4). *Reviews the bilayer plasma membrane and membrane proteins and the process of endocytosis.*

BROWN, M. S., AND J. L. GOLDSTEIN, 1984. How LDL receptors influence cholesterol and atherosclerosis, *Scientific American* 251 (5).

DAUTRY-VARSAT, A., AND H. F. LODISH, 1984. How receptors bring proteins and particles into cells, *Scientific American* 250 (5). *The life cycle of coated pits.*

GOODSELL, D. 1992. A look inside the living cell. *American Scientist* 80 (5). *What the interior of a cell might look like if we could see the molecules themselves.*

SHARON, N., AND H. LIS, 1993. Carbohydrates in cell recognition. *Scientific American* 268 (1). *On the cell-surface markers that play a role in cell adhesion, development, and the immune response.*

STOSSEL, T. P. 1994. The machinery of cell crawling. *Scientific American* 271 (3). *On how cells move in amoeboid fashion.*

TAYLOR, D. L., M NEDERLOF, F. LANNI, AND A. S. WAGGONER, 1992. New vision of light microscopy. *American Scientist* 80 (4). *On modern methods of creating contrast.*

UNWIN, N., AND R. HENDERSON, 1984. The structure of proteins in biological membranes, *Scientific American* 250 (2).

CHAPTER 5

INSIDE THE CELL

The cell membrane creates a general-purpose chemical environment favorable to life—a comfortable organic home in a hostile world. A human home may have, in addition to a heating and cooling system to maintain a suitable general temperature, an oven for tasks that require unusually high temperatures, a refrigerator with freezer for low-temperature needs, and other specialized tools and appliances to help it operate more efficiently. Cells too need special structures to create the unusual internal conditions necessary for particular tasks, as well as a set of tailor-made tools. Many of these subcellular entities are enclosed in their own membranes, and are called *organelles*. The term is also applied to the more complex of the other subcellular units. This chapter examines these specialized cellular components.

THE ROLES OF SUBCELLULAR COMPARTMENTS

■ THE NUCLEUS: THE CELL'S CONTROL CENTER

The largest internal compartment in most organisms (it is absent in bacteria) is the membrane-bounded *nucleus* (Fig. 5.1). The nucleus is the control center of the cell; its genetic material issues the "instructions" that guide the life processes of the cell.

Though bacteria too possess genetic material that controls the cell's activities, they lack a membrane-bounded nucleus. Similarly, this group has few of the subcellular structures found in other organisms. These differences are so fundamental that bacteria are classified into separate kingdoms of their own. They are called *procaryotes* ("before any nucleus"), whereas all other organisms are *eucaryotes* ("having a true nucleus"). The characteristics of procaryotic cells will be discussed later.

The eucaryotic nucleus contains two large structures, the chromosomes and the nucleolus. Both are embedded in a matrix of filaments that seem to provide structure and paths along which processing and transport can occur (Fig. 5.2). The entire nucleus is bounded by a closely associated pair of membranes called the *nuclear envelope*.

The *chromosomes* (Fig. 5.3) are long, threadlike bodies clearly visible only when they condense in preparation for cell division; at

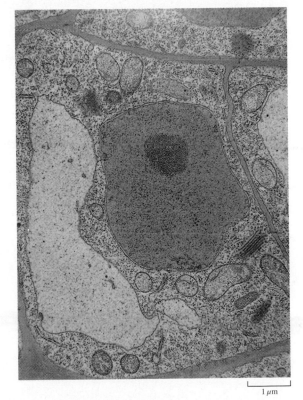

1 μm

5.1 A plant cell nucleus

The nucleus of most cells is about one-third the diameter of the whole cell, thus occupying 3–4% of cellular volume. In this case, the nucleus is the large orange body at the center, with the darker nucleolus embedded in it.

5.2   Internal matrix of the nucleus

The nucleus contains a meshwork of fibers along which molecules are transported.

5.3   Chromosomes in a dividing cell of *Trillium*

The separate chromosomes in the two nuclei of this spring-flowering plant can easily be distinguished.

all other times the chromosomes are in an uncondensed conformation that, when stained, appears as a dark, hazy material called chromatin. The chromosomes are composed of DNA and protein; the DNA is the substance of the basic units of heredity, called *genes*, while the protein provides spool-like supports, or cores, on which the DNA is wound to form *nucleosomes* (Fig. 5.4). The genes are duplicated whenever a cell divides, and a copy is passed to each new cell. The genes determine the characteristics of cells and control the day-to-day activities of living cells.

The hereditary information carried by the genes is written in the sequence of the nucleotide building blocks of the DNA molecules. The genes remain in the nucleus, while most of the processes they control take place in the cytoplasm. The mechanism for conveying the genetic information from the nucleus to the cytoplasm is transcription (detailed in Chapter 9): the nucleotide sequence in the DNA gives rise to a corresponding nucleotide sequence in RNA. This RNA sequence apparently moves along a nuclear filament on which it is modified into messenger RNA (mRNA), which leaves the nucleus and moves to the sites of protein synthesis in the cytoplasm. There amino acids are linked by peptide bonds in a sequence corresponding to that of the mRNA nucleotides to form proteins, including enzymes (Fig. 5.5). This process is known as translation (discussed in Chapter 9). The sequence of amino acids—the primary structure of a protein—determines the three-dimensional conformation of the protein and the biological activity that this conformation bestows. The genes encode all the information necessary for the synthesis of the enzymes regulating the myr-

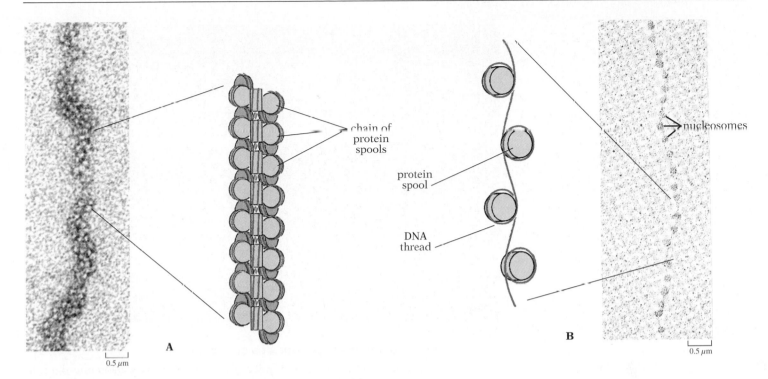

chain of protein spools

protein spool

DNA thread

nucleosomes

A

0.5 μm

B

0.5 μm

5.4 DNA on nucleosomes

The chromosomal DNA of eucaryotes is wound on protein spools, or cores, to form structures called nucleosomes. Normally the spools adhere to one another in a regular way, giving the DNA the appearance of a piece of yarn (A). When the DNA is treated to break the connections between nucleosomes, the individual protein spools and the thread of DNA can be seen (B).

5.5 Flow of information from nucleus to cytoplasm

Information encoded in a region of the chromosome (a gene) is transcribed by an enzyme complex to create an RNA copy. This messenger RNA is then exported to the cytoplasm, where it is decoded by a ribosome and the protein specified by the gene is synthesized.

folded protein

section of chromosome

gene being transcribed in nucleus

RNA transcript (mRNA) moves to cytoplasm

transcript being translated in cytoplasm

1 μm

5.6 The nuclear envelope of a corn root cell

The large structure filling the upper left quarter of this electron micrograph is the nucleus. The unlabeled arrow indicates a point where the endoplasmic reticulum and the double nuclear membrane interconnect. ER, endoplasmic reticulum; G, Golgi apparatus; M, mitochondrion; N, nucleus; NE, nuclear envelope; P, pore in nuclear envelope; W, cell wall.

iad interdependent chemical reactions that determine the characteristics of cells and organisms.

The other structure in the nucleus is the ***nucleolus***, one or more dark-staining, generally oval areas usually visible within the nuclei of nondividing cells. Nucleoli are, in fact, simply specialized parts of the chromosome; like the rest of the chromosome, they are composed of DNA and protein. The DNA of the nucleoli includes multiple copies of the genes from which a type of RNA called ribosomal RNA (rRNA) is transcribed. After the rRNA is synthesized, it combines with proteins, and the resulting complex detaches from the nucleolus, leaves the nucleus, and enters the cytoplasm, where it becomes a part of the protein-synthesizing complexes called ***ribosomes***.

The nuclear envelope helps maintain a chemical environment in the nucleus different from that of the surrounding cytoplasm. It also appears to provide anchorage for the two ends of each chromosome. Unlike the cell membrane, the complete nuclear envelope consists of distinct inner and outer membranes, with space enclosed between them (Figs. 5.1, 5.6).

This double-membrane envelope is interrupted by fairly large and elaborate pores at points where the outer and inner membrane are continuous (Figs. 5.6–5.8). Nevertheless, the membrane is highly selective: many substances that can cross the cell membrane into the cytoplasm cannot pass through the nuclear envelope into the nucleus. Even molecules much smaller than the pores fail to

1μm

5.7 Freeze-fractured cell from the tip of an onion root

At upper right of this electron micrograph is the surface of the nuclear envelope, with numerous pores; a typical nucleus has a few thousand pores. A variety of vesicles can be seen in the cytoplasm.

move through these openings, yet some large molecules pass readily. These macromolecules are primarily substances produced on the genes (such as mRNA) that are moving out of the nucleus, proteins moving into the nucleus to be incorporated into nuclear structures or to catalyze chemical reactions in the nucleus, and various substances from the cytoplasm that move into the nucleus and help regulate gene activity.

This highly selective and directional exchange through the nuclear pores between the nucleus and the cytoplasm is based on "passwords," chemical signal sequences attached to one end of the various molecules that tell the pores whether to allow the larger molecule through, and if so in which direction. Certain viruses (including HIV, the AIDS virus) have broken the molecular code used by nuclear pores; they are able to smuggle in a copy of their chromosome, which can then be incorporated into the cell's DNA library.

The nuclear envelope has another interesting elaboration: it is directly connected to an extensive cytoplasmic membrane system called the endoplasmic reticulum (Fig. 5.6).

■ THE ENDOPLASMIC RETICULUM: SYNTHESIS AND SORTING

Attempts to separate subcellular organelles from the cytoplasm often make use of centrifugation. Cells are first lysed (broken) by detergents that disrupt the lipids in the membrane, and then they

0.5μm

5.8 The rough endoplasmic reticulum

This electron micrograph of a thin section of a pancreatic cell from a bat shows many flattened cisternae of rough ER; the ribosomes lining the RER membranes can be clearly distinguished. In the lower right portion of the micrograph is part of the nucleus (N); note the very prominent pores (P) in the double membrane of the nuclear envelope. A mitochondrion (M) is at the top.

are layered on top of a test tube containing a viscous solution of, say, glucose; finally, the samples are spun in a centrifuge to sort out the cell parts by weight. Each component travels through the solution at a characteristic rate, the densest moving most rapidly.

Using these techniques, Albert Claude isolated some unusual cytoplasmic components in 1938. These "microsomes" (as he called them) made up 15–20% of the total cell mass and had a very high nucleic acid content. They also contained a high percentage of the cytoplasmic phospholipids. In 1945 Keith R. Porter found that the microsomes were the fragments of a complex network of membranes in the cytoplasm plus associated ribosomes. This system, which Porter named the *endoplasmic reticulum (ER)* (*reticulum* being Latin for "network"), exists in large part to synthesize new membrane phospholipids, to package proteins into vesicles for transport to other parts of the cell, and to store calcium ions.

There are two anatomically distinct forms of endoplasmic reticulum, each with a different role in synthesis and packaging. One, the *rough endoplasmic reticulum (RER)* exists as flattened, fluid-filled, membrane-enclosed sacs called *cisternae*. The RER probably forms a series of interconnected sheets, so that all the cisternae interconnect. The rough appearance of the RER membrane arises from its association with the numerous protein-synthesizing ribosomes (Fig. 5.8). Ribosomes also bind to the outer membrane of the nuclear envelope, to which the RER is connected.

The association of the ribosomes with the RER membrane appears to be necessary for the proteins they produce to penetrate the membrane to the cisternae or to become embedded in the membrane. Most or all of the proteins to be packaged and transported by the ER are synthesized by the ribosomes of the RER. Proteins synthesized on free ribosomes in the cytoplasm are apparently released to function as enzymes in the *cytosol* (the more fluid part of the cytoplasm). The mRNA from the nucleus that is to produce enzymes for the ER binds first to free ribosomes. The mRNA, however, signals the ribosome (with the help of signal-recognition particles) to bind to special channels in the RER. Once the ribosome binds, the translation begins and the enzyme being synthesized on the ribosome is threaded through the channels into the lumen (the interior) of the RER.

As soon as a *signal sequence* in this mRNA is synthesized as part of the growing polypeptide chain, and this sequence is threaded into the lumen of the ER, enzymes in the RER membrane attach a small sugar complex (an oligosaccharide) that will aid in subsequent sorting (Fig. 5.9): some proteins remain in the ER, whereas others are exported to a variety of sites. Modifications of this molecular "mailing label" specify where the protein is to go. Errors in the tagging system can have serious consequences.

Beyond its importance in intracellular transport, the RER functions as a cytoplasmic framework, providing catalytic surfaces for some of the biochemical activity of the cell. Its complex folding provides an enormous surface for such activity.

The other form of endoplasmic reticulum, the *smooth endoplasmic reticulum (SER)*, not only lacks ribosomes but also has a much more tubular appearance (Fig. 5.10) and a very different set of characteristic membrane proteins. The most obvious functions

5.9 Oligosaccharide mailing label attached to proteins in the ER

A 14-sugar side chain is attached to nearly all proteins synthesized on the RER, and serves as a "mailing label." Proteins lacking this label remain in the RER. When the four terminal sugars are removed, the protein is exported to the Golgi apparatus in vesicles. The size of the tag in this drawing is exaggerated compared with the protein. (G, glucose; M, mannose; N, *N*-acetyl-glucosamine)

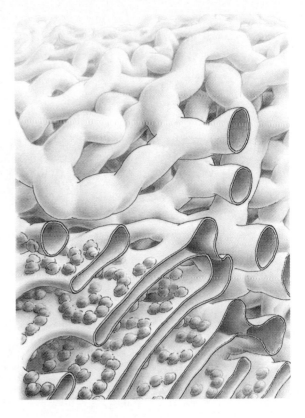

0.2 µm

5.10 Rough and smooth endoplasmic reticulum

The endoplasmic reticulum from a steroid-producing cell of a guinea pig testis, shown in this electron micrograph, consists of a complex system of membranes including the RER, with associated ribosomes, and SER. The relationship between RER and SER varies from cell to cell; the sketch shows an association more typical of a cell in which macromolecules are synthesized in the RER and transported to the SER. This representation assumes that the RER and SER are physically continuous; they may actually be separate organelles that communicate via vesicles.

of the SER are to synthesize membrane phospholipids and to package certain proteins in the cisternae into membrane-bounded vesicles for transport to other locations in the cell. The vesicles also carry membrane-mounted proteins with them. Though most of the proteins in the cisternae and the SER membrane are synthesized first in the RER, it is not yet clear how they get to the SER. It may be that the RER and SER are continuous, as was once universally assumed; alternatively, the RER may send membrane-embedded proteins and cisternal proteins to the SER via vesicles.

Only liver, muscle, and hormone-secreting cells have large amounts of SER. The SER of liver cells holds enzymes that detoxify many poisons, including barbiturates, amphetamines, and morphine. Poisons that appear in the bloodstream are transported to the liver. There the rapid synthesis of more phospholipids can cause the surface area of the liver cells' SER to double. The positioning of detoxification enzymes in the ER takes maximum advantage of the cell's potential for compartmentalization to protect the cytosol from the poisons. The products of the detoxification process are probably packaged in vesicles that have been formed from the ER itself, and are then transported to other organelles, where they are further broken down.

The SER also stores and releases calcium ions, whose concentration, as we will see, is important in controlling the building or disassembly of elements of the cytoskeleton. The SER in muscles holds enormous quantities of calcium ions, which are used to coordinate contraction. The relative sizes of some of the subcellular components discussed so far are shown in Figure 5.11.

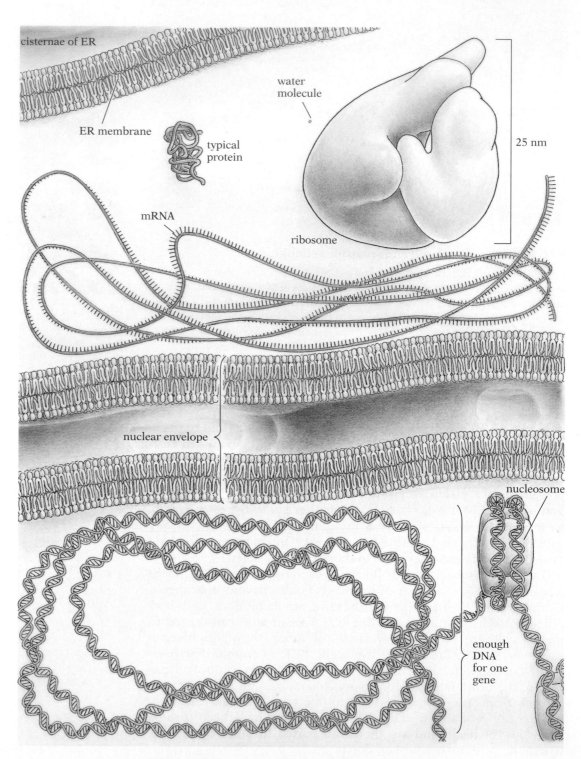

5.11   Relative sizes of subcellular components

■ THE GOLGI APPARATUS: SORTING AND PACKAGING

In 1898 Camillo Golgi discovered a different system of membrane-delimited cellular compartments, arranged approximately parallel to one another; this system is now called the **Golgi apparatus** (Fig. 5.12).

The Golgi apparatus is particularly prominent in cells involved in the secretion of chemical products; as the level of secretory activity of these cells changes, corresponding changes occur in the morphology of the organelle. For example, in certain cells of the pancreas of guinea pigs, a zymogen (the inactive precursor of an enzyme) synthesized on the ribosomes moves into the channels of the ER; it reaches the Golgi apparatus in vesicles that have budded off from the ER. The zymogen arrives at the compartment nearest the nucleus (the most highly curved cisterna in Figure 5.12, known as the *cis* compartment), and then moves from one layer to the next until it reaches the farthest layer (the *trans* compartment). Movement between layers is by means of vesicles that bud off from one cisterna and then fuse with another, farther from the nucleus. During this *cis-trans* movement the zymogen is modified, and in the final cisterna it is concentrated and stored; it is eventually released from the cell via secretory vesicles that are produced by this outer compartment of the Golgi and move to the cell surface. In this way the Golgi apparatus stores, modifies, and packages secretory products.

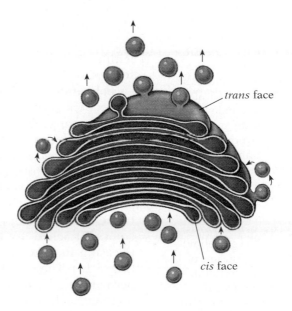

0.05μm

5.12 The Golgi apparatus

Electron micrograph of Golgi apparatus from an amoeba. Vesicles forming at the ends of some of the cisternae can be seen in the EM and in the interpretive drawing. The movement is from the *cis* layer, through the various intermediate compartments, to the *trans* cisterna. *Cis* (meaning "this side of") and *trans* ("other side of") are defined with respect to the nucleus. The Golgi sorts, modifies, relabels, and packages molecules into vesicles for further transport. In this EM, the *trans* compartment and the vesicles they have released are labeled with an electron-dense stain.

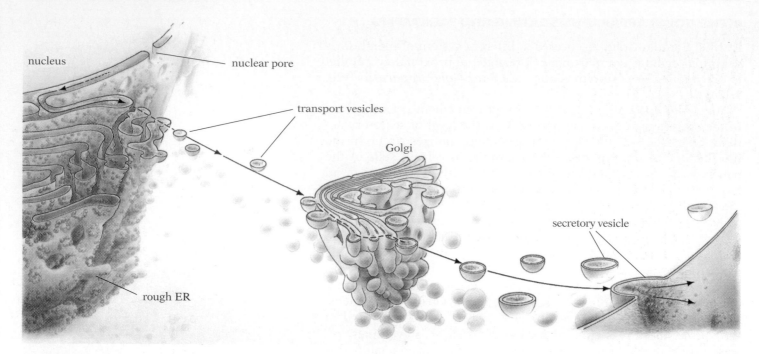

5.13   One path of phospholipid movement in the cell

Structural molecules of cellular membranes are constantly on the move. The path traced here shows the movement of a phospholipid from a point of synthesis in the nuclear envelope until it is incorporated into the cell membrane. (Phospholipids are also synthesized on the ER.) The lipid first moves through the lumen of the RER and (either directly or via vesicles) into the SER. There it receives a specific carbohydrate marker, which makes it a glycolipid, and becomes part of a vesicle transporting proteins to the Golgi apparatus; it is then incorporated into the membrane of that organelle. Subsequently the same molecule, perhaps with a new or modified carbohydrate marker, moves to the outer layer of the Golgi, where it becomes part of a secretory vesicle that is carried to the plasma membrane. There, in the process of exocytosis, the vesicle membrane (including the glycolipid molecule) fuses with the plasma membrane of the cell. Other membrane molecules follow different paths. (Note that the most highly curved Golgi compartment in this illustration is the *trans* cisterna, whereas in Fig. 15.12 it was the *cis* compartment. This serves to emphasize the point that *cis* and *trans* in this flexible organelle are defined by their relative distance from the nucleus.)

The Golgi apparatus is also the major director of macromolecular transport in cells. Though no protein synthesis takes place in the Golgi, polysaccharides are synthesized there from simple sugars and attached to proteins and lipids to create glycolipids and glycoproteins. Some of these are transported to the glycocalyx as part of vesicle membranes. In addition, proteins already marked with carbohydrate groups (true for nearly all proteins transported to the Golgi apparatus from the ER) usually have their sugar-based carbohydrate tags modified there. How the modified tags aid in subsequent sorting and packaging is still a mystery.

Traffic through the Golgi is not entirely one way: some proteins synthesized into the RER get into Golgi-directed vesicles by error; these misplaced molecules are captured in the Golgi, packaged into vesicles, and returned to the sender.

The secretory vesicles produced by the Golgi apparatus probably play an important role in adding surface area to the cell membrane. When one of these vesicles moves to the cell surface, it becomes attached to the plasma membrane and then ruptures, releasing its contents to the exterior in the process of exocytosis (Fig. 5.13). The membrane of the ruptured vesicle may remain as a permanent addition to the plasma membrane, or it may eventually migrate back to the Golgi apparatus or some other organelle as part of an empty vesicle. Recycling of membrane phospholipids back to the Golgi and ER is essential if the outer membrane is not to grow indefinitely.

■ LYSOSOMES: DIGESTIVE ORGANELLES

First described in the 1950s by Christian deDuve, **lysosomes** are membrane-enclosed bodies that function as storage vesicles for many powerful digestive (hydrolytic) enzymes (Fig. 5.14). The lysosome membrane contains an ionic pump that maintains a highly

0.1 μm

5.14 Lysosomes in a connective-tissue cell from the vas deferens of a rat

The small dark body at upper right is a primary lysosome. The much larger body at left is a secondary lysosome (digestive vacuole) formed by fusion of a primary lysosome with a phagocytic or pinocytic vesicle. (The dark appearance of the lysosomes results from staining for acid phosphatase, a digestive enzyme whose presence is used as the definitive test for these organelles.)

acidic internal environment. The membrane permits desirable reaction products to pass through to the cytosol, but it is impermeable to the hydrolytic enzymes and capable of withstanding their digestive action. If the lysosome membrane is ruptured, the hydrolytic enzymes—no longer safely confined—are released into the surrounding cytoplasm and break down the interior of the cell.

Lysosomes act as the digestive system of the cell, enabling it to process some of the bulk material taken in by endocytosis. Their hydrolytic enzymes are synthesized in the form of zymogens in the RER, packaged into transport vesicles in the SER, and carried to the Golgi apparatus. There, receptor proteins on the inside of the Golgi membrane form coated pits to attract and trap these zymogens, which are recognized by their characteristic carbohydrate markers. When a region has received an appropriate supply of the zymogens, it buds off from the Golgi membrane as a lysosome, and the enzymes are activated. Proteins mounted on the exterior of the lysosome's membrane serve as recognition sites to assure that the enzymes it carries are delivered to the appropriate target. The targets of these primary lysosomes include endocytotic vesicles (endosomes) newly arrived from the cell surface, and secondary lysosomes, also known as digestive vacuoles, which are the digestive vesicles already produced by the fusion of primary lysosomes and endocytotic vesicles. When digestion is complete, the useful products pass into the cytosol, while the residue is discharged by exocytosis, and the vesicle membrane and any receptors it contains are reincorporated into the cell membrane (Fig. 5.15).

Several diseases are caused by lysosome disorders. In human inclusion cell disease, for instance, the enzyme that adds a distinctive sugar to the mailing label is defective; as a result most of the 40 or so hydrolytic enzymes are mismarked for secretion. Undigested "food" builds up in cells, resulting in general debilitation, followed by death when one or more of the body's organs fail. In the devastating nervous-system disorder known as Tay-Sachs disease, lipid-digesting lysosomes lack a particular enzyme. When these deficient lysosomes fuse with lipid-containing vesicles, they do not fully digest their contents. The resulting defective secondary lysosomes

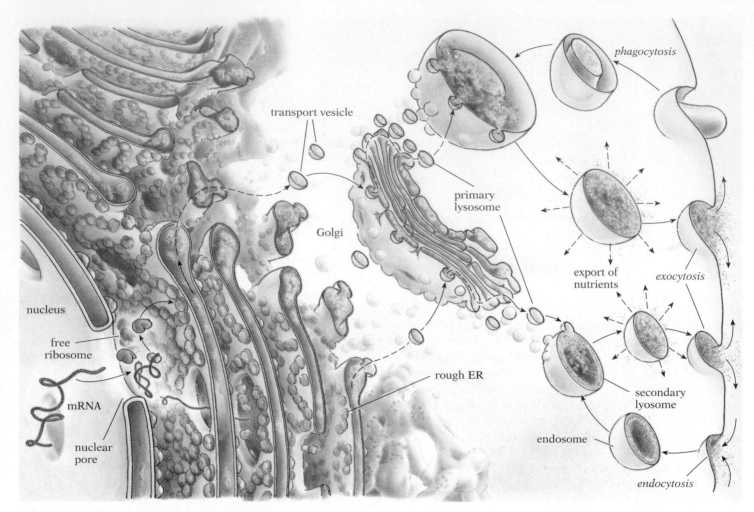

5.15   The hydrolytic enzyme cycle

The role of the ER and the Golgi apparatus in cellular digestion is representative of their functions in the cell. DNA in the cell nucleus produces the mRNA that encodes the hydrolytic enzymes. The mRNA is transported to ribosomes, which then bind to the RER. The enzymes are synthesized there, passed into the ER, and marked with signal sequences for later transport. The enzymes are next collected by receptors in the SER and packaged into transport vesicles. Markers on the outside of the transport vesicles cause them to fuse with the membrane of the Golgi apparatus, where new vesicles (primary lysosomes) are created. These packets of specific mixtures of hydrolytic enzymes fuse with appropriately marked endocytotic vesicles (endosomes) or with secondary lysosomes (produced by previous fusions of this kind) and help digest the contents. Useful products of this digestion pass through the membranes of the secondary lysosomes into the cytosol, while the residue is disposed of through exocytosis. The same pattern of synthesizing and marking enzymes in the ER, transporting them to the Golgi apparatus, sorting, chemically modifying, and repackaging them there, and then dispatching them to various intracellular targets seems to be the general strategy for managing macromolecules in cells.

can accumulate and block the long thin parts of nerve cells responsible for transmitting nerve impulses.

The elaborate organization of membrane movement in cells is vulnerable to many infectious diseases. Semliki Forest virus (SFV), for instance, infects a wide range of hosts from invertebrates to humans. The virus carries a marker on its surface that mimics a substance for which one kind of coated pit has specific receptors. SFV is carried into the cell by endocytosis, after which its chromosome escapes into the cytosol. The virus uses the host cells' ribosomes,

ER, Golgi apparatus, and the whole system of carbohydrate tags to reproduce itself and direct the subsequent exocytosis of its viral progeny. The virus thus ensures its own propagation by exploiting much of the specialized membrane machinery of the cell.

■ PEROXISOMES: DETOXIFICATION ORGANELLES

Cells also contain a variety of small organelles known collectively as microbodies. The most thoroughly understood are the *peroxisomes* (Fig. 5.16). Like lysosomes, peroxisomes contain an assortment of powerful enzymes. However, where the enzymes of the lysosomes are hydrolytic (water-splitting), these enzymes catalyze condensation reactions. Examples are the oxidative removal of amino groups from amino acids, the detoxification of alcohol, the oxidation of the dangerous compound hydrogen peroxide into water and oxygen, the breakdown of fatty acids into useful building blocks, and the reactions involved in the production of macromolecules used in respiration and other biochemical pathways.

Peroxisomes are independent organelles that multiply by fissioning: a cell that fails to inherit at least one peroxisome from the cytoplasm of its parent cell cannot make its own from scratch and will die. Peroxisomes are not self-sufficient, however: their membrane phospholipids are synthesized in the SER, recognized and picked up by transfer proteins, carried through the cytoplasm, and deposited in the peroxisomal membrane; the precursors of their enzymes are produced in the cytoplasm (rather than the ER) and transported to these tiny organelles. Just as proteins synthesized on the ER must be supplied with labels to assure accurate delivery, so too proteins built in the cytoplasm must have signals to assure proper routing, whether to peroxisomes or other destinations. In the cytoplasm, however, the signals are embedded in the amino acid sequence at one end of the protein rather than in a carbohydrate side chain.

■ MITOCHONDRIA: CELLULAR POWER STATIONS

Mitochondria carry out respiration, the chemical reactions that extract energy from food and make it available for innumerable energy-demanding activities. Each mitochondrion is bounded by a double membrane; the outer membrane is smooth, while the inner membrane has many inwardly directed folds called *cristae* (Fig. 5.17). Energy-rich reactants are concentrated in this organelle, and here, with the help of the appropriate enzymes, they are combined with oxygen to generate water, carbon dioxide, and the energy that runs the cell. The operation of mitochondria is fully described in Chapter 6.

■ CHLOROPLASTS: SOLAR ENERGY RECEPTORS

Plastids are large cytoplasmic organelles found in the cells of plants but not in those of fungi or animals. Plastids are clearly visible through an ordinary light microscope. There are two principal

0.1 μm

5.16 A peroxisome

This electron micrograph shows part of a tobacco-leaf cell, which has been treated with stain to reveal the crystalline core of a peroxisome and the presence of oxidizing enzyme within it. In mammals, peroxisome enzymes neutralize hydrogen peroxide, alcohol, and other potentially harmful substances.

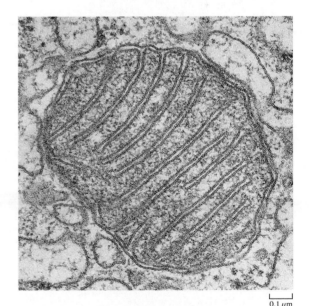

0.1 μm

5.17 A mitochondrion from a rat epithelial cell

Note the double outer membrane and the numerous cristae, which can be seen to arise as folds of the inner membrane in this electron micrograph.

granal thylakoids

stromal thylakoids

A

stroma

0.5μm

cell wall   cytoplasm   chloroplast

vacuole

nucleus

vacuolar
membrane

chloroplast membrane

B

plasma membrane

5μm

5.18 Chloroplasts

(A) Stacks of disklike thylakoids, forming grana, can be seen in this chloroplast of a corn leaf. (B) Numerous chloroplasts lie close to the perimeter of a mature leaf cell of timothy grass.

categories: ***chromoplasts*** (colored plastids) and ***leucoplasts*** (white or colorless plastids).

Chloroplasts are chromoplasts that contain the green pigment ***chlorophyll***, along with various yellow or orange pigments. Chapter 7 fully describes how the radiant energy of sunlight is trapped in the chloroplasts by molecules of chlorophyll and is then used in the manufacture of complex organic molecules (particularly glucose) from simple inorganic raw materials such as water and carbon dioxide.

Typical chloroplasts are bounded by two concentric membranes and have a complex internal membranous organization (Fig. 5.18). The fairly homogeneous internal proteinaceous matrix is called the ***stroma***. Numerous flat compartments, called ***thylakoids***, are embedded in it. In most higher plants these thylakoids come in two varieties: separate thylakoids that run through the stroma, and stacks of platelike thylakoids known as ***grana*** (Fig. 5.18). Most of the important reactions of photosynthesis occur on membranes of the thylakoids.

■ LEUCOPLASTS: STORAGE ORGANELLES IN PLANTS

The colorless plastids, or leucoplasts, are primarily used to store starch, oils, and protein granules. Plastids filled with starch, called *amyloplasts*, are particularly common in seeds and in storage roots and stems, such as carrots and potatoes. Starch is an energy-storage compound in plants, and is deposited as a grain or group of grains in plastids (Fig. 5.19); no starch is found in other parts of the cell.

All types of plastids form from small colorless bodies called proplastids. Once formed, many kinds of plastids can be converted into other types under appropriate conditions. For example, under certain conditions leucoplasts exposed to light develop chlorophyll and are converted into chloroplasts, complete with the characteristic internal thylakoids, which develop from invaginations of the inner membrane of the plastid.

■ VACUOLES: MULTIPURPOSE ORGANELLES

Membrane-enclosed, fluid-filled spaces called *vacuoles* are found in both animal and plant cells, though they have their greatest development in plant cells. There are various kinds of vacuoles, with a corresponding variety of functions. In some protozoans, specialized vacuoles, called contractile vacuoles, play an important role in expelling excess water and some wastes from the cell. Many protozoans also have food vacuoles, chambers that contain food particles. They are similar to the vesicles formed by many cells when they take in material by endocytosis.

The distinction between vesicles and vacuoles, both of which are membrane-bounded, is hard to draw with any precision, particularly since vesicles may fuse with or bud off from vacuoles. The most obvious differences are in permanence, activity, and size: vesicles are relatively short-lived transport vehicles, while vacuoles tend to be long-lived; vesicles usually move quickly, while vacuoles are relatively static; finally, vesicles are usually small, while vacuoles are most often quite large. These hazy distinctions are underscored by the observation that many plant vacuoles develop from lysosomes.

In most mature plant cells, a large vacuole occupies much of the volume of the cell. The immature cell usually contains many small vacuoles. As the cell matures, the vacuoles take in more water and become larger, eventually fusing to form the very large definitive vacuole of the mature cell (Fig. 5.20). This process pushes the cyto-

0.5μm

5.19 Leucoplasts from the root tip of *Arabidopsis*

Because they contain numerous prominent starch grains, these leucoplasts from the small desert plant *Arabidopsis* are called amyloplasts.

5.20 Development of the plant-cell vacuole

The immature cell (left) has many small vacuoles, which arise from lysosomes. As the cell grows (middle), these vacuoles fuse and eventually form a single large vacuole, which occupies most of the volume of the mature cell (right). Note the cytoplasm has been pushed to the periphery.

5.21   An actin microfilament

This portion of an actin microfilament shows the helically intertwined chains of protein subunits.

plasm to the periphery of the cell, where it forms a relatively thin layer.

The plant vacuole contains a liquid called cell sap, which is primarily water with a variety of substances dissolved in it. Since the cell sap is generally hypertonic relative to the external medium, the vacuole tends to take in water by osmosis. As the vacuole swells, the vacuolar membrane (or tonoplast, as it is often called) pushes outward against the cytoplasm, which transmits the pressure to the cell wall. The wall is strong enough to limit the swelling and prevent the cell from bursting; The outward push of the vacuolar membrane thus maintains cell turgidity and stiffness.

Many substances of importance in the life of the plant cell are stored in the vacuoles, among them high concentrations of soluble organic nitrogen compounds, including amino acids; vacuoles also store sugars, various organic acids, and some proteins. The vacuoles also function as dumping sites for noxious wastes. Enzymes secreted into them degrade some of these wastes into simpler substances that can be reabsorbed into the cytosol and reused.

THE CYTOSKELETON: STRUCTURE AND MOVEMENT

As we have seen, the cell is full of specialized, membrane-lined organelles that mediate much of cellular chemistry. Cells also need a variety of important protein-based components that help organize movement, not only within the cell but by the cell itself, and that aid in defining and controlling cell shape.

■ MICROFILAMENTS: CELLULAR ROPES AND CABLES

Molecules of a protein called **actin** polymerize spontaneously under conditions of elevated Ca^{++} and Mg^{++} concentration, or normal levels of K^+ and Na^+. The resulting long, extremely thin polymers, helically intertwined, form actin microfilaments (Fig. 5.21), a general-purpose cabling that accounts for 2–20% of the protein in most cells. A filament spanning the entire length of a cell can form in a matter of minutes.

Actin microfilaments can play a purely structural role; when cross-linked for strength, they are a component of the cytoskeleton, the complex weblike array of molecules that helps maintain cell shape. This network is densest just under the cell membrane (Fig. 5.22), to which it is anchored by special proteins. Parallel, cross-linked arrays of actin microfilaments provide reinforcement for various stiff cellular protuberances, including the dense forests

0.02μm

5.22   Actin in a rat fibroblast cell

Actin microfilaments appear to run between distinct anchor points on the membrane, providing structure and a scaffolding for movement.

5.23 Microfilaments in the microvilli

Cross-linked into tight bundles, actin microfilaments in the microvilli provide rigidity.

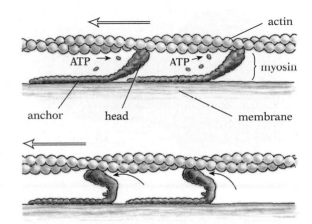

5.24 Interaction of actin and myosin

Some (perhaps all) nonmuscle myosin is anchored in membranes and, when fueled by ATP, can bind to actin microfilaments and undergo a conformational change. The resulting movement of the myosin head (described in detail in Chapter 37) moves the membrane (to which the anchor is permanently connected) and the microfilament (to which the head is temporarily attached) past one another in opposite directions. Myosin may also be able to anchor itself to vesicles or actin filaments, and so move them along the microfilament network. Myosin can create movement of 1–9 μm/sec.

of rodlike projections known as microvilli that line the intestine (Fig. 5.23).

In addition to its structural role, actin is involved in cell movement through interactions with a second cytoskeleton protein, myosin. *Myosin* consists of a two-part head plus an anchor. The tip of the head can bind to actin, while the base can accept energy from ATP, which causes the head to twist relative to the anchor. This twisting is used to pull actin cables (Fig. 5.24).

In muscle cells (described in detail in Chapter 37) the anchor is actually a long tail that binds to other myosin tails to create a highly specialized bundle that makes coordinated contractions possible. In nonmuscle cells, the myosin anchor is relatively short and is bound to cell or vesicle membranes or to actin filaments.

The best-understood case of myosin-mediated movement in nonmuscle cells occurs during cell division, when the daughter cells are created. Actin filaments are formed along the midline of the parent cell, and myosin molecules (perhaps mounted on the membrane) pull on the actin to "pinch off" the two cells from one another.

Similar actin-myosin interactions are responsible for cell movement (Fig. 5.25). At the trailing edge of the cell the membrane is being constantly tightened; the myosin pulls the actin cables, thus squeezing the cytoplasm toward the front. Vesicles produced from the rear membrane are sent forward to enable the leading edge to grow. Myosin-mediated movement of actin filaments is believed to cause extension of the pseudopodium at the front, though the details of this process are still a mystery. While most of the cytoplasm in the cell, a region often called ectoplasm, contains highly cross-

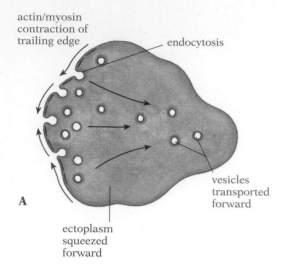

actin/myosin
contraction of
trailing edge

endocytosis

vesicles
transported
forward

A

ectoplasm
squeezed
forward

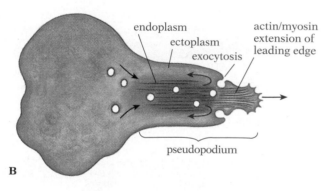

endoplasm

ectoplasm

exocytosis

actin/myosin
extension of
leading edge

pseudopodium

B

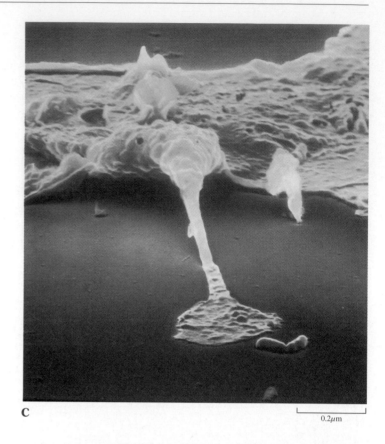

C

0.2μm

5.25 Amoeboid movement

(A) As the trailing edge of the cell contracts, squeezing the cytoplasm forward, the cytoplasm moving into the pseudopodium
loses its actin cross-linking, and becomes more fluid (endoplasmic). This endoplasmic core slides into the pseudopodium
between the peripheral layers of stationary ectoplasm. (B) The
pseudopodium extends as peripheral actin filaments (straight
lines) push it forward and stiffen its cortex; vesicles transported
from the rear provide more membrane. Endoplasm is pushed
down the middle of the pseudopodium. As the core endoplasm
moves forward, ectoplasm at the rear of the pseudopod and
from the rest of the cell is converted into endoplasm, while
endoplasm at the front of the pseudopod is converted into ectoplasm. (C) In this dramatic example of actin-based cell movement, a phagocyte (an amoeboid cell in the blood that captures
foreign material by endocytosis) is extending a pseudopodium
toward a bacterium, which it has detected on the basis of a
waste chemical released by the bacterium. (D) Actin bundles
are visible in the tip of the pseudopodium of this migrating
nerve cell.

D

5.26 Cross-linked actin filaments
This electron micrograph shows the internal structure of the leading edge of a moving white blood cell.

linked actin filaments and is relatively rigid (Fig. 5.26), the cytoplasm flowing into the leading edge, the endoplasm, is unlinked, and so is much more fluid. Once this endoplasm has been squeezed into the extension provided by the myosin-moved actin filaments, the actin becomes cross-linked again; other regions of ectoplasm are then converted into endoplasm to feed further movement.

■ MICROTUBULES: CELLULAR GIRDERS AND RAILWAYS

Microtubules can be thought of as heavy-duty versions of microfilaments. They are long, hollow, cylindrical structures (Fig. 5.27) that, like microfilaments, form spontaneously in response to the

A   0.2μm   B   0.2μm

5.27 Microtubules

(A) Longitudinal section, from bovine brain. (B) Cross section, from hamster spermatid.

vesicle

kinesin:
light chains
heavy chains

ATP

microtubule

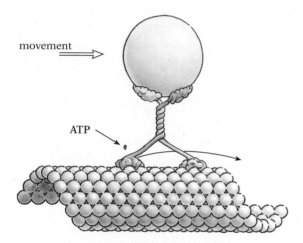

movement

ATP

5.29 Kinesin-mediated movement

Kinesin has two identical heavy chains (which bind to the microtubule and "walk" along it in a particular direction) and two identical light chains (which bind the vesicle to the heavy chains).

0.1 μm

5.30 Dynamin-mediated movement

A mitochondrion from the giant amoeba *Reticulomyxa* is being moved along a microtubule. The dynamin molecule attaching the organelle to the microtubule is clearly visible.

5.28 Structure of portion of a microtubule

The subunits of tubulin, each consisting of two proteins (shown colored and white), are helically stacked to form the wall of the tubule, which is usually 13 subunits in circumference.

proper ionic signals. Molecules of the globular protein **tubulin**, each molecule consisting of two proteins (α and β), polymerize to form a helical stack (Fig. 5.28). Microtubules radiate from **microtubule organizing centers** (usually near the Golgi) and play a critical role in general cell structure, in vesicle and organelle movement, and in cell division. During cell division prominent arrays of microtubules radiate from **centrosomes**, areas near structures called centrioles at each end of the cells in many species, to form a basketlike arrangement (the spindle), which is instrumental in moving the chromosomes to the locations of the new nuclei.

Special proteins (called kinesins) move vesicles away from the organizing center; one end binds to the vesicle while the other, which has two "feet," steps along the microtubule at speeds of 1–3 μm/sec (Fig. 5.29). Other kinesin-like proteins move mitochondria and other cargo. Yet other molecules (especially dynamin) move various loads *toward* the organizing center (Fig. 5.30). The positioning of the organizing center of this elaborate transportation network near the cellular mailroom—the Golgi—is no accident. Like microfilaments, microtubules also help provide shape and support for the cell and its organelles as part of the cytoskeleton. In fact, the microtubules appear to actively stretch the ER out into the farthest reaches of the cell: when the connections between microtubules and ER are broken, the ER collapses into a compact clump around the nucleus.

■ INTERMEDIATE FILAMENTS: A CELLULAR LATTICE?

There are numerous tubular fibers larger in diameter than actin but smaller in diameter than microtubules, with (in general) a much lower degree of organization. Although there are several chemically distinct varieties of these fibers, they are lumped together as **intermediate filaments** (Fig. 5.31). (Had their hollow structure been evident when they were named, they would doubtless have been called "intermediate tubules" instead.) Other intermediate filaments are associated with the cell membrane (where they provide structure and aid in the formation of cell-to-cell junctions) and the nuclear envelope (Fig. 5.32).

The collection of intermediate filaments in the cytoplasm, which is most prominent in cells subject to mechanical stress, is often referred to as a lattice (Fig. 5.33). The information we have about the cytoplasmic lattice can be interpreted in two quite different ways. For example, the proteins in the cytosol are concentrated on these

5.31 Structure of intermediate filaments

Top: The basic subunit is a protein dimer formed
from two identical molecules coiled together.
Bottom: each dimer is paired with another,
slightly offset, dimer to form a tetramer. The
tetramers bind head to tail to create strands.
Eight strands bound together in the form of a
hollow tube generate the structure typical of
intermediate filaments.

5.32 Nuclear lamina

This meshlike array of intermediate filaments
lies just inside the inner membrane of the nuclear envelope in frog eggs.

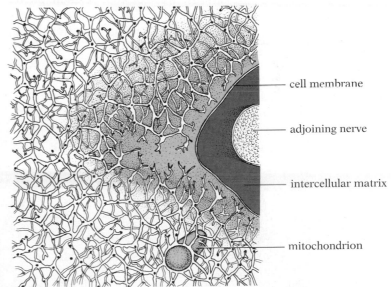

5.33 Cytoplasmic lattice

The cytoplasmic lattice is thought by many researchers to be composed of a
network of fibers that help give shape to the cell and hold various cellular
organelles in position. The hypothesis suggests that fibers are anchored to
the cell membrane, as shown in the interpretive drawing, and are also
linked to cellular microtubules and microfilaments (not shown). Other
researchers argue that the lattice is merely an artifact of the fixation technique used to prepare the specimen for the EM.

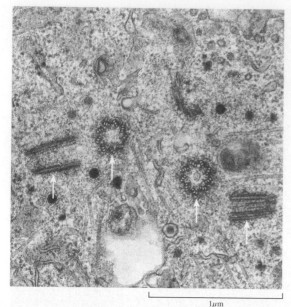

5.34 Centrioles

This electron micrograph shows newly replicated centrioles (arrows). Since the centrioles of each pair lie at right angles to each other, the sectioning of the specimen results in one from each pair being cut longitudinally and one being cut in cross section.

fibers, while the surroundings are basically aqueous. Perhaps in living cells the proteins are mounted on the fibers and organized into arrays for the sequential processing of reactants along an enzymatic pathway, or perhaps they merely stick to the fibers when the specimen is undergoing fixation for viewing. Similarly, the free ribosomes of the cytoplasm are concentrated at lattice intersections, perhaps in living cells, perhaps as a result of fixation.

■ CENTRIOLES AND BASAL BODIES: STRUCTURAL ORGANIZERS

Centrioles are found in pairs, oriented at right angles to each other, just outside the nuclei of many kinds of cells (Fig. 5.34). From the neighborhood of these organelles projects an array of microtubules. When centrioles are present, they are the focus of the microtubule spindle during cell division. In cross section, centrioles display a uniform structure, with nine triplets arranged in a circle and each triplet composed of three fused microtubules (Fig. 5.35B). Centrioles do not appear to be necessary for cell division; they may simply be the templates for making basal bodies.

Basal bodies anchor the many hairlike cilia and flagella (discussed next) to the cell membrane. As Figure 5.35A indicates, basal bodies have exactly the same structure as centrioles, and are probably the same organelle since basal bodies can become centrioles and vice versa. During cell division in the mobile alga *Chlamydomonas*, for example, the basal bodies of the two flagella abandon their posts and migrate to the poles of the cell, where they take up their positions in the spindle. Similarly, the basal body of the flagellum in many kinds of sperm becomes one centriole of the egg after fertilization. Also, the centrioles of cells that differentiate to line the oviduct multiply to become the basal bodies of the cilia

5.35 Basal bodies and centrioles

(A) Basal bodies, such as the three shown in cross section in this electron micrograph of a protozoan, are essentially identical to centrioles in structure. (B) Centrioles are composed of nine triplet microtubules.

whose rhythmic beating moves eggs from the ovary to the site of fertilization.

■ CILIA AND FLAGELLA: CELL MOVERS

Some cells of both plants and animals have one or more movable hairlike structures projecting from their free surfaces. If there are only a few of these appendages and they are relatively long in proportion to the size of the cell, they are called *flagella*. If there are many and they are short, they are called *cilia* (Fig. 5.36). The basic structure of flagella and cilia in eucaryotes is the same, and the terms are often used interchangeably. Both usually function either in moving the cell or in moving liquids (or small particles) across the surface of the cell. They occur commonly on unicellular and small multicellular organisms and on the male reproductive cells of most animals and many plants, and can be the principal means of locomotion. They are also common on the cells lining many internal passageways and ducts in animals, where their beating aids in moving materials through the passageways. In the trachea they reach densities of a billion per square centimeter.

The flagella and cilia of eucaryotic cells are an extension of the cell membrane, containing a cytoplasmic matrix, with 11 groups of microtubules embedded in the matrix. Almost invariably, nine of these groups are fused pairs arranged around the periphery of the cylinder, while the other two are isolated microtubules lying in the center (Fig. 5.37). Each cilium and flagellum is anchored to the cell by a basal body; the similarities between these organelles and the

10 μm

5.36 Ciliated surface of the trachea of a hamster

Coordinated movements of the many cilia shown in this SEM function to sweep dust particles and other foreign material from the respiratory surfaces to the mouth.

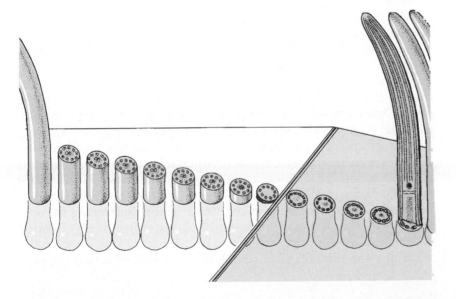

5.37 Cross section of cilia from the protozoan *Tetrahymena*

Most of this electron micrograph of an oblique section of surface tissue reveals the "9 + 2" arrangement of microtubules in cilia; the two cross sections at the far right show the nine triplet microtubules characteristic of basal bodies. The interpretive drawing shows the plane of the section and the position of the microtubules in a lateral section of an adjacent cilium.

0.2 μm

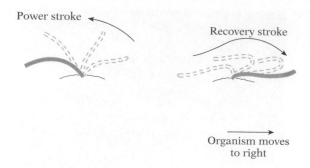

Power stroke

Recovery stroke

Organism moves
to right

5.38 Stroke cycle of a cilium

In the power stroke the stalk is extended fairly
rigidly and swept back by bending at its base. The
recovery stroke brings the cilium forward again,
as a wave of bending moves along the stalk from
its base. Note that at no time during the recovery
stroke is much surface opposed to the water in
the direction of movement.

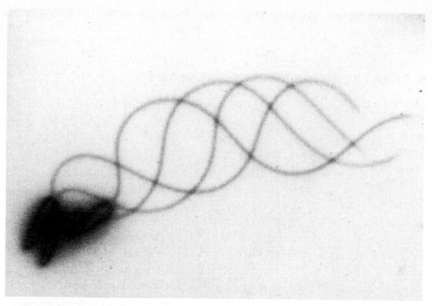

5.39 Flagellum movement in a sea urchin spermatozoon

Successive waves of bending that move along the flagellum, from its base
toward its tip, push against the water and propel the sperm cell forward.
This photograph was taken using four light flashes 10 msec (milliseconds)
apart. In some protozoans, the flagella have lateral "hairs" that reverse the
thrust; as a result, the waves move in the opposite direction, and so pull
the cell along.

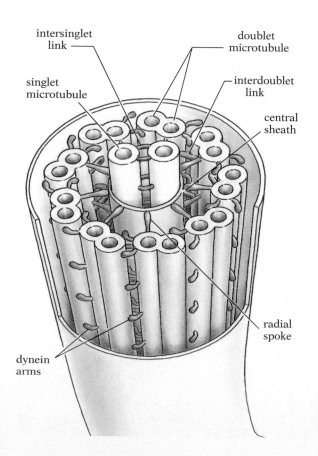

intersinglet
link

doublet
microtubule

singlet
microtubule

interdoublet
link

central
sheath

radial
spoke

dynein
arms

basal body in the arrangement of microtubules is obvious. The
beating of cilia and flagella (Figs. 5.38 and 5.39) depends on two
clawlike arms that allow each of the nine outer doublets to interact
with an adjacent doublet (Fig. 5.40).

In a series of definitive experiments in the 1970s, Ian Gibbons
showed that a protein called *dynein*, found in cilia, can extract energy from ATP. He also found that when all the dynein was extracted from cilia, the doublets were left armless, and when the
dynein was restored, the arms reappeared—a clear demonstration
that the arms are the site of ATP hydrolysis for the cilia. Current
models of ciliary movement postulate that the arms provide the
basis for a rachetlike mechanism that enables ciliary microtubules
to "walk along" or slide over one another. The models also postulate that because of the shear resistance within the cilium, due in
large part to the radial spokes that bind all the doublets to the central sheath, the sliding of some doublets past others brings about a
bending of the ciliary stalk (Fig. 5.41).

Despite the remarkable resemblance in mode of action between

5.40 Internal structure of a cilium

The nine microtubule doublets are arranged around two singlet
microtubules in the center of the cilium. Movement is generated when the
dynein arms of one subset of doublets begin to "walk" along the length of
other doublets. The doublets moving the most thus force the other doublets
to bend. The intersinglet links may prevent lateral bending, and so explain
why ciliary beats are constrained to a fixed plane.

A

B   0.2µm

5.41   Bending in cilia produced by microtubules sliding past each other

(A) The microtubules on the concave side of the bend (green, right) have slid tipward. Because the tubules are all interconnected, their changes in relative position can be accommodated only if the stalk of the cilium bends. (B) Rows of outer dynein arms are visible in this electron micrograph of two *Tetrahymena* cilia that have been stripped of their outer cell membranes. The cilia were first frozen and then fractured obliquely to their axes. The surrounding matrix was subsequently etched back to expose the internal structures.

the tubulin-dynein system and the actin-myosin system, the two are unrelated: the amino acid sequences in tubulin and dynein show no similarities to those of actin and myosin. These two systems evolved their similar mechanisms for producing ratchet-driven sliding motion independently; this is a truly impressive example of what is called convergent evolution—the independent evolution of two very similar solutions to the same problem.

ARE EUCARYOTES COOPERATIVES OF PROCARYOTES?

There are two basic kinds of cells: eucaryotes (most organisms) and procaryotes (bacteria). Nearly all the organelles we have surveyed are unique to eucaryotes. Yet there is good reason to believe that some of these organelles may have evolved from procaryotes. To examine this argument we must first summarize eucaryotic organization.

■ HOW A "TYPICAL" EUCARYOTIC CELL IS ORGANIZED

The "typical" cell is a useful fiction; in fact, plant and animal cells differ from one another in dramatic ways, cells of particular plants or animals differ from those of other plants or animals almost as much, and within the body of any one plant or animal the various cells often differ strikingly in shape, size, and function. Now that the number of known cellular components has grown so large and their great variability has been so well demonstrated, it becomes even more obvious that no single diagram, or even series of diagrams, can really portray a typical cell. Nevertheless, to help summarize and visualize the arrangement of the organelles discussed in the preceding pages, two diagrams of relatively unspecialized cells (Figs. 5.42 and 5.43) are given here.

rough endoplasmic reticulum

smooth endoplasmic reticulum

glycocalyx

vacuole

Golgi apparatus

nucleus

nucleolus

primary lysosome

secondary lysosome

plasma membrane

centrioles

mitochondrion

5.42 A "typical" animal cell

Not every animal cell includes all the organelles
shown here, nor are all the substructures that
may occur in an animal cell represented. While
the diagrams here and in Figure 5.41 indicate
relative size, some of the organelles in the electron micrographs are much enlarged in comparison to others, in order to show internal detail.

endoplasmic
reticulum

plasmodesma

nucleolus   nucleus

chloroplast

plasma
membrane

cell wall

mitochondrion

leucoplast

vacuole

Golgi apparatus

5.43 A "typical" plant cell

The organelles shown here do not occur in every plant cell, and some plant-cell substructures are not represented. The small amount of inner membrane compared with that of animal mitochondria is typical of plants.

5.44   Part of a bacterial cell

The light area in the center of the cell in this electron micrograph is called the nucleoid; it contains DNA but is not bounded by a membrane. Note the prominent cell wall, with the plasma membrane visible just inside it, and the absence of membrane-bounded cellular organelles.

■ WHAT MAKES PROCARYOTES DIFFERENT?

Procaryotic cells lack most of the cytoplasmic organelles present in eucaryotic cells (Fig. 5.44). As their name implies, they have no nuclear membrane; they also lack other membranous structures such as an endoplasmic reticulum, a Golgi apparatus, lysosomes, peroxisomes, and mitochondria. (Many of the functions of mitochondria are carried out by the inner surface of the plasma membrane.) Many photosynthetic bacteria, however, do have separate membranous vesicles or lamellae containing chlorophyll.

Procaryotic cells contain a large DNA molecule, which, though not tightly associated with proteins as DNA is in eucaryotic cells, is nonetheless a true chromosome (Fig. 5.45). Often there are also small independent pieces of DNA, called plasmids. Unlike eucaryotic chromosomes, which are usually linear, the procaryotic chromosome and plasmids are ordinarily circular. Like eucaryotic chromosomes, the procaryotic chromosome bears, in linear array, the genes that control both the hereditary traits of the cell and its ordinary activities. The DNA functions by directing protein synthesis on ribosomes via mRNA, in the way already described for eucaryotic cells. Ribosomes are prominent in the cytoplasm of both eucaryotic and procaryotic cells. Those of procaryotic cells, however, are somewhat smaller than those of eucaryotic cells.

Some bacterial cells possess hairlike organelles, used in swimming, that have traditionally been called flagella, but they lack microtubules; instead they are composed of a single kind of protein, flagellin. Flagellin forms a stiff helix that is rotated like a propeller by a special structure in the membrane at its base. Table 5.1 gives a summary of some of the most important differences between procaryotic and eucaryotic cells.

5.45   A disrupted cell of *Escherichia coli*

Exposure to detergent releases the DNA of the common intestinal bacterium *E. coli*, most of which can be seen outside the cell in this electron micrograph. Though not readily recognizable here, the main chromosome and the small accessory chromosomes (plasmids; arrow points to one) of a bacterium each form a circle rather than individual strands in the cell. The mottled appearance of the surface of the bacterium results from the effects of alcohol, drying, and shadowing with platinum.

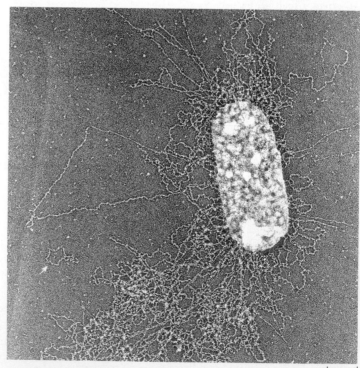

Table 5.1 A comparison of typical procaryotic and eucaryotic cells, and certain eucaryotic organelles

CHARACTERISTIC	PROCARYOTIC CELLS	EUCARYOTIC CELLS	MITOCHONDRIA AND CHLOROPLASTS
Size	1–10 μm	10–100 μm	1–10 μm
Nuclear envelope	Absent	Present	Absent
Chromosomes	Single, circular, with no nucleosomes	Multiple, linear, wound on nucleosomes	Single, circular, with no nucleosomes
Golgi apparatus	Absent	Present[a]	Absent
Endoplasmic reticulum, lysosomes, peroxisomes	Absent	Present[a]	Absent
Mitochondria	Absent	Present[a]	
Chlorophyll	Not in chloroplasts	In chloroplasts	
Ribosomes	Relatively small	Relatively large	Relatively small
Microtubules, intermediate filaments, microfilaments	Absent	Present	Absent
Flagella	Lack microtubules	Contain microtubules	

[a] Certain parasites and anaerobic organisms lack mitochondria, having lost them either because high-energy compounds are available directly from the host cell, or because they cannot use the oxygen-dependent enzymatic pathways unique to mitochondria. One kingdom of primitive eucaryotes (discussed in Chapter 21) lacks mitochondria, ER, and Golgi.

■ THE ENDOSYMBIOTIC HYPOTHESIS: A PROCARYOTIC CORPORATION

Scientific opinion today is moving toward the view that at least two organelles found only in eucaryotes—mitochondria and chloroplasts—are the descendants of procaryotic organisms that took up residence in the "hospitable" precursors of eucaryotes. There are several lines of evidence for this *endosymbiotic hypothesis* (from *endo-*, "within," and *symbiosis*, "state of living together"); some of these are based on the features summarized in Table 5.1.

1 Many symbiotic associations of procaryotes and eucaryotes are known. For instance, many present-day photosynthetic bacteria live inside eucaryotic hosts, providing them with food in return for shelter. Similarly, certain nonphotosynthetic bacteria live symbiotically within eucaryotes, extracting and sharing energy from foods that their hosts cannot themselves metabolize. Such associations provide a clear starting point for the evolution of *obligate* symbioses, in which two organisms depend on one another for their mutual survival.

2 Both mitochondria and chloroplasts contain their own ribosomes and chromosomes; the chromosomes code for their ribosomal RNA and ribosomal proteins, and for some (though not all) of their enzymes. Mitochondria and chloroplasts also build their own membranes.

3 The organelle chromosomes resemble those of procaryotes in that they are circular, not wound on special protein spools, and not enclosed by a nuclear envelope.

4 The internal organization of the organelle genes is similar to that of procaryotes but very different from that of eucaryotes.

5 Cell division in eucaryotes involves a spindle apparatus, whereas bacteria, mitochondria, and chloroplasts divide by fissioning.

6 The ribosomes of mitochondria and chloroplasts are more similar to those of procaryotes than to the ribosomes in the cytosol of the cells in which they live.

In summary, the endosymbiotic hypothesis assumes that a bacterium with the unique metabolic capabilities exhibited by mitochondria was captured through endocytosis about 1.5 billion years ago, resisted digestion, lived symbiotically inside its host, and divided independently of its host. Later, some of the symbiont's genes were moved to the host nucleus, which took control from its guest. From this partnership all present-day plants, animals, fungi, and unicellular eucaryotic organisms must have evolved. Subsequently, according to this hypothesis, various photosynthetic bacteria met the same fate, the result being the chloroplasts found in algae and plants.

Some current speculation, less well supported but thought-provoking, holds that peroxisomes, basal bodies and centrioles, and nematocysts (the devices by which jellyfish and similar creatures harpoon potential prey and careless swimmers) may also have had an endosymbiotic origin.

CHAPTER SUMMARY

THE ROLES OF SUBCELLULAR COMPARTMENTS

THE NUCLEUS: THE CELL'S CONTROL CENTER The nucleus is bounded by a double membrane, the nuclear envelope, and contains the chromosomes. (p. 121)

THE ENDOPLASMIC RETICULUM: SYNTHESIS AND SORTING The RER binds the ribosomes synthesizing proteins destined for the ER. Some of these proteins are used in the ER; the rest are transferred to the SER, where they are packaged into vesicles for transport elsewhere. The ER also synthesizes new membrane. (p. 125)

THE GOLGI APPARATUS: SORTING AND PACKAGING The Golgi apparatus modifies and sorts molecules arriving in vesicles, and dispatches new vesicles including secretory vesicles. (p. 129)

LYSOSOMES: DIGESTIVE ORGANELLES Lysosomes contain digestive enzymes; they fuse with endocytosed vesicles containing food. (p. 130)

PEROXISOMES: DETOXIFICATION ORGANELLES Peroxisomes contain powerful oxidative enzymes. (p. 133)

MITOCHONDRIA: CELLULAR POWER STATIONS Mitochondria are complex organelles that carry out respiration. (p. 133)

CHLOROPLASTS: SOLAR ENERGY RECEPTORS Chloroplasts are complex organelles that carry out photosynthesis. (p. 133)

LEUCOPLASTS: STORAGE ORGANELLES IN PLANTS Leucoplasts store high-energy compounds for later use. (p. 134)

VACUOLES: MULTIPURPOSE ORGANELLES Vacuoles help give plant cells support and can store some molecules. (p. 135)

THE CYTOSKELETON: STRUCTURE AND MOVEMENT

MICROFILAMENTS: CELLULAR ROPES AND CABLES Actin microfilaments give structural support and, in conjunction with myosin, are involved in cell movement. (p. 136)

MICROTUBULES: CELLULAR GIRDERS AND RAILWAYS Microtubules give cells support and provide channels along which vesicles can move. Many radiate from an organizing center, the centrosome. (p. 139)

INTERMEDIATE FILAMENTS: A CELLULAR LATTICE? Intermediate filaments are tubules with at least some structural role. (p. 140)

CENTRIOLES AND BASAL BODIES: STRUCTURAL ORGANIZERS Centrioles are found near the centrosomes, and generate the basal bodies found at the base of cilia and flagella. (p. 142)

CILIA AND FLAGELLA: CELL MOVERS Cilia and flagella are mounted on the membrane and are used either to propel a cell or move material past it. (p. 143)

ARE EUCARYOTES COOPERATIVES OF PROCARYOTES?

HOW A "TYPICAL" EUCARYOTIC CELL IS ORGANIZED Eucaryotes have the organelles discussed in this chapter. (p. 145)

WHAT MAKES PROCARYOTES DIFFERENT? Procaryotes lack the organelles described in this chapter, though some are organized like mitochondria or chloroplasts. (p. 148)

THE ENDOSYMBIOTIC HYPOTHESIS: A PROCARYOTIC CORPORATION This hypothesis proposes that many organelles were either once free-living procaryotes or internal components of procaryotes. Some organelles even have their own DNA organized along procaryotic lines and use procaryote-like ribosomes. (p. 149)

STUDY QUESTIONS

1 Trace the pathway of information flow from the chromosomes to the outer membrane during the secretion of a digestive enzyme. (pp. 121–133)

2 Are the carbohydrate markers placed on proteins that are destined to be mounted in the cell membrane ever exposed to the cytosol? (pp. 129–131)

3 What would happen to a cell if it lost its Golgi apparatus? its nucleolus? its peroxisomes? (pp. 123–124, 129–130, 133)

4 List five important differences between procaryotes and eucaryotes. How do these contrasts bear on the endosymbiotic hypothesis? (pp. 148–150)

SUGGESTED READING

ALBERTS, B., D. BRAY, J. LEWIS, M. RAFF, K. ROBERTS, AND J. D. WATSON, 1994. *Molecular Biology of the Cell*, 3rd ed. Garland, New York. *This massive tome is the most complete and up-to-date summary of cell biology available.*

ALLEN, R. D., 1987. The microtubule as an intracellular engine, *Scientific American* 256 (2).

DEDUVE, C., 1986. *The Living Cell*. Scientific American Library, New York. *This two-volume set provides a well-illustrated, up-to-date tour of the cell.*

GOVER, D. M., C. GONZALEZ, AND J. W. RAFF, 1993. The centrosome. *Scientific American* 268 (6). *On how these organelles help determine cellular structure.*

ROTHMAN, J. E., 1985. The compartmental organization of the Golgi apparatus. *Scientific American*, 253 (3). *An incisive analysis of the fine structure of this important organelle.*

SIMMONS, K., H. GAROFF, AND A. HELENIUS, 1982. How an animal virus gets into and out of its host cell, *Scientific American* 246 (2). (Offprint 1511) *A fascinating account of how Semliki Forest virus subverts the membrane system of its many vertebrate hosts.*

STOSSEL, T. P., 1994. The machinery of cell crawling. *Scientific American* 271 (3).

WEBER, K., AND M. OSBORN, 1985. The molecules of the cell matrix, *Scientific American* 253 (4). *Reviews the structure and function of microfilaments and microtubules.*

CHAPTER 6

RESPIRATION

Energy is essential to life. It is used by cells to fuel growth and reproduction, and even just to maintain the status quo. On the earth at present, stored energy is usually released through aerobic respiration, in which, ultimately, oxygen and glucose are combined to yield CO_2 and water; the energy can then be used to run the cell's enzyme-mediated reactions (Fig. 6.1). These reactions, collectively known as **metabolism**, belong to one of two phases—**anabolism**, the processes by which complex organic molecules like glucose are assembled, and **catabolism**, the processes by which living things extract energy from food. This chapter will examine the cell's strategy of energy extraction and the machinery it uses.

THE FLOW OF ENERGY

■ HOW METABOLISM EVOLVED

Metabolism evolved under conditions very different from those that exist over most of the earth today. In particular, the ancient atmosphere had no oxygen. Relict anaerobic species found today in bogs, volcanic pools, and on the ocean bottom provide a glimpse of

6.1   Summary of biological energy flow

Today virtually all the energy for life originates in the sun, where hydrogen is converted by fusion into helium, and light is produced. In the process of photosynthesis, green plants convert the radiant energy of sunlight into chemical energy, which is most often stored initially in glucose. When cells in most organisms need energy, the glucose is broken down and some of its chemical energy is recovered by the process of aerobic respiration; the resulting product—ATP—supplies the energy in a more manageable form, making it available for muscular contraction, nerve conduction, active transport, and other work. Aerobic respiration utilizes oxygen, a by-product of photosynthesis, and photosynthesis utilizes carbon dioxide and water, by-products of respiration. With each of the transformations shown, much energy is lost as waste heat.

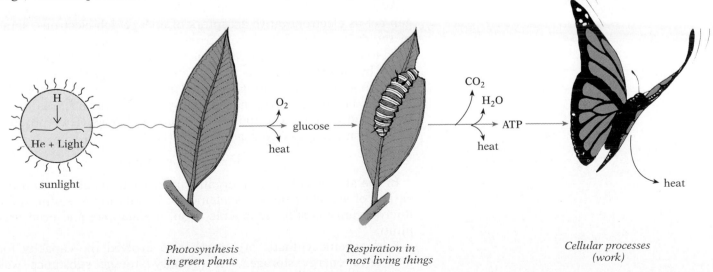

Photosynthesis in green plants   *Respiration in most living things*   *Cellular processes (work)*

0.1 μm

6.2 Chemosynthetic bacteria

These chemosynthetic bacteria, *Methanothermus fervidus*, live at temperatures up to 97°C in Icelandic hot springs. They combine H_2 and CO_2 to produce methane and energy.

the earliest metabolic pathways. The first organisms appeared about 3.5 billion years ago and subsisted on a limited supply of energy-rich organic molecules generated by inorganic processes. However, natural selection strongly favored organisms capable of synthesizing their own food from scratch. To create such organic compounds, this second group of early organisms used energy from inorganic sources, a process known as chemosynthesis.

Some early chemosynthetic organisms combined naturally occurring CO_2 with energy-rich H_2 to produce methane and water:

$$CO_2 + 4H_2 \longrightarrow CH_4 + 2H_2O + energy$$

As a result of this reaction, in which carbon and (especially) oxygen act as electronegative acceptors of energy-rich electrons, some of the energy stored in the electrons of H_2 is released and can be used to do work in the cell. Methane ("swamp gas") is still produced today by anaerobic bacteria in bogs and hot springs (Fig. 6.2).

Another energy-liberating reaction uses sulfur as an acceptor to liberate the energy in H_2; indeed, many present-day bacteria thrive in the sulfur-rich waste found in sewers and bogs. They produce the foul-smelling gas hydrogen sulfide (H_2S):

$$H_2 + S \longrightarrow H_2S + energy$$

In the absence of oxygen, organisms evolved the ability to use a variety of less electronegative elements: in addition to sulfur, some microorganisms still use arsenic, iron, manganese, and even uranium.

Early chemosynthetic organisms also evolved the capacity for long-term energy storage. The primary storage substance was probably glucose, or polysaccharides that could be converted into

EXPLORING FURTHER

THE HIGH ELECTRONEGATIVITY OF OXYGEN: ITS ROLE IN ENERGY TRANSFORMATIONS

Life depends on the efficient management of electrons. With the aid of highly specialized enzymes, each cell harvests the energy of electrons as they move from a position of relatively great potential energy to one of less potential energy. In biochemical pathways, an electron's potential energy can be unlocked in two ways. In photosynthesis, as we will see, a photon can energize an electron to a higher energy level within an atom—from level L to M, for example—and then this extra potential energy can be captured and used later. More often, however, biological reactions make use of the difference in potential energy between the energy of an outer electron in one atom and a vacancy with a lower potential energy in another atom. When an electron is shifted between atoms, energy is released. This movement need not be—indeed, usually is not—between different electron energy levels; instead, because the energy of, for instance, the L level of carbon is significantly higher than that of the L level of oxygen (Fig. A), movement of an electron from one to the other is highly exergonic. The same holds for electrons in covalent bonds within a molecule: the shared electrons in a C-H bond, for example, have about 11 kcal/mole more potential energy than those in an O-H bond. Transferring a hydro-gen from a carbon to an oxygen, either within a single covalently bonded molecule or between two different molecules, thus releases a substantial amount of energy.

The difference in potential energies of outer-shell electrons accounts for the differences in electronegativities (see Fig. B). Of the elements commonly found in living things, oxygen is the most electronegative; hence, moving electrons to oxygen releases more free energy than shifting them to any other element. In the early stages of the evolution of life, free oxygen was exceedingly rare, so electrons usually had to be shifted to other electron acceptors. Even when oxygen that was part of a larger molecule could be used, the result was not optimal: though a substitution of electrons can result in a net liberation of energy—replacing a C-O bond with an O-H bond, for instance, is exergonic—it releases only about 25% of the energy available from the direct formation of an O-H bond from free oxygen (Fig. B). Using sulfur as an alternative electron acceptor, to produce an S-H bond rather than an O-H bond, is also exergonic but yields less than 30% of the electron's potential energy. Some bacteria still use sulfur, nitrogen, or even carbon as their electron acceptors.

Nucleus | Carbon | Oxygen

A   Relative energy levels

The corresponding electron levels in two elements never have the same energy. As a result, an electron moving from the L level of carbon to the L level of oxygen will lose (liberate) energy in the process.

B   Electronegativity and potential energy

The electronegativity chart has been inverted to show differences in electron potential energy—the essential factor controlling whether a reaction is endergonic (requiring that electrons move up on the graph) or exergonic (in which electrons move down to lower energy values). Only the six most important elements in living tissue are shown. Note that hydrogen is the least electronegative of these elements; electron transfers from hydrogen are therefore inevitably exergonic. Oxygen, on the other hand, is the most electronegative; transfers of electrons to it always release energy.

glucose. When needed, some of the energy of the glucose was extracted, as energy still is today, through glycolysis—a series of reactions that we will examine in a moment.

About 3 billion years ago certain organisms acquired a primitive ability to capture the sun's energy directly and use it to synthesize glucose and other important organic compounds. Then, about 2.5 billion years ago, the forerunner of aerobic respiration evolved. This pathway, which includes glycolysis as its first stage, extracts large amounts of energy from the end products of glycolysis, but it must use molecular oxygen (O_2) to be maximally efficient. Since molecular oxygen was in short supply, less electronegative acceptors (sulfur, nitrogen, and carbon) had to suffice, and so cells wasted much of the potential energy in the chemical bonds of storage compounds.

Oxygen became increasingly available when the descendants of the early photosynthetic organisms began producing it as a by-product of a more advanced form of photosynthesis; thus, about 2.3 billion years ago, oxygen began to accumulate in the atmosphere. Since oxygen (in the form of ozone, O_3) filters out destructive X rays and ultraviolet light from the sun, organisms could now emerge from beneath the radiation shielding provided by earth and water to colonize the surface of the land. In addition, the abundance of oxygen made aerobic respiration the dominant catabolic pathway.

■ HOW ENERGY IS TRANSFERRED IN REACTIONS

Sugars like glucose are energy-rich compounds, while carbon dioxide is energy-poor. Through aerobic respiration glucose is "burned" to produce CO_2 and energy for the cell:

$$C_6H_{12}O_6 + 6O_2 + 6H_2O \longrightarrow 6CO_2 + 12H_2O + \text{energy}$$

As the colored arrows indicate, the essential molecular rearrangement that occurs in this reaction is the removal of all of the carbon-bound hydrogens from the sugar and their rebonding to oxygen. Energy is released in this reaction because any transfer of hydrogen electrons from bonds with carbon to bonds with oxygen is exergonic. The enzymes of aerobic respiration use some of this energy to do work in the cell; the rest of the energy is lost as heat. This transfer of hydrogen atoms from carbon to oxygen is an example of *oxidation*. Stated more generally, a substance is oxidized when it loses electrons (almost always hydrogen electrons)—and sometimes the rest of the atom to which the electrons are attached—to a more electronegative substance (usually oxygen); as a result, oxidation inevitably liberates energy.

The converse process—storing energy in a compound—is called, oddly enough, *reduction*. Photosynthesis, for example, works to reduce CO_2, basically reversing aerobic respiration: solar energy is used to power the synthesis of high-energy sugar from energy-poor CO_2. The reduction of CO_2 occurs through the redeployment of hydrogens from low-energy bonding with electronegative oxygen to high-energy bonding with carbon:

$$6CO_2 + 12H_2O + \text{light} \longrightarrow C_6H_{12}O_6 + 6O_2 + 6H_2O$$

Though reduction in cellular reactions usually involves the transfer of hydrogen atoms from oxygen to carbon, reduction in general occurs whenever an electron (with or without the rest of its atom) is moved to a less electronegative atom. This broader definition is important because we will see cases in both photosynthesis and aerobic respiration in which oxidation and reduction entail the movement of isolated electrons, or involve elements other than oxygen and hydrogen.

We can see that oxidation and reduction are mirror-image processes, one lowering, the other raising an electron's potential energy. Since all reactions involve making or rearranging bonds, it follows that if one atom or molecule gains energy, another must have lost energy—that is, whenever one substance is reduced, another is oxidized. We can follow this so-called **redox** reaction from the perspective of an electron, as it moves from one atom to another:

$$A^{e^-} + B \longrightarrow A + B^{e^-}$$

has been   has been
oxidized   reduced
(lost energy)   (gained energy)

More often though, the electron undergoing a change in potential is part of an atom that is rebonded; no ions are created. Just as the most common reduction reactions involve the removal of oxygen or addition of hydrogen, oxidation usually involves the addition of oxygen or removal of hydrogen:

$$A + BO \longrightarrow AO + B$$

has been   has been
oxidized   reduced

or

$$AH + B \longrightarrow A + BH$$

has been   has been
oxidized   reduced

In cells, removal or addition of an electron derived from hydrogen is the most frequent mechanism of redox reactions. A summary of redox reactions appears in Table 6.1.

It now becomes clear why the synthesis of sugar from carbon dioxide constitutes reduction of the CO_2: hydrogen obtained by splitting water molecules is added to the CO_2, thereby storing energy that is later recovered through respiration. The main product of this subsequent respiration is ATP, the major energy currency of the cell.

■ ATP: THE CURRENCY OF ENERGY EXCHANGE

One of the essential substances of life is the compound **adenosine triphosphate**, or **ATP**, which plays a key role in nearly every energy transformation in living things (Fig. 6.3).

Table 6.1 Redox reactions

OXIDATION	REDUCTION
Adds oxygen	Removes oxygen
Removes hydrogen	Adds hydrogen
Removes electron(s)	Adds electron(s)
Liberates energy	Stores energy

Adenine   Ribose P_1 P_2 P_3

ADENOSINE   PHOSPHATES

6.3 The ATP molecule

Adenosine triphosphate (ATP) is composed of an adenosine unit (a complex of adenine and ribose sugar) and three phosphate groups arranged in sequence. The last two phosphates are attached by so-called high-energy bonds (wavy lines). The cell stores energy by adding a phosphate group to ADP (adenosine diphosphate) to make ATP, and later recovers some of this energy by hydrolyzing ATP into ADP and inorganic phosphate. Occasionally ADP is hydrolyzed to make AMP (adenosine monophosphate), one of the nucleotides of which RNA is composed. The widely used term *high-energy bonds* is misleading: the energy is actually stored within the phosphate group itself.

The ATP molecule is composed of a nitrogen-containing compound (adenosine) plus three phosphate groups bonded in sequence:

$$\text{adenosine} - Ⓟ \sim Ⓟ \sim Ⓟ$$

According to convention, Ⓟ stands for the entire phosphate group, and the wavy lines between the first and second and the second and third phosphate groups represent so-called high-energy bonds.[1]

Actually, it is often only the terminal phosphate bond of ATP that is involved in energy conversions. The exergonic reaction by which this bond is hydrolyzed and the terminal phosphate group removed leaves a compound called **adenosine diphosphate,** or **ADP** (adenosine plus two phosphate groups), and inorganic phosphate (symbolized by P_i):

$$\text{ATP} + H_2O \xrightarrow{\text{enzyme}} \text{ADP} + P_i + \text{energy}$$

If both the second and third phosphate groups are removed from ATP, the resulting compound is **adenosine monophosphate,** or **AMP**.

New ATP can be synthesized from ADP and inorganic phosphate

---

[1] The widely used term *high-energy bond* does not mean that energy is stored in the phosphate bonds themselves. Instead, it means that the bonds between the negatively charged phosphate groups are very unstable (and so are thermodynamically inexpensive to break) and that substantially more energy is released when the phosphate groups form new bonds.

if adequate energy is available to force a third phosphate group onto the ADP. Addition of phosphate is termed *phosphorylation*:

$$ADP + P_i + energy \xrightarrow{enzyme} ATP + H_2O$$

ATP is often called the universal energy currency of living things. Cells initially store energy in the form of carbohydrates, such as glucose, and of lipids, such as fat. But the amount of energy in even a single glucose molecule—670 kcal/mole—is inconveniently large for driving most reactions. The hydrolysis of ATP to ADP releases a more useful amount of energy: 7.3 kcal/mole. Glucose molecules are the hundred-dollar bills of the cellular economy, while ATP molecules are the everyday denominations, sufficient for making and breaking covalent bonds. Though some other compounds can supply energy, ATP is the one most often used by cells in the various kinds of work they perform: it powers synthesis of more complex compounds, muscular contraction, nerve conduction, active transport across cell membranes, light production, and so on. The price of the work is paid through the energy-releasing hydrolysis of ATP to ADP.

HOW CELLS HARVEST ENERGY

The energy stored in lipids and carbohydrates is liberated gradually through a series of reactions, each catalyzed by its own specific enzyme. Certain steps release small amounts of energy that are transferred to ATP by the phosphorylation just described. Let us look at the most important steps in the catabolism of glucose, the central energy-liberating pathway in all organisms.

■ RESPIRATION WITHOUT OXYGEN

Glycolysis The complete catabolism of glucose involves five stages, divided between anaerobic and aerobic series of reactions. Stage I, *glycolysis*, is the anaerobic portion of the process, and breaks glucose down to pyruvic acid. Glycolysis is the most ancient part of the pathway; it evolved long before free oxygen became available. We will trace the steps of glycolysis in detail because they illustrate how reaction pathways are organized into small, enzyme-mediated steps and are made to work through the careful management of free energy changes. It is these principles and the overall sequence of steps (summarized in Fig. 6.4) that are important, not the exact thermodynamics and the structural details of the molecular intermediates.

Glucose is a stable compound, one with little tendency to break down spontaneously. If its energy is to be harvested, glucose must first be activated by the investment of a small amount of energy. The first steps of glycolysis, therefore, are preparatory, enabling the later steps to extract the stored energy.

The energy for initiating glycolysis (Fig. 6.4) comes from ATP. The initial reaction, like the succeeding ones, is made possible by a step-specific enzyme, which binds (via weak bonds) to the reactants, activates them (by causing the electrons of the bound molecules to redistribute themselves), and then joins or rearranges the

PREPARATORY STEPS

Step

glucose

ATP → hexokinase → 1

ADP

phosphoglucoisomerase 2

ATP → phosphofructokinase → 3

ADP

aldolase 4

phosphotriose isomerase 5

2 PGAL

to oxidation steps
(right-hand column)

OXIDATION WITH FORMATION OF ATP

Step

2 P_i   2 NAD_ox   glyceraldehyde 3-phos-   6
phate dehydrogenase
2 NAD_re + 2H^+

2 ADP
phosphoglycerate kinase   7
2 ATP

phosphoglyceromutase   8

enolase   9
2 H_2O

2 ADP
pyruvate kinase   10
2 ATP

2 pyruvic acid

in most plants and   *in animals and*
many microorganisms   *some microorganisms*

2H^+   2H^+   2 NAD_re + 2H^+
lactate   11
dehydrodenase
2 CO_2   2 NAD_ox

2 acetaldehyde   2 lactic acid

2 NAD_re
alcohol
dehydrodenase   12
2 NAD_ox

2 ethanol

6.4   Glycolysis and fermentation

The entire reaction series for glycolysis is shown to illustrate how biochemical pathways work; there is no reason to try to memorize each step. Glucose in solution normally exists as a ring structure, but the straight-chain form is adopted here for clarity. Energy to initiate the breakdown of glucose is supplied by two molecules of ATP (Steps 1 3). The resulting compound is then split into two molecules of PGAL (5). This completes the preparatory reactions. Next, the PGAL is oxidized by the removal of hydrogen, and inorganic phosphate is added to each of the three-carbon molecules (6). A series of reactions then results in the synthesis of four new molecules of ATP, for a net gain of two (7–10). The pyruvic acid produced by this anaerobic breakdown can be further oxidized in most cells if O_2 is present (by reactions not shown here). In the absence of sufficient O_2, the pyruvic acid may be converted to lactic acid in some kinds of organisms, or to CO_2 and ethanol in others (11–12). At each step a particular enzyme catalyzes a specific redistribution of electrons and thereby brings about changes in bonding. This step-by-step strategy is the only one by which enzymes can guide reactions. (In each diagram, bonds to be altered by enzymatic actions in the next step are shown in green. The fate of particular oxygens and hydrogens cannot always be traced from step to step because they are sometimes incorporated into units, such as phosphate groups, for which full diagrams are not given.) Figure 6.4 provides a more schematic summary of glycolysis.

reactants before releasing the products. In the first reaction, a molecule of ATP donates its terminal phosphate group to the glucose.

$$(1) \quad C\!-\!C\!-\!C\!-\!C\!-\!C\!-\!C + ATP \xrightarrow{\text{enzyme}} C\!-\!C\!-\!C\!-\!C\!-\!C\!-\!C\!-\!\textcircled{P} + ADP$$

$$\underset{\text{glucose}}{} \qquad \qquad \underset{\text{glucose-6-phosphate}}{}$$

$$(\Delta G = \text{-4.0 kcal/mole})$$

(The simplified equations given here show only the carbon skeleton; the more complete molecular structures are shown in Fig. 6.4.) The name of the product tells us that it is a glucose with a phosphate group attached to its sixth carbon atom.

Let's look carefully at what has happened in this reaction. An enzyme, hexokinase, has bound glucose and ATP, catalyzed the transfer of a phosphate group to the glucose, and released the products. The overall change in free energy of the electrons rearranged in this reaction is -4.0 kcal/mole; the free energy is liberated primarily as heat. As always, the minus sign means this is an exergonic (downhill) reaction.[2] In fact, the reaction has an equilibrium constant (K_{eq}) of about 1000—that is, the products outnumber the reactants by 1000 to 1. The free energy for this reaction comes from ATP: the energy available from the terminal phosphate of the ATP is 7.3 kcal/mole. Only 4.0 kcal/mole is liberated to drive this first step of glycolysis, and the other 3.3 kcal/mole from the ATP is stored in the electrons of the product—the activated glucose.

---

[2] As noted in Chapter 3, we use ΔG (rather than the formal symbol $\Delta G'^0$) for the change in the free energy of a reaction under standard conditions. In addition, we divide the ΔGs value simply between products and liberated free energy, even though, in fact, a more complicated but poorly understood division is probably occurring. The resulting approximation is reasonably accurate. The equilibrium constant (K_{eq}) is being treated for simplicity as though there is only a single product. This is possible because glycolysis has relatively little effect on the concentration of water, ATP, ADP, and so on; the concentration of these compounds is maintained through various homeostatic mechanisms in the cell.

The next step in glycolysis converts glucose-6-phosphate into the nearly identical compound fructose-6-phosphate:

$$(2) \quad \text{C—C—C—C—C—C—}\textcircled{P} \xrightarrow{\text{enzyme}} \text{C—C—C—C—C—C—}\textcircled{P}$$

glucose-6-phosphate \qquad\qquad fructose-6-phosphate

$$(\Delta G = +0.4 \text{ kcal/mole})$$

The positive ΔG indicates that this is an uphill, endergonic (energy-requiring) reaction, one that cannot proceed spontaneously. How is it, then, that glycolysis does not grind to a halt? The answer lies in the operation of ***coupled reactions:*** two reactions that share a common intermediate molecule—in this instance, glucose-6-phosphate, the product of Step 1 *and* the reactant of Step 2—can proceed as a single reaction. The 1000:1 ratio of products to reactants in Step 1 supplies an enormous number of reactant molecules to Step 2; this abundance insures that, even though the equilibrium constant of the second reaction is only about 0.2 (that is, a 1:5 ratio of products to reactants), many product molecules are nevertheless created to feed Step 3.

From the perspective of thermodynamics, these two steps can be treated as one reaction. The -4.0 kcal/mole liberated by Step 1 is combined with the +0.4 kcal/mole consumed by Step 2 to yield a net ΔG of -3.6 kcal/mole. Taken together, the two steps are strongly exergonic, and so the reaction proceeds. The glycolytic pathway is a series of such coupled reactions, in which exergonic steps push or pull endergonic steps, with the favorable *net* free-energy change of the steps taken together enabling the sequence of reactions to proceed.

The conversion of glucose-6-phosphate into fructose-6-phosphate also provides a good illustration of how enzymatic pathways work. Recall that when a substrate binds to an enzyme by means of weak bonds, a slight shift in the electron distribution of the substrate is induced, which lowers the activation energy required for a particular change in bonding and thereby catalyzes the reaction. The result of the redistribution of electrons can be seen in Step 2, in which two hydrogens, bonded to the fifth carbon and its oxygen, are transferred to the first carbon and its oxygen. This trivial change is essential to prepare for the next step in glycolysis. Once the change has taken place, the substrate no longer "fits" the enzyme, and so the substrate drifts away to be captured by the next enzyme in the series. Each step in the glycolytic pathway is therefore very small and is mediated by a highly specific enzyme.

After the formation of fructose-6-phosphate in Step 2, another molecule of ATP is consumed, to add a phosphate to the other end of the molecule. Of the ATP's energy, 3.9 kcal/mole is stored in the product, while the remaining 3.4 kcal/mole is liberated as heat; hence the reaction is exergonic:

$$(3) \quad \text{C—C—C—C—C—C—}\textcircled{P} + \text{ATP} \xrightarrow{\text{enzyme}}$$

fructose-6-phosphate

$$\textcircled{P}\text{—C—C—C—C—C—C—}\textcircled{P} + \text{ADP}$$

fructose-1,6-bisphosphate

$$(\Delta G = -3.4 \text{ kcal/mole})$$

Since this reaction in its turn is coupled to the previous one (with which it shares the intermediate compound fructose-6-phosphate), we can add the free energy of the separate steps along the way. The overall reaction chain so far has liberated 7.0 kcal/mole, and so it has a highly favorable K_{eq} of more than 10^5.

Next, the fructose-1,6-bisphosphate is split between the third and fourth carbons, forming two essentially similar three-carbon molecules (Step 4). One is **PGAL** (phosphoglyceraldehyde), and the other, an intermediate compound, is usually converted immediately to PGAL, in Step 5. (The cell can also use PGAL to synthesize fat if conditions warrant.) PGAL, a phosphorylated three-carbon sugar, is a key intermediate in both glycolysis and photosynthesis.

(4–5) $\text{P—C—C—C—C—C—C—P} \xrightarrow{\text{enzyme}} 2 \text{ C—C—C—P}$
            fructose-1,6-bisphosphate   PGAL

$$(\Delta G = +7.5 \text{ kcal/mole})$$

To this point, instead of releasing energy from glucose to form new ATP molecules, glycolysis has actually cost the cells two ATPs. Indeed, Steps 4 and 5 are so unfavorable energetically that the net change in free energy is now +0.5 kcal/mole. For subsequent reactions to proceed, significant amounts of free energy must be liberated to pull the reactants past the five preparatory steps.

The next reaction actually involves two separate molecular changes, which we summarize for simplicity in one step. The first change is oxidation of the PGAL by reduction of **NAD** (nicotinamide adenine dinucleotide). (The characteristic function of NAD is temporary storage of high-energy electrons; it transports energy from one pathway to another, or from one step in a pathway to another step, elsewhere in the pathway.) Each NAD_{ox} accepts two hydrogens, keeping one and the electron of the other to produce NAD_{re} and an H^+ ion. The second change is phosphorylation of the PGAL:

(6) $2 \text{ C—C—C—P} + 2 \text{ NAD}_{ox} + 2 P_i \xrightarrow{\text{enzyme}}$

$$2 \text{ P}\sim\text{C—C—C—P} + 2 \text{ NAD}_{re} + 2 \text{ H}^+$$
$$(\Delta G = +3.0 \text{ kcal/mole})$$

The oxidation phase of this reaction, taken alone, is strongly exergonic, while the phosphorylation phase is strongly endergonic. Since the two processes occur together, the energy that would have been released by the oxidation (more than 100 kcal/mole) is conserved in the reduced NAD (NAD_{re}) and the phosphorylated PGAL. The consequence of Step 6 is to make the net change in free energy for the overall reaction chain even more unfavorable (+3.5 kcal/mole), but the next downhill reaction, to which Step 6 is coupled, begins once again to turn the balance, as the high-energy phosphate bond is broken; some of the free energy is harvested by transferring the phosphate group to the ADP, to form ATP:

(7) $2 \text{ P}\sim\text{C—C—C—P} + 2\text{ADP} \xrightarrow{\text{enzyme}} 2 \text{ C—C—C—P} + 2\text{ATP}$
$$(\Delta G = -9.0 \text{ kcal/mole})$$

At this point, then, the cell regains the two ATP molecules invested to activate the glucose in Steps 1 and 3, and the overall net change

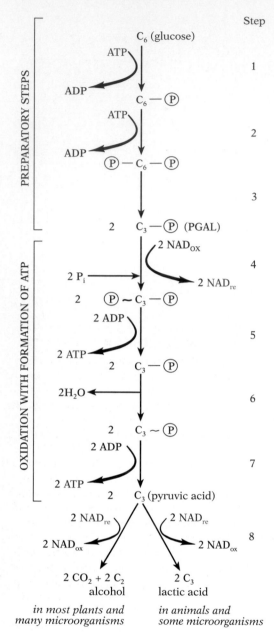

Step

6.5 Summary of glycolysis and fermentation

in free energy is again favorable: -5.5 kcal/mole. Moreover, a great deal of energy has been stored in NAD_{re}.

Next comes a reaction that energizes the remaining phosphate groups:

$$(8) \quad 2 \, C-C-C-\textcircled{P} \xrightarrow{\text{enzyme}} 2 \, C-C-C\sim\textcircled{P} + H_2O$$
$$(\Delta G = +1.5 \text{ kcal/mole})$$

After a reaction that rearranges the substrate for a ΔG of -0.4 kcal/mole (Step 9), these energized phosphate groups are transferred to ADP; the products are ***pyruvic acid*** (also referred to as pyruvate) and ATP:

$$(10) \quad 2 \, C-C-C\sim\textcircled{P} + 2 \, ADP \xrightarrow{\text{enzyme}} \underset{\text{pyruvic acid}}{2 \, C-C-C} + 2 \, ATP$$
$$(\Delta G = -15.0 \text{ kcal/mole})$$

Obviously this last reaction is overwhelmingly favorable; indeed, the K_{eq} is greater than 10^9. Moreover, a profit of two ATPs is generated by this step of glycolysis. Because the two ATP molecules used in Steps 1 and 3 have already been regained (in Step 7), the two additional molecules formed here represent a net gain in ATP for the cell. The highly exergonic last step results in an overall ΔG of -19.4 kcal/mole. It is the liberation of this energy, primarily as heat, that causes this series of coupled reactions to proceed.

Figure 6.5 summarizes the essential steps of glycolysis; Figure 6.6 shows graphically both the thermodynamics of the reaction steps and the changes in free-energy content at each successive step in glycolysis, from glucose to pyruvic acid (as well as in the process of fermentation, which will be discussed shortly).

We can now summarize the most important features of glycolysis:

1 Each molecule of glucose (a six-carbon compound) is broken down into two molecules of pyruvic acid (a three-carbon compound).

2 Two molecules of ATP are used to initiate the process. Later, four new molecules of ATP are synthesized, for a *net* gain of two molecules of ATP from each glucose.

3 Two molecules of NAD_{re} are formed.

4 Glycolysis can occur whether or not O_2 is present. Glycolysis is found in the cytoplasm of all living cells.

Fermentation We have seen that in glycolysis two molecules of NAD_{ox} are reduced to NAD_{re}, and that NAD functions in the cell as a shuttle for high-energy electrons. Thus NAD is only a temporary acceptor of electrons, promptly passing its extra electrons to some other compound and then going back for another load. The cell has only a limited supply of NAD molecules, and these must be used over and over again. If the NAD_{re} molecules formed in glycolysis could not quickly unload electrons (that is, be oxidized back into NAD_{ox}), all the cell's NAD would soon be tied up. With Step 6 of glycolysis thus blocked, the process would come to an end.

As we noted in the discussion of Step 6, more than 50 kcal/mole of free energy is stored in each of the two molecules of NAD_{re} produced during the glycolysis of a glucose molecule. For the cell to harvest this energy, the NAD_{re} must transfer its electrons to a lower energy level in some more electronegative acceptor molecule. In most cells, molecular oxygen is the ultimate acceptor of the transferred electrons. Under anaerobic conditions, however, with no oxygen present, the pyruvic acid formed by glycolysis must accept the electrons from NAD_{re}. This reduction of pyruvic acid results in the formation of *lactic acid* in animal cells and some unicellular organisms, and of *ethanol* (ethyl alcohol) and carbon dioxide in most plants and many unicellular organisms:

(11)   pyruvic acid + NAD_{re} + H^+ $\xrightarrow{\text{enzyme}}$ lactic acid + NAD_{ox}

or

(11)   pyruvic acid $\xrightarrow{\text{enzyme}}$ acetaldehyde + CO_2

(12)   acetaldehyde + NAD_{re} + H^+ $\xrightarrow{\text{enzyme}}$ ethanol + NAD_{ox}

Thus, under anaerobic conditions, NAD shuttles back and forth, picking up electrons (becoming NAD_{re}) in Step 6 and giving up the electrons (becoming NAD_{ox}) in Step 11 or 12. The process that begins with glycolysis and ends with the transformation of pyruvic acid into ethanol or lactic acid is called **fermentation**.[3]

Fermentation enables a cell to continue synthesizing ATP by breaking down nutrients under anaerobic conditions. However, because the electrons transferred from NAD_{re} remain at a relatively high energy level in the reduction of pyruvic acid, fermentation extracts only a very small portion (about 2%) of the energy present in the original glucose.

Fermentation by yeast cells and other microorganisms is the basis for the industries that produce breads and alcohol. Bacterial fermentation is also essential to the production of most cheeses, yogurt, and a variety of other dairy products.

■ **RESPIRATION WITH OXYGEN**

In eucaryotic cells the more efficient process of energy extraction that occurs in the presence of abundant molecular oxygen, aerobic respiration, takes place exclusively in the mitochondria. When O_2 is present, it acts as the electron acceptor; mitochondrial enzymes move the transient electrons to the oxygen, thereby releasing the free energy of NAD_{re}:

$$O_2 + 2\ NAD_{re} + 2H^+ \longrightarrow 2H_2O + 2\ NAD_{ox}$$
$$(\Delta G = -52.4\ \text{kcal/mole})$$

6.6   Changes in free energy at successive steps in glycolysis and fermentation

The graph shows two very different free-energy summaries. The first (red) traces only the free-energy changes resulting from each reaction; these values correspond to the ΔG values cited in the text, and show the net change of -19.4 kcal/mole that accompanies glycolysis. The black line represents the energy in the chemical intermediates; it includes the cost of the ATPs invested in Steps 1 and 3 to activate the glucose. The large drops in free energy correspond to steps in which energy is removed from those intermediates and stored in ATP or NAD_{re}. In the five preparatory steps, which convert glucose into PGAL, the free-energy content of the intermediate substances is slightly increased owing to the investment of two molecules of ATP. In Step 6 there is a sharp drop in free energy associated with the formation of two molecules of NAD_{re}. There are also major drops in Steps 7 and 10, each associated with the formation of two molecules of ATP. The space between the two free-energy summaries indicates the cellular energy profit; the process runs at a loss until Step 6.

[3] The term *fermentation* has been used in countless ways in the scientific literature. It is often restricted to the breakdown of glucose to ethanol. It is also applied to the production of either ethanol or lactic acid by microorganisms—lactic acid production in animal cells being called glycolysis. Both these uses lead to confusion between the terms *fermentation* and *glycolysis*, and both tend to obscure the general occurrence of the same basic fermentation process in all living cells. Accordingly, *fermentation* is here applied to any process in which glucose is catabolized and organic molecules are used as the electron acceptors of glycolysis. The glycolytic pathway to pyruvic acid is taken both as a preparatory reaction sequence leading to the Krebs citric acid cycle when sufficient oxygen is present, and as the initial portion of fermentation in the absence of sufficient oxygen.

COUPLED REACTIONS

Many important reactions are endergonic; outside energy is required to build DNA, RNA, proteins, and other molecules, to form structures like cell membranes, and to activate energy sources such as glucose and fat so their stored energy can be utilized. This outside energy is supplied by a coupling of the unfavorable endergonic reactions to one or more strongly exergonic reactions. The first two reactions of glycolysis provide an example of how coupling works:

$$ \text{(1)} \quad \text{glucose} + \text{ATP} \xrightarrow{K_{eq} = 610} \text{glucose-6-phosphate} + \text{ADP} $$
$$ (\Delta G = \text{-4.0 kcal/mole}) $$

$$ \text{(2)} \quad \text{glucose-6-phosphate} \xrightarrow{K_{eq} = 0.54} \text{fructose-6-phosphate} $$
$$ (\Delta G = \text{+0.4 kcal/mole}) $$

Reaction 1 is downhill: the products eventually outnumber the reactants by 610:1, while reaction 2 is uphill, as the reactant eventually outnumbers the product by about 2:1. Because one of the products of reaction 1 is the reactant of reaction 2, the reactions are coupled—they share an *intermediate compound*, glucose-6-phosphate. The table, a molecular scorecard, shows these coupled reactions set in motion by the combination of 10^5 molecules of glucose and an equal number of molecules of ATP (*a*). The equilibrium constant of reaction 1 (610) means that the reaction will proceed until a

ratio of roughly 99,836 molecules of glucose-6-phosphate to 164 of glucose has been reached (*b*). Since glucose-6-phosphate is a reactant of reaction 2, some of it will be converted into fructose-6-phosphate. The equilibrium constant of this reaction is 0.54, so about 35,007 molecules of product will then be produced (*c*).

With the consequent reduction in the total amount of glucose-6-phosphate, reaction 1 is no longer at equilibrium, so additional glucose and ATP react to produce glucose-6-phosphate (*d*), only to have some of this intermediate promptly converted into fructose-6-phosphate (*e*). At this point equilibrium is reached, with the ratios of product to reactant for both reactions 1 and 2 approximating the corresponding equilibrium constants. The addition of more glucose and ATP (which happens almost continuously in cells) or the removal of fructose-6-phosphate (which also happens continuously, as it is converted into another compound in the third step of glycolysis) will cause more molecules to move along this reaction pathway (*f–j* and *k–n*).

Thus favorable and unfavorable reactions are coupled in the cell: reactants are pushed through the uphill steps as long as appropriate intermediate compounds are present and the *net* ΔG is negative. In fact, cells function by coupling literally thousands of reactions into long chains in this manner.

REACTANTS	REACTION 1 ($K_{eq} = 610$)	INTERMEDIATE COMPOUND	REACTION 2 ($K_{eq} = 0.54$)	PRODUCT
glucose + ATP	$\longrightarrow$	glucose-6-phosphate	$\longrightarrow$	fructose-6-phosphate
a 100,000		0		0
b 164		99,836		0
c 164		64,829		35,007
d 106		64,887		35,007
e 106		64,866		35,028
Addition of 10^4 molecules of glucose and 10^4 molecules of ATP				
f 10,106		64,866		35,028
g 123		74,849		35,028
h 123		71,349		38,528
i 117		71,355		38,528
j 117		71,353		38,530
Removal of 10^4 molecules of fructose-6-phosphate				
k 117		71,353		28,530
l 117		64,857		35,026
m 106		64,868		35,026
n 106		64,864		35,030

The free energy is then utilized in the formation of ATP. Moreover, the pyruvic acid (which still has 590 kcal/mole of free energy at Step 10) can be broken down to yield additional energy for the synthesis of still more ATP. (If lactic acid has already been formed, it can be reconverted into pyruvic acid—thus regenerating the lost NAD_{re}—when sufficient oxygen becomes available.)

The process of aerobic breakdown of nutrients with accompanying synthesis of ATP is called *aerobic respiration*. Whereas anaerobic respiration comprises fermentation (glycolysis, followed by the transfer of electrons to pyruvic acid, producing lactic acid or ethanol), aerobic respiration consists of a longer sequence of events: glycolysis (Stage I of aerobic respiration), oxidation of pyruvic acid to acetyl CoA, the reactions of the Krebs citric acid cycle, a series of reactions involving an electron-transport chain, and finally the processes that culminate in the synthesis of ATP. It is the transfer of electrons from NAD_{re} to highly electronegative oxygen atoms—a transfer mediated by the electron-transport chain—that ultimately produces most of the ATP from glucose.

Oxidation of pyruvic acid (Stage II of aerobic respiration) The net effect of the aerobic oxidation of pyruvic acid is to break down the three-carbon pyruvic acid to CO_2 and the two-carbon compound acetic acid, which is connected by a high-energy bond to a coenzyme called CoA; the complete compound is acetyl-coenzyme A, or *acetyl-CoA*. When a molecule of pyruvic acid is oxidized to acetyl-CoA and CO_2, hydrogen is removed and a molecule of NAD_{re} is formed. Since two molecules of pyruvic acid were formed from each glucose molecule, two molecules of NAD_{re} are formed here. This complicated series of reactions can be summarized by the following equation:

$$2 \text{ pyruvic acid} + 2 \text{ CoA} + 2 \text{ NAD}_{ox} \longrightarrow$$

$$2 \text{ acetyl} \sim \text{CoA} + 2CO_2 + 2 \text{ NAD}_{re} + 2H^+$$

The Krebs cycle (Stage III of aerobic respiration) The acetyl-CoA is next fed into a complex series of reactions called the Krebs citric acid cycle, after the British scientist Sir Hans Krebs, who was awarded a Nobel Prize for working out this system. Each of the two two-carbon acetyl-CoA molecules formed from one molecule of glucose is combined with a four-carbon compound (oxaloacetic acid) already present in the cell, to form a new six-carbon compound called *citric acid*. Each of the citric acid molecules is then oxidized to a five-carbon compound plus CO_2. The five-carbon unit, in turn, is oxidized to a four-carbon compound plus CO_2. This four-carbon compound is then converted into the four-carbon compound: oxaloacetic acid. It can now pick up more acetyl-CoA, forming new citric acid and beginning the cycle again (Fig. 6.7). The cycle is simplified in Figure 6.8.

We see, then, that two carbons are fed into the Krebs cycle as the acetyl group, and two are released as CO_2. Since each glucose molecule being oxidized yields two molecules of acetyl-CoA, two turns of the cycle are required, and four carbons are released as CO_2 during this stage of glucose breakdown. With the two carbons already released as CO_2 during the oxidation of pyruvic acid to acetyl-CoA, all six carbons of the original glucose are accounted for.

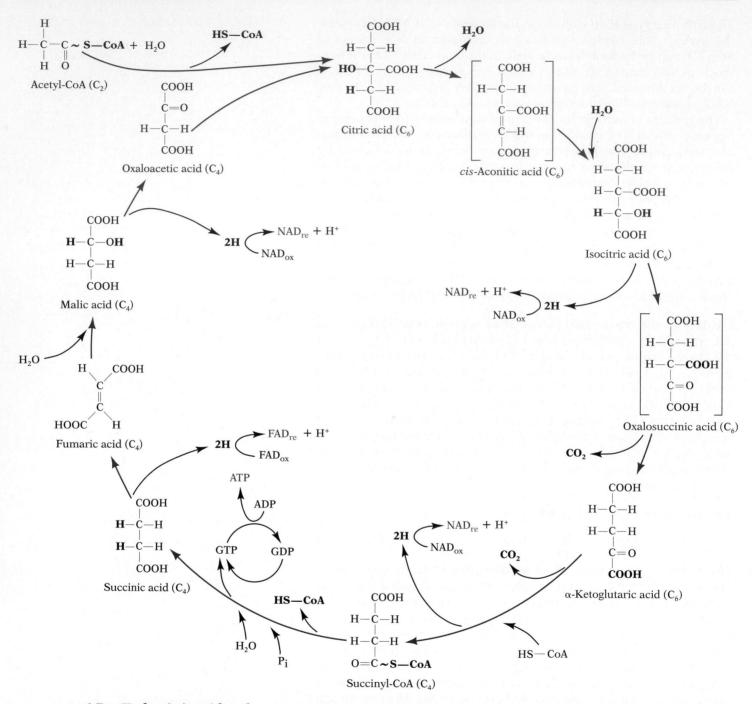

6.7 Krebs citric acid cycle

The complete cycle is shown here only to illustrate the characteristic complexity of metabolic pathways; there is no point in trying to learn all the reactions involved. As shown at upper left, the acetyl group (two carbons) from acetyl-CoA enters the cycle by combining with oxaloacetic acid (four carbons) to form citric acid (six carbons). During subsequent reactions two of the carbons are removed as CO_2 (between oxalosuccinic acid and α-ketoglutaric acid, and between α-ketoglutaric acid and succinyl-CoA), and a total of eight hydrogens are removed. These hydrogens are picked up by NAD (or by the related acceptor molecule FAD). One molecule of ATP is synthesized (bottom). Finally, oxaloacetic acid is regenerated and can combine with a new acetyl group to start the cycle over again. The cycle is completed twice for each molecule of glucose oxidized. (The atoms removed at each step are shown in boldface in the structural formulas. The two substances in brackets—*cis*-aconitic acid and oxalosuccinic acid—are enzyme-bound intermediates that seldom exist as free compounds.)

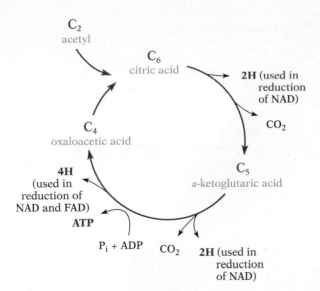

C₂
acetyl

C₆
citric acid

2H (used in reduction of NAD)

CO₂

C₄
oxaloacetic acid

4H (used in reduction of NAD and FAD)

ATP

C₅
a-ketoglutaric acid

Pᵢ + ADP   CO₂   2H (used in reduction of NAD)

6.8 Simplified version of the Krebs citric acid cycle

The two carbons of the acetyl group combine with a four-carbon compound to form citric acid, a six-carbon compound. Removal of one carbon as CO_2 leaves a five-carbon compound. Removal of a second carbon as CO_2 leaves a four-carbon compound, which can combine with another acetyl group and start the cycle over again. In the course of the cycle, one molecule of ATP is synthesized and eight hydrogens are released, which are used in the reduction of NAD and FAD. Since one molecule of glucose gives rise to two acetyl units, two turns of the cycle occur for each molecule of glucose oxidized, with production of four molecules of CO_2, two molecules of ATP, and 16 hydrogens.

The oxidative breakdown of each molecule of acetyl-CoA via the Krebs cycle also involves the removal of eight hydrogens, which are picked up by NAD_{ox} (or by FAD_{ox}, the oxidized form of a related electron-carrier protein called flavin adenine dinucleotide); four units of reduced carrier are thus formed (Fig. 6.7). Since the breakdown of one molecule of glucose leads to two turns of the Krebs cycle, a total of eight molecules of reduced carrier (six NAD_{re} and two FAD_{re}) are formed during this stage of the breakdown of glucose. Two molecules of ATP are also synthesized in two turns of the Krebs cycle.

Figure 6.9 summarizes the yield of ATP, NAD_{re}, FAD_{re}, and CO_2 from the three stages of the breakdown of glucose.

The electron-transport chain (Stage IV of aerobic respiration)
So far we have seen a net gain of only four new ATP molecules (two in glycolysis and two in the Krebs cycle). These represent just a small fraction of the energy originally available in the glucose. Of the remainder, some is liberated during the first three stages (mostly as heat, which is essential for the reactions to proceed); the rest is stored in the high-energy intermediates NAD_{re} and FAD_{re}. Twelve of these molecules are synthesized in the breakdown of each molecule of glucose (Fig. 6.9).

How is this energy used to synthesize ATP? We said earlier that under aerobic conditions the regeneration of NAD_{ox} from NAD_{re} is

6.9 Summary of the most important products of Stages I, II, and III in the complete breakdown of one molecule of glucose

Stage I (glycolysis) begins with the expenditure of two molecules of ATP to activate glucose and produce two molecules of PGAL. The two PGAL molecules are then broken down to two molecules of pyruvic acid, in a process that first pays back the two ATP molecules originally invested and then yields two molecules each of ATP and NAD_{re} (red). Stage II yields two molecules each of CO_2 and NAD_{re}. Stage III, in which the two molecules of acetyl-CoA are fed into the Krebs cycle and further broken down, yields four CO_2, two ATP, six NAD_{re}, and two FAD_{re} molecules. (The H^+ ions liberated in the production of NAD_{re} and FAD_{re} are not shown.)

I (Glycolysis)

gluclose

2 ATP →

fructose-1,6-bisphosphate

2 PGAL

2 ATP ←   → 2 ATP

→ 2 NAD_re

2 pyruvic acid

II

→ 2 CO₂

→ 2 NAD_re

2 acetyl-CoA

III (Krebs cycle)

2 citric acid

→ 4 CO₂
→ 2 ATP
→ 6 NAD_re
→ 2 FAD_re

6.10 Respiratory electron-transport chain

The reactions summarized in this diagram take place on the inner membrane of the mitochondrion. NAD_{re} donates two electrons and a proton to the electron-transport chain; a second hydrogen ion is drawn from the medium. The electrons are passed from one acceptor substance to the next, step by step down an energy gradient from their initial high energy level in NAD_{re} to their final low energy level in H_2O. (When available, molecules of FAD_{re} can also donate their electrons to the electron-transport chain; since these electrons have less energy than those of NAD_{re}, they enter lower down on the chain, at Q.) Each successive acceptor molecule is cyclically reduced when it receives the electrons, and then oxidized when it passes them on to the next acceptor molecule. At three sites along the chain, some of the free energy released is used to pump H^+ ions into the compartment outside the inner membrane. Later the H^+ gradient generated by the electron-transport chain is used in the synthesis of ATP. The electron acceptors are a flavoprotein (FP); coenzyme Q; cytochromes a, a_3, b, b_2, c, and o; and two proteins containing iron and sulfur, FeS_a and FeS_b. For simplicity, steps have been combined wherever possible. A more complete sequence is shown as part of Figure 6.12.

achieved by the passage of electrons from NAD_{re} to O_2, with oxygen thus acting as the ultimate acceptor of electrons:

$$O_2 + 2NAD_{re} + 2H^+ \longrightarrow 2H_2O + 2NAD_{ox}$$

The NAD_{re} does not, however, pass its electrons directly to the oxygen, as this summary equation suggests. The electrons and their associated protons reach their ultimate targets indirectly. In particular, the hydrogen electrons are passed down a "respiratory chain" of electron-transport compounds, many of which are iron-containing enzymes called **cytochromes** (Fig. 6.10). The electrons move to ever-lower energy levels with each transfer. As we will see, the energy extracted in the electron-transport chain is then used for the production of ATP.

The anatomy of respiration The elaborate internal structure of the mitochondrion plays a crucial role in respiration. The extensively folded inner membrane divides each mitochondrion into two compartments (Fig. 6.11). This membrane separates two very different chemical environments and serves as scaffolding for gates,

A

B

500 Å

C

0.5 μm

6.11 Structure of a mitochondrion

(A) A typical animal mitochondrion. (B) In this interpretive drawing, much of the outer membrane has been cut away, and the interior has been sectioned to show how the inner membrane folds into cristae. The mitochondria of metabolically very active cells have more cristae than those of less active cells. The inner compartment is within the inner membrane, while the outer compartment is the space between the two membranes. (C) A close-up of the inner membrane. As shown in Figure 6.12, much of the activity of cellular respiration, such as electron transport, occurs across the inner membrane between the inner and outer compartments. The knoblike structures are the F_1 enzyme complexes responsible for synthesizing ATP; as such, they are often called ATP synthases.

pumps, and organized enzyme arrays. While the enzymes of the Krebs cycle are contained in the fluid of the inner compartment, the enzymes of the electron-transport chain are found in the inner membrane.

The anatomy of the electron-transport chain is shown in Figure 6.12. The major raw material, NAD_{re}, was generated by the sequence of events already described: glycolysis (which takes place outside the mitochondrion in the cytosol) ⟶ pyruvic acid (which is transported through both the outer and inner membranes to the inner compartment of the mitochondrion) ⟶ acetyl–CoA ⟶ Krebs cycle ⟶ NAD_{re} (as well as CO_2, ATP, and FAD_{re}). The electron-transport chain must now harvest this energy.

The high-energy compound NAD_{re} carries its electrons to the inner membrane. Here NAD_{re} is oxidized by losing a hydrogen and an electron to a flavoprotein (FP), which is reduced; simultaneously, FP accepts a H^+ ion from the medium of the inner compartment. The hydrogen from NAD_{re} is split into a hydrogen ion (H^+) and an electron (e^-). The H^+ ion, along with the H^+ accepted from the medium, is deposited in the outer compartment; the two elec-

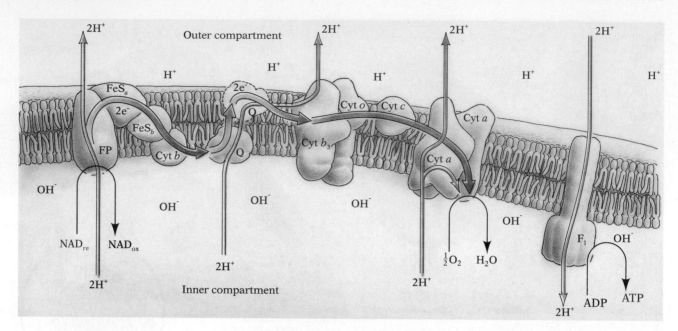

6.12   Anatomy of Stages IV and V of aerobic respiration

Glycolysis (not shown) takes place in the cytosol of the cell, and supplies pyruvic acid, which is converted into acetyl-CoA in the inner compartment of the mitochondrion. The Krebs cycle (not shown) operates in the inner compartment, producing CO_2, ATP, NAD_{re}, and FAD_{re}. Mounted in the inner membrane of the mitochondrion, which separates the inner and outer compartments, are the enzymes of the electron-transport chain. They extract the energy from NAD_{re} and FAD_{re}, and use it to pump H^+ ions from the inner to the outer compartment. Some of the ions are carried across by Q, which shuttles between the inner and outer surface of the membrane. The energy of the resulting electrochemical gradient is then used by the enzyme complex F_1 (right) to make ATP from ADP. The ATP is then exported from the cytoplasm (not shown). Approximately two H^+ ions must pass through the F_1 complex to create one ATP from ADP. Since the energy of the two electrons donated by each NAD_{re} molecule is used to transport six H^+ into the outer compartment, the oxidation of an NADre produces about three ATPs. Two electrons from a molecule of FAD_{re}, which have less energy than those of NAD_{re}, can enter the electron-transport chain at Q; the oxidation of the lower-energy compound FAD_{re} produces about two ATPs.

trons are passed immediately to another enzyme in the sulfur-containing complex FeS_a. The result is that the H^+ concentration of the outer compartment is raised, the H^+ concentration of the inner compartment is lowered, high-energy-level electrons (two from each NAD_{re}) are inserted into the transport chain (center) where they will be used to do work, and NAD_{ox} is returned to the citric acid cycle to be "recharged."

The transfer of electrons from one acceptor to another proceeds because free energy is liberated (that is, lost by electrons) at each step. For each electron reaching enzyme Q (which shuttles back and forth across the membrane), another H^+ enters the chain from the inner compartment and is deposited in the outer compartment. (At this stage, the lower-energy electrons of FAD_{re} can also enter the electron-transport chain.) The electrons, whatever their origin, now move through the cytochrome series (left) to cytochrome a_3. This last enzyme group uses the energy of the electrons to split molecular oxygen (O_2) and catalyze the following reaction:

$$1/2 O_2 + 4H^+ + 2e^- \xrightarrow{\text{cytochrome } a_3} H_2O + 2H^+$$

This reaction reduces by four the number of hydrogen ions in the inner compartment. Two of these hydrogen ions are exported to the outer compartment, while the other two are incorporated into the water molecule. About 100 electrons can pass through each transport chain each second.

Chemiosmotic synthesis (Stage V of aerobic respiration) All the energy stored in the NAD_{re} has now been used up without generating any new ATP. However, an electrostatic and osmotic concentration gradient has built up across the inner membrane: the inner compartment, having lost H^+, has become negatively charged, while the outer compartment—the space between the inner and outer membrane—is filled with H^+ ions and therefore is positively charged. The result is a molecular battery that uses the difference in charge across the inner mitochondrial membrane to make ATP from ADP. Most of the H^+ ions return from the outer to the inner compartment by way of the F_1 complex (Fig. 6.12), passing down a ***chemiosmotic gradient***. As they do so, the energy of this electrochemical gradient—the combined osmotic and electrostatic gradients—is harvested (Figure 6.13). The abundant energy generated by this elegant system is stored in the phosphate groups of ATP. The same H^+ gradient probably also powers the pump that exports ATP to the cytosol. Peter Mitchell won the Nobel Prize in 1978 for working out this indirect route by which mitochondria convert the energy of NAD_{re} into ATP.

It is a continuing challenge to explain how such efficient, highly refined examples of biological engineering as the electron-transport chain might have evolved. How do the first elements in such a complicated array, which seem useless without the rest, ever come into existence? To answer the question, biologists look at a wide variety of species for preadaptations—fragments of the system as we know it now that are employed in other ways and which might have been present in ancestral organisms and available to natural selection for "remodeling."

Consider the most intricate element of the mitochondrial system, the F_1 complex, which uses the H^+ gradient to make ATP. Relict species of anaerobic bacteria living today still lack an electron-transport chain but nevertheless have an F_1 assembly that works in reverse. These anaerobes burn ATP to pump H^+ out of their cells, eliminating the acidic by-products of fermentation. There is similar evidence to suggest that the cytochrome complexes evolved later in ancient bacteria, where they served similar detoxification functions. Natural selection must then have favored combining these various systems into a rudimentary electron-transport chain using sulfur, carbon, or nitrogen as the ultimate electron acceptor.

Summary of respiration energetics Now that we have looked at all five stages in the aerobic breakdown of glucose, we can summarize their combined energy yield. The overall flow of energy in aer-

6.13 Operation of the F1 complex

The F_1 complex is composed of at least 18 subunits. The core group carries the H^+ to one of three protein pairs of the head region where ATP is formed; each of the three pairs is undergoing one of the three steps in the cycle: ADP and phosphate binding, ATP formation, and ATP release. Many researchers believe that the head rotates during ATP synthesis.

HOW WE KNOW:

TESTING THE MITCHELL HYPOTHESIS

Convincing evidence for Peter Mitchell's brilliant chemiosmotic hypothesis comes from several researchers, including (besides Mitchell) Efraim Racker, Andre T. Jagendorf, and Peter C. Hinkle. Much of their research employs artificially reconstructed vesicles. The technique takes advantage of the tendency of phospholipids in an aqueous solution to form spherical liposomes. When mitochondria are exposed to ultrasound (sound of very high frequency), both the outer and inner membranes break into sheetlike fragments; the inner membranes subsequently "heal" by forming spheres about 100 nm (nanometers) in diameter, as shown in the accompanying figure. These submitochondrial vesicles display 9-nm spheres that are thought to be parts of the ATP-synthesizing F_1 complexes. Researchers disrupt the mitochondria, let the submitochondrial vesicles form in one solution—trapping that chemical milieu inside the vesicles—and then transfer the vesicles to a different chemical environment. In this way, the chemistry of the two sides of the membrane can be altered at will.

According to Mitchell's chemiosmotic battery model, in a normal mitochondrion the necessary gradient exists because the H^+ concentration is higher on the outside of the inner membrane than on the inside. If the model is correct, the inside-out patches of membrane should function when the H^+ concentration is higher on the inside. And in fact, when the vesicles are created with more H^+ inside than out, they begin producing ATP, but when the H^+ concentration is higher on the outside, they do nothing. The model also predicts that if these vesicles are supplied with NAD_{re} on the outside, they will increase the gradient by raising the H^+ concentration on the inside. This too happens. In addition, the model predicts that the insertion into these vesicles of channels permeable to H^+ should short-circuit the battery by allowing H^+ to cross the membrane freely, and should thereby prevent ATP synthesis. Again, this is exactly what is observed.

inner membrane   outer membrane

A   normal mitochondrion

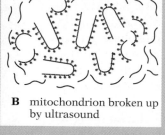

B   mitochondrion broken up by ultrasound

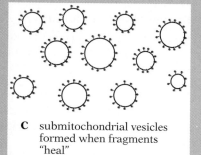

C   submitochondrial vesicles formed when fragments "heal"

1 μm

Formation of submitochondrial vesicles

Submitochondrial vesicles are made by breaking the mitochondrion (A) into pieces with ultrasound (B), and then allowing the bits to "heal" (C). By controlling the medium in which vesicle formation takes place, and the solution into which the resulting inside-out membrane spheres are then put, the researcher can create virtually any desired combination of internal and external chemical environments.

obic respiration is summarized in Figure 6.14. As we have seen, glycolysis generates two molecules of ATP per molecule of glucose, along with two NAD_{re}; the conversion of pyruvic acid into acetyl-CoA yields another two NAD_{re}; the Krebs citric acid cycle produces an additional two ATPs, six NAD_{re}, and two FAD_{re}; finally, and most critically, the charging of the mitochondrial battery—the expulsion of H^+ across the inner membrane by means of the electron-transport chain (driven by oxidation of the NAD_{re} and FAD_{re})—stores enough energy to synthesize roughly another 28 ATPs. Each NAD_{re}

glucose

2 ATP ⟶

fructose-1,6-bisphosphate

STAGE I
Glycolysis

2 TGAL

2 ATP ⟵   ⟶ 2 ATP

2 NAD$_{re}$ ⟵

2 pyruvic acid

STAGE II
Pyruvic acid
to acetyl-CoA

⟶ 2 CO$_2$

2 NAD$_{re}$ ⟵

2 acetyl-CoA

2 citric acid

STAGE III
Krebs cycle

⟶ 4 CO$_2$

6 NAD$_{re}$ ⟵   2 ATP

2 FAD$_{re}$ ⟵

STAGE IV
Respiratory
electron-
transport
chain

6 O$_2$ ⟶   ⟶ 12 H$_2$O

H$^+$ ⟵

STAGE V
Chemiosmotic
synthesis
of ATP

F$_1$
complex ⟶   28 ATP

6.14 The ATP yield from complete breakdown of glucose to carbon dioxide and water

In the first three stages, four molecules of ATP are directly synthesized, along with ten of NAD$_{re}$ and two of FAD$_{re}$. In Stage IV the NAD$_{re}$ and FAD$_{re}$ are fed to the electron-transport chain, and their stored energy is utilized to create a difference in charge across the inner membrane of the mitochondrion. Finally, in Stage V the F$_1$ enzyme complex uses the energy of this chemiosmotic gradient across the membrane to make approximately 28 more molecules of ATP.

molecule provides the energy to pump enough H$^+$ ions for the synthesis of almost three ATPs (10NAD$_{re}$ ⟶ 25ATP), while each FAD$_{re}$ pumps enough H$^+$ to power the synthesis of two ATPs (2FAD$_{re}$ ⟶ 3ATP).

Now it is clear why atmospheric oxygen is so important to life as we know it: aerobic respiration extracts 16 times as much energy from glucose as does anaerobic metabolism (glycolysis followed by fermentation). Metabolic poisons like cyanide (which binds irreversibly with a cytochrome and thereby blocks the electron-transport chain) are fatal because the cell is suddenly denied 92% of its normal energy. However, parts of our bodies *can* operate anaerobically for short periods. During intensive exercise, for instance, muscles often need so much energy that the oxygen supplied from breathing is insufficient. In such cases glycolysis and fermentation provide the needed energy for a time, but they are inefficient and fatigue soon results. Later, the oxygen debt is paid back by deep breathing or panting, and the lactic acid that accumulated in the muscles as a result of fermentation is removed to the liver and converted back into glucose.

We can calculate the overall efficiency of aerobic respiration: a molecule of glucose has a free energy of about 670 kcal/mole, while a molecule of ATP stores about 7.3 kcal/mole. Since the 32 ATPs that are generated therefore represent about 234 kcal/mole, the cell has retained 35% of the energy originally stored in the glucose; the other 65% is released, primarily as heat. This liberated energy is essential to shuttling the reactants smoothly through the various chemical chains. Moreover, some of the inevitable "waste" heat can be used to raise an organism's internal temperature so that the chemistry of life runs more efficiently.

■ HOW FATS AND PROTEINS ARE METABOLIZED

Cells can extract energy in the form of ATP, not only from the carbohydrates we have focused on so far but also from the two other major categories of nutrients: fats and proteins. As Figure 6.15 shows, early steps in the breakdown of fats and proteins create products that can be fed into the enzyme pathways we have already discussed.

Catabolism of fats begins with their hydrolysis to glycerol and fatty acids. The glycerol (a three-carbon compound) is then converted into PGAL and fed into the glycolytic pathway at the point where PGAL normally appears. The fatty acids, on the other hand, are broken down to a number of two-carbon fragments, which are converted into acetyl-CoA and fed into the respiratory pathway farther along. Since fats are more completely reduced compounds than carbohydrates (that is, they have a higher proportion of hydrogens), their full oxidation yields more energy per unit weight; 1 g of fat yields more than twice as much energy as 1 g of carbohydrate.

The amino acids produced by hydrolysis of proteins are catabolized in a variety of ways. After removal of the amino group (deamination) in the form of ammonia (NH_3), some amino acids are converted into pyruvic acid, some into acetyl-CoA, and some into one or another compound of the citric acid cycle. Complete oxidation of a gram of protein yields roughly the same amount of energy as that of a gram of carbohydrate.

Such compounds as pyruvic acid, acetyl-CoA, and the compounds of the citric acid cycle, which are common to the catabolism of several different types of substances, not only play a crucial role in the oxidation of energy-rich compounds to carbon dioxide and water but also function in the anabolism of amino acids, sugars, and fats. They serve as biochemical crossroads, at which several enzyme pathways intersect. By investing energy, the cell can reverse the direction in which substances move along some of these pathways; for example, the PGAL and acetyl-CoA produced at different points in the breakdown of carbohydrate can be moved up the pathways to glycerol and fatty acids, for use in building fats. Similarly, many amino acids can be converted into carbohydrate via the common intermediates in their metabolic pathways. Not all pathways are two-way, however; most higher animals lack a reversible pyruvic acid–to–acetyl-CoA enzyme, as well as any alternative enzyme, and so cannot convert fatty acids into carbohydrate.

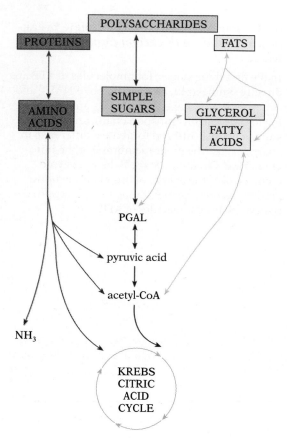

6.15   Relationships of the catabolism of proteins and fats to the catabolism of carbohydrates

CHAPTER SUMMARY

THE FLOW OF ENERGY

HOW METABOLISM EVOLVED Early organisms harvested energy from inorganic molecules. Later they evolved the ability to store energy as glucose for later use. The evolution of photosynthesis allowed massive conversion of low-energy inorganic molecules into high-energy organic compounds. (p. 153)

HOW ENERGY IS TRANSFERRED IN REACTIONS Electron energy is transferred in chemical reactions; molecules losing energy are oxidized, while those gaining energy are reduced. (p. 156)

ATP: THE CURRENCY OF ENERGY EXCHANGE Most energy transfers between biochemical pathways in the cell are made via ATP. (p. 157)

HOW CELLS HARVEST ENERGY

RESPIRATION WITHOUT OXYGEN During glycolysis the cell harvests some of the energy in glucose, storing it in ATP and NAD_{re}. In the absence of oxygen, NAD is oxidized during fermentation, and its energy is lost. (p. 159)

RESPIRATION WITH OXYGEN In the presence of oxygen, the pyruvic acid produced by glycolysis is oxidized and enters the Krebs cycle, which extracts additional energy and generates CO_2. Much of the extracted energy is in the form of high-energy electrons in NAD and FAD, which enter the electron-transport chain where they create a strong electrochemical gradient in the mitochondrion. The energy in this gradient is then used to synthesize ATP. Aerobic respiration extracts 16 times more energy from glucose than does fermentation. (p. 165)

HOW FATS AND PROTEINS ARE METABOLIZED Fats and proteins are metabolized by being converted into molecules found at some step in aerobic respiration; they enter the pathway at these points. (p. 176)

STUDY QUESTIONS

1 What is the source of energy in the glucose molecule that is extracted in fermentation? How exactly can you localize it? (pp. 159–165)

2 In the absence of oxygen, which element is the best to use as an electron acceptor in order to extract the most energy: carbon, nitrogen, or sulfur? Why? (p. 155)

3 Why are there two separate entry points into the electron-transport chain, one for electrons from NAD and one for electrons from FAD? (pp. 170–172)

4 What is the purpose of the space between the inner and outer membranes of the mitochondrion? Why not have just one membrane, as the procaryotic precursor of this organelle must have had? (pp. 169–173)

5 Why does reaction coupling work? How does it help explain the operation of fermentation? (pp. 159–166)

SUGGESTED READING

CLOUD, P., 1983. The biosphere, *Scientific American* 249 (3). *On the combined evolution of the earth, life, and the atmosphere, with particular emphasis on the role of oxygen concentration.*

HINKLE, P. C., AND R. E. MCCARTY, 1978. How cells make ATP, *Scientific American* 238 (3). *A difficult but rewarding explanation of how ATP is made.*

MORAN, L. A., K. G. SCRINGEOUR, H. R. HORTON, R. S. OCHS, AND J. D. RAVEN, 1994. *Biochemistry*, 2nd ed. Prentice Hall, Englewood Cliffs, N.J. *Traces in great detail the biochemical pathways of respiration, and their regulation.*

TUNNICLIFFE, V. 1992. Hydrothermal-vent communities of the deep sea. *American Scientist* 80 (4). *How chemosynthetic bacteria manage to extract energy from sulfurous 250°C water at deep sea vents, and support an exotic ecosystem of other creatures.*

CHAPTER 7

PHOTOSYNTHESIS

The earliest forms of life on earth survived by metabolizing high-energy inorganic molecules. Without oxygen, these anaerobic species could harvest only a tiny fraction of the available energy, and they were also in danger of extinction when their limited supply of inorganic food ran out. For eons, life on earth was a marginal phenomenon.

About 3 billion years ago, prospects suddenly changed: some primitive organisms evolved the ability to capture the sun's energy. Photons energized the electrons in special pigments, and these activated molecules passed their energy to other molecules, leading eventually to the synthesis of glucose and other organic molecules (Fig. 7.1).

This primitive form of photosynthesis—the transformation of light into chemical energy—freed organisms from their dependence on inorganic food. Further evolution resulted in a more efficient form of photosynthesis that also produced molecular oxygen as a by-product. As oxygen began to accumulate in the atmosphere 1.5–2 billion years ago, its high electronegativity allowed high-efficiency aerobic respiration to become the main mechanism for extracting energy from food. Thus the two major chemical path-

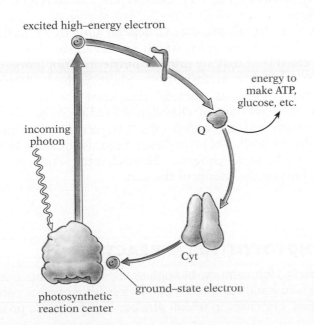

7.1 Energetic basis of life

The basic energy equation of life begins when a photon from the sun excites an electron in a pigment molecule by moving the electron up to a higher level. The energy thus acquired ultimately excites an electron in an acceptor molecule. Subsequently, as the excited electron is returned to a low-energy level in a pigment molecule by way of a series of acceptors, its free energy is used for generating an energy-storage compound (usually glucose or ATP). The pathway shown here is the oldest and most basic form of photosynthesis.

ENERGY PRODUCTION   METABOLISM

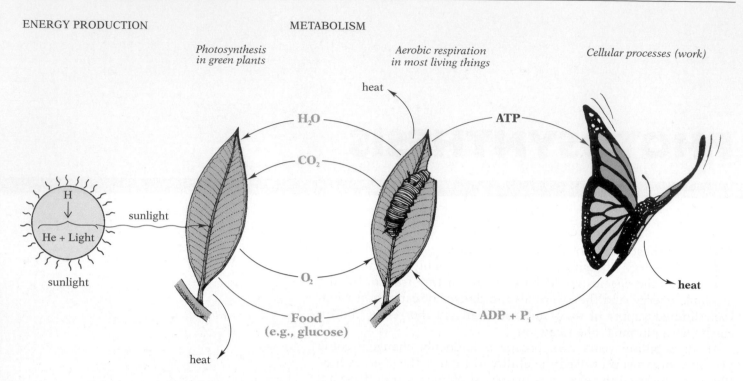

Photosynthesis in green plants   *Aerobic respiration in most living things*   *Cellular processes (work)*

7.2 Energy flow in aerobic pathways

With the evolution of the modern form of photosynthesis approximately 2.3 billion years ago, oxygen, the electronegative element essential to the efficient operation of catabolic processes, began to accumulate in the atmosphere. This led (about 1.5 billion years ago) to the spread of organisms capable of high-efficiency aerobic respiration, and to the unification of the two great metabolic pathways—photosynthesis and respiration—that is characteristic of life processes today. The pathways are unified by shared products and by-products: organic food, H_2O, CO_2, and O_2. Red arrows trace the flow of energy. Note that both the caterpillar and the leaf respire.

ways—photosynthesis and respiration—were joined through shared by-products (Fig. 7.2). Life became the dominant feature of the earth.

Today, virtually all organisms depend directly or indirectly on photosynthesis. *Autotrophs*, which are organisms (such as plants) that are capable of making organic nutrients from inorganic materials, depend almost exclusively on photosynthesis. *Heterotrophs* are organisms (such as animals) that must obtain organic nutrients from the environment; they depend indirectly on photosynthesis since they consume autotrophs, heterotrophs that have eaten autotrophs, or both. Photosynthesis, then, is life's single most important biochemical process. This chapter explores how autotrophs harvest the energy of the sun.

THE PHOTOSYNTHETIC REACTION

By the early 19th century, biologists knew that plants use light to combine water and carbon dioxide to generate organic compounds and oxygen. For cases in which glucose is the organic product, the

HOW WE KNOW:
THE INGREDIENTS OF PHOTOSYNTHESIS

For most of recorded history scientists had no idea that the sun supplies the surface of the earth with virtually all its energy or that green plants trap that energy and produce the invisible gas we breathe. Only in 1772 did the English clergyman Joseph Priestley demonstrate that plants affect air in such a way as to reverse the effects of burning or of breathing. As he reported it:

> One might have imagined that, since common air is necessary to vegetable, as well as to animal life, both plants and animals had affected it in the same manner, and I own I had that expectation, when I first put a sprig of mint into a glass-jar, standing inverted in a vessel of water; but when it had continued growing there for some months, I found that the air would neither extinguish a candle, nor was it at all inconvenient to a mouse, which I put into it.
>
> Finding that candles burn very well in air in which plants had grown a long time, and having had some reason to think, that there was something attending vegetation, which restored air that had been injured by respiration, I thought it was possible that the same process might also restore the air that had been injured by the burn-

ing of candles.

> Accordingly, on the 17th of August, 1771, I put a sprig of mint into a quantity of air, in which a wax candle had burned out, and found that, on the 27th of the same month, another candle burned perfectly well in it. This experiment I repeated, without the least variation in the event, not less than eight or ten times in the remainder of the summer.

Priestley was the first to show that plants produce oxygen, though he did not realize that this was happening. Seven years later, the Dutch physician Jan Ingenhousz demonstrated the necessity of sunlight for oxygen production, and he also showed that only the green parts of the plant could photosynthesize. Just five years later a Swiss pastor, Jean Senebier, showed that the process depended on a gas that he called "fixed air" (carbon dioxide). Finally, in 1804, another Swiss researcher, Nicolas Théodore de Saussure, found that water is necessary for the photosynthetic production of organic materials. Thus, in a span of only 32 years, the qualitative formula of photosynthesis was worked out:

$$\text{light} + \text{water} + \text{carbon dioxide} \longrightarrow \text{organic matter} + \text{oxygen}$$

formula of the most important reaction for life on earth is:

$$6CO_2 + 12H_2O + \text{light} \longrightarrow 6O_2 + C_6H_{12}O_6 + 6H_2O$$
$$(\Delta G = -1300 \text{ kcal/mole})$$

What the formula does not reveal, however, is how this reaction actually occurs—that is, how a special combination of enzymes and cellular architecture manages to harvest energy from light. Nor does it explain why water appears on both sides of the equation. In fact, it does not even tell us whether the oxygen we breathe comes from H_2O or CO_2.

For many years, most scientists believed that light energy splits carbon dioxide (CO_2), and that the carbon is then combined with water (H_2O), to form the group —CH_2O, from which carbohydrates like glucose ($C_6H_{12}O_6$) are in turn constructed. According to this view, the oxygen released by the plant during photosynthesis comes from CO_2. In 1930, however, C. B. van Niel showed that photosynthetic bacteria that use hydrogen sulfide (H_2S) instead of water as a raw material for photosynthesis, give off sulfur instead of oxygen as a by-product. It seemed clear that these bacteria must be using sulfur from H_2S as their electron acceptor instead of the oxygen from CO_2; thus the oxygen produced by plants during photosynthesis must come from H_2O rather than CO_2. This supposition was later proven by using a heavy isotope of oxygen (^{18}O instead of the usual ^{16}O) to label the oxygen in water; ^{18}O$_2$ rather than ^{16}O$_2$ was liberated. Thus we can now trace the fates of all the atoms involved in generating the basic organic building block —CH$_2$O (shown below in parentheses because it is a component rather than an independent molecule):

$$CO_2 + 2H_2O + light \xrightarrow{chlorophyll} O_2 + (-CH_2O) + H_2O$$

The reason water appears on both sides of the equation is that the water produced by the photosynthetic process is new; it is not the water used as a raw material.

Now that we have sorted out the fates of the various atoms, we can look at how the energy for making —CH$_2$O is harvested. The process occurs in two basic steps: "light" reactions convert and store the energy from light in specialized energy-transfer molecules like ATP, while "dark" reactions use that stored energy to convert (or fix) carbon dioxide into carbohydrates such as glucose.

HOW THE LIGHT REACTIONS STORE LIGHT ENERGY

Light is used to drive two processes: the temporary storage of energized electrons in electron acceptors and the phosphorylation of energy-storage compounds (usually ADP). The term ***photophosphorylation*** is often used to describe the light-dependent conversion of ADP into ATP:

$$ADP + P_i + energy \xrightarrow{enzyme} ATP + H_2O$$

■ HOW CHLOROPHYLL CAPTURES THE ENERGY OF LIGHT

Visible light constitutes one small region of the spectrum of electromagnetic radiation (Fig. 7.3). Each form of radiation in this spectrum has a specific wavelength and energy content. These two characteristics are inversely related: the longer the wavelength, the smaller the energy content. Within the narrow band visible to humans, the shortest light waves produce the sensation of violet and

7.3 Portion of the electromagnetic spectrum

Light visible to humans constitutes only a very small portion of the total spectrum. Within the visible spectrum, light of different wavelengths stimulates different color sensations in us. Besides vision and photosynthesis, other radiation-dependent biological processes rely on this same small portion of the electromagnetic spectrum (sometimes extended into the ultraviolet or infrared). Light of wavelengths shorter than about 300 nm is absorbed by the atmosphere, while wavelengths longer than 1000 nm have too little energy to drive biological reactions.

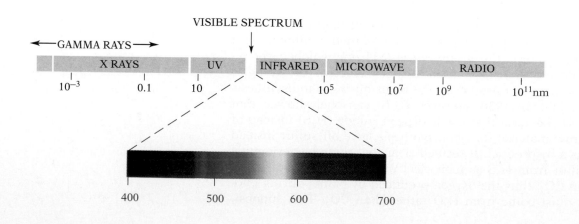

7.4 Molecular structure of chlorophyll

The structure of chlorophyll *a* is shown. The structure of the other major type of chlorophyll in green plants, chlorophyll *b*, differs only in the side group shown in the inset: a formyl group (—CHO) is substituted for the methyl group (—CH$_3$) of the *a* form. The electron that is excited by light energy is in the "head" region, near the magnesium atom. Note the similarity of chlorophyll's prosthetic group to the heme of hemoglobin.

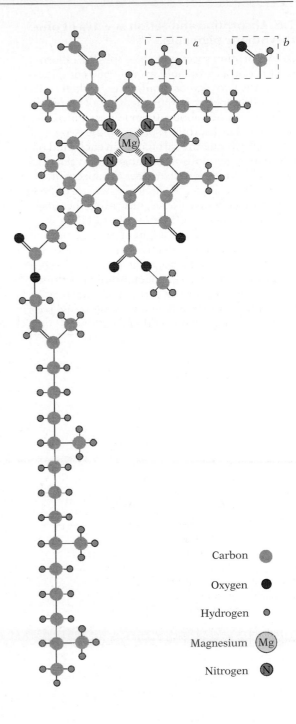

Carbon ●

Oxygen ●

Hydrogen ○

Magnesium (Mg)

Nitrogen (N)

the longest produce the sensation of red. The combination of all visible wavelengths creates the sensation of white. Electromagnetic waves shorter than those of violet light, such as ultraviolet (UV) radiation, X rays, and gamma rays, are invisible to us; waves longer than those of red light—infrared, microwave, and radio or TV rays, for instance—are also invisible.

Not all light is equally effective for photosynthesis. The all-important green pigment ***chlorophyll***, which traps light energy and helps convert it into chemical energy, captures only certain wavelengths. There are several slightly different kinds of chlorophyll; we will look first at the light-capturing ability of the most widespread form: chlorophyll *a* (Fig. 7.4).

Light falling on an object may pass through the object (be transmitted), be absorbed by it, or be reflected from it (Fig. 7.5). We can see transmitted and reflected light, but not absorbed light. Since chlorophyll is green, then it follows that chlorophyll cannot be absorbing much green light or there would be little radiation of that color reaching our eyes. On the other hand, chlorophyll must be absorbing radiation of other wavelengths within the visible part of the spectrum, or the light transmitted or reflected to us would appear white. Thus chlorophyll must be absorbing some light at the red and blue ends of the spectrum.

Chlorophyll can be extracted from a leaf and exposed to light to determine more precisely the amount of absorption at each wave-

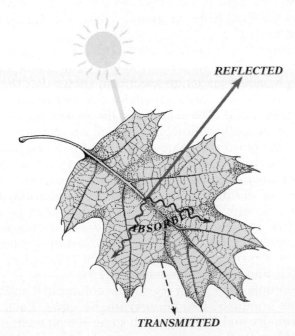

7.5 Light striking a leaf

Light striking an object, such as a leaf, may be reflected, absorbed, or transmitted.

7.6 Absorption and action spectra of photosynthetic pigments

(A) Absorption spectra of chlorophyll *a*, chlorophyll *b*, and a carotenoid from a P680 complex—one of the two pigment complexes in plants. (The absorption spectra in the other type of complex, P700, are shifted about 20 nm to the right.) Taken together, the absorption spectra of the two chlorophylls and the carotenoid in each kind of complex cover more of the range of wavelengths available to the plant than does the spectrum of chlorophyll *a* alone. (B) Action spectrum of P680-dominated photosynthesis, indicating the relative effectiveness of different colors of light in driving photosynthesis. Light of intermediate wavelengths is more effective in driving photosynthesis than would be predicted on the basis of the absorption spectrum of chlorophyll *a* alone. Other pigments, such as carotenoids, absorb light of these intermediate wavelengths and pass the energy to the photosynthetic pathway.

length. The resulting absorption spectrum of chlorophyll *a* (Fig. 7.6A) shows that light in the violet, blue-violet, and red regions is absorbed, while light from the green, yellow, and orange regions is mostly transmitted.

But chlorophyll *a* is not the whole story: if we measure which wavelengths contribute to photosynthesis in an intact leaf, the resulting graph, called an action spectrum, shows that there is high activity in parts of the spectrum where chlorophyll *a* absorbs very little light (Fig. 7.6B). The reason for this difference is that other pigments—principally yellow and orange carotenoids, as well as other forms of chlorophyll—are also present in green plants. These molecules absorb light and then pass the energy to the chlorophyll *a*. **Accessory pigments** like the carotenoids thus enable the plants to use more of the light than is trapped by chlorophyll *a* alone. Most of the chlorophylls and pigments are bound by proteins in a highly ordered structure called a light-harvesting complex (LHC). The most common form, LHC II, has three identical subunits, each consisting of a protein, seven molecules of chlorophyll *a*, five of chlorophyll *b*, and two carotenoids (Fig. 7.7).

In the chloroplasts of green plants, the chlorophyll and accessory pigments are organized into **photosynthetic units**. Each unit contains between five and 10 LHCs, with some 300 pigment molecules, including chlorophyll *a*, another variety of chlorophyll (*b* in green

A   B

7.7 How chlorophyll is mounted in the membrane

The antennas are organized as five to ten protein-bound units called light-harvesting complexes (LHC). Each LHC includes three subunits, viewed here from the stromal side of the membrane (A) and from the plane of the membrane (B). Each of these subunits, in turn, has approximately seven molecules of chlorophyll *a* (blue), five molecules of chlorophyll *b* (green), and several carotenoids (red); all are held in a protein matrix (tan). (The tails of the chlorophylls have been omitted for clarity.)

plants, *c* in brown algae, and *d* in red algae), and carotenoids. One group of four pigment molecules in each unit is distinct from all the rest; it is part of a complex that acts as a ***reaction center*** where energy conversion occurs. The other pigment molecules function somewhat like antennas for collecting light energy and transferring it to the reaction center.

When a photon strikes a chlorophyll (or a carotenoid) molecule and is absorbed, its energy is transferred to an electron of the pigment molecule; the excited electron moves up to a higher, relatively unstable energy level (Fig. 7.8). Not just any photon will do: because electrons occupy discrete energy levels, the photon must have a particular amount of energy to excite the pigment electron from its ground state to the higher level. This explains the well-

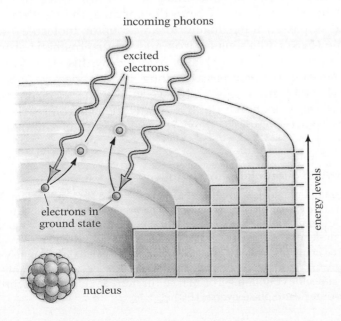

incoming photons

excited electrons

electrons in ground state

nucleus

energy levels

7.8 Effect of light on chlorophyll

When a photon is absorbed by a chlorophyll molecule, the photon's energy raises an electron to a higher energy level. This simplified representation shows a typical distribution of electrons at their lowest available energy levels, and the distribution as it is altered by absorption of a photon of red light, which raises one electron from its ground state in the first energy level to its first excited state in the second energy level. As we saw in Figure 7.6A, the absorption spectrum of a particular pigment can have several peaks. These exist because electrons can be excited from their ground state to their second excited state, two energy levels higher. In this illustration, another electron at the first level is shown being excited up to the third level—its second excited state— by a blue photon, which has enough more energy than a red photon that it can lift electrons two levels rather than one.

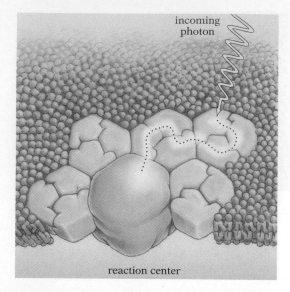

incoming photon

reaction center

7.9 Flow of excited state within a photosynthetic unit

A photon strikes one of the antenna pigments in one of the three subunits of an LHC, raising an electron in the pigment to a higher energy level. This excited state is then passed from one pigment molecule to another in a random sequence (dotted pathway) until it eventually reaches the reaction center where it is trapped.

change in potential energy

incoming photon

FeS

Fd

Cyt b_6

Cyt f

PC

30

20

10

Free energy (kcal/mole)

antenna molecules   P700   reaction center

defined absorption peaks in Figure 7.6A: since the energy of a photon is related to its wavelength, photons with the proper amount of energy come only from light within a certain range of wavelengths (Fig. 7.6B).

An excited and therefore unstable electron will return spontaneously to its inactive state almost immediately, giving up its absorbed energy. Isolated chlorophyll in a test tube, for instance, promptly loses part of the energy it captures by reemitting it as visible light—a process known as fluorescence. (The rest of the energy is lost as heat.) Chlorophyll molecules alone, separated from their photosynthetic units, are incapable of converting light energy into chemical energy. But in the functioning chloroplast, once light energy has raised an electron in an antenna molecule to a high-energy state, the excited state is passed from one pigment molecule to another and eventually reaches the reaction-center complex, which traps it (Fig. 7.9).

The excited state is captured in the reaction center because the free energy of this complex is *lower* than that of the antenna molecules. Hence, once the excited state has reached the reaction center trap, it cannot easily escape. In this molecule, the energized electron that characterizes the excited state is passed quickly to an acceptor molecule and enters a series of redox reactions that convert the energy into a form more readily used by the cell.

■ HARVESTING THE ENERGY OF EXCITED ELECTRONS

There are two general pathways by which the energy from excited electrons is harvested: the cyclic pathway and the noncyclic pathway. The cyclic pathway involves only one of the two types of photosynthetic units found in most plants; the noncyclic pathway involves both. The cyclic pathway is the simpler and more ancient of the two.

Cyclic electron flow and photophosphorylation In cyclic photophosphorylation the reaction center is designated P700 because its chlorophyll absorbs maximally at 700 nm. Once the reaction center has received energy from the antenna, the electron thus activated is passed immediately to an acceptor molecule—a redox enzyme (FeS) containing an iron and sulfur prosthetic group. The acceptor is reduced and P700 is oxidized by this electron transfer. The electron is then passed along by a series of membrane-mounted molecules (Fig. 7.10) very like the electron-transport chain of the mitochondrion. Eventually the electron reaches a molecule of plastocyanin (PC), where it waits until it can fill an open-

7.10 Cyclic photophosphorylation

A photon of light strikes a pigment molecule in the antenna system. The excited state eventually reaches the P700 complex, which is thereby energized. An energized electron from P700 then begins passage from one acceptor molecule to the next, each more electronegative than the previous one. Free energy is released at each step, and the electron ultimately returns to the ground state at which it began in P700. The electron acceptors are FeS, a molecule containing iron and sulfur; ferredoxin (Fd); cytochrome b_6; cytochrome f; and plastocyanin (PC).

ing in the P700 complex. When another reaction-center electron is energized and transferred to FeS, the slot opens and the first electron fills it, thereby completing the cycle.

When a light-energized electron leaves a chlorophyll molecule, it is energy-rich; when it finally returns, it is energy-poor. The transition is gradual. As the electron is passed from transport molecule to transport molecule in the chain, it releases some of its extra energy with each transfer. Hence when the electron finally falls back into the chlorophyll, it has discharged all its extra energy, but it has not released it all at once, as happens with isolated chlorophyll in a test tube. Instead, the energy has been released in a series of small portions of manageable size. The process is called *cyclic photophosphorylation* because the electron is returned to the chlorophyll and some of its energy is used (indirectly, as we will see) to phosphorylate ADP into ATP.

Cyclic photophosphorylation was the first elaborate photosynthetic pathway to evolve, and it is still the dominant form found in bacteria. As Figure 7.10 indicates, however, the cyclic system is not very efficient: of the 25 kcal/mole of energy gained when P700 is excited by a photon, only the energy liberated in the passage from cytochrome b_6 to cytochrome f—3.4 kcal/mole—is actually available to the cell. The energy released in the other steps is wasted. Though 3.4 kcal/mole is better than nothing (particularly since photons are free), most photosynthesis now follows a highly modified, noncyclic pathway, which is more efficient under many conditions. In most organisms today cyclic photophosphorylation merely augments the noncyclic pathway.

Noncyclic electron flow and photophosphorylation Like the cyclic process, noncyclic electron flow and photophosphorylation (Fig. 7.11) begins when a photon of light strikes an antenna molecule of chlorophyll and raises an electron to an excited state, which is later trapped by the reaction center P700. As before, an ener-

7.11 Noncyclic photophosphorylation

As in cyclic photophosphorylation, electrons move along the pathway one at a time. However, for convenience in illustrating reactions at the two ends of the noncyclic pathway, the passage of a pair of electrons is shown throughout the diagram. One of the two essential light events occurs when a photon of light strikes a pigment molecule in Photosystem II; the resulting excited state eventually reaches P680, which donates an energized electron to substance Q. The electron moves down the Photosystem II electron-transport chain from Q to PQ to cytochrome f to PC; as in cyclic photophosphorylation, the energy released in the passage from PQ (plastoquinone) to cytochrome f is used by the cell (by the chemiosmotic process described later). At PC the electron waits to fill an opening in the P700 complex, which occurs when a photon strikes Photosystem I and P700 transfers an energized electron to the Photosystem I electron-transport chain. This electron passes from FeS to Fd and then, with the help of the enzyme FAD, to NADP. Energy released in the step down from FAD is used by the cell, and for each two electrons transported, one $NADP_{re}$ is generated to supply energy for carbon fixation. The vacancy in P680 that was created by the passage of an electron to Q is filled by an electron released in the splitting of water (lower left) by a manganese-containing enzyme (Mn), with one molecule of water yielding two electrons, along with two H^+ ions and one atom of oxygen. As indicated by the changes in relative free energy, the electron that has moved through Photosystem II still has much of its original energy when it enters Photosystem I. The primitive cyclic pathway (which connects Fd to PQ via cytochrome b_6) is also shown.

gized electron is then led away from the P700 complex by electron acceptors, the first two being FeS and ferredoxin (Fd). But here the similarity to cyclic photophosphorylation ends. Instead of continuing down the cyclic transport chain, the electron is passed from Fd to a different acceptor molecule: **nicotinamide adenine dinucleotide phosphate,** or **NADP,** which is closely related to the NAD in mitochondria. The antenna molecules and the P700 reaction center, together with the electron-transport chain from FeS to Fd, constitute **Photosystem I**.[1]

Unlike the reduced electron-acceptor molecules in cyclic photophosphorylation, $NADP_{re}$ does not promptly pass along the electrons it receives from the electron-transport chain to another acceptor molecule. Instead, it retains a pair of energized electrons and their associated protons. Eventually $NADP_{re}$ acts as a high-energy electron donor in the reduction of CO_2 to carbohydrate, a process known as carbon fixation. Thus electrons move from the chlorophyll, through an electron-transport chain to NADP, to carbohydrate (via other intermediate compounds).

Since energized electrons from P700 are retained by $NADP_{re}$ and eventually incorporated into carbohydrate, it follows that Photosystem I is left short of electrons. These "electron holes" are filled, indirectly, by electrons derived from water through a second light-driven process. This second light event involves a different type of photosynthetic unit, which, like Photosystem I, contains about 300 molecules of chlorophyll including about 200 molecules of chlorophyll *a*. In this case the reaction-center complex is called P680 because it absorbs light maximally at 680 nm. The antenna molecules and the P680 reaction center, plus its special set of electron-transport molecules, constitute **Photosystem II**.

When light of a proper wavelength strikes a pigment molecule of Photosystem II, the energy is passed around within the photosynthetic unit until it reaches the P680 trap. Excited P680, in turn, donates a high-energy electron to an acceptor, designated Q (Fig. 7.11). Substance Q then passes the electron to a chain of acceptor molecules, which transport the electron, by some of the same steps we saw in cyclic photophosphorylation, down an energy gradient to the electron hole in the P700 molecule of Photosystem I. As the electron moves down the transport chain, some of the energy released along the way is used by the cell indirectly to synthesize ATP (via the chemiosmotic pathway described later).

Thus the electron holes created in Photosystem I by the first light event are refilled by electrons moved from Photosystem II by the second light event. However, this process alone simply shifts the holes from Photosystem I to Photosystem II. This difficulty is resolved when P680 (with the aid of a manganese-containing enzyme which we shall designate simply as Mn) pulls replacement electrons away from water, leaving behind free protons and molecular oxygen:

$$2H_2O \longrightarrow 4e^- + 4H^+ + O_2$$

$$\downarrow$$

$$P680$$

---

[1] Some workers do not consider the electron-transport chain to be part of the photosystem.

HOW WE KNOW:
THE P680 REACTION CENTER

Painstaking research by H. Michel, J. Deienhofer, and R. Huber (for which they shared a Nobel prize) has revealed the structure of a photosynthetic reaction center in purple bacteria. In the details lies the explanation of how the center complex manages to capture and keep electron energy that is transferred freely between the conventional chlorophyll molecules in the antenna. The reconstruction of a P680 chloroplast reaction center shown here is based on the structure of the evolutionary precursor in photosynthetic bacteria; the two are thought to be very similar. A photon is shown activating an antenna chlorophyll in the chloroplast membrane immediately adjacent to the reaction center (1); this molecule transfers its energy to one of the two central reaction-center chlorophylls (2). The energized electron is first moved to a peripheral reaction-center chlorophyll (3) and then to a pheophytin molecule (4); from there it moves immediately to a bound quinone (5), which is the molecule Q of the electron-transport chain. The entire process is then repeated (white arrow), with the resulting high-energy electron being transferred to the same quinone used by the first electron (6). Now two electrons are waiting to leave.

The electrons leave the reaction-center complex by way of the membrane quinone, PQ (7,8). The electrons missing from the P680 chlorophylls are replaced (10) by the enzyme Mn when it splits a water molecule (9). Because the movement of only one electron is pictured here, the water-splitting reaction is shown as involving only half a water molecule. The eight molecules of the reaction-center complex are bound into a unit by a large protein, shown here in outline.

The oxygen from H_2O is released as a gaseous by-product. The protons, as we will see, also play an important role.

To summarize, the electrons involved in the second light event move from water to the P680 reaction center of Photosystem II, to Q, to the transport chain of Photosystem II, and to Photosystem I

(Fig. 7.11). Thus the sequence from water through the two photosystems to carbohydrate is as follows:

$$H_2O \longrightarrow \boxed{P680} \longrightarrow \text{Photosystem II transport chain} \longrightarrow \boxed{P700} \longrightarrow$$

$$\text{Photosystem I transport chain} \longrightarrow \boxed{NADP_{re}} \longrightarrow \text{carbohydrate}$$

This sequence shows that the electrons necessary to reduce carbon dioxide to carbohydrate come from water, and that the movement of electrons from water to carbohydrate involves the cooperation of a series of redox reactions similar to those occurring in mitochondria.

Since electrons are not passed in a closed loop, this series of reactions is termed noncyclic electron flow, which gives rise to ***noncyclic photophosphorylation***. The whole process results in the formation of both ATP and $NADP_{re}$ as well as the release of molecular oxygen. Though the events of NADP reduction and photophosphorylation have traditionally been known as the light reactions of photosynthesis, only two steps—the capture of photons by the antenna complexes, followed by the export of the electron from the reaction centers—are directly light-dependent.

The details given above apply only to photosynthesis by plants, algae, and the cyanobacteria. Other bacteria possess different forms of chlorophyll and can use light energy in the synthesis of ATP and $NADP_{re}$, but they do not use water as the source of electrons. As mentioned earlier, some use hydrogen sulfide (H_2S) and give off sulfur instead of oxygen; others use nitrogen-based compounds or organic acids. Even ordinary plants can be induced to use a source of electrons other than water. If oxidation of water is blocked by chemical inhibitors and a strong electron donor is provided as a substitute, noncyclic electron flow and photophosphorylation can continue without production of oxygen. However, oxygen is the most electronegative element among these alternatives, and its source—water—is abundant. It is not surprising that the most successful photosynthetic organisms, the plants, evolved using photosynthesis based on water, which is ubiquitous in the chemistry of life.

■ THE ROLE OF CHLOROPLAST STRUCTURE

There are several physical and chemical similarities between the electron-transport strategies of photosynthesis and those of respiration. For instance, the functioning of the respiratory electron-transport chain depends on ordered arrays of redox molecules embedded in the highly folded inner membrane of the mitochondrion; we will see something very similar in the chloroplast. We will also find that transport of H^+ across compartment membranes, which creates an electrochemical gradient, supplies the energy for the synthesis of ATP.

Like the mitochondrion, the chloroplast is a membrane-bounded organelle with its own genome, which probably derives from an ancient endosymbiotic bacterium. However, unlike the mitochondrion, which has a heavily folded inner membrane containing electron-transport chains, the chloroplast has an inner membrane that is relatively smooth and flat, and follows the contours of the outer membrane (Fig. 7.12); the inner membrane has many selective gates and pumps but no electron-transport complexes. It encloses the main volume, or *stroma*, of the chloroplast. The antenna pigments, reaction centers, and electron-transport-chain molecules of the chloroplast are in fact embedded in a third membrane inside the stroma, which forms a series of flattened, interconnected compartments known as *thylakoids*.

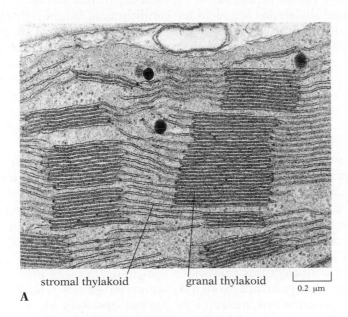

stromal thylakoid   granal thylakoid   0.2 μm

A

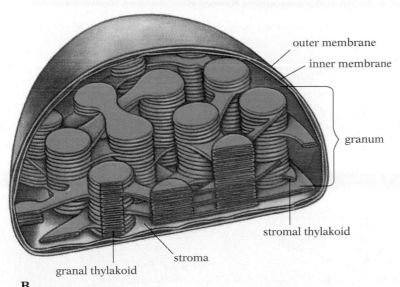

outer membrane

inner membrane

granum

stromal thylakoid

stroma

granal thylakoid

B

7.12 Structure of a chloroplast

(A) Electron micrograph of a section of a chloroplast of timothy grass showing several grana; note the continuity between granal and stromal thylakoids. (B) This cutaway view of a typical chloroplast shows the inner and outer membranes lying close together, enclosing the large compartment known as the stroma. Inside the stroma is a third distinct membrane, which forms the interconnected compartments called thylakoids. The stacks of flat, disklike thylakoids are grana. Chlorophyll molecules and most of the electron-transport-chain molecules are located in the thylakoid membrane.

7.13 · Anatomy of photophosphorylation

The enzyme molecules of photophosphorylation are shown here for simplicity as a linear chain; in the membrane the chain is thought to be folded back on itself, so that cytochrome b_6 connects Fd and PQ, thereby making possible cyclic photophosphorylation (not shown). The antenna molecules of the two photosystems are omitted. This representation makes it clear how electrons from water are moved along the electron-transport chains, and how the thylakoid battery is charged. Since the pH scale is logarithmic (see Fig. 2.14, p. 31), the values shown indicate that the concentration of H^+ ions is 10,000 times greater within the thylakoid than on the other side of the membrane. Note that PQ migrates from one side of the membrane to the other to transport H^+ ions. The CF_0-CF_1 complex, which uses the energy of the electrochemical-potential gradient to make ATP, is also shown.

Thylakoids occur in two distinct but interconnected arrangements, each with its own functional specializations (Fig. 7.12). When loose in the stroma, the proportion of the surface area of the ribbonlike stromal thylakoids that is in contact with stromal fluids is essentially 100%. When stacked neatly in piles called **grana**, the densely packed, chlorophyll-laden membranes of granal thylakoids are highly efficient at capturing photons, but only a small proportion of the granal membrane is exposed to stromal fluids.

Stromal thylakoids are ideal for conditions of bright light, while granal thylakoids are well adapted to take advantage of moderate light. Like the inner mitochondrial membrane, the thylakoid membrane makes possible an electrochemical-potential gradient, which, functioning like a battery, supplies energy for the synthesis of ATP. However, in the chloroplast the H^+ ions accumulate in the innermost compartment of the organelle—the interior of the thylakoids—while the outer compartment, the stroma, becomes more alkaline.

Let's trace the sequence of events in photophosphorylation in the chloroplast. A photon is absorbed in the antenna complex of Photosystem II, exciting an electron to a higher energy level. This energy is transferred to the reaction-center complex. After another electron follows the same path, the pair is fed to an acceptor, plastoquinone (PQ), in the membrane (Fig. 7.13). The resulting electron vacancies in the normal chlorophylls in the reaction center are filled with electrons from water. The entire process involves abstracting four electrons from two water molecules (catalyzed by Mn) to produce one molecule of oxygen and four H^+ ions, and takes about a nanosecond (10^{-9} sec).

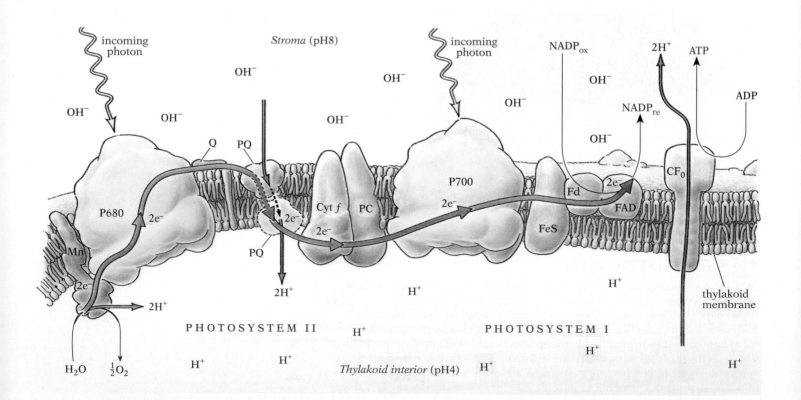

The two pairs of energized electrons that left the reaction center via the acceptor (Q) are delivered, one electron at a time, to PQ, a molecule that can shuttle between the stromal side of the membrane and the thylakoid side. We will follow only one pair, and thus in our electrical accounting consider only two of the four H^+ ions produced when the two molecules of water were split. Some of the energy of each electron pair is used here to move two H^+ ions from the stroma to the thylakoid interior. Each electron now moves, one at a time, to cytochrome f and then to PC until there is a vacancy in P700.

When a photon is absorbed by an antenna molecule in Photosystem I and excites an electron, the excitation energy is passed to P700. The energized electron of P700 is immediately passed to an acceptor (FeS). The resulting vacancy in P700 can now be filled by an electron from Photosystem II waiting in PC. A second photon absorption is required to create a vacancy for the other electron in the pair we are following.

The two energized electrons of Photosystem I, meanwhile, move one at a time to another acceptor (Fd), from which each can follow one of two pathways. Under normal conditions, the most likely fate for the electrons is to help reduce $NADP_{ox}$. This process requires both electrons and consumes two H^+ ions from the stroma, so that in all, four H^+ ions are now gone from the stroma while four H^+ ions have been added to the thylakoid: two via PQ and two when the water molecule was split. The alternative (not shown in Fig. 7.13) is the cyclic pathway (dashed line in Fig. 7.11). The cyclic alternative represents a less efficient use of energy, but it is favored when $NADP_{ox}$ is in short supply or when the cell is more in need of ATP than of $NADP_{re}$.

The result of this highly ordered flow of electrons is twofold. On the one hand, it generates a supply of the high-energy carrier molecule $NADP_{re}$, whose role in carbohydrate synthesis we will examine shortly. On the other, it helps create, through the flow of H^+ ions, a powerful electrochemical gradient, which is then used to generate ATP: as in the mitochondrion, the membrane in the thylakoid contains numerous enzyme complexes (here called CF_0-CF_1) that can utilize the energy of the gradient to phosphorylate ADP.

The organization and anatomy of the electron-transport chains of photophosphorylation and of respiration are very similar. In fact, there is little doubt that the anabolic and catabolic mechanisms have a common evolutionary origin. Current evidence indicates that a low-efficiency anaerobic pathway similar to glycolysis followed by fermentation evolved first, enabling chemosynthetic cells to store energy in carbohydrates for later use. Then followed a higher-efficiency membrane-embedded electron-transport chain, which used electron acceptors like sulfur, carbon, and nitrogen to yield greater supplies of free energy for the cell.

Although primitive kinds of photosynthesis had evolved earlier, a major modification of the electron-transport system used in metabolism gave rise to cyclic phosphorylation. When the more versatile noncyclic system evolved, providing energy for carbon fixation, atmospheric oxygen was produced. For all its potential advantages, oxygen was toxic to much of the anaerobically evolved chemistry of the cell; the evolution of oxygen-tolerant enzymes must have

7.14 Light and dark reactions

Photosynthesis consists of two physically separate but interlocking sets of reactions. The light reactions use light to generate the energy intermediates ATP and $NADP_{re}$. These reactions, as we have seen, take place in the thylakoid membrane. The dark reactions—carbon fixation—use these energy intermediates to turn carbon dioxide into carbohydrates like glucose. Noncyclic photophosphorylation, summarized in the left half of this figure, produces roughly six ATPs in excess of those needed for the synthesis of a single molecule of glucose. Cyclic photophosphorylation goes on simultaneously to provide additional ATP for the cell, but since it does not produce $NADP_{re}$ for carbon fixation, we have ignored it here.

been the next step. With this, the evolution and rapid spread of the respiratory pathway using oxygen as its electron acceptor became possible.

HOW THE DARK REACTIONS FIX CARBON

So far, we have seen how the energy of photons is captured and used to make ATP and $NADP_{re}$, but not how this energy is used in turn to transform the low-energy substance CO_2 into high-energy compounds like glucose. The entire process of photosynthesis is frequently divided into the light reactions of electron-H^+ transfer and photophosphorylation on the one hand, and the dark reactions of carbon fixation on the other.

■ CARBOHYDRATE SYNTHESIS BY THE CALVIN CYCLE

Carbohydrates contain much chemical energy, while CO_2 contains very little. The reduction of CO_2 to form glucose proceeds by many small uphill steps, each catalyzed by a specific enzyme. The energy for this stepwise synthesis of carbohydrates from CO_2 comes from light via ATP and $NADP_{re}$ (Fig. 7.14). The series of dark reactions is called the **Calvin cycle** in honor of Melvin Calvin, who first worked out the many steps involved.

The Calvin cycle begins as CO_2 combines with the five-carbon sugar **ribulose bisphosphate** (**RuBP**) to form an unstable six-carbon compound, which is promptly broken into two three-carbon molecules of phosphoglyceric acid (PGA). Each molecule of PGA is

---

HOW WE KNOW:
THE CALVIN CYCLE

Since there are many sequential steps in the reduction of CO_2 to carbohydrates, and since many of the intermediate compounds occur also in other processes, leading to different end products, how was the exact sequence of steps ever discovered? The tool that made such discoveries possible was a radioactive isotope of carbon, ^{14}C. Samuel Ruben and Martin D. Kamen at the University of California, Berkeley, who discovered this isotope at about 1940, immediately recognized its potential as a research tool in photosynthesis. They showed that plants exposed to carbon dioxide containing the radioactive isotope (that is, $^{14}CO_2$ instead of the normal $^{12}CO_2$) incorporated the isotope into a variety of compounds.

In 1946, Melvin Calvin and his associates, also at Berkeley, began an intensive long-term investigation of carbon dioxide fixation in photosynthesis, using ^{14}C as their principal tool. They exposed algal cells to light in an atmosphere of $^{14}CO_2$ for a few seconds and then killed the cells by immersing them in alcohol. The alcohol also inactivated the enzymes that cat-

alyze the reactions of photosynthesis. With the enzymes paralyzed, whatever amount of each intermediate compound existed in the cell at the moment of inactivation was, in effect, locked in. Calvin and his coworkers could then determine which of these locked-in intermediate compounds contained ^{14}C.

How long the algal cells were exposed to the $^{14}CO_2$ before being killed determined the number of compounds in which ^{14}C was detected: when the time was very short, the ^{14}C reached only the first few compounds in the synthetic sequence; when the time was longer, the isotope moved through more steps in the sequence and appeared in a great variety of compounds. Thus the order in which these compounds appeared in cells reflected their place in the reaction chain. After years of painstaking research, Calvin, who in 1961 was awarded the Nobel Prize for his critically important investigations, worked out the sequence of reactions now called the Calvin cycle.

6 ATP

6 ADP

6 C$_3$
PGA

[3 C$_6$]
*unstable
intermediate*

6 C$_3$ — Ⓟ   **6 NADP$_{re}$**

6 NADP$_{ox}$

3 CO$_2$

6 C$_3$ — Ⓟ
PGAL

3 C$_5$
RuBP
(3 × C$_5$ = 15 C)

5 C$_3$ — Ⓟ
PGAL
(5 × C$_3$ = 15 C)

3 ADP

1 PGAL *for
synthesis of
glucose*

3 ATP

7.15 Synthesis of carbohydrate by the Calvin cycle

Each CO$_2$ molecule combines with a molecule of ribulose bisphosphate (RuBP), a five-carbon sugar, to form a highly unstable six-carbon intermediate, which promptly splits into two molecules of a three-carbon compound called PGA. Each PGA molecule is phosphorylated by ATP and then reduced by NADP$_{re}$ to form PGAL, a three-carbon sugar. Thus each turn of the cycle produces two molecules of PGAL. Five of every six new PGAL molecules formed are used in the synthesis of more RuBP by a complicated series of reactions (not shown here) driven by ATP. The sixth new PGAL molecule can be used in the synthesis of glucose. The path of carbon from CO$_2$ to glucose is traced by blue arrows. Since it takes three turns of the cycle to yield one PGAL for glucose synthesis, the diagram begins with three molecules of CO$_2$; it would require a total of six turns to produce one molecule of glucose, a six-carbon sugar. Note that the cycle is driven by energy from ATP and NADP$_{re}$, both formed by the light reactions of photophosphorylation.

then phosphorylated by ATP and reduced by hydrogen from NADP$_{re}$. The resulting energy-rich three-carbon compound is PGAL (phosphoglyceraldehyde). This compound, which is also an intermediate in glycolysis, is a true sugar and, in a sense, is the stable end product of photosynthesis. Because PGAL is a three-carbon compound, as are the intermediate compounds leading to its formation, the Calvin cycle is often called *C$_3$ photosynthesis* (Fig. 7.15).

Five of every six molecules of PGAL are used in the formation of new RuBP (by a complicated series of reactions powered by ATP), with which more CO$_2$ can be processed. The sixth molecule can be combined by a series of steps with another molecule of PGAL (produced in another turn of the cycle) to form the six-carbon sugar glucose. Thus, it takes six carbon dioxide molecules and six turns of the Calvin cycle to produce one molecule of glucose.

Though glucose has traditionally been considered the end product of photosynthesis, free glucose is not present in significant amounts in most higher plants. Some of the PGAL produced by the Calvin cycle is at once utilized in the formation of lipids, amino acids, and nucleotides. Even when glucose is synthesized, it is normally used almost immediately as a building-block unit for double sugars (disaccharides like sucrose), starch, cellulose, or other polysaccharides. Carbohydrates are generally stored in higher plants in the form of starch, which, because it is insoluble in water, has much less osmotic activity than glucose. An excessive accumulation of carbohydrates in the form of sugars would raise the osmotic concentration of the cytoplasm relative to the environment and severely upset the osmotic balance between the cell and the surrounding fluid. The result would be the intake of too much water by the cell, leading to extreme swelling.

■ THE MYSTERY OF PHOTORESPIRATION

One property of the Calvin cycle is perplexing in that it has no obvious biological function: RuBP carboxylase, the enzyme that catalyzes the carboxylation of ribulose bisphosphate (that is, the addition of CO_2 to RuBP) at the start of the Calvin cycle, can also catalyze the oxidation of RuBP by molecular oxygen (the addition of O_2 to RuBP). In other words, CO_2 and O_2 are alternative substrates that compete with each other for the same binding sites on this enzyme. When the concentration of CO_2 is high and that of O_2 is low, carboxylation is favored and carbohydrate synthesis by the Calvin cycle proceeds. When the reverse conditions prevail—when the concentration of CO_2 is low and that of O_2 is high—oxidation is favored. Higher-than-normal temperatures also favor the oxidation pathway.

Oxidation of the high-energy photosynthetic intermediate RuBP leads to regeneration of low-energy CO_2, a wasteful process known as *photorespiration*. Since it does not result in synthesis of ATP, as other types of respiration do, it appears to short-circuit the Calvin cycle to no purpose. The situation could be worse, however. By means of a complex series of reactions involving chloroplasts, mitochondria, and peroxisomes, plant cells salvage much of the energy they stand to lose from the breakdown of RuBP. Only one of every three carbons entering photorespiration is actually lost as CO_2. Still, photorespiration squanders useful energy.

Because photorespiration predominates over photosynthesis at low concentrations of CO_2, plants that depend exclusively on the Calvin cycle for CO_2 fixation cannot synthesize carbohydrates unless the CO_2 concentration in the air is above a critical level (commonly about 50 ppm, or parts per million); even at normal levels much of the production of photosynthesis is undercut by concurrent photorespiration. At atmospheric CO_2 concentrations, net photosynthesis by such plants could be increased by as much as 50% if photorespiration could be stopped. We will return to this question presently.

THE LOGIC OF LEAF DESIGN

As we have seen repeatedly, life depends both on the precisely catalyzed reactions of various biochemical pathways, with the accompanying interplay of their products, and on the particular anatomy of cells and their organelles. Nowhere is this intimate relationship between form and function more obvious than in the specialized tissues responsible for photosynthesis.

■ THE ANATOMY OF ORDINARY LEAVES

Photosynthesis can occur in all green parts of the plant, but in most plants the leaves expose the greatest area of green tissue to the light and are therefore the principal organs of photosynthesis.

Figure 7.16 shows leaves of a variety of familiar land plants. Most leaves consist of a stalk, or *petiole*, and a flattened *blade*. However, some leaves, particularly those of the grasslike plants, lack petioles, the base of the blade being attached directly to the stem. The blades of most leaves are broad and flat and contain a

VENATION

LEAF TYPE

blade

petiole

Pinnate
venation

Simple

Palmate
venation

Pinnately
compound

Parallel
venation

Palmately
compound

7.16 Leaf types

A leaf usually consists of a blade and a petiole, which sometimes has stipules at its base. Veins run from the petiole into the blade. The main veins may branch in succession off the midvein (pinnate venation); or they may all branch from the base of the blade (palmate venation), or they may be parallel. The blade may be simple, or it may be compound—that is, divided into leaflets that may be pinnately or palmately arranged. Leaves with parallel venation (lower left) are characteristic of the monocot group, plants whose seedlings have only one "seed leaf" or cotyledon; they include grasses, grains, and spring bulbs, for example. The other leaves are from the dicot group (containing two cotyledons).

7.17 Leaf Anatomy

This scanning electron micrograph shows a cross section of a typical dicot leaf. Note that the mesophyll cells are loosely packed and have many spaces between them. The round structure in the center is a vein.

upper epidermis

palisade mesophyll

vein

spongy mesophyll

lower epidermis

complex system of veins. Because of the flatness of the blade, the leaf exposes to the light an area that is very large in relation to its volume.

When we examine a transverse section of a leaf under the microscope (Fig. 7.17, 7.18), we see that the outer surfaces are formed by layers of epidermis, usually only one but sometimes two or more cells thick. A waxy layer, the *cuticle*, usually covers the outer surfaces of both the upper and the lower epidermis, but is generally

A C_3 leaf

bundle-sheath cell

cuticle
upper epidermis
palisade mesophyll

veinlet

spongy mesophyll

chloroplast

lower epidermis

stoma
guard cell

B C_4 leaf

bundle-sheath cell

mesophyll

7.18 Anatomy of C_3 and C_4 (Kranz) leaves

In a C_3 leaf the palisade mesophyll cells typically form a layer in the upper part of the leaf; the corresponding mesophyll cells in a C_4 leaf are usually arranged in a ring around the bundle sheath. While the bundle-sheath cells of C_4 leaves have chloroplasts (dark green), those of C_3 leaves usually lack them.

thicker on the upper side. The chief function of the epidermis is to protect the internal tissues of the leaf from excessive water loss, from invasion by fungi, and from mechanical injury. Most epidermal cells do not contain chloroplasts.

The entire region between the upper and lower epidermis constitutes the **mesophyll**. The mesophyll is usually divided into two fairly distinct parts: an upper palisade mesophyll, consisting of cylindrical cells arranged vertically, and a lower spongy mesophyll, composed of irregularly shaped cells (Fig. 7.18A). The cells of both parts of the mesophyll are very loosely packed and have many intercellular air spaces between them. These spaces are interconnected and communicate with the atmosphere outside the leaf by way of holes in the epidermis called **stomata** ("mouth" in Greek; *stoma* is the singular). The size of the stomatal openings is regulated by a pair of modified epidermal cells called **guard cells**.

A conspicuous system of veins (also called vascular bundles) branches into the leaf blade from the petiole (Fig. 7.16). The veins form a structural framework for the blade and also act as transport pathways. Each vein contains cells of the two principal vascular tissues, xylem and phloem. Each vein is also usually surrounded by a **bundle sheath**, composed of cells packed so tightly together that there are few spaces between them. In most cases the branching of the veins is such that no mesophyll cell is far removed from a vein; typically the veins have a combined length of 100 cm per square centimeter of leaf blade.

■ LEAVES WITH KRANZ ANATOMY

As early as 1904, German plant anatomists observed that the leaf anatomy of some plants of tropical origin—plants associated with bright, hot, but especially dry habitats—showed a combination of features not generally found in plants native to the temperate zones. This unusual complex of features came to be called **Kranz anatomy**. *Kranz*, which means "wreath" in German, refers to the ringlike arrangement of photosynthetic cells around the leaf veins of these plants.

The bundle-sheath cells of plants with Kranz anatomy (also called C_4 plants, for reasons we will see shortly) contain numerous chloroplasts, whereas those of other plants (C_3 plants) often do not. In plants with Kranz anatomy the mesophyll cells that correspond to the palisade layer tend to be clustered in a ringlike arrangement around the veins, just outside the bundle sheaths (Fig. 7.18B). These mesophyll cells contain numerous chloroplasts, but the spongy mesophyll cells outside the rings often have few chloroplasts or none at all. In Kranz plants the chloroplasts of the bundle-sheath cells and mesophyll cells usually differ in a number of ways. In the bundle-sheath cells the chloroplasts are bigger, they accumulate large amounts of starch in the presence of light, and the grana are few and poorly developed; in the mesophyll cells the chloroplasts are smaller, they usually do not accumulate much starch in the presence of light, and they have numerous large grana (Fig. 7.19).

Though there is considerable variation in the structural details among species exhibiting Kranz anatomy, the common trends sug-

7.19 Two kinds of chloroplasts in a C_4 leaf

At left in this electron micrograph of a segment of a corn leaf is part of a bundle-sheath cell; its chloroplasts have small grana and contain starch grains (light areas). At right and bottom are parts of two mesophyll cells; their chloroplasts are smaller and contain numerous grana but no starch grains.

gest a special adaptation to conditions of high temperature, intense light, low moisture, and low CO_2 and high O_2 concentrations—all conditions far from optimal for plants that depend entirely on the Calvin cycle for CO_2 fixation. All these various structural specializations serve a common biochemical adaptation: the C_4 pathway of photosynthesis.

■ AVOIDING PHOTORESPIRATION: C₄ PHOTOSYNTHESIS

As Figure 7.20A shows, corn, which originated in the tropics and has Kranz anatomy, can carry out photosynthesis at very low concentrations of CO_2; bean plants, which are native to the temperate zone, fail to photosynthesize at CO_2 levels below about 50 ppm because of photorespiration. At CO_2 concentrations of 200 or 300 ppm, where corn approaches its maximum photosynthetic capacity, beans still perform well below their potential capacity. Again because of photorespiration, the concentration of O_2 that inhibits photosynthesis is far lower for beans than for corn (Fig. 7.20B). Kranz anatomy is correlated with the ability of a plant to function even when the CO_2 level in the leaf is low and O_2 concentration is high.

Low CO_2 and high O_2 result from exposure to great heat, dryness, and brilliant light, as in a desert or on a dry savanna. Under such conditions, when the moist walls of the mesophyll cells risk losing too much water by evaporation through the stomata, the guard cells close the stomata almost completely. Water loss is thus reduced, but now gases can no longer move freely between the atmosphere and the air spaces inside the leaf. As CO_2 is used up in photosynthesis, the nearly closed stomata prevent the supply inside the leaf from being fully replenished, and the CO_2 concentration in the air spaces around the mesophyll cells falls. Under such conditions, non-Kranz plants like the bean will carry out so much photorespiration that their ability to synthesize carbohydrate from CO_2 will be greatly reduced. By contrast, corn and other Kranz

7.20 Comparison of photosynthetic efficiency in C₃ and C₄ plants

(A) Corn can fix carbon at CO_2 concentrations as low as 1 ppm, and it carries out photosynthesis at a very high rate at concentrations of 200–300 ppm. (A normal concentration of CO_2 in the atmosphere is about 330 ppm.) By contrast, beans perform no net carbon fixation at CO_2 concentrations below about 50 ppm, and their rate of photosynthesis at concentrations of 200–300 ppm is not very high. (B) Photosynthesis in corn shows no inhibition at all at O_2 concentrations below 65%, whereas the photosynthetic rate of beans falls steadily as the O_2 concentration rises. (A normal concentration of O_2 in the atmosphere is about 21%). Both A and B are for a temperature of 20°C and a light intensity of 2000 foot-candles. Obviously, C_4 photosynthesis is superior under these conditions. However, when the temperature or illumination varies (and while the O_2 and CO_2 concentrations remain at normal levels), a very different picture emerges. For example, as the temperature drops (C), C_3 plants clearly perform more efficiently, so they have the advantage in cooler (that is, more temperate) climates.

plants *can* synthesize carbohydrate under dry conditions; they are thus able to survive in climates that would be fatal to other plants. Even under less extreme conditions they can often carry out photosynthesis at a higher rate than other plants.

In the late 1960s M. D. Hatch and C. R. Slack worked out the biochemical pathway of photosynthesis unique to Kranz plants. They found that in the mesophyll cells—the cells richest in chloroplasts that are arranged in rings around the veins—CO_2 is combined, not with ribulose bisphosphate as in the Calvin cycle, but rather with a three-carbon compound called phosphoenolpyruvate or PEP, to form a four-carbon (C_4) compound. The enzyme that catalyzes this carboxylation of PEP, unlike the one that catalyzes the carboxylation of RuBP in the Calvin cycle, does not have O_2 as an alternative substrate and thus is not inhibited by high O_2 concentrations. Thus it enables Kranz plants (which are therefore also called C_4 plants) to fix CO_2 under conditions when photorespiration would predominate over photosynthesis in C_3 plants, which use only the Calvin cycle.

Curiously enough, the C_4 compound formed in the mesophyll cells is not used directly for growth or nutrition by the plant. Instead, it is passed in reduced form into the bundle-sheath cells (which in C_4 plants are very well developed and contain chloroplasts), where it is decarboxylated: the C_4 compound is broken down to CO_2 and a C_3 compound (Fig. 7.21). The C_3 residue moves back to the mesophyll cells, where it is reconverted into PEP and starts the C_4 cycle over again. The CO_2 remains in the bundle-sheath cells, where it is picked up by the RuBP carboxylase in these chloroplasts and incorporated into carbohydrate through the conventional Calvin cycle.

In both C_3 and C_4 plants the ultimate assimilation of CO_2 into carbohydrate is by the Calvin cycle. The difference is that in C_3 plants the Calvin cycle is the only pathway of CO_2 fixation, whereas in C_4 plants there is another, preliminary fixation pathway. It might seem strange that a plant would have evolved a special mechanism for fixing CO_2 as C_4 in the mesophyll, only to combine it with a mechanism for promptly breaking off the CO_2 again and refixing it as carbohydrate in the bundle-sheath cells. However, this "laundering" of CO_2 by C_4 plants gives them an advantage over C_3 plants under conditions of high temperature and intense light, when stomatal closure results in low CO_2 and high O_2 in the air spaces inside the leaf. Under such conditions, C_3 plants are unable to use CO_2 effectively because O_2 competes for RuBP. In contrast, C_4 plants *can* fix CO_2, because the mesophyll cells, acting as CO_2 pumps, can elevate the CO_2 concentration in the bundle-sheath cells to a level at which carboxylation of ribulose bisphosphate (leading into the Calvin cycle) exceeds its oxidation. The Kranz anatomy, with its concentric rings of mesophyll and bundle-sheath cells, makes possible the compartmentalization on which the process of CO_2 pumping depends.

The combination of Kranz anatomy and C_4 photosynthesis has evolved independently in a variety of unrelated plants, including corn, sugarcane, sorghum, crabgrass, saltbush, and portulaca. It is an impressive illustration of the intimate relationship between structure and function in living systems.

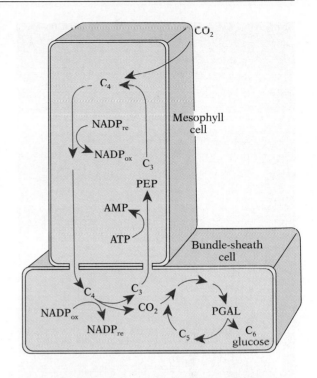

7.21 Hatch-Slack pathway of C_4 photosynthesis

The path of carbon is traced by red arrows. A mesophyll cell absorbs CO_2 from the intercellular air spaces (top). The CO_2 combines with PEP, a three-carbon compound, to form a four-carbon compound. After reduction by $NADP_{re}$, the C_4 substance moves into an adjacent bundle-sheath cell. There it is oxidized by $NADP_{ox}$ and split into a C_3 compound and CO_2. The C_3 compound moves back to the mesophyll cell and is converted (by a reaction probably driven by energy from ATP) into PEP. The CO_2 is fed into the Calvin cycle in the bundle-sheath cell and is incorporated into carbohydrate.

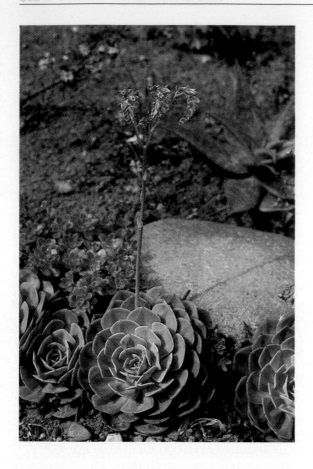

■ AVOIDING PHOTORESPIRATION: CRASSULACEAN ACID METABOLISM

Another remarkable dry-climate variation of photosynthesis is found in many succulents—plants that store water in fleshy leaves (Fig. 7.22)—and a few other plants, including pineapple, Spanish moss, and some cacti. Like C_4 plants, these organisms avoid water loss in their hot environment by closing their stomata during the day and opening them at night. The CO_2 necessary for photosynthesis is captured and stored at night in the form of malic acid and isocitric acid, and then slowly released in the cells during the day, to be fixed by C_3 photosynthesis. This process is called ***crassulacean acid metabolism (CAM)***, named for the genus of plants in which the process was discovered. Though CAM seems to have evolved independently of C_4 photosynthesis, it serves the same purpose. The essential difference is that Kranz plants solve the CO_2-buildup problem by segregating the two steps of carbon fixation anatomically, while CAM plants separate them temporally. The existence of these variations illustrates again how natural selection can lead to several solutions of the same problems.

7.22 Stonewort, a member of the Crassulaceae
The thick, fleshy leaves of this plant are typical of succulents.

CHAPTER SUMMARY

THE PHOTOSYNTHETIC REACTION
Photosynthesis combines water and carbon dioxide:
$$CO_2 + 2H_2O + light \longrightarrow O_2 + (—CH_2O) + H_2O \text{ (p. 180)}$$

HOW THE LIGHT REACTIONS STORE LIGHT ENERGY

HOW CHLOROPHYLL CAPTURES THE ENERGY OF LIGHT Chlorophyll absorbs red and blue light, exciting an electron to a higher energy level; accessory pigments extend the action spectrum. Groups of chlorophyll act as antennas, passing their excited electrons to a reaction center, from which the electrons enter an electron-transport chain. (p. 182)

HARVESTING THE ENERGY OF EXCITED ELECTRONS The cyclic pathway, used exclusively by anaerobic photosynthesizers but available as a backup in aerobic autotrophs, transfers excited electrons through an electron-transport chain back to the antenna, extracting energy used to create an electrochemical gradient in the chloroplast. In the noncyclic pathway the excited electron is passed through the electron-transport chain of Photosystem II to the antenna of Photosystem I, where it fills a vacancy created when an electron excited there passes into its electron-transport chain; the Photosystem I electron ends up in $NADP_{re}$. The vacancy in the antenna of Photosystem II is filled with an electron obtained from the splitting of water, which produces oxygen as a by-product. Energy extracted in the electron-transport chain is used to create an electrochemical gradient in the chloroplast. (p. 186)

THE ROLE OF CHLOROPLAST STRUCTURE The electrochemical energy is stored as a H^+ gradient across the thylakoid membrane, and is used to synthesize ATP. (p. 190)

HOW THE DARK REACTIONS FIX CARBON

CARBOHYDRATE SYNTHESIS BY THE CALVIN CYCLE The Calvin cycle uses ATP and the high-energy electron in $NADP_{re}$ to convert CO_2 into PGAL, which is then converted into other compounds, especially glucose. (p. 194)

THE MYSTERY OF PHOTORESPIRATION In the presence of too much oxygen, the enzyme that begins the fixation of CO_2 can instead bind O_2 and "unfix" CO_2. (p. 195)

THE LOGIC OF LEAF DESIGN

THE ANATOMY OF ORDINARY LEAVES Leaves have veins to bring in raw materials and export products, a cuticle and adjustable stomata to minimize water loss, and an internal mesophyll, where gases are exchanged and photosynthesis occurs. (p. 196)

LEAVES WITH KRANZ ANATOMY Leaves of Kranz plants have much of their photosynthetic mesophyll packed tightly around the veins; very little cell surface is exposed for gas exchange. Kranz anatomy is an adaptation for C_4 photosynthesis. (p. 199)

AVOIDING PHOTORESPIRATION: C4 PHOTOSYNTHESIS C_4 plants synthesize a four-carbon intermediate in mesophyll cells using an enzyme that is not sensitive to oxygen; this intermediate is then moved to bundle-sheath cells, where it is converted into a three-carbon molecule and CO_2. The three-carbon molecule is moved back to the photosynthetic cell, and the CO_2 is fixed via the Calvin cycle in the bundle-sheath cells. C_4 photosynthesis is most common in plants adapted for arid or hot conditions, where the need to conserve water by keeping the stomata closed leads to a buildup of waste oxygen in the leaves. (p. 200)

AVOIDING PHOTORESPIRATION: CRASSULACEAN ACID METABOLISM CAM metabolism solves the same problem by capturing CO_2 in intermediate compounds at night, when temperatures are lower; the CO_2 is released during the day for use in the Calvin cycle. (p. 202)

STUDY QUESTIONS

1 Make a list of the analogies between the electron-transport chain in respiration and that in photosynthesis. Be sure to include not only the strategies used by the two processes but also the enzymes and the chemicals involved in each. (pp. 169–173; 186–190)

2 Distinguish between C_3 photosynthesis, C_4 photosynthesis, and CAM photosynthesis. Why hasn't C_4 photosynthesis taken over? (pp. 195–196; 199–202)

3 If the predictions of some climatologists are borne out, human activity (especially the burning of trees and fossil fuels) will increase the CO_2 concentration in the atmosphere significantly and, through the greenhouse effect, raise the average temperature of the earth. What might this do for the various strategies of photosynthesis? (pp. 200–202)

4 Compare the dark reactions of photosynthesis with the Krebs citric acid cycle. Is there any hint of a common evolutionary origin? (pp. 167–169; 194–195)

SUGGESTED READING

GOVINDJEE, AND W. J. COLEMAN, 1990. How plants make oxygen, *Scientific American* 262 (2). *On the operation of the water-splitting enzyme of noncyclic photosynthesis.*

MILLER, K. R., 1979. The photosynthetic membrane, *Scientific American* 241 (4). *An excellent discussion relating the chemiosmotic theory of chloroplast function to the structure of thylakoid membranes as shown by freeze-etch microscopy.*

YOUVAN, D. C., AND B. L. MARRS, 1987. Molecular mechanisms of photosynthesis, *Scientific American* 256 (6). *An account of the molecular events occurring during the first 200m seconds following photon absorption.*

PART II

THE PERPETUATION OF LIFE

◀ **Molecule of inheritance,** the double helix of DNA is vividly portrayed in this painting by the noted scientific illustrator Irving Geis. The atomic-scale "ball and stick" model—based on single-crystal X-ray analysis—represents the dioxyribose backbones of the molecule as blue and the nucleotide bases as red; weak hydrogen bonds are indicated by dotted lines. The white ribbons accentuate the molecule's overall double-helical structure.

CHAPTER 8

THE STRUCTURE AND REPLICATION OF DNA

The processes of life are guided by an elaborate and precise series of information transfers. The genes of an organism's DNA contain all the information necessary to build and operate the organism— information that orchestrates its development from a single unspecialized cell into a complex corporation of specialized tissues and organs, and also directs events ranging from glycolysis and the Krebs cycle to the organism's behavior. This chapter is concerned with how the genetic information as a whole is duplicated and transferred to each new cell, whether it be a cell added as the organism grows, or one of the **gametes** (sex cells, like sperm and egg) that unite to form an entirely new organism (Fig. 8.1C).

Genes are most conveniently defined as sequences of DNA that encode proteins (or, occasionally, encode special sequences of the nucleic acid RNA that are used directly in the construction of ribosomes and in other ways). The genes are organized in a linear sequence on one or more chromosomes. The chromosomes and the DNA in organelles, as well as the tiny chromosome-like elements called plasmids (discussed in Chapter 10), together comprise the **genome**.

Modern molecular genetics begins with the study of DNA structure; we will look first at the structure of the genetic material and then go on to examine the process of chromosome duplication, the essential step of which is **replication**. This process, which includes self-repair of any replication errors, enables each daughter cell produced during cell division to receive its own copy of each chromosome (and thus a complete set of genes) from the parental cell.

WHAT ARE GENES MADE OF?

■ THE UNUSUAL CHEMISTRY OF CHROMOSOMES

The special chemistry of the nucleus came to light in 1868, when Friedrich Miescher showed that protein-digesting enzymes, though they destroy nearly everything else in cells, leave the contents of the nucleus largely intact. This meant that the nucleus contains large quantities of an unusual compound, now called nucleic acid.

A

Instructions in the DNA are used to build cellular components and direct biochemical events. In most eucaryotes, there arc two copies of each chromosome.

B

Chromosomes are duplicated prior to cell division and a complete copy of each pair is passed to each new cell.

C

When gametes (sex cells) are formed, they receive only one chromosome from each pair. The normal number of chromosomes is restored by fusion of a male and female gamete.

8.1 Transfer of genetic information

The set of instructions in the DNA is transcribed and used to direct events within the cell (A), is duplicated and passed on to daughter cells during normal cell division (B), and is halved and passed to gametes in preparation for sexual reproduction (C).

8.2 Feulgen staining

Feulgen staining was used to distinguish the DNA in this sample of liver cells.

Table 8.1 Griffith's results

BACTERIA INJECTED	REACTION OF MICE
Live strain R	Survived
Live strain S	Died
Dead strain S	Survived
Live strain R + dead strain S	Died

In 1914 Robert Feulgen devised a method of selectively staining nucleic acids a brilliant crimson, and showed that nuclear DNA is restricted to the chromosomes (Fig. 8.2).

Staining and other techniques show that, in a given organism, all the **somatic cells** (all the cells except the gamete-producing **germ cells**) ordinarily contain the same amount of DNA. This is true even though cells from such different tissues as liver, kidney, heart, nerve, and muscle differ drastically in the amounts of other substances they contain. Furthermore it became clear that egg and sperm cells contain only half as much DNA as the somatic cells. Since cell division distributes a complete set of genes to every somatic cell, regardless of its eventual role, while the process of gamete production distributes to every sperm and egg cell exactly half the amount of genetic material found in the somatic cells, these two discoveries indicated that either nucleic acid or the protein intimately associated with it is the essential material of genes.

■ GENES: DNA OR PROTEIN?

In 1928 Fred Griffith showed that mice injected with a virulent (disease-producing) strain of pneumococci, the bacteria that cause pneumonia, inevitably die. This S strain (so named because on artificial media it grew as smooth colonies) was compared with a nonvirulent R (for rough colonies) strain, which did not kill mice, and heat-killed S-strain bacteria, which were also harmless (Table 8.1). However, mice injected with a mixture of live strain-R and heat-killed strain-S bacteria died. How could a mixture of dead and nonvirulent live bacteria have killed the mice? Griffith found the dead mice full of live strain-S bacteria! Somehow the live strain-R bacteria had been transformed into live strain-S bacteria by material from the dead strain-S cells.

In 1943 O. T. Avery, Colin MacLeod, and Maclyn McCarty showed that the essential agent in the process of bacterial **transformation** was DNA (Fig. 8.3). However, this did not prove that

8.3 Rough versus smooth pneumococci

Rough (untransformed) colonies are on the left; smooth (transformed) colonies are on the right. (Photograph from the 1994 paper by Avery, MacLeod, and McCarty.)

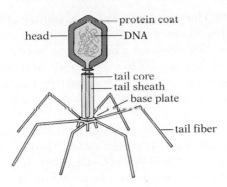

8.4   A complex bacteriophage

genes are made of DNA since the DNA from the virulent strain might merely have activated the protein-based genes in the nonvirulent strain.

In 1952 Alfred D. Hershey and Martha Chase showed that one of the viruses that attacks the bacterium *Escherichia coli*, which is abundant in the human digestive tract, does so by means of its DNA. Bacteria-destroying viruses are called **bacteriophage,** or *phage* for short (from the Greek *phagein*, "to eat"). These tiny parasites subvert the cellular machinery of host organisms to reproduce their own genes. They consist primarily of a protein coat and a nucleic acid core. Some phage viruses have a head region (containing the genetic material) and an elongate tail region made up of a hollow core, a surrounding sheath, and six distal fibers (Fig. 8.4). When such a phage attacks a bacterial cell, proteins on the tip of its tail fibers and its base plate bind to proteins on the wall of the bacterial cell (Fig. 8.5). The DNA of the phage enters the cell, but the protein coat is left behind. Later the host cell releases dozens of hundreds of new viruses.

0.1 µm

8.5   Bacteriophage replication

Each bacteriophage attaches to the bacterial cell wall by its tail fibers and base plate, and injects its genetic material into the cell. Once inside, the genetic material takes over the metabolic machinery of the cell and puts it to work making new phage.

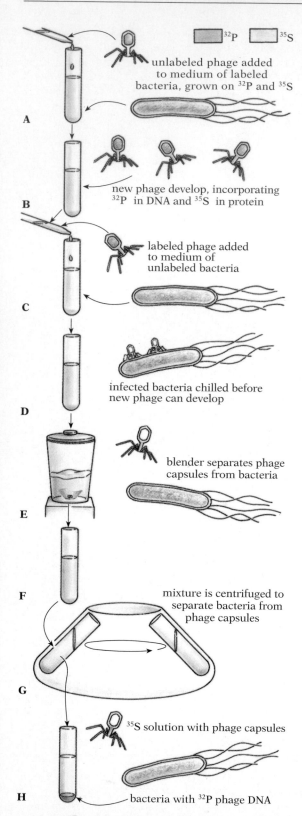

8.6   Hershey-Chase experiment

For details of this experiment, which demonstrated that DNA rather than protein carries genetic information, see the text.

The conclusive Hershey and Chase experiment took advantage of differences in the chemistry between proteins and nucleic acids: proteins have sulfur but not phosphorus, whereas nucleic acids have phosphorus but not sulfur. They cultured phage on bacteria grown on a medium containing radioactive phosphorus (^{32}P) and radioactive sulfur (^{35}S) (Fig. 8.6A). The phage incorporated the ^{35}S into their proteins and the ^{32}P into their DNA (Fig. 8.6B). Hershey and Chase then infected nonradioactive bacteria with the radioactive phage. They allowed the phage time to bind to the bacteria and inject hereditary material (Fig. 8.6C–D). Next they agitated the bacteria in a blender to detach what remained of the phage from their surfaces. They then centrifuged the mixture to separate the infected bacteria (reduced to a pellet at the bottom of the sample tube) from the phage coats (suspended in the fluid left behind; Fig. 8.6E–G). The fluid contained a substantial amount of ^{35}S but little ^{32}P, an indication that only the empty protein coat had been left outside the bacterial cell. The bacterial fraction contained much ^{32}P but no ^{35}S, meaning that only DNA had been injected into the bacteria by the phage (Fig. 8.6H). From these results they concluded that DNA alone was sufficient to transmit all the genetic information necessary to produce new viruses.

■ THE MOLECULAR STRUCTURE OF DNA

As detailed in Chapter 3, DNA is composed of building blocks called nucleotides, each of which is itself composed of a five-carbon sugar bonded to a phosphate group and a nitrogenous base (Fig. 8.7). By convention, the five carbons are designated by num-

8.7   Diagram of a nucleotide from DNA

A phosphate group and a nitrogenous base are attached to deoxyribose, a five-carbon sugar. The carbons in deoxyribose are designated 1'–5', as shown here, though the numbers are not normally included in molecular diagrams. The phosphate group is bound to the 5' carbon of the sugar, while a hydroxyl group is bound to the 3' carbon. Of the four different nitrogenous bases, cytosine is illustrated here. RNA nucleotides have a sugar identical to the one shown except that the 2' carbon has a hydroxyl group bound to it.

8.8 Portion of a single chain of DNA

5′ end of chain

5′

G

T

C

3′

A

3′ end of chain

Nucleotides are linked together by bonds between their sugar and phosphate groups. The nitrogenous bases (G, guanine; T, thymine; C, cytosine; A, adenine) are side groups. In this diagram P represents the main components of each phosphate group—the phosphorus atom with its hydroxyl group and the double-bonded oxygen; only the oxygen atoms in the connecting chain are shown separately.

bers and primes: 1′ to 5′. There are four kinds of nucleotides in DNA, which differ from one another in their nitrogenous bases. Two of the bases, *adenine* and *guanine*, are purines, which are double-ring structures; the other two, *cytosine* and *thymine*, are pyrimidines, which are single-ring structures (Fig. 8.8).

The nucleotides in a particular DNA molecule are arranged in a specific sequence; the sugars are held together by the phosphate groups that link the 3′ carbon of one sugar to the 5′ carbon of the next; the nitrogenous bases are arranged as side groups off the chains (Fig. 8.8). DNA molecules ordinarily exist as double-chain structures, with the two chains, or strands, held together by hydrogen bonds between their nitrogenous bases. Such bonding can occur only between cytosine and guanine or between thymine and adenine (Fig. 8.9). Thus the sequence of bases in one strand determines the complementary sequence in the other (Fig. 8.10). Notice that the polarities of the two strands are opposite: one runs from 5′ to 3′, while the other goes from 3′ to 5′.[1] Finally, the ladderlike double-chain molecule is coiled into a double helix (Fig. 8.11), and stabilized further by hydrogen bonds aligned with the chains; these bonds between separate rungs of the nucleotide ladder are

[1] This strand "polarity" has nothing to do with the unequal distribution of charge that gives rise to polar molecules.

Thymine   Adenine

Cytosine   Guanine

PYRIMIDINES   PURINES

8.9 Bonding of nitrogenous bases in nucleotides

Because of the differing electronegativities of oxygen, hydrogen, nitrogen, and carbon, the nitrogenous bases of DNA have polar segments. The spacing and polarity of these segments allow thymine to form hydrogen bonds (red striped bands) with adenine, and cytosine to bond with guanine. (The asterisk marks the point where each base attaches to a sugar.)

8.10   Portion of a DNA molecule uncoiled

The molecule has a ladderlike structure, with the two uprights composed of alternating sugar and phosphate groups and the cross rungs composed of paired nitrogenous bases. Each cross rung has one purine base (a pentagon attached to a hexagon) and one pyrimidine base (a hexagon). When the purine is guanine (G), the pyrimidine with which it is paired is always cytosine (C); when the purine is adenine (A), the pyrimidine is thymine (T). Adenine and thymine are linked by two hydrogen bonds, guanine and cytosine by three. Note that the two chains run in opposite directions: the free phosphate is linked to the 5' carbon at the upper end of the left chain and to the corresponding carbon at the lower end of the right chain.

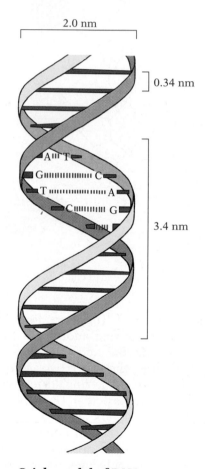

8.11   Watson-Crick model of DNA

The molecule is composed of two polynucleotide chains held together by hydrogen bonds between their adjacent bases. The double-stranded structure is coiled in a helix. The width of the molecule is 2.0 nm, the distance between adjacent nucleotides is 0.34 nm, and the length of one complete coil is 3.4 nm. Interactions between bases within each chain (not shown) help stabilize the molecule in the helical shape shown here.

analogous to those that stabilize the α helices of proteins (see Fig. 3.23, p. 62).

Determining the structure of so complicated—and important—a molecule as DNA had become an irresistible challenge to many scientists. X-ray diffraction analysis of DNA by Rosalind Franklin

8.12   X-ray diffraction image of DNA

As Crick had discovered, the X-shaped cross radiating from the center of this pattern is diagnostic of a helix. The position of the strong bands at the top and bottom indicate a periodicity of 0.34 nm. More subtle patterns suggest a 3.4-nm period as well.

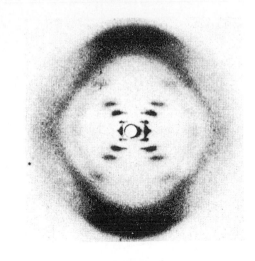

produced the first sharp X-ray diffraction patterns (Fig. 8.12). Francis H. C. Crick, who had developed mathematical methods for interpreting X-ray patterns of protein helices, used the Franklin photographs to deduce that DNA is a helix with three major periodicities: repeating patterns of 0.34, 2.0, and 3.4 nm.

James D. Watson (Fig. 8.13) and Crick decided to develop a model of the structure of the DNA molecule by combining (1) what was known about the chemical content of DNA, (2) the information gained from Crick's analysis of the X-ray diffraction studies, and (3) data on the exact distances between bonded atoms in molecules, the angles between bonds, and the sizes of atoms. Watson and Crick built scale models of the component parts of DNA and then attempted to fit them together in a way that would agree with the information from all these sources.

They were certain that the 0.34-nm periodicity corresponded to the distance between successive nucleotides in the DNA chain, the 2.0-nm periodicity to the width of the whole chain, and the 3.4-nm periodicity to the distance between successive turns of the helix. They found that a single chain of nucleotides coiled in a helix 2.0 nm wide with turns 3.4 nm long would have a density only half as great as that of DNA; from this they concluded that DNA molecules must have two nucleotide chains rather than one. They tried several arrangements of their scale model; the one that best fitted all the data had the two nucleotide chains twisted together, running in opposite directions, fitting within a hypothetical cylinder of appropriate diameter, with the purine and pyrimidine bases oriented toward the interior (Fig. 8.11).

With the bases oriented in this manner, hydrogen bonds between the bases of opposite chains could supply the force to hold the two chains together and to maintain the helical configuration. The DNA molecule, when unwound, would have a ladderlike structure, with the ladder uprights formed by the two long chains of alternating sugar and phosphate groups, and each of the cross rungs formed by two nitrogenous bases linked loosely by hydrogen bonds (Fig. 8.10).

Watson and Crick realized that each cross rung must be composed of one purine base and one pyrimidine base: two purines opposite each other occupied too much space (because each had two rings, for a total of four), and two pyrimidines opposite each other did not come close enough to bond properly (because each had only one ring). This left four possible pairings: A–T, A–C, G–T, and G–C. Further examination revealed that only A–T and G–C pairs had the right geometry to form hydrogen bonds. This pairing quite unexpectedly explained an earlier finding by the biochemist Erwin Chargaff. He had noted that while the proportion of cytosine in DNA varies from species to species, it is constant between individuals of a given species, and always matches the proportion of gua-

8.13   Watson describing the Watson-Crick model of DNA

This photograph was taken at a 1953 seminar at the Cold Spring Harbor Laboratory.

nine in that species. Similarly, the DNA of any given species always contains exactly equal amounts of adenine and thymine nucleotides. (Conversely, the amounts of adenine and thymine within a species usually differ from the amounts of guanine and cytosine.) Watson and Crick completed their deduction of DNA structure in 1953.

HOW DNA IS REPLICATED

■ THE TEMPLATE HYPOTHESIS

To be the genetic substance, DNA must have built into it the information necessary to replicate itself. One of the most satisfying things about the Watson-Crick model is that the basic strategy of replication is obvious from the structure of DNA.

The Watson-Crick model Since the DNA of all organisms is alike in being a polymer composed of only four different nucleotides, the essential distinction between the DNA of one gene and the DNA of another gene must be the sequence in which the four possible types of base-pair cross rungs (A–T, T–A, G–C, and C–G) occur. The basic question of genetic replication is, then, What tells the cell's biochemical machinery how to put these nucleotide building blocks together in exactly the sequence characteristic of the DNA already present in the cell?

Watson and Crick pointed out that if the two chains of a DNA molecule are separated by rupturing the hydrogen bonds between the base pairs, each chain provides all the information necessary for synthesizing a new partner identical to its previous partner. Since an adenine nucleotide must always pair with a thymine nucleotide, and a guanine always with a cytosine, the sequence of nucleotides in one chain (strand) precisely specifies what the sequence of nucleotides in its complementary strand must be. Thus, separating the two chains of a DNA molecule and using each chain as a template, or mold, against which to synthesize a new partner for it would result in two complete double-chained molecules identical to the original molecule (Fig. 8.14).

Experimental support Satisfying as it was, this template theory of DNA replication was pure speculation when first put forward by Watson and Crick. Since then, convincing evidence has come from the work of a number of investigators.

In 1957 Arthur Kornberg and his associates developed a method for synthesizing DNA in a test tube. They extracted an enzyme complex from the bacterium *E. coli* that catalyzes the synthesis of DNA—a ***DNA polymerase*** (an enzyme that creates a DNA polymer)—and combined it with a radioactive ATP-activated supply of the four nucleotides as raw material. (The activation consisted of a high-energy phosphate group that provides the energy for the later reaction steps; the isotope used was ^{14}C.) DNA was added to serve both as a primer—a starting point for the expected reactions—and as a potential template. Soon new DNA containing ^{14}C appeared.

8.14 Replication of DNA

As the two polynucleotide chains of the old DNA (yellow and blue) uncoil, new polynucleotide chains (green) are synthesized on their surfaces. The process produces two complete double-stranded molecules, each of which is identical in base sequence to the original double-stranded molecule. For simplicity, replication is shown here beginning at the end of the molecule; in actuality it begins at specific internal points.

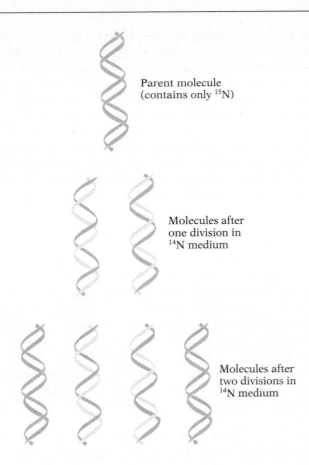

Parent molecule
(contains only ^{15}N)

Molecules after
one division in
^{14}N medium

Molecules after
two divisions in
^{14}N medium

8.15   Results of the Meselson-Stahl experiment

The parent DNA molecule (blue chains) contained only heavy nitrogen. After one division, the DNA had an intermediate density, an indication that half the nitrogen in each molecule was heavy and half was light (normal); the two heavy parental chains had separated, and each had acted as the template for synthesis of a complementary normal chain (yellow and green). Even after several additional duplications, the two original heavy parental chains remained intact, though in two separate molecules of DNA.

The ratio of adenine and thymine to guanine and cytosine in the new DNA was precisely the same as that in the primer DNA.[2]

Direct support for the Watson-Crick model of replication came in 1958 from work by Matthew S. Meselson and Franklin W. Stahl. Meselson and Stahl grew *E. coli* for many generations on a medium whose only nitrogen source was the heavy isotope ^{15}N. Eventually all the DNA in these bacteria contained the heavy isotope instead of the normal isotope ^{14}N. Then the researchers changed the nitrogen source abruptly from ^{15}N to ^{14}N. Cell samples were removed at regular intervals thereafter, and the DNA was extracted from them and subjected to a complicated procedure designed to separate DNA of different densities.

The experiment showed that when cells containing only heavy (^{15}N) DNA were allowed to undergo one division in the normal (^{14}N) medium, the DNA of the new cells was intermediate in density between heavy DNA and normal DNA—the nitrogen in the DNA of the new cells was half ^{15}N and half ^{14}N. This is precisely what would be expected if the two chains of the heavy parental DNA separated and acted as templates for the synthesis of new partners from nucleotides containing only ^{14}N (Fig. 8.15). Each

---

[2] Ironically, we now know that there is more than one kind of DNA polymerase in cells, and that the one Kornberg isolated, now called DNA polymerase I to indicate its priority of discovery, is primarily used for repairing damage to chromosomes and cleaning up loose ends created during replication (see the *Exploring Further* box, pp. 216–217). The second DNA polymerase found is involved in creating the nucleolus (see pp. 124 and 286). The main enzyme complex responsible for replication, DNA polymerase III, was discovered last.

new DNA molecule should be composed of one heavy chain from the parent and one lighter (normal) chain newly synthesized; the new molecule would thus have an intermediate density.

To prove that all the ^{15}N really was in one chain of the intermediate-density DNA and all the ^{14}N was in the other chain, Meselson and Stahl subjected the DNA to a treatment that breaks the hydrogen bonds between the bases and separates the chains. As expected, this procedure produced single chains of heavy and normal density; the two isotopes had not been distributed randomly throughout the DNA molecule but had been localized, each in one of the chains, just as the Watson-Crick theory predicted.

■ THE DIFFICULTIES OF REPLICATION

The process of replication is complex and requires precision. Hydrogen bonds that stabilize the helical shape and link the two chains of the DNA molecule must be broken, and the chains must be separated; complementary nucleotides must be paired with the nucleotides of each existing chain; the new nucleotides must be covalently linked to form a chain; and so on. Every step is managed by specific enzymes, and takes place quickly and accurately. In *E. coli*, for instance, a complex of several enzymes, collectively known as DNA polymerase III, adds an average of 500 base pairs per second to the new strand, with only one error in every billion pairs copied.

The basic task of DNA replication is the same in procaryotes and eucaryotes: the process in bacteria and the process in humans display many similarities. For example, though the chromosomes differ in shape—bacterial chromosomes are circular while ours are linear—in both groups, replication begins at particular spots in the DNA and proceeds in both directions away from the initiation site; replication in eucaryotes never begins at an end. In both procaryotes and eucaryotes, replication enzymes can move along the parental strands only from the 3′ end to the 5′ end, generating complementary strands running from 5′ to 3′ (Fig. 8.16). This means that one chain is copied continuously while the other must be replicated "backward," in discontinuous segments (as described in the *Exploring Further* box). Both groups of organisms have mechanisms that locate and correct errors; both have special mechanisms that prevent the chains from tangling.

Along with these general similarities come differences in detail. For example, because eucaryotic chromosomes are packaged on protein spools to form nucleosomes (see Fig. 5.4, p. 123), replication proceeds at a rate of only 50 base pairs per second; it takes time for the DNA to unwind from the spools and, after a duplicate set of spools has been synthesized, to be rewound. In addition, because eucaryotic chromosomes are much larger than procaryotic chromosomes, replication is initiated at many independent sites on each chromosome simultaneously. Otherwise, it might take weeks or even months for a complex eucaryotic cell to divide.

Another major difference between procaryotes and eucaryotes comes in their strategies for *segregation*, or sorting newly replicated chromosomes into two daughter cells. Procaryotes accomplish this task by attaching their single chromosome to the cell

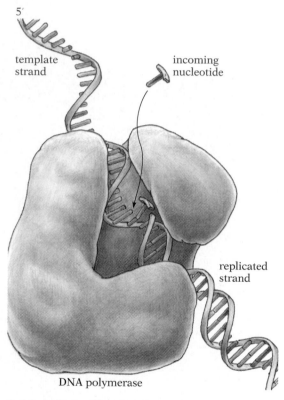

5′

template strand

incoming nucleotide

replicated strand

DNA polymerase

8.16   DNA polymerase

The approximate shape of DNA polymerase complex is known from X-ray crystallography. It is shown here replicating one strand of DNA, moving from upper left to lower right. Not shown is a ring-shaped protein that forms from two subunits and surrounds the DNA just behind the polymerase; by attaching itself to the replication enzyme, this protein keeps the polymerase firmly attached to the DNA it is copying.

membrane; the replicated chromosome also attaches to the cell membrane near the same anchor point. After replication, a membrane grows in between these two anchor points, enclosing one copy in each daughter cell (Fig. 8.17A). For eucaryotes, which typically have 10–20 times as many genes, and two copies of each of their several chromosomes, the problem is more difficult. In these organisms, a special spindle structure aligns each replicated pair along the plane of the cell's coming division. Microtubules run from each pole to one member of each replicated pair, and serve to draw each new cell's complement of chromosomes away from the midline as cell division proceeds (Fig. 8.17B).

Another unique feature of nearly all eucaryotes is sexual reproduction. Most eucaryotes reproduce by means of gametes (eggs and sperm, for example), which contain only one copy of each chromosome; fusion of two gametes restores the normal number. Clearly, the pattern of replication or segregation necessary to produce gametes must be different from that used in normal cell division (Fig. 8.17C). Sexual reproduction serves, at least in part, to promote variation between parent and offspring. This variation occurs because the two copies of each chromosome in an individual are not absolutely identical, since one of the chromosomes came from the female gamete—the egg—while the other was contributed by the male gamete.

Although each member of a chromosome pair contains the same kinds of genes in the same order, the base sequence of a gene on one chromosome is often slightly different from that of the same gene on the other, and thus may encode a product with slightly different characteristics. Alternative forms of a gene are called *alleles*. The precise set of alleles in one gamete of an individual usually differs from that in most of the other gametes, and usually differs substantially from the set of alleles in gametes produced by other individuals. The result is that when two gametes from different in

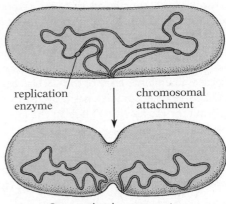

replication enzyme

chromosomal attachment

A Segregation in procaryotes

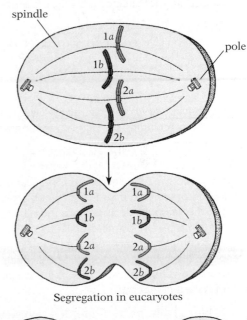

spindle

pole

B Segregation in eucaryotes

8.17 Alternative patterns of segregation after DNA replication

(A) In procaryotes, the chromosome remains attached to the cell membrane. After cell replication, the two copies are separated by the growth of the new membrane. (B) In most eucaryotes, there are two copies of each chromosome before replication, and each copy is used to produce a replica of itself; in this example, there was copy *a* and *b* of chromosome 1 and chromosome 2, each of which was then replicated to create the four pairs shown. As cell division proceeds, each pair of replicates becomes aligned along a plane, and one member of each pair is drawn to each end of the cell. (C) In preparation for the production of gametes, each copy of a chromosome is replicated, and then each two replicated pairs join to form a group of four—a tetrad. Two rounds of cell division follow, generating four gametes, each with only one of the two chromosomes in each pair.

C Gamete production

EXPLORING FURTHER

REPLICATION OF THE *E. COLI* CHROMOSOME

The process of replication in *E. coli* illustrates the complex series of steps necessary to make an accurate copy of a chromosome. Highly specific enzymes are responsible for each event. Like all enzymes they recognize specific reactants by their complementary shape and pattern of polar charges. And as enzymes often do, they work together to rearrange the bonds in various reactants and produce a final product—in this case, a complete copy of the circular bacterial chromosome. Enzymes in complex pathways work nearly simultaneously, but for clarity we will discuss the events of this process in sequence.

1 In *E. coli*, the process begins when a protein, DNA B, recognizes an initiation site on the chromosome by its particular sequence of bases and binds to it (top figure).

2 Next, molecules of the enzyme group known as DNA gyrases (or topoisomerases, enzymes that change the shape of molecules) begin to relax the supercoiling of the chromosome on each side of the DNA B protein.

3 As the two DNA gyrase molecules move away from the initiation site, two molecules of the **rep** enzyme unzip the double helix, using energy from ATP to break the hydrogen bonds that hold the bases together.

4 Single-strand binding proteins (SSB) then form a scaffolding, which holds the two strands apart and prevents them from rebinding to each other spontaneously (Fig. A).

5 The last of the five steps that must occur before the actual replication of DNA can begin involves a primer enzyme (the primase). Because the replication enzyme can only add bases to the end of an incomplete strand as it reads the template strand, there must be at least a few new bases in place. The primase binds to the initiation site and adds a complementary sequence about 10 bases long. Oddly enough, the primer is made of RNA, which must later be replaced by DNA; we will see how this substitution is accomplished presently.

6 Next, **DNA polymerase III**, a complex consisting of several proteins bound together, begins to replicate one of the two DNA strands by binding to it and adding complementary bases, creating, as its name implies, a polymer of DNA nucleotides. A complex is necessary because DNA polymerase must catalyze several different reaction steps and must be able to use four different nucleotides as reactants, depending on what it "reads" from the strand it is copying (the template strand). Presumably there are four

active sites, one each for adenine, cytosine, guanine, and thymine. Once the complex has read the template strand and brought the complementary nucleotide into place, it catalyzes the binding of the nucleotide to the growing complementary strand. The nucleotides that are added have been activated by the addition of a high-energy phosphate group from ATP, which provides the energy for this step.

Because of its enzymatic specificity, DNA polymerase III can add nucleotides only to the 3′ end of a nucleotide strand—the end without the phosphate group. This creates a serious problem: since the two strands of the double helix have opposite polarities, with one running from 5′ to 3′ and the other from 3′ to 5′ (see Fig. 8.10), they must be copied in opposite directions. Only one strand can be copied by DNA polymerase III, following along behind the rep enzyme as it unzips the DNA (Fig. A); the other strand must somehow be copied backward. The DNA formed by the DNA polymerase that follows the rep enzyme is known as the **leading strand**, while the DNA synthesized backward is known as the **lagging strand**. The latter is formed bit by bit in a looping "backstitch" pattern.

7 Backstitching begins when the primer enzyme synthesizes short lengths of complementary RNA nucleotides at intervals along the single-stranded DNA (Fig. A).

8 The short segments created in Step 5 provide the 3′ free end for the DNA polymerase III; from here the polymerase works backward, copying the strand, until it reaches the preceding RNA primer segment (Fig. B). One strand, then, is copied continuously while the other is copied in sections, known as Okazaki fragments, which are 1000–2000 bases long and are flanked by the RNA primers.

9 Now a series of enzymes must patch the fragments together into a continuous strand. The repair complex **DNA polymerase I** removes the anomalous 10-base RNA primer segments and replaces then with DNA (Fig. C).

10 Finally, the fragments are welded together by **DNA ligase** (Fig. D), and the new strand is finished.

Current evidence suggests that the two DNA polymerase IIIs—the one on the leading strand and the one on the lagging strand—are physically attached. This means that the section (shown here as linear) that includes the SSB proteins and primer, running from the rep enzyme and the DNA polymerase III on the lagging strand, is actually a loop.

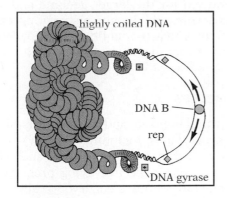

Step	Symbol	Substance	Function
1		DNAB	finds and marks initiation site
2		DNA gyrase	relaxes coiling
3		rep	separates DNA strands
4		SSB	hold strands apart
5,7		RNA primase	primes lagging strand for replication
6,8		DNA polymerase III	synthesizes complementary DNA
9		DNA polymerase I	erases primer and replaces with DNA
10		DNA ligase	welds gaps

dividuals fuse and begin developing into an organism, the resulting individual is likely to be at least somewhat different from either of its parents, and even from its siblings.

Despite the differences required for the larger genome and sexual reproduction of most eucaryotes, the biochemistry of procaryotic and eucaryotes replication is nearly identical. The *Exploring Further* box on page 216 summarizes the process in bacteria.

■ REPLICATION IN ORGANELLES

As described in Chapter 5, mitochondria and chloroplasts share several characteristics with procaryotes, including (usually) circular chromosomes. The most interesting interpretation of such similarities is that these organelles were once free-living procaryotes that took up symbiotic residence in primitive eucaryotic cells.

The discovery that organelles must have their own genes came in 1909, when Carl Correns traced variegation, which is marked by the appearance of white patches on green leaves, in the bushy, evening-blooming plant called four o'clocks (*Mirabilis jalapa*). He discovered that variegation is transmitted through the cytoplasm of the maternal gamete. Correns correctly guessed that chloroplasts must contain genes and replicate themselves, and that variegation can occur when one or more of the genes involved in chlorophyll synthesis is defective. Since plants cannot survive without some functioning chloroplasts, the mutations must be carried by some but not all of the chloroplasts supplied by the seed-producing plant. (Pollen grains, the male gametes, lack chloroplasts.) During the course of rapid cell division in growing leaves, certain cells, by chance, wind up with only chloroplasts carrying the mutant gene; cells derived from the first cell that lacks at least one chlorophyll-producing chloroplast appear as an elongated streak on the leaf.

The first evidence of genes in another organelle—the mitochondrian—did not come until 1938, when T. M. Sonneborn discovered that some strains of *Paramecium aurelia* carry a cytoplasmic gene that produces a poison (originally called the kappa factor) that kills other strains. The poison is produced by a mitochondrial gene. Mitochondria, like chloroplasts, replicate themselves and are inherited from the female (egg-producing) parent in nearly all species.

We now know that DNA replication and organelle division in mitochondria and chloroplasts occur out of phase with chromosome replication in the nucleus, though the pace of cell division regulates organelle division in some way. The DNA in organelles is almost always circular (Fig. 8.18). Each organelle contains several copies of its DNA, and this DNA is more similar to procaryotic DNA than to the nuclear DNA of eucaryotic cells: it lacks the his-

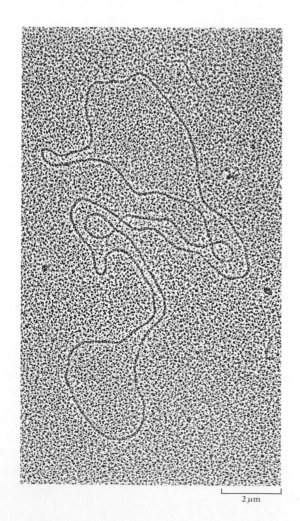

2 μm

8.18   Organelle DNA

Electron micrograph of mitochondrial DNA from a liver cell of a chicken.

tone nucleosome cores characteristic of nuclear DNA, for example, and is attached to the inner membrane of the organelle much as the procaryotic chromosome is attached to the cell membrane. However, it differs from bacterial DNA in having considerably fewer base pairs: the *E. coli* chromosome encodes about 3000 products, while the genes of a typical animal mitochondrion encode about 40. Since these gene products are insufficient to carry out organelle synthesis and operation (even in bacteria at least 90 gene products are required just for replication, transcription, and translation), most of the genes necessary for organelle function must at some point in evolution have been transferred to the cell's nucleus.

■ HOW ERRORS ARE REPAIRED

The precise replication of DNA is essential for normal cell function. *Mutations*—random changes in genes—are far more likely to disrupt a pathway than to improve it. If the delicate architecture of an essential enzyme is altered, it usually becomes less active, thus slowing the entire reaction chain of which it is a part. Since the genetic instructions for synthesis of the thousands of enzymes, structural proteins, regulatory proteins, and so on are each hundreds or thousands of bases long, an error rate as low as one in a thousand bases is far too great. Special enzymes have evolved in both procaryotes and eucaryotes to detect and repair mutations. Some enzymes keep uncorrected replication errors at a very low level, while other enzymes locate and repair most of the damage that occurs to the DNA *between* periods of replication. Repair enzymes attach themselves to faulty or damaged areas of the DNA that display the patterns of spacing and polar charges to which their active sites bind; the enzymes loosen specific bonds and catalyze the formation of new ones. Low rates of repair are correlated with susceptibility to cancer and premature aging.

The need for proofreading The initial error rate of the replication enzymes of both procaryotes and eucaryotes is about 3 in every 100,000 pairs.[3] If the errors produced at this rate were left uncorrected, the result would be a mistake in roughly 3% of each cell's proteins—perhaps 1000 changed proteins in every human cell after each replication.

Fortunately, the DNA polymerase complex includes one or more enzymes that successively proofread each base, clipping out mistakes. Other enzymes in the polymerase complex then substitute a rematched base for the excised unit; the complex then moves on without checking again. This second pass reduces the chance of an error to roughly 1 in 10^9. For humans, with our 50,000 functional genes, each of which has an average of about 1500 bases on each

---

[3] This value, like much of what is known about how DNA polymerases work, is obtained from studies of polymerase complexes with one or more inactive component enzymes. From the change in the operation of the complex, the function of the inactive enzyme can often be deduced. In this case replication proceeds when the enzymes under study are inactive, but most errors are not corrected.

of its two strands, this corresponds to an error in a gene somewhere in the chromosomes once in every 10 cell divisions.

Repairing mutations The integrity of the genetic message is also threatened by alterations in base sequences induced by heat, radiation, and various chemical agents. The rate at which these mutations occur is astoundingly high: thermal energy alone, for instance, breaks the bonds between roughly 5000 purines (adenine and guanine) and their deoxyribose backbones in each human cell every day. Cytosine is chemically converted to uracil (a nucleotide normally found only in RNA, and misread by DNA replication and transcription enzymes) at a rate of about 100 per cell per day. Ultraviolet (UV) radiation from sunlight fuses together adjacent thymines at a high rate in exposed epidermal cells (see Fig. 9.20, p. 242). And yet, because of the continuous operation of repair enzymes, the rate at which mutations actually accumulate in cells is even lower than the rate of uncorrected errors in replication.

The strategy for repairing mutations is basically the same as for replication errors: enzymes locate and bind to the faulty sequences and clip out the flaws, and the intact complementary strand guides repair. A remarkable array of specific enzymes is involved. For instance, chromosomes are scanned for chemically altered bases by fully 20 different enzymes, each specific for a particular class of problems. Another five or so enzymes are specific for finding faulty covalent bonding between a base and some other chemical, or between adjacent bases on one strand; the fusion of thymines induced by UV radiation is the most common error of this type. Other enzymes bind at the sites of missing bases, like the purines that are easily lost to thermal energy. In all, some 50 enzymes locate and correct errors.

Although elaborate, no enzymatic repair system is perfect, so some mutations manage to survive. The results can range from neutral effects like blue eyes (which result from a defective pigment gene) to genetic diseases. Mutations survive for several reasons: Some errors are missed in the repair process, and others are repaired incorrectly. When a mutation occurs just before or during replication, there may not be time for detection and repair. Still other mutations are not overlooked by the repair system, but are actually created by it.

One well-understood class of mutations generated by repair enzymes is *misalignment deletion*. Small deletions (loss of a few base pairs) tend to occur at susceptible sequences. The basis of this differential susceptibility is the relative weakness of the base pairing between adenine and thymine: they are connected by only two hydrogen bonds, whereas cytosine and guanine are connected by three. The hydrogen bonds are continually breaking and re-forming, and it frequently happens, particularly in regions rich in adenine-thymine pairs, that by chance all the bonds in a small segment of DNA are broken simultaneously. The bonds re-form spontaneously, but occasionally the rebonding is incorrect. For example, when a region with an extended series of adenine-thymine pairs undergoes a transient separation, there is a small chance that the two strands will be misaligned when they pair again. DNA repair enzymes may then remove the unpaired bases and thereby in-

3′ ···—G—C—T—T—T—T—T—T—G—G— ··· 5′
5′ ··· —C—G—A—A—A—A—A—A—C—C— ··· 3′

A

3′ ···—G—C—T—T—T—T—T—T—G—G— ··· 5′
5′ ··· —C—G—A—A—A—A—A—A—C—C— ··· 3′

B

3′ ···—G—C T—T—T—T—T—G—G— ··· 5′
5′ ··· —C—G—A—A—A A—A—C—C— ··· 3′

C

repair
enzyme

3′ ···—G—C T—T—T—T—T—G—G— ··· 5′
5′ ··· —C—G—A—A—A A—A—C—C— ··· 3′

D

3′ ···—G—C—T—T—T—T—T—G—G— ··· 5′
5′ ··· —C—G—A—A—A—A—A—C—C— ··· 3′

E

8.19 Misalignment deletion

(A) A sequence with an extended series of A–T pairs is especially susceptible to deletion. (B) When, by chance, a series of hydrogen bonds breaks simultaneously, the two strands of the helix separate transiently. (C) When they pair again, there is a small chance that they will be misaligned. (D) If the resulting distortion is detected by DNA repair enzymes, the two unpaired bases will be cut out by the enzymes . (E) The loose ends will then be reattached with DNA ligase, leaving each strand of DNA one nucleotide short.

troduce a misalignment deletion (Fig. 8.19). The repair is likely to alter the meaning of the gene by causing errors during translation of the gene's mRNA.

Another common mutation involves regions in genes that normally have the same three-base sequence repeated end to end several times (for example, . . . CAGCAGCAGCAGCAG . . .). Sometimes, either during replication or repair, extra copies of the triplet are added. When the number is large enough, the gene becomes defective: a massive CAG repeat is the cause of Huntington's disease, while one kind of mental retardation results from a hugely expanded CGG repeat.

Finally, some mutations simply cannot be detected by the repair enzymes. An example of this occurs in bacterial DNA when a cytosine that was modified by the addition of a methyl group ($-CH_3$) loses its amino group ($-NH_2$) after replication, producing a thymine (Fig. 8.20). The thymine is now mismatched with guanine, the partner of its predecessor. Since thymine is a normal base, there is no way for the repair enzymes to determine whether the incorrect base is the thymine in one chain or the guanine in the other.

The methylated-cytosine problem arises because the methylation of certain cytosines can be a useful adaptation. In bacteria, enzymes known as **endonucleases**, present in the cytoplasm, are often able to break up the DNA of invading viruses. (As detailed in Chapter 10, the DNA-chopping ability of certain endonucleases is used in recombinant DNA research.) If the enzymes digest invading DNA, what keeps them from also cutting bacterial chromosome? In the bacterial DNA, cytosines in those sequences

Cytosine   Methylated   Thymine
            cytosine

8.20 Deamination of methylated cytosine

Cytosine is sometimes methylated by enzymes. When a methylated cytosine undergoes deamination (loss of its amino group), usually as the result of oxidation by a mutagenic chemical, the result is a normal thymine. This change cannot be corrected reliably because the repair system cannot determine whether it is the thymine or the guanine (the base in the complementary strand to which the cytosine was originally paired) that is incorrect.

vulnerable to the endonucleases are methylated, so the endonucleases cannot bind to them. However, bacterial DNA protected in this way from endonuclease action is susceptible to mutation by conversion of cytosines into thymines. (There is active methylation of specific DNA sequences in eucaryotes as well, though its function there is to help regulate genes.)

CHAPTER SUMMARY

WHAT ARE GENES MADE OF?

THE UNUSUAL CHEMISTRY OF CHROMOSOMES DNA is the one substance whose quantity does not usually vary from one somatic cell to another, although it is reduced by half in gametes. (p. 205)

GENES: DNA OR PROTEIN? Chromosomes are composed of DNA and protein. Only the DNA of viruses needs to enter their host in order to begin multiplying; thus DNA is the genetic material. (p. 206)

THE MOLECULAR STRUCTURE OF DNA DNA has four bases: adenine, guanine, cytosine, and thymine. They are long chains of sugars linked by phosphate groups; pairs of chains are attached to one another through polar bonding between complementary bases—adenine to thymine, guanine to cytosine—to form a double helix. (p. 208)

HOW DNA IS REPLICATED

THE TEMPLATE HYPOTHESIS Replication occurs by separating the two strands of a double helix and using each strand as a template for synthesizing a complementary strand. (p. 212)

THE DIFFICULTIES OF REPLICATION The strands must be kept separate while the various enzymes do their work. The polymerase can only synthesize DNA in one direction, so one strand must be replicated in small backward steps, which must then be joined. In eucaryotes the nucleosomes must be removed and a new set synthesized. The daughter chromosomes must be segregated into separate progeny cells. (p. 214)

REPLICATION IN ORGANELLES Organelles have their own very small chromosomes, similar to bacterial DNA in structure and replication. (p. 218)

HOW ERRORS ARE REPAIRED Mutations occur in the DNA, either because of the action of chemicals or UV light on the chromosomes, or through errors in replication. Proofreading by the polymerase minimizes replication errors. Many kinds of special enzymes patrol the chromosomes looking for spontaneous mutations and repairing them. (p. 219)

STUDY QUESTIONS

1 Compare and contrast ATP and the nucleotide having adenine for its nitrogenous base (that is, adenosine monophosphate, or AMP). (pp. 158, 209)

2 Bacteria have only one copy of each gene, whereas most eucaryotes have two. If most serious mutations destroy the activity of the product encoded by a gene, what effect does having two copies of a gene have on the chance of suffering a complete loss of a gene?

3 The bacterial chromosome has about 4 million base pairs. Assuming all bases occur with equal frequency, calculate the approximate total energy of the hydrogen bonds holding the two strands together. (p. 35)

4 Assuming one ATP molecule is needed to prime each of the 8 million bases added during *E. coli* replication, how many glucose molecules must be metabolized to fuel a single such cell division? (Ignore the energy needed for the other steps.) (p. 175)

5 In the 1958 experiment of Meselson and Stahl, what pattern of DNA densities would be observed halfway through the first replication? halfway through the second replication? (pp. 212–214)

SUGGESTED READING

HOWARD-FLANDERS, P., 1981. Inducible repair of DNA, *Scientific American* 245 (5). *On how cells recognize when DNA has been damaged, how they "switch on" genes to generate repair enzymes, and how the enzymes work.*

LINDAHL, T., 1993. Instability and decay of the primary structure of DNA. *Nature*, 362, 709–714. *An excellent overview of spontaneous mutation, repair, and their connection with cancer and aging.*

RADMAN, M., AND R. WAGNER, 1988. The high fidelity of DNA replication, *Scientific American* 259 (2). *On how the proofreading component of DNA polymerase works.*

WANG, J. C., 1982. DNA topoisomerases, *Scientific American* 247 (1). *On the enzymes responsible for untangling DNA during replication.*

YUAN, R., AND D. L. HAMILTON, 1982. Restriction and modification of DNA by a complex protein, *American Scientist* 70, 61–69. *On how some endonucleases can cut DNA at a specific site if it is fully unmethylated, or finish methylating (and thereby protect) the same site if it is partially methylated.*

CHAPTER 9

TRANSCRIPTION AND TRANSLATION

In this chapter we will look at how information is actually encoded in DNA, and how the information is transcribed into ribonucleic acid (RNA). We will also see how a fraction of this RNA is used directly in structural and enzymatic roles, while the rest—messenger RNA (mRNA)—is translated into protein. Finally, we will look at some of the consequences of transcription and translation errors.

HOW TRANSCRIPTION WORKS

With a few exceptions, every cell has a full set of chromosomes, which are replicated, repaired, and passed on at every cell division. But at any given moment, only a small proportion of the thousands of genes in the cell are active. Which genes are needed depends not only on what the cell is doing—dividing, growing, resting, moving—but also on the cell's environment and, if it is part of a complex multicellular organism, on its specialty. Transcription enzymes must recognize the small number of active genes in a cell and ignore the rest, and must know where genes begin and end. (Replication, which is an all-or-none process, does not make any distinction between genes.) In Chapter 11 we will take up the important question of *gene expression*: how genes are activated and inactivated according to the needs of the cell.

■ THE NEED FOR MESSENGER RNA

Genetic information is stored in the DNA of chromosomes, while the proteins encoded by genes are synthesized on ribosomes. How is information conveyed from the genetic library to the cellular building sites? By the early 1940s cells in tissues such as the vertebrate pancreas, where protein synthesis is particularly active, were known to contain large amounts of RNA. Radioactive-tracer experiments demonstrated that eucaryotic RNA is synthesized in the nucleus and then moves into the cytoplasm (Fig. 9.1). These two pieces of evidence suggested that RNA might be the chemical messenger between DNA and the ribosomes.

Though RNA and DNA are very similar, they differ in three important ways, described in Chapter 3: (1) The sugar in RNA is ribose, whereas in DNA it is deoxyribose (Fig. 9.2). (2) RNA has

A

B

25 μm

9.1 Radioactive-tracer experiment showing the movement of messenger RNA

(A) Cells were placed briefly in a medium containing radioactive RNA nucleotides. After 15 min the labeled nucleotides were incorporated into messenger RNA in the nucleus, where they appear as dark spots. (B) 75 min later, however, the labeled mRNA was found in the cytoplasm, demonstrating that mRNA moves from the nucleus to the cytoplasm.

Uracil   Thymine

Ribose   Deoxyribose

9.2 Comparison of uracil in RNA and thymine in DNA

The five-carbon sugars of RNA and DNA differ only at the site shown in color, where deoxyribose lacks an oxygen atom that is present in ribose. Uracil differs from thymine only in that it lacks a methyl group ($-CH_3$).

uncopied DNA strand

copied DNA strand

mRNA   RNA polymerase

A

B $\overline{0.2\,\mu m}$

9.3 Transcription of a gene

(A) As the RNA polymerase complex moves along the DNA, it catalyzes transcription of only one of the two DNA strands. (B) An electron micrograph shows three polymerases transcribing a gene in *E. coli*.

uracil where DNA has thymine;[1] (3) RNA is ordinarily single-stranded, whereas DNA is usually double-stranded.

These differences aside, DNA acts as a template for the production of RNA. The synthesis proceeds in essentially the same way as that of new DNA: the two strands of a DNA molecule uncouple, and RNA is synthesized along one of the DNA strands by RNA polymerase (Fig. 9.3). For every adenine in the DNA template, a uracil ribonucleotide, rather than a thymine, is added to the growing RNA strand; for every thymine in the DNA, an adenine ribonucleotide is added; for every guanine, a cytosine; and for every cytosine, a guanine (Fig. 9.4). In short, the synthesis of RNA—the process of *transcription*—operates exactly like replication; the resulting strand of RNA, which is complementary to the transcribed DNA, will later act as an intermediary in protein synthesis.

There are three major types of RNA: ***messenger RNA (mRNA)***—the type just discussed—carries the information necessary to specify the sequence of amino acids in a protein to the ribosomes where proteins are synthesized; ***ribosomal RNA (rRNA)*** forms part of the ribosomes themselves; and ***transfer RNA (tRNA)*** brings amino acids to the ribosomes during protein synthesis.

■ CREATING MESSENGER RNA

Transcription of DNA into RNA is accomplished by a six-protein enzyme complex called ***RNA polymerase***, which both binds to the DNA and opens up the helix. Like DNA polymerase, the transcription complex moves from the 3' end to the 5' end of the DNA, and synthesizes a strand of the opposite polarity (5' to 3'), in this case a strand composed of ribonucleotides. In procaryotes, a single kind of RNA polymerase is responsible for all RNA synthesis. In eucaryotes, RNA polymerase II is responsible for the synthesis of mRNA. Two other RNA polymerases synthesize rRNA, tRNA, and other RNAs that serve structural and, in some cases, enzymatic roles; all of these RNA polymerases work in much the same way.

Transcription in procaryotes   RNA polymerases know where to start and stop synthesizing the messenger because specific control sequences in DNA mark the beginning and end of a gene. The first pair of these is called the ***promoter***; active sites on the RNA polymerase bind specifically to regions with the base sequences of the promoter. The promoter in *E. coli*, reading from 5' to 3' on the complementary strand, usually begins with TTGACA or a very similar sequence.[2] This is followed by roughly 17 bases with another

---

[1] The presence of uracil in place of thymine in RNA makes it possible for the active site of a wide variety of enzymes to distinguish DNA from RNA, and so to be selective in their activity. DNA polymerase, for instance, does not attempt to replicate or transcribe messenger RNA, nor can enzymes in the cytoplasm that digest the DNA of invading viruses chew up the cell's own RNA.

[2] Though the RNA polymerase works from the template strand of the DNA, the custom is to refer to the sequences in question as they exist on the complementary strand. This method of reference is convenient because the RNA sequences produced in transcription are identical to those of the corresponding regions of the complementary DNA strand—except, of course, that U is substituted for T. Thus, since the first three DNA bases transcribed are usually GTA (Fig. 9.5A), they are rendered as CAU in the mRNA, which corresponds to CAT in the complementary

9.4   Synthesis of RNA by transcription of a DNA template

The sugar in RNA (ribose, r) is slightly different from that in DNA (deoxyribose, d); in addition, the uracil (U) in RNA takes the place of the thymine in DNA. Transcription is accomplished by the enzyme complex RNA polymerase (not shown).

function; then comes TATATT or a very similar sequence (Fig. 9.5A). The polymerase recognizes these two regions in the DNA and is large enough to bind simultaneously to both.

The exact promoter sequences shown in Figure 9.5 are called *consensus sequences*, since each base shown is the one most commonly found at the promoter location in question in procaryotes. (Much less is known about eucaryotic promoters, but the sequence TATA appears to be important in them as well.) The promoter se-

9.5 Transcription signals in *E. coli* DNA

(A) RNA polymerase binds to a region of DNA with the sequences shown. Once the polymerase is bound, transcription begins with the GTA signal (corresponding to the CAT sequence on the complementary strand) and proceeds until a termination signal is encountered. (B) The termination signal consists of a "self-complementary" sequence—one that can bind to itself—and four to eight adenines (corresponding to thymines on the complementary strand). (C) At this point, the tail of the mRNA forms a short double-helix hairpin loop, and the polymerase stops synthesizing. Note that RNA polymerase inserts a uracil wherever DNA polymerase would add a thymine.

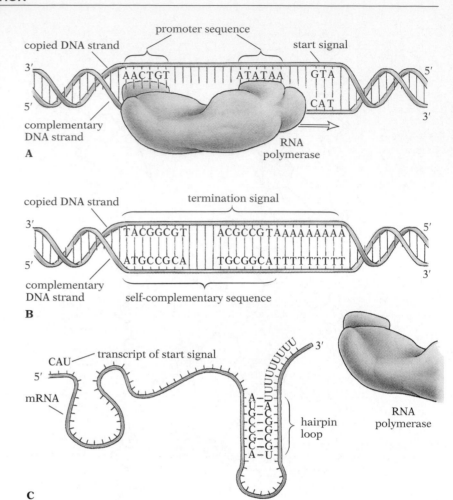

quences of most genes differ in one way or another from the consensus sequences; more than a hundred variations have been discovered in *E. coli*, for example. Minor variations in the promoter sequences cause the polymerase to bind less strongly and less often to some of them, so some genes are transcribed less frequently than others.

Once bound, the polymerase does not begin synthesizing mRNA immediately. Instead, it must find the start signal—often CAT—located about seven bases beyond the binding point, toward the 3′ end of the complementary strand. In most procaryotes, synthesis of mRNA continues until the polymerase encounters a termination signal. The termination signal has two components. First, there is a region with a base sequence that allows the corresponding bases in the tail of the mRNA to pair off and bind together to form a small loop, known as a **hairpin loop** (Fig. 9.5C). This is followed on the copied strand by a run of four to eight adenines.

When the polymerase moves into the region of the adenine run, the hairpin loop forms in the RNA just produced and, apparently from the physical stress it puts on the enzyme complex, slows or temporarily halts transcription. Two things now happen that terminate the production of the RNA. First, because the loop sequence has pulled away from the DNA, that portion of the gene is able to re-form its double helix, adding yet more strain to the complex. Second, the weak bonding between the run of adenines in the DNA and the uracils of the RNA copy (only two hydrogen bonds between each base pair) is unable to anchor the RNA to the gene for long; the pause and the stress together allow the transcribed copy and its polymerase to drift away from the chromosome.

Transcription in eucaryotes: editing of mRNA The mechanism of mRNA synthesis just described is found in procaryotes, mitochondria, and chloroplasts. Transcription in eucaryotic nuclei is more complicated. One complication is that eucaryotic mRNA is "tagged" on both ends: a cap of 7-methylguanosine is added at the front (5′) end, while a tail of 100–200 adenines is affixed to the 3′ end. Among other things, these modifications are used later to identify the molecule as mRNA. The result—the **primary transcript**—is still not a usable messenger.

The dramatic difference between a primary transcript and functional mRNA came in 1977 as a total surprise: Phillip A. Sharp discovered that though most primary transcripts are about 6000 bases long, the mRNA actually transcribed is only about one-third that length. Large specific regions *within* the primary transcript of most eucaryotic messengers must be removed in the nucleus to create a functional mRNA molecule. The regions of the primary RNA transcript that survive this processing and operate during protein synthesis (as well as the parts of the gene that gave rise to these sections) are called **exons** (because they are expressed), while the intervening sequences of the primary transcript, that are removed in the nucleus (and the corresponding regions of the gene) are referred to as **introns** (Fig. 9.6).

Despite their early removal, many introns are necessary to the functioning of RNA: most mRNA transcribed from artificially manufactured genes that lack introns fails to get into the cytoplasm, while mRNA from genes with some introns intact is often processed correctly and slips through the nuclear pores to the cytoplasm.

Intron removal is a formidable task (Fig. 9.7). Some genes have as many as 50 introns, and a mistake of even a single base in the excision process can render the mRNA useless. The precise beginnings and ends of introns on the primary transcripts are marked by signals so that they can be recognized and removed.[3] The intron-boundary signals are recognized by a short bit of RNA found in curious RNA-protein complexes known as the small nuclear ribonucleoprotein particles—snRNPs (sometimes pronounced "snurps"), for short. At least four different kinds of snRNPs cooperate in most splicing. The RNA in these particles is like ribosomal RNA in that it is used directly, and has both an enzymatic and a structural role.

The best known snRNP has a base sequence complementary to the boundary between the end of exons and the beginning of introns. SnRNP complexes catalyze the breakage of the bond between the primary-transcript nucleotides at each intron-exon

9.6   Visible evidence of introns

A gene for an egg protein, ovalbumin, was denatured to create single-stranded DNA, and then mixed with mature mRNA transcribed from this gene. Because the mRNA sequence is complementary to the DNA strand from which it is transcribed, the two molecules are able to form a double-stranded hybrid. Because the introns in the primary transcript have been removed by editing, the intron regions in the DNA form unmatched loops, revealing that this gene has six introns. The scale in the photograph is 1 μm.

---

[3] The sequence at the beginning of the intron is AG-GUAAGU; the hyphen marks the exon-intron boundary. At the end of the intron it is CAG-G.

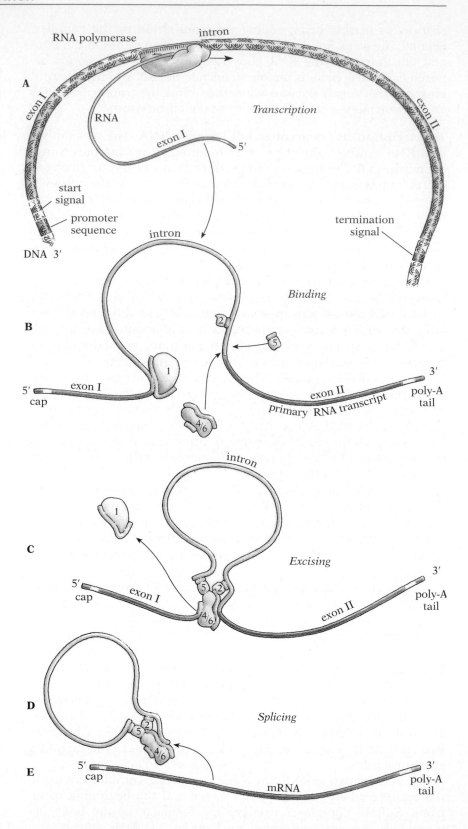

9.7 Messenger RNA processing in eucaryotes

(A) RNA polymerase binds to a promoter sequence and moves along the DNA strand, synthesizing a complementary RNA strand. (B) A 7-methylguanosine cap is added at the beginning of the transcript by another enzyme. After the RNA polymerase transcribes the termination signal, the RNA between the last exon and the sequence of four to eight adenines is removed; a separate enzyme extends the poly-A tail by 100–200 bases, completing the primary RNA transcript. (B–E) With the aid of several small nuclear ribonucleoprotein particles (snRNPs), any introns are removed and the resulting transcript is spliced. For clarity, only one intron is shown. The process begins with recognition of the beginning of the intron by snRNP 1 (B). Next, snRNPs 2, 4/6, and 5 come together to form a complex at the end of the intron (C). The leading end of the intron is cut (D) and bound to the complex, which then cuts the trailing end of the intron and splices the two exons together (E). At this point the mature mRNA is ready for export to the cytoplasm.

0.1 μm

9.8 A snRNP-intron complex

boundary. The intron drifts away (Fig. 9.8) and is thought to be digested by other enzymes. Other components of the snRNP complex then splice the two exons. Once all the introns have been removed, the mature mRNA is exported to the cytoplasm. This splicing system probably evolved from a self-splicing mechanism, still seen in a few genes in certain species: the intron itself folds into a catalytic form, excises itself, and splices the exons together (Fig. 9.9).

HOW THE MESSAGE IS TRANSLATED

Messenger RNA contains the sequence of bases that specifies the identity and order of the amino acids that will be linked to form the protein encoded by the gene. This sequence, which is preceded by a short "leader" and followed by a "trailer" (analogous to the blank margins at the top and bottom of a page), is decoded on a ribosome, a process called ***translation***: information is translated from one molecular "language" to another. We will look first at the two languages, and then at the process of translation.

■ THE GENETIC CODE

The need for codons When Watson and Crick discovered that DNA is essentially a linear array of four nitrogenous bases, it became clear that the unique sequence of amino acids found in a particular protein must be encoded by *groups* of bases. After all, if each DNA base designated a particular amino acid, there would have to be 20 different bases instead of four. If the bases were taken two at a time—AA, AC, AG, AT, CA, CC, and so on—only 16 combinations would be possible. Since 20 amino acids are commonly present in proteins, the information has to be encoded in sets of three or more bases, called ***codons***.

9.9 A self-splicing intron

The intron portion of the transcript has formed internal pair bonds and folded into an enzymatic conformation that brings the end of one exon adjacent to the beginning of the next exon. In this drawing the 5' end of the intron has just been cut free of the transcript.

Crick and his associates established the length of the codon in 1961. Bacteria were treated with compounds that insert or delete nucleotides from DNA. Crick reasoned that adding or deleting a single nucleotide would render a message meaningless because it would disrupt the translation process by shifting the apparent starting point in subsequent coding units. Suppose the message is

THE BIG RED ANT ATE ONE FAT BUG

Deletion of the first *E* will make it

THB IGR EDA NTA TEO NEF ATB UGX

A single deletion redefines all subsequent codons; if it occurs near the start of the message, the amino acid sequence of the protein is completely altered. Two deletions would have the same effect, but if the number of deletions corresponded to the codon length, and if they all occurred near the beginning, translation would produce an enzyme that would probably retain some activity:

THE GRD ANT ATE ONE FAT BUG

Crick and his associates made different numbers of nucleotide deletions, and determined for each concentration whether or not active enzymes were produced. Subjecting these results to elaborate statistical analyses, they concluded that codons were three bases long.

Deciphering the code The next problem to be solved in understanding translation was to determine the exact relationships between codons and amino acids. This was a long and difficult process, and was not completed until H. G. Khorana developed a technique for generating specific RNA sequences. Table 9.1 on page 242 summarizes the genetic code.

As the genetic dictionary of the table makes clear, all but two amino acids (methionine and tryptophan) are represented by more than one codon. The synonymous codons—those coding for the same amino acid—usually have the same first two bases, but differ in the third; thus CCU, CCC, CCA, and CCG all code for proline. In fact, U and C are always equivalent in the third position, and A and G are equivalent in 14 of 16 cases. Notice also that four codons—AUG, UAA, UAG, and UGA—serve as "punctuation," marking the beginning and end of the actual message within the mRNA.[4]

■ HOW RIBOSOMES WORK

Translation occurs on ribosomes. Ribosomes have a large and a small subunit, each a complex of ribosomal RNA (rRNA), enzymes, and structural proteins (Fig. 9.10). When not carrying out protein synthesis, the ribosomal subunits are separate. In procaryotic ribo-

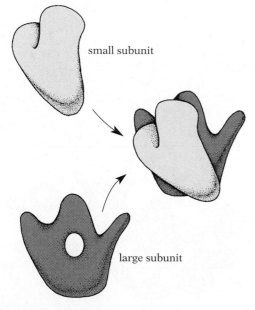

small subunit

large subunit

9.10 A ribosome

A functional ribosome is formed when the two kinds of independent subunits join. This joining can take place only after mRNA binds to the small subunit. The tunnel in the large subunit is about 2.5 nm in diameter.

---

[4] Certain minor inconsistencies in the genetic code have been discovered. For example, in some ciliates the codons UAA and UAG, which are normally termination signals, instead code for glutamine; in some bacteria and most (perhaps all) eucaryotic mitochondria, UGA, which can also be a termination signal, codes for tryptophan.

paired

folded

somes the large subunit is composed of two molecules of rRNA and roughly 35 proteins; the small subunit has one molecule of rRNA and approximately 20 proteins. Eucaryotic ribosomes are somewhat larger. In both eucaryotes and procaryotes, the rRNA itself is highly structured: long segments of complementary base pairs are held together in a helical chain by hydrogen bonds, creating a complex pattern of arms and loops (Fig. 9.11).

In procaryotes, the 5′ end of the mRNA binds to the small subunit, which is then able to bind the large subunit (Fig. 9.12). The binding of the mRNA involves base pairing between a section of the rRNA and a binding signal—usually AGGAGGU, near the end of the mRNA in the leader segment. This signal binds in the small subunit's cleft (Fig. 9.12B–C), along with the initiation codon (AUG) located just a few bases farther along on the mRNA. Special proteins called initiation factors aid in this process. When the binding of the large subunit is complete, translation can begin. AUG codes for methionine, so methionine is always the first amino acid incorporated during translation. Another special enzyme later removes the methionine from the end of most polypeptides.

Because procaryotes have no nuclear membrane, and the process of transcription is not segregated from the ribosomes and other translational machinery in the cytoplasm, ribosomes are able to bind one end of an mRNA molecule and begin translating it into

9.11 Ribosomal RNA

Ribosomal RNA has many stretches of complementary base pairs (color), which allow the chain to fold back on itself and form double-stranded helical arms. The open loops represent areas in which the bases are not complementary. This arrangement probably plays an important role in ribosomal function. The rRNA shown here is found in the small ribosomal subunit of a bacterium.

protein while the RNA polymerase is still transcribing the rest of the message from the chromosome. (The same situation occurs in chloroplasts and mitochondria.) While the first ribosome to bind is translating later parts of the message, additional ribosomes can bind and begin translation (Fig. 9.13).

9.12   Synthesis of a polypeptide chain by a ribosome

Free ribosomes exist as two separate subunits (A–B). A signal sequence on the mRNA binds to the small subunit, causing a change that allows it to bind the large subunit (B–C). The ribosome then translates the mRNA into a polypeptide chain, producing a protein (D). The growing polypeptide chain is thought to be threaded out through the tunnel in the large subunit. The mRNA is much longer than shown here, and additional ribosomes can bind and begin translation while the first ribosome is still at work.

9.13   Simultaneous transcription and translation

In procaryotes and in eucaryotic organelles, translation can begin as soon as the first part of a message has been transcribed from the chromosome, and several ribosomes can be involved simultaneously. The complex of mRNA and two or more ribosomes is often called a polysome. In the micrograph, the two thin horizontal lines are DNA; the lower strand is being transcribed and, simultaneously, the RNA transcripts are being translated by ribosomes (dark spots). Note that the transcripts become longer from left to right as transcription of the gene proceeds. The growing polypeptide chains are not visible. The boxed area corresponds to the interpretive sketch.

9.14 Nuclear pores

(A) The nuclear envelope contains many specialized pores, which provide a pathway from the nucleus to the cytoplasm for mRNA. The density of these portals is indicated by the micrograph of a small section of the nuclear membrane from a frog egg. (B) The probable structure of these octagonal pores includes a central plug, which appears to have two sets of eight protein arms, one on the inner face of the nuclear envelope, the other on the outer face. (C) The arms seem to act like an iris, opening to a diameter of 20 nm to admit correctly tagged macromolecules, and closing to 9 nm after they pass.

In eucaryotes, four steps intervene between transcription and translation. We have already mentioned three of these: modification of the transcript at the 5' end, modification at the 3' end, and splicing of the RNA after the removal of introns (see Fig. 9.7). The fourth step is movement through the pores of the nuclear envelope (Fig. 9.14) into the cytosol, where unbound ribosomes are available to do translation. Most RNA is translated by ribosomes in the cytosol, but the mRNA for certain proteins is always translated on the rough endoplasmic reticulum (ER). The mRNA of this type begins the process of translation in the normal way, by binding to a small ribosomal subunit in the cytosol and thereby making possible the binding of the large subunit and the start of translation. However, translation ceases almost immediately; the leading end of the partially synthesized protein contains a sequence that binds to a signal-recognition complex; this binding causes translation to stop.

The combined mRNA, recognition complex, and ribosome assembly then binds to the rough ER. The partially synthesized protein is inserted into a channel of the ER, translation resumes, and the growing chain of amino acids is fed into the lumen of the ER. At the end of translation, the ribosomal subunits release the mRNA, separate, and drift free of the ER. As detailed in Chapter 5, proteins synthesized in the rough ER are destined for noncytoplasmic roles: they may be soluble, and either drift free in the ER lumen, or be collected together in the ER and packaged into secretory or other vesicles (Fig. 9.15). Alternatively, they may become mounted in the ER membrane, either to remain there to perform their function, or to be packaged into vesicle membranes or into the cell membrane itself. Most proteins synthesized on ribosomes in the cytosol, by contrast, remain in the cytosol.

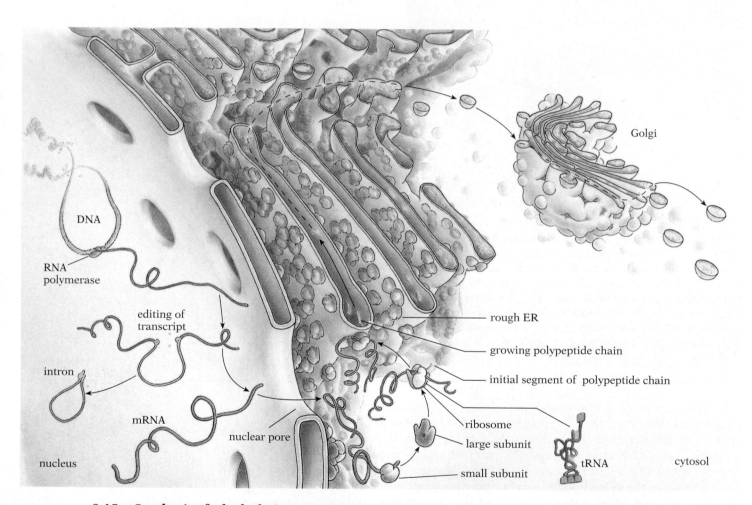

9.15   Synthesis of a hydrolytic enzyme

Most mRNA is translated freely in the cytosol, but when protein products are destined for the ER lumen, the ER membrane, or secretion, translation of mRNA occurs on the rough ER. A gene is transcribed, and the transcript is processed to form mRNA. Next, the mRNA passes through the nuclear pore (shown here without its plug) and binds to a small ribosomal subunit. A large ribosomal subunit then binds. Translation begins and continues until a signal sequence on the growing polypeptide chain binds to a signal-recognition complex that interrupts translation. The ribosome complex then binds to the rough ER and translation resumes, with the growing polypeptide chain passing into the lumen of the ER. Finally, the newly synthesized protein, in this case a hydrolytic enzyme, is packaged into a vesicle for transport to the Golgi apparatus.

■ THE NEED FOR TRANSFER RNA

In 1957 Mahlon Hoagland and his associates demonstrated that each amino acid becomes attached to some form of RNA *before* it arrives at the ribosome, where it is added to the growing polypeptide. This kind of RNA acts to transfer amino acids to the ribosomes; Hoagland and his colleagues called it transfer RNA. Each tRNA molecule binds a single molecule of amino acid, activates it with energy from ATP, and transports it to a ribosome.

At least one form of tRNA is specific for each of the 20 common amino acids; the amino acid arginine, for example, combines only with tRNA specialized to transport arginine. All forms of tRNA share certain structural characteristics: a length of 73–93 nucleotides, a structure consisting of a single chain that is folded into a cloverleaf shape, with internal base pairing (Fig. 9.16), and a CCA termination sequence. When an amino acid binds to its specific tRNA, it always binds at the CCA end.

Specific enzymes (aminoacyl tRNA synthetases) match and bind each amino acid with its appropriate tRNA (Fig. 9.17). The tRNA then carries its amino acid to a ribosome that has bound a strand of mRNA. There the tRNA becomes attached to the mRNA by complementary base pairing: each tRNA has an exposed portion containing an unpaired triplet of bases called an **anticodon**, which is complementary to an mRNA codon for its particular amino acid. For example, the mRNA codon CCG codes for the amino acid proline. One type of tRNA for proline has the anticodon sequence GGC, which is complementary to CCG. When a molecule of the proline tRNA with an attached molecule of proline approaches a molecule of mRNA, its exposed GGC triplet can bind to the mRNA only where the mRNA has a CCG triplet.

■ THE TRANSLATION CYCLE

The actual binding of tRNA molecules and linkage of the amino acids they carry is managed in a stepwise fashion by the large subunit of the ribosome, but only after completion of the preliminaries: first, the tRNA with the anticodon sequence for methionine is bound to the small subunit; in eucaryotes as well as procaryotes this sequence is complementary to the initiation codon on the mRNA, which can now also bind to the small subunit; finally, the large subunit is bound to the small one. As translation then proceeds, the ribosome moves along the mRNA, with the part of the message being translated lying in the groove between the two subunits; two adjacent tRNA binding sites—the **P-site** and the **A-site**—bring the appropriate nucleotide sequences together.

When the cycle begins, the anticodon of the first tRNA is in register with the initiation codon of the mRNA at the P-site (Fig. 9.18:4). The adjacent A-site then brings a molecule of tRNA with

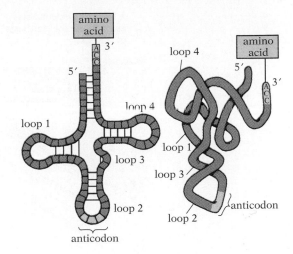

9.16 Structure of a molecule of tRNA

The single polynucleotide chain folds back on itself, forming five regions of complementary base pairing, four loops with unpaired bases, and an unpaired terminal portion to which the amino acid can attach. The unpaired triplet that acts as the anticodon is on loop 2. Loop 4 is thought to function in binding the tRNA to the ribosome (probably by binding to rRNA), and loop 1 probably binds an activating enzyme. The molecule is shown flattened at left; its tertiary structure is diagrammed at right.

9.17 Matching of an amino acid with its tRNA

The matchmaking enzyme responsible for loading a particular amino acid (glutamine in this case) onto its tRNA is shown in this computer-generated image in blue. It binds to the tRNA (red and yellow) at both the anticodon (bottom) and the acceptor arm (top center), the two regions of the tRNA responsible for its specificity. The green molecule is ATP.

1 DNA in nucleus acts as template for synthesis of mRNA

DNA

2 mRNA goes to cytoplasm, where it complexes with ribosomes

amino acids in cytoplasm

amino acids

3 tRNA carries amino acid to mRNA

tRNA

mRNA

4 tRNA bearing methionine couples briefly with mRNA on ribosome at P-site. Another tRNA, bearing the appropriate anticodon, pairs with the codon at the vacant A-site.

5 The amino acid that was attached to the tRNA at the P-site is detatched and bound to the amino acid on the next tRNA.

6 The first tRNA moves off to pick up more amino acid, and the ribosome moves along the mRNA.

9.18 Translation cycle on the ribosome

After messenger RNA is created by transcription (1), it enters the cytoplasm (2), where it binds to a small ribosomal subunit. After a large subunit binds, translation can begin. At this point the tRNA with the anticodon for methionine (which was bound to the small subunit before the two subunits are combined) is already paired with the initiation codon on the mRNA at the P-site (4). Another tRNA (3), bearing the appropriate anti-codon, pairs with the codon at the vacant A-site (5). Next the amino acid that was attached to the tRNA at the P-site is detached and bound to the amino acid on the tRNA at the A-site. The P-site tRNA then drifts free (6). Finally, the messenger is moved so that the remaining tRNA occupies the P-site, and the next codon is brought into position at the A-site, ready to accept another complementary anticodon (7). In these drawings part of the small subunit is shown cut away to reveal the mRNA, which lies in the groove between the subunits. The size of the tRNAs is greatly exaggerated.

the appropriate anticodon together with the next codon triplet on the mRNA (Fig. 9.18:5). Once the new tRNA is bound to the A-site, the enzyme peptidyl transferase moves the amino acid from the P-site tRNA and binds it to the amino acid at the A-site. The tRNA at the P-site, having relinquished its amino acid, is released (Fig. 9.18:6), and the cycle is completed as the ribosome moves along the mRNA by one codon, thereby bringing the codon that was at the A-site (and its tRNA, with the growing polypeptide chain) to the P-site (Fig. 9.18:7). As a result, the codon to be translated next now occupies the A-site.

The translation cycle adds about 15 amino acids per second to the polypeptide chain, with an error rate of about one mistake in every 30 chains. It comes to an end when the ribosome encounters one of the termination codons and, with the aid of a protein known as the release factor, causes the completed polypeptide to be released.

The ribosome complex's potential for producing protein is enormous. Since the average protein chain is 300–500 amino acids long, a ribosome can produce a chain in about 25–35 sec. Because many ribosomes can read a single mRNA at the same time, new peptide chains can be generated from each mRNA about every 3 sec. A typical active eucaryotic cell has on the order of 300,000 mRNA molecules in circulation; if sufficient raw materials are available, some 100,000 proteins can be synthesized every second. Cells specialized for secretion or rapid growth are far more active than this.

■ TRANSCRIPTION AND TRANSLATION IN ORGANELLES

Just as predicted by the endosymbiotic hypothesis (discussed in Chapters 5 and 19), transcription and translation in organelles are remarkably similar to the corresponding processes in procaryotes. For example, since there is no nuclear membrane to segregate a strand of mRNA from the ribosomes during its synthesis in an organelle, translation can begin at one end before transcription of the DNA has been completed at the other, just as it does in a procaryote (see Fig. 9.13). The resulting transcript receives neither the 7-methylguanosine cap nor the poly-adenine (poly-A) tail characteristic of mRNA produced in the nucleus.

Another similarity can be seen in the translation of the initiation codon, usually AUG. Eucaryotic tRNA pairs this codon with an unmodified methionine, whereas procaryotes begin translation with a tRNA carrying a special methionine—one to which a formyl group has been attached; organelles also begin translation with a formylated methionine. The rRNAs and ribosomal proteins of organelles are much closer in size and composition to those of procaryotes than they are to those elsewhere in the same cell (Fig. 9.19). Indeed, the ribosomes of *E. coli* and chloroplasts are interchangeable.

Despite their genetic similarity to procaryotes, organelles have many fewer genes. Mitochondrial DNA, for instance, usually encodes about 40 products, including two of the three rRNAs of the organelle ribosomes, all of their various tRNAs, one of the dozens of ribosomal proteins, one of the nine proteins of the F_1 complex

9.19 Proteins from chloroplast and cytoplasmic ribosomes of eucaryotes

The proteins from large and small subunits of organelle and cytoplasmic ribosomes have been dissociated and separated by molecular weight (vertical axis) and charge (horizontal axis). It is evident that organelle ribosomes (left) have fewer proteins than the cytoplasmic ribosomes of eucaryotes (right); furthermore, few if any of the proteins from the two sources are alike in weight and charge, an indication that they also differ in composition.

(which makes ATP, utilizing energy stored by the respiratory electron-transport chain), and several other organelle proteins. The other products used by mitochondria—the third rRNA; other ribosomal proteins; the replication, transcription, and translation enzymes; the electron-transport and other F_1 proteins, and so on—must be encoded by nuclear genes and transported to the organelles. The nucleus, therefore, contains not only all the genes for the cytoplasmic ribosomes, but nearly all the genes for mitochondrial ribosomes as well.

The abbreviated chromosome of mitochondria is probably an evolutionary consequence of intracellular competition. During early development, when a growing organism's cells are multiplying rapidly, any mitochondrion that can reproduce faster than others in the cell will contribute a greater proportion of mitochondria to the next generation of cells, including the ones destined to produce gametes in the adult. In time, natural selection favors such mitochondria, which come to dominate their slower-growing counterparts. The key to reproductive speed lies largely in reducing the number of genes that must be replicated. Selection, therefore, has favored mitochondria that have transferred as many genes as possible to the nucleus. The more puzzling question is why any genes at all are left in organelles.

HOW ERRORS ARISE

Mutations—alterations in the sequence of bases in the DNA—can change the information content of genes and thus produce new alleles. As discussed in Chapter 8, genes are subject to various mutational events. Here we will examine briefly some of the consequences of these events.

■ TYPES OF MUTATION

Two types of mutation that are particularly harmful are **additions** and **deletions** of single bases. These usually result in the production of inactive enzymes: at the point of the insertion or deletion the ribosomes begin to translate incorrect triplets, and the original meaning of subsequent codons is lost. Synthesis of the polypeptide chain may even be stopped too early if the shift in reading the codons has created a spurious termination signal, or stopped too late if the stop signal is misread. Alterations that result in the misreading of subsequent codons are called **frameshift mutations**.

Another type of mutation is **base substitution** (also called point mutation), in which one nucleotide is replaced by another. For example, a codon that normally has the base composition CGG may be changed to CAG, which codes for a different amino acid. Base substitution is not as serious as addition or deletion, however, because in most codons a change in the third base does not alter the meaning (see Table 9.1). Roughly 30% of all base substitutions therefore have almost no effect.[5] Moreover, even when an amino acid is changed, the substitution usually has relatively little effect if, for example, one nonpolar amino acid replaces another. On the other hand, exchanges that occur between polar and nonpolar amino acids, or between positively charged and negatively charged amino acids, are likely to cause trouble. And alterations that create new prolines or cysteines, or remove old ones, are often serious since they may change the tertiary structure of the protein. All in all, just over half of all base substitutions are likely to affect protein function strongly.

Another very common type of mutation is **transposition**. This phenomenon (discussed in detail in Chapter 10) results from the insertion of long stretches of DNA from one part of the genome into the middle of another. The majority of easily detectable spontaneous mutations in *Drosophila*, for example, result from transposition.

[5] There is some effect from a third-base mutation because the tRNAs for some alternative codons for an amino acid may be less numerous, less active, or less accurate in binding to the codon. In genes that are transcribed and translated at a high rate, selection favors the version of the codon that yields the best translation performance; for less frequently translated genes, there is little effect from a third-base substitution.

Table 9.1 The genetic code (messenger RNA)

FIRST BASE IN THE CODON	SECOND BASE IN THE CODON				THIRD BASE IN THE CODON
	U	C	A	G	
U	Phenylalanine	Serine	Tyrosine	Cysteine	U
	Phenylalanine	Serine	Tyrosine	Cysteine	C
	Leucine	Serine	*Termination*	*Termination*	A
	Leucine	Serine	*Termination*	Tryptophan	G
C	Leucine	Proline	Histidine	Arginine	U
	Leucine	Proline	Histidine	Arginine	C
	Leucine	Proline	Glutamine	Arginine	A
	Leucine	Proline	Glutamine	Arginine	G
A	Isoleucine	Threonine	Asparagine	Serine	U
	Isoleucine	Threonine	Asparagine	Serine	C
	Isoleucine	Threonine	Lysine	Arginine	A
	Methionine[a]	Threonine	Lysine	Arginine	G
G	Valine	Alanine	Aspartic acid	Glycine	U
	Valine	Alanine	Aspartic acid	Glycine	C
	Valine	Alanine	Glutamic acid	Glycine	A
	Valine	Alanine	Glutamic acid	Glycine	G

[a] Also *Initiation* when located at leading end of mRNA.

9.20   A thymine dimer within a DNA molecule

Two adjacent thymine nucleotides are bonded to each other covalently, so they cannot form hydrogen bonds with the adenine nucleotides of the complementary strand. Such a mutation, often induced by UV light, will inactivate the DNA if it is not repaired. Some species that normally live in the dark (for instance, the intestinal bacterium *Salmonella*) lack the enzyme that can repair thymine dimers.

■ ENVIRONMENTAL SOURCES OF MUTATION

High-energy radiation, including both ultraviolet light and ionizing radiation such as X rays, cosmic rays, and radioactive decay, can cause mutations. In addition, a variety of chemicals are mutagenic. A normal spontaneous mutation rate for a single gene is one mutation in every 10^6-10^8 replications, but the rate can be greatly increased by unusual exposure to mutagenic agents.

Ionizing radiation at various wavelengths sometimes induces simple base substitutions, but it also frequently produces large deletions of genetic material. It accomplishes the latter either directly by colliding with the DNA and causing breaks to occur, or indirectly by splitting water to produce highly reactive free radicals (single, unbonded, strongly electronegative oxygen atoms), which combine with and break the DNA.

Ultraviolet (UV) light most often exerts its mutagenic effect by causing abnormal bonding of thymine bases. When two adjacent thymine units absorb UV light and are thus energized, they can bond to each other, forming a ***thymine dimer*** (Fig. 9.20). An unrepaired dimer inactivates a strand of DNA; not only can no mRNA be transcribed from it, but DNA replication cannot take place.

Some mutagenic chemicals produce their effects by directly converting one base into another. For example, nitrous acid (HNO_2), a powerful mutagen found in cigarette smoke, deaminates (removes the NH_2 group from) cytosine, changing it into uracil; it also converts adenine into hypoxanthine and guanine into xanthine. Other chemical mutagens, called base analogues, are themselves some-

9.21 Ames test

Each of these Petri dishes has vast numbers of histadine-requiring bacteria in a layer of agar. The agar contains everything the bacteria need to grow except histadine. (A) Each white spot is a colony descended from a single bacterium in which a spontaneous mutation has restored the ability to synthesize histadine. (B–D) The large white disk in the center contains a mutagenic substance that is diffusing out from the disk: furylfuramide (B), aflatoxin (C), and 2-aminofluorene (D). The mutagenic potential of these chemicals is indicated by the large number of colonies able to grow as a result of a genetic change that restored their capacity to synthesize histadine.

times incorporated into nucleic acids in place of one of the normal bases. An example is 5-bromouracil, an analogue of thymine. When a strand of DNA contains a unit of 5-bromouracil, it is prone to errors of replication, because the 5-bromouracil will sometimes pair with guanine rather than with the requisite adenine. Virtually all of these mutations are detected and repaired before they can exert any effect.[6]

There is a close correlation between the mutagenicity of a chemical and its carcinogenicity—its cancer-inducing activity. This strongly suggests that many cancers are caused, at least in part, by mutations in somatic cells. Thus we see that alterations of the DNA are important not only in germ cells, where they may affect future offspring, but also in somatic cells, whose metabolism or growth they may disrupt, causing disease or degeneration.

The connection between mutagenicity and carcinogenicity is the basis for the test developed by Bruce N. Ames. Environmental pollutants, reagents used in industrial processes, proposed new drugs, and food additives are screened by the Ames test for potential carcinogens. The compound to be tested is added to a special mutant strain of *Salmonella* requiring the amino acid histidine as a nutrient. When the mixture of bacteria and chemical is incubated on a medium deficient in histidine, any cells that undergo a mutation that reverses their original mutation regain the ability to synthesize histidine and to grow on the deficient medium. A count of these so-called revertant colonies, each derived from one cell with a reverse mutation, provides a rough measure of the mutagenic potential of the chemical being tested (Fig. 9.21). An increase in the

---

[6] By flooding a cell with 5-bromouracil, it is possible to overwhelm the repair enzymes. This strategy is often used in cancer chemotherapy to kill rapidly dividing tumor cells. An unfortunate side effect is that some highly active normal cells—those producing hair, for instance—are also killed.

mutation rate in the bacteria over the normal spontaneous level predicts the likely cancer-inducing potency of the chemical in humans.

Many chemicals, though not themselves mutagens or carcinogens, are transformed in the body, especially by detoxification enzymes in the liver, into derivatives that are mutagenic and carcinogenic.[7] For this reason newer versions of the Ames test add liver homogenate to the bacterial culture, so that the bacteria will be exposed both to the original chemical and to its metabolic derivatives.

[7] As described in Chapter 31, the liver operates on the principle that unfamiliar chemicals are bad, and alters them with a variety of enzymes that prepare them for excretion; unfortunately, a few substances (like the pesticide contaminant dioxin) are thereby turned into carcinogens.

CHAPTER SUMMARY

HOW TRANSCRIPTION WORKS

THE NEED FOR MESSENGER RNA Messenger RNA (mRNA) carries genetic information from specific genes on the chromosome to sites of protein synthesis. (p. 225)

CREATING MESSENGER RNA RNA polymerase transcribes one strand of the DNA in a gene, creating an mRNA. The polymerase binds to a promoter region, begins transcription at a nearby start sequence, and stops at a termination sequence. In eucaryotes this creates a primary transcript whose introns must be removed before the mature mRNA can leave the nucleus. (p. 226)

HOW THE MESSAGE IS TRANSLATED

THE GENETIC CODE Nucleic acid base triplets (codons) encode the 20 common amino acids as well as signals for initiating and ending translation. (p. 231)

HOW RIBOSOMES WORK Ribosomes consist of two complex subunits, each a combination of rRNA and proteins. The smaller subunit binds to the mRNA, and then to the larger subunit. Translation begins with the first codon for methionine. The messengers for products destined for the ER have a sequence that causes their ribosomes to bind to the rough ER early in translation. (p. 232)

THE NEED FOR TRANSFER RNA Each amino acid is carried to the ribosomes while attached to a tRNA specific to that kind of amino acid; each tRNA has an anticodon sequence complementary to an mRNA codon for that amino acid. (p. 237)

THE TRANSLATION CYCLE The ribosome has adjacent P- and A-sites that bind sequential codons. The tRNA binds to the A-site; its amino acid binds to the amino acid attached to the tRNA already bound to the P-site; the P-site tRNA then detaches and the mRNA is shifted so that the codon formerly at the A-site is now attached to the P-site. (p. 237)

TRANSCRIPTION AND TRANSLATION IN ORGANELLES These two processes in organelles are very similar to those of procaryotes: there are no introns or nuclear membrane, and ribosomes can bind to mRNA while it is still being transcribed. (p. 239)

HOW ERRORS ARISE

TYPES OF MUTATION Additions and deletions of single bases create frameshift mutations that change the meaning of subsequent codons. Base substitutions can change individual codons. Transpositions rearrange pieces of genes. (p. 241)

ENVIRONMENTAL SOURCES OF MUTATION A variety of chemicals as well as ionizing radiation can induce mutations. Many cancers result from mutations. (p. 242)

STUDY QUESTIONS

1 Compare and contrast replication and transcription. (pp. 212–218, 226–229)

2 Compare the several types of RNA used in eucaryotic cells. (pp. 226, 229–233, 237–239)

3 Compare and contrast transcription and translation in procaryotes with the more complicated series of events in eucaryotes. How does the process in organelles fit into the picture? Is there any obvious reason for these differences? (pp. 225–236, 239–241)

4 Make a list of all the signal sequences (and the function of each) we have encountered so far, whether in DNA, RNA, or peptides. What others are likely to exist that have not been discussed yet? (pp. 226–239, 242)

5 In all, it takes four high-energy phosphate bonds to catalyze the several steps necessary to add each amino acid to a growing peptide chain. Given that the average protein contains 300–500 amino acids (the lower figure being more typical of procaryotes), how expensive (in terms of glucoses metabolized) is a protein, assuming that the amino acids are freely available? (p. 175)

SUGGESTED READING

CECH, T. R., 1986. RNA as an enzyme, *Scientific American* 255 (5). *On the enzymatic properties of ribosomal and especially snRNP RNA, and the possibility that RNA originally served the functions now taken over by DNA and protein.*

DARNELL, J. E., 1985. RNA, *Scientific American* 253 (4). *Reviews transcription, processing, translation, and transcriptional control.*

ALBERTS, B., D. BRAY, J. LEWIS, M. RAFF, K. ROBERTS, AND J. D. WATSON, 1994. *Molecular Biology of the Cell*, 3rd ed. Garland, New York. *Excellent detailed study of transcription and translation.*

LAKE, J. A., 1981. The ribosome, *Scientific American* 245 (2). *On the three-dimensional structure of the ribosome and the details of translation.*

STEITZ, J. A., 1988. "Snurps," *Scientific American* 258 (6). *On the enzymes that remove the introns from eucaryotic primary transcripts.*

CHAPTER 10

MOBILE GENES AND GENETIC ENGINEERING

Life depends on a dynamic balance between stability and change. Stability is essential for DNA replication, as well as for the transcription and translation of genes to produce RNA and protein. Change also is critical: natural selection can work only on the variation in a species, and a static genome would lead to the inevitable extinction of any species unable to evolve in the face of changing climate, resources, or competition.

Our emphasis in this chapter is on large-scale, dynamic alterations that can occur in the genome. By *large scale* we mean changes that involve an entire gene or major parts of it. Mutations involving single-base changes in functional genes play a smaller role in evolution, because such "tinkering" is unlikely to alter the gene product dramatically.[1] By contrast, the consequences of large-scale chromosomal changes range from cancerous growth in the individual to the potential for major evolutionary change in the species. Even for bacteria, whose sensitivity to mutational change is great and whose opportunities for major gene exchange and movement are limited compared with those of most eucaryotes, mutational alterations are thought to account for only a few percent of the evolutionary changes observed in the species.

Much of the work being done in genetic engineering today makes use of the natural agents of genetic mobility, especially the curious "accessory" chromosomes discussed presently. We will also look at viruses, whose unique reproductive strategies allow them to move new genes into target cells. Finally, we'll examine how gene mobility is orchestrated from within the genome, and how the enzymes responsible for mobile genes facilitate the techniques of genetic engineering, with their enormous potential for improving everything from food crops to human health.

THE DISCOVERY OF BACTERIAL PLASMIDS

Bacteria have one copy of their DNA instructions on a large circular chromosome; like all other organisms with only one set of instructions, they are classified as **haploid.** By contrast, sexually

---

[1] As described in Chapter 11, single-base changes in the sequences that *control* gene expression can have a significant effect on cell physiology by altering the *amount* of product synthesized.

A

1 μm

B

0.2 μm

C

0.2 μm

10.1   Electron micrographs of plasmids from *E. coli*

(A) A plasmid (arrow; top, center) is visible next to the main chromosome of this lysed *E. coli.* (B) The isolated plasmid seen here carries a gene encoding a product that confers resistance to the antibiotic tetracycline. (C) Plasmids are replicated independently of the main chromosome. The bacterial enzyme complex DNA polymerase can be seen copying each of these plasmids.

reproducing organisms, which have two copies of their genetic instructions in every cell except in their gametes, are **diploid;** gametes fuse to restore the duplicate set prior to fertilization. In 1946, however, Joshua Lederberg and Edward L. Tatum found that mating (of a sort) nevertheless does take place in bacteria. This led to the discovery of **plasmids,** small circular accessory molecules of DNA (Fig. 10.1), which are also found in eucaryotes ranging from yeast to mammals.

■ DO BACTERIA MATE?

Lederberg and Tatum isolated two mutant strains of *E. coli,* one of which was unable to synthesize the amino acid methionine or the vitamin biotin, while the second could not make the amino acids threonine or leucine. The mutants could be grown only on a medium that supplied the nutrients they could not synthesize. However, when the two mutant strains were mixed on a minimal medium (one lacking all four critical nutrients), a few healthy colonies grew on the deficient medium. The individuals in these colonies had somehow inherited both the first strain's ability to synthesize threonine and leucine and the second strain's ability to synthesize methionine and biotin.

The survival of these bacteria resulted from a **recombination** of traits from the two original mutant strains: genetic material from two different individuals had become combined in a single genome. Recombination occurs whenever two haploid gametes fuse to form a fertilized egg, which is thereby rendered diploid;

bacteria, however, are never diploid, and in most species a stiff cell wall precludes fusion. Bacterial recombination usually depends on **conjugation.** When two bacterial cells lie close to each other and other conditions are favorable, a narrow cytoplasmic bridge, or pilus, may be built from one to the other (Fig. 10.2). Genetic material passes through the bridge from one cell to the other.

Conjugation can occur only between cells of different mating types, designated F^+ ("male") and F^- ("female"). Unlike F^- cells, F^+ cells have a DNA plasmid containing a **sex factor** in their cytoplasm. Under most circumstances this factor replicates within non-dividing bacteria, and a copy can be transferred from F^+ cells to F^- cells. The products of conjugation are always F^+. The population never becomes a pure culture of F^+ cells because conjugation is very time-consuming; thus an F^- cell can undergo one or more fissions while an F^+ cell engages in one conjugation. Hence, the proportion of F^+ cells can actually decline with conjugation.[2] Another problem limits the F^+ population: the pilus (which only F^+ cells can build) provides an attachment point for certain viruses, which thus kill only F^+ bacteria.

F^+ strains usually include a very few cells that, when they conjugate, may fail to transfer an intact sex factor to the F^- cells; instead they transfer chromosomal DNA. As a result, conjugation initiated by these high-frequency recombinant (*Hfr*) cells does not usually convert F^- cells into F^+ or *Hfr* cells. The difference between F^+ and *Hfr* strains is that in *Hfr* strains the plasmid is incorporated into the bacterial chromosome (Fig. 10.3). During conjugation, a copy of the main chromosome of the donor cell is moved through the pilus, but this delicate bridge rarely remains unbroken long enough for the sex factor (part of which is at the end of the chromosome) to be transferred. *Hfr* conjugation is important because it vastly increases chromosomal recombination.

The sex factor is unusual because at times it is free in the cytoplasm as a tiny circle of DNA, while at other times it is incorporated into the chromosome and behaves like other chromosomal genes. In a later section we will look at similar behavior in viruses and in movable elements in eucaryotic chromosomes called transposons.

■ AUTONOMOUS PLASMIDS

Most bacterial cells contain some autonomous plasmids, which, unlike the sex-factor plasmid, are never integrated into the main chromosome. Some such plasmids carry only one or two genes, while others may be as much as one-fifth the size of the main chromosome and carry many genes.

Plasmids with genes for resistance to various antibiotics, including streptomycin, tetracycline (Fig. 10.1B) and ampicillin, are particularly important to modern medicine. When bacteria with resistance plasmids are exposed to one of these antibiotics, the plasmids begin replicating, and a cell that originally had only two

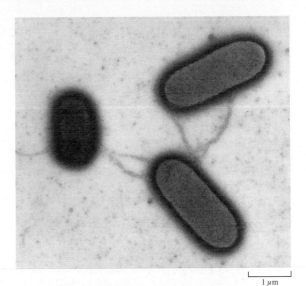

$\llcorner$ 1 μm $\lrcorner$

10.2   Conjugating bacteria

Long cytoplasmic bridges, or pili, connect the lower cell with two others with which it is conjugating simultaneously. The pili also provide attachment sites for certain viruses that can only infect F^+ (or "male") cells.

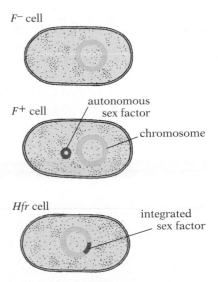

10.3 F^-, F^+, and *Hfr* cells of *E. coli* compared

F^- cells lack the sex factor. In F^+ cells the sex factor is a plasmid free in the cytoplasm. In *Hfr* cells the sex factor is integrated into the chromosome.

---

[2] This reproductive disadvantage is kept to a minimum because F^+ bacteria typically initiate conjugation only when the population is growing very slowly, perhaps because of a shortage of some critical nutrient.

or three plasmids may soon have a thousand or more. Plasmids may also carry genes that make the bacteria virulent, as in the case of the pneumococci that cause pneumonia. As we will see, recombinant DNA technology makes it possible to insert new genes into both autonomous plasmids and plasmids that are integrated into the host genome.

■ TRANSFORMATION

Some species of bacteria can take up pieces of DNA from their surroundings. (Many species, for instance, can be induced to take up foreign DNA when the cell walls of the bacteria are digested away and the bacteria are then maintained in an isotonic medium.) Imported DNA, often a fragment of the chromosome of a dead bacterium, may recombine with a similar region of the host genome. This necrophilic version of procaryotic sex, called *transformation*, is a very important tool in genetic engineering.

HOW VIRUSES MOVE GENES

Viruses are tiny intracellular parasites with a miniature chromosome, a protective capsule, and (rarely) an enzyme. Their various reproductive strategies provide avenues for both natural and artificial transfer of genes.

■ HOW VIRUSES REPRODUCE

Lytic viruses In the 1940s Max Delbrück elucidated the reproductive cycle of a common bacteriophage. When he mixed phage with a culture of bacteria, the viruses seemed to vanish, but half an hour or so later a hundred times as many phage as he had started with suddenly appeared in the culture. The phage had entered the host cells, taken control of the cellular machinery, replicated themselves, and then caused the host cells to burst (lyse), releasing a new generation of infectious phage. The new viruses then attacked other bacterial cells and started the *lytic cycle* over again (Fig. 10.4A). Many techniques of genetic engineering take advantage of the willingness of viral-coat proteins to encapsulate nonviral DNA: the new genes to be delivered to other cells are introduced into infected cells; new viral caspules are built automatically around small pieces of DNA in the host, but in this case many of these genetic sequences are the nonviral fragments; and then the resulting viruses are used to infect new hosts.

Retroviruses A variety of different viral strategies are now understood. For example, many disease-causing viruses in animals have an RNA chromosome, which is both replicated and transcribed (Fig. 10.4B). The most extreme departure from the standard DNA- and RNA-based strategies is found in *retroviruses.* These parasites have an RNA chromosome, and carry an enzyme, *reverse tran-*

A

B

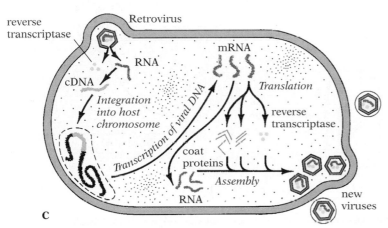

C

10.4 Three common strategies of viruses

Viruses exploit their hosts in a variety of ways. The most frequent strategy is for a viral genome consisting of DNA to be both replicated and transcribed, and the transcripts translated, all by means of host enzymes. Transcription and translation generate coat proteins, enzymes that modify host-cell function, and enzymes that later lyse the host. (A) Other DNA viruses utilize replication enzymes produced through transcription and translation. (B) For RNA viruses the transcription step is unnecessary, but most encode a special enzyme, RNA replicase, to catalyze replication of the RNA. Some RNA viruses lyse their hosts, as shown, but many others escape from their hosts without lysing them. (C) Retroviruses are an unusual kind of RNA virus. Each retrovirus carries an enzyme, reverse transcriptase, that catalyzes the formation of a cDNA copy of the viral RNA; this copy is then incorporated into a chromosome of the host (often an animal cell). The host's enzymes then take over and accomplish replication (not shown), transcription, and translation of the viral genes. As the drawing indicates, retroviruses may enter the host cell by a kind of endocytosis, losing the protein coat only after entry; similarly, retroviruses generally do not lyse their hosts but instead leave by extrusion, a process whereby the virus is enveloped in a segment of the cell membrane as it emerges from the cell.

scriptase, that catalyzes the formation of a DNA copy of the RNA (known as *cDNA* because it is complementary to the RNA from which it is copied); thus they reverse the usual flow of information from DNA to RNA (Fig. 10.4C). Retroviral genes are usually inserted into the host chromosome, where they may become permanent residents.

Temperate viruses Some viral chromosomes act in a manner analogous to the *F*⁺ plasmid: they can both enter and leave the host chromosome. This ability was first discovered when certain apparently normal strains of bacteria were exposed to ultraviolet (UV) light or X rays or various chemicals used to induce mutations. The bacteria would lyse within 1 hour, releasing large numbers of infectious phage: the bacteria had been carrying dormant viruses. Cells that harbor inactive viruses are said to be *lysogenic.*

Viruses whose reproductive cycle includes an inactive stage inside their hosts are called *temperate* rather than virulent. Virulent phages invariably kill their hosts; temperate phages may or may not kill their hosts, depending on a variety of conditions. When they do not kill their hosts, their injected DNA usually becomes associated with the host chromosome at a particular location, and the *lysogenic cycle* begins (Fig. 10.5).

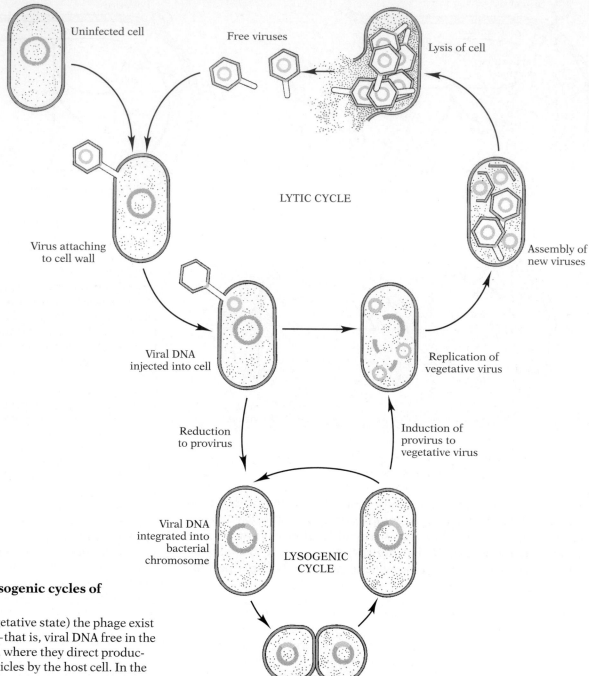

LYTIC CYCLE

Uninfected cell

Free viruses

Lysis of cell

Virus attaching
to cell wall

Assembly of
new viruses

Viral DNA
injected into cell

Replication of
vegetative virus

Reduction
to provirus

Induction of
provirus to
vegetative virus

Viral DNA
integrated into
bacterial
chromosome

LYSOGENIC
CYCLE

Reproduction of lysogenic bacteria

10.5 Lytic and lysogenic cycles of bacteriophage

In the lytic cycle (vegetative state) the phage exist only as free viruses—that is, viral DNA free in the host cell's cytoplasm, where they direct production of new viral particles by the host cell. In the lysogenic cycle the phage DNA is integrated into the host cell's chromosome, and only on occasion is it induced to break loose and initiate viral replication. (Bacteriophages are much smaller relative to their hosts than is indicated here.)

While integrated into the bacterial chromosome, the viral DNA, like the sex factor of *Hfr* strains, behaves as part of that chromosome. It is replicated with the rest of the chromosome, and it can be transferred from one cell to another during conjugation. Its genes can undergo recombination with bacterial genes, and they can even produce visible effects in the host bacterium, such as modifications of the morphology of the colony, changes in the properties of the cell wall, and changes in the production of enzymes. For example, diphtheria bacteria produce a toxin that causes the disease only if they are carrying a specific type of viral gene. Viral genes often make the bacterium in which they reside immune to further infection by the same type of virus, a phenomenon similar to the ability of the sex-factor genes to block the entry of additional F^+ DNA. Thus the DNA of temperate viruses can exist in an autonomous state, replicating independently and eventually destroying the cell, or it can exist in the integrated state, as a **provirus,** functioning and replicating as a part of the chromosome. Several human ailments, including cold sores and shingles, are caused by lysogenic viruses engaging in a bout of lytic reproduction.

■ VIRAL TRANSFER OF NONVIRAL GENES

Not only can some viral genes move into chromosomes, but the reverse can also occur. The phenomenon is best understood in bacteria. When temperate viruses are in the autonomous state and have put the bacterial cell to work making more viruses, small fragments of the bacterial chromosome may become enclosed in the new viral coats. If a temperate virus carrying bacterial DNA in this manner infects a new host, it injects bacterial DNA into the new host. Sometimes the injected bacterial genes undergo recombination with the new host's genes. The virus has thus acted as a vehicle for transferring genes from one bacterial cell to another, a process called **transduction** (Fig. 10.6).

Some of the genes that produce important observable effects in the human body were probably moved into human chromosomes by viruses. These genes may have been transferred from other humans or even from other species. This can happen because some viruses (notably those responsible for influenza) can infect other hosts, as in the "swine flu" epidemic of the 1980s. As described in the next chapter, some genes transported by viruses may be involved in cancer. Viruses may also have been involved in moving organelle genes into the nucleus of eucaryotes. In short, by moving entire genes, viruses can play a crucial role in creating the variation on which natural selection works in organisms.

HOW GENES ARE MOVED WITHIN THE GENOME

While the movement of genes from one organism into the chromosomes of another is important, equally significant alterations can be caused by moving genes around *within* a genome. Since each gene's transcription is controlled by nearby sequences (including the promoter) that are not necessarily moved with the gene, the timing and degree of transcription can be significantly altered. Gene movement, therefore, can lead to dramatic changes.

■ THE BEHAVIOR OF TRANSPOSONS

Genetic units that move about in the genome, either by removing themselves to new locations or by duplicating themselves for inser-

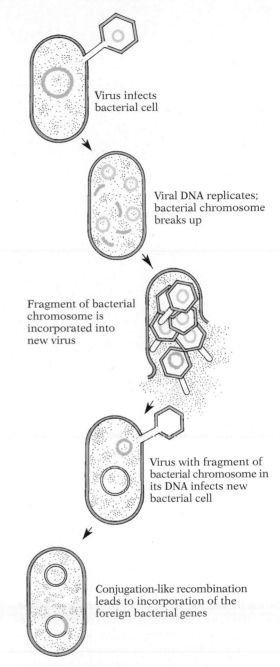

Virus infects bacterial cell

Viral DNA replicates; bacterial chromosome breaks up

Fragment of bacterial chromosome is incorporated into new virus

Virus with fragment of bacterial chromosome in its DNA infects new bacterial cell

Conjugation-like recombination leads to incorporation of the foreign bacterial genes

10.6   Model of transduction by bacteriophage

tion elsewhere, are called *transposons*. They are found in all cells, procaryotic and eucaryotic, and can also insert themselves into plasmids. Transposition in eucaryotes was discovered more than four decades ago by Barbara McClintock. She found that in corn certain genetic elements will occasionally move, particularly after cells are subjected to trauma, such as exposure to intense UV radiation. These movements produced kernels with unusual colors that could not have resulted from normal recombination. Her results were so completely at odds with the prevailing concept of a static gene that the mobile genetic elements she had discovered were considered some sort of abnormality. By 1983, however, when she was awarded the Nobel Prize, many such transposons had been discovered, and their possible role in evolution was beginning to be recognized.

Transposons can consist of one or several genes, or just a control element, and can move in several ways, none of which is fully understood. Transposons within a chromosome are flanked by a pair of identical sequences, some of which are actually part of the transposon. Some transposons move from one site on the chromosome to another; others leave a copy behind; still others remain fixed but dispatch copies to other sites. A transposon that has physically moved leaves the flanking sequences as telltale scars (Fig. 10.7A); those that leave duplicates behind are copied in full by

10.7 A model of transposition events

While integrated into a chromosome a transposon exists as a stretch of DNA flanked by characteristic sequences. At least one of these flanking sequences moved into the chromosome as part of the transposon, while the other may already have been there, and served as a target site. The movement of the transposon may involve either excision from the chromosome, leaving one or more target sequences behind (A), or duplication, with one copy of the transposon then left behind at the original location (B). In both cases, the mobile transposon probably exists as a circle, which can be integrated into any other part of the genome with (or, for certain transposons, without) the target sequence (C). The integration is frequently accompanied by a duplication, which produces another mobile transposon to continue the cycle (D).

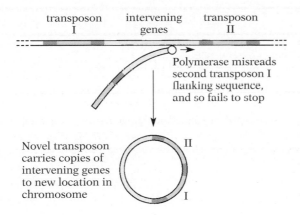

transposon I   intervening genes   transposon II

Polymerase misreads second transposon I flanking sequence, and so fails to stop

Novel transposon carries copies of intervening genes to new location in chromosome

10.8   Model for capture of chromosomal genes by transposons

Chromosomal genes lying between two transposons are occasionally incorporated into a single hybrid transposon. A mutation in the second flanking sequence makes this incorporation inevitable.

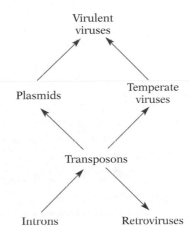

DNA polymerase (Fig. 10.7B). Those that dispatch a copy can employ either DNA or RNA. In the case of an RNA copy the "retrotransposon" encodes reverse transcriptase. When such an RNA copy escapes to the cytoplasm, it is translated; the reverse transcriptase thus produced proceeds to make a cDNA copy for insertion back into the chromosome. Transposons are by no means rare: approximately 5% of the human genome consists of copies of a single kind—the L1 retrotransposon, present in at least 50,000 partial copies; this particular transposon, when it moves into a specific position on the X chromosome, appears to cause the most common form of hemophilia.

In many cases the receiving chromosome carries particular sequences that serve as targets for the mobile transposon. This sequence can pair with the one copy of the flanking sequence carried on the transposon. A special enzyme, usually encoded by the transposon itself, enables the transposon to recognize and act on the target sequence in the host chromosome. The transposon can then, with the aid of another enzyme usually encoded by the transposon, incorporate itself into the host chromosome. Other transposons have enzymes able to effect insertion anywhere in the genome. Like viruses and plasmids, transposons can sometimes pick up additional genes from the main chromosome (Fig. 10.8). There is an obvious similarity between transposons, temperate viruses, and those plasmids capable of incorporating themselves into a chromosome: each generally encodes the enzymes it needs to orchestrate its movements. The possible evolutionary relationships between these genetic entities are outlined in Figure 10.9.

■ OTHER WAYS TO MOVE GENES

Although of less use in genetic engineering, there are other ways genes can move. These alternative mechanisms are important in understanding chromosomal organization, cancer, immunology, and evolution. Perhaps the most intriguing case is the occasional instance of retroinsertion, in which an mRNA is mistakenly bound by reverse transcriptase; it can then serve as a template for the creation of novel cDNA, which may then be incorporated back into the chromosome. The human genome has many such cDNAs, which are easily recognized from their lack of introns (missing in

10.9   Possible evolutionary relationships between mobile genetic entities

Although it is not yet possible to trace the evolution of mobile genetic elements with certainty, it seems likely that the development of self-splicing introns may have been the first step. (The likely adaptive value of early introns is discussed in Chapter 17.) The next step would have been the evolution of a system to allow reinsertion of excised introns, producing transposons. Transposons could then have given rise to three new entities: (1) retroviruses, from the evolution of RNA transposons; (2) plasmids, from defective transposons, trapped in their independent circular state; and (3) the less benign entities we know as temperate viruses, from transposons that developed the ability to move between cells. Virulent viruses could have evolved from either plasmids or temperate viruses. Though not indicated here, at least some of these steps are reversible: temperate viruses, for example, could lose their ability to switch from one cell to another, and thus return to the status of transposons.

Table 10.1   Sources of large-scale genetic change

PHENOMENON	MECHANISM	POSSIBLE CONSEQUENCES
INSERTION	Transformation	New genes from a dead cell imported from surrounding medium and incorporated into chromosome of bacterium
	Transduction	New genes accidently picked up from previous host and imported into cell by a virus
	Plasmid insertion	Existing gene or genes become integrated into genome, and subject to novel controls
	Lysogenic insertion	Novel genes of temperate phage inserted into host genome
	Retroviral insertion	cDNA copy of novel genes of retrovirus inserted into host genome
	Intron insertion	Excised introns inserted into genome, mainly at exon-exon junctions in cDNA insertions
DUPLICATION	Retroinsertion	cDNA copies of transcribed host DNA incorporated into genome, providing duplicate copies of genes
	Breakage and fusion	Part of one chromosome breaks off and fuses to the end of another during gamete formation; some gametes may obtain duplicate copies of genes on the broken fragment (discussed further in Chapter 16)
	Unequal crossing over	Chromosomes may be misaligned during a complex process called crossing over (described in Chapters 12 and 16); some gametes may obtain duplicate copies of some genes
GENE MOVEMENT	Transposition	Chromosomal DNA moved with genome, or both duplicated and moved

mRNAs after processing) and their telltale caps and poly-A tails (absent in normal genes, but added to mRNA during processing).

Genes can also be moved when chromosomes break and the pieces are reattached in the wrong places, or when errors occur during meiosis. Finally, introns removed during mRNA processing can sometimes act as templates for cDNA, and then be incorporated back into the genome. The range of possible mechanisms for gene movement is summarized in Table 10.1.

GENETIC ENGINEERING

■ ARTIFICIAL TRANSFORMATION

The agents of gene mobility, especially reverse transcriptase and plasmids, play a major role in *recombinant DNA* techniques for inserting selected genes from one kind of organism into cells of another kind. These techniques have great scientific, medical, and commercial potential. Specific genes can be isolated, introduced into bacteria, and used to produce large quantities of a desirable gene product such as insulin. Alternatively, new genes can be introduced into plants or animals that might benefit from them—genes for disease resistance in plants, for instance, or genes for growth hormones in domestic animals. However, recombinant DNA techniques have become a subject of public debate because of concern over the possibility of accidentally producing new pathogens, or developing "genetic monsters."

Plasmids, like DNA from bacterial chromosomes, can transform bacterial cells: if bacteria of certain species are placed in a medium containing free plasmids, some of the cells will pick up the plasmids. Thus nonresistant bacteria can be transformed into resistant ones by exposure to a medium in which bacteria with plasmids for antibiotic resistance have been killed. Recombinant DNA technology makes use of this transforming potential of plasmids: purified plasmids are modified by the addition of foreign genetic material, and when bacterial cells then pick up the modified plasmids, they acquire the foreign genes.

Let's follow the procedure in more detail: Bacterial cells containing plasmids are broken up, and their DNA is extracted. The DNA is then centrifuged to separate the plasmids from the main bacterial chromosomes. The purified plasmids are next exposed to a restriction endonuclease, one of a class of procaryotic enzymes that cleave the DNA circle at a particular nucleotide sequence (Fig. 10.10). The plasmid DNA is now linear, with "sticky" ends (unpaired bases) where it has been cleaved. The plasmid DNA is next mixed with fragments of foreign DNA prepared with the same restriction endonuclease and therefore equipped with sticky ends that are complementary to those of the plasmid DNA. In such a mixture, under the appropriate environmental conditions (temperature, pH, and so on), the plasmid DNA and the foreign DNA spontaneously anneal by complementary base pairing, re-forming a circle in the process (Fig. 10.11). The backbones (chains of alter-

endonuclease *Eco*RI

endonuclease *Eco*RI

A

B

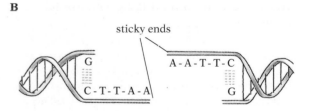

sticky ends

C

10.10 Endonuclease action

(A) Restriction endonucleases bind to a pair of target sequences—GAATTC in the case of the enzyme *Eco*RI. (B) They break a particular phosphate bond in the backbone of each strand of the DNA. (C) The resulting cut ends with unpaired bases are called sticky ends because under favorable conditions they will pair with cut ends having the complementary sequence of bases. (D) A more detailed look at a restriction endonuclease. One *Eco*RI attaches to a target sequence; the other endonuclease is omitted for clarity. The blue portion of the enzyme is important for recognizing the target sequence; the binding is indicated by dashed red lines. The solid red portion of the enzyme is involved in weakening the bonds in the DNA; the arrow indicates where the helix is cut. More than 100 endonucleases are known, each with its own specific target sequence. The normal function of endonucleases is to digest infecting viral DNA.

D

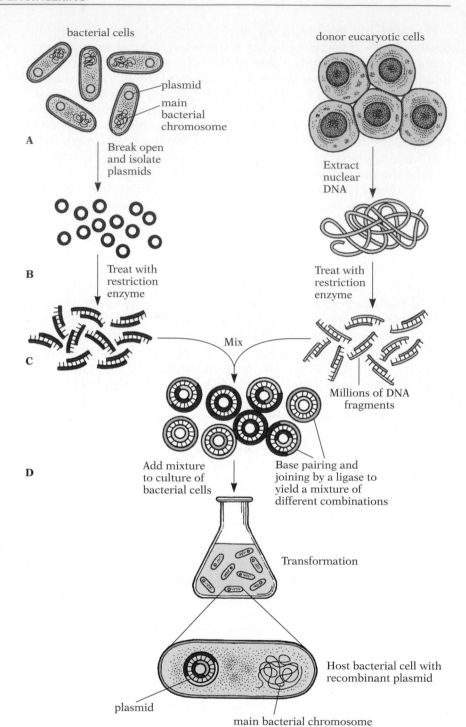

10.11 A recombinant DNA technique using transformation

Plasmids (shown here in red), removed from donor bacterial cells (A), are cut by an endonuclease (B). They are then mixed with fragments of DNA (blue and yellow) from other cells, produced with the same endonuclease (C). The plasmid DNA and the foreign DNA can join at their sticky ends by complementary base pairing, re-forming a circle in the process (D). The DNA ends are then sealed by treatment with a ligase. The modified plasmids, some bearing foreign genes, are added to a medium containing live bacterial cells. Some of the cells pick up the plasmids bearing foreign genes and are thus transformed. Alternatively, the plasmids can be packaged into viral capsules and inserted in the bacterial cells by transduction (not shown).

nating phosphate and sugar groups) of the DNA circle can then be sealed with the enzyme DNA ligase. The end products are plasmids that contain a graft of foreign genetic material.

These hybrid plasmids are then mixed with bacterial cells (usually treated with a calcium salt to make them more permeable). The bacterial cells pick up the modified plasmids. For more efficient transfer, transduction can be used: the hybrid plasmids are mixed with viral coat proteins, which will self-assemble around the plasmid if it is not too large; these plasmid-carrying viruses are

A

bacteriophages (phages)

donor cells

Extract DNA from phages

Extract nuclear DNA

B

Cut with restriction enzyme

Cut with restriction enzyme

Mix

recombinant molecules

C

phage proteins

Add phage proteins; recombinant DNA is then "packaged" in proteins to form new bacteriophages

D

Add new bacteriophages to a culture of bacteria

E

bacteria

phage

F

Bacteriophages attack bacterial cells and inject DNA into the cell

10.12 A recombinant DNA technique using transduction

The first steps of the transduction technique (A–B) are similar to those of the transformation approach (see Fig. 10.11), except that viral DNA may be used instead of a bacterial plasmid. After the recombinant molecules are obtained, viral proteins are added, and the hybrid DNA is packaged into new viral particles (C). The viruses are then added to the host culture (D) where they infect the target cells (F).

used to infect target cells (Fig. 10.12). In general, plasmids containing genes that confer antibiotic resistance are used; the experimenter then treats the bacterial cells exposed to the recombinant plasmids with antibiotic, and only the bacteria that have incorporated hybrid plasmids survive. As the plasmids reproduce, the foreign DNA is copied, or ***cloned***.

One disadvantage of this so-called genomic-cloning approach is that the desired combination of plasmid and foreign DNA is not the sole result. After endonuclease treatment, the foreign DNA is left as a mixture of various fragments, of which only one may be of interest. These genomic-clone fragments are frequently used as

10.13   Screening for cells with recombinant DNA

There are several ways to locate colonies with recombinant plasmids carrying a desired gene. (A) In many of them the treated cells are plated into a dish and allowed to grow. (B) A piece of special filter paper is pressed against the dish, and thus picks up a few cells from each colony. (C) One method then requires the cells on this filter-paper "replica" to be exposed to antibodies specific to the gene product being sought. (D) After washing away the antibody molecules that fail to bind to the cells containing the desired gene DNA of the killed cells, chemicals are added that stain the antibodies. This stain indicates the location of the colonies in the original dish containing the desired gene; these colonies are removed from the dish and cultured. (The way in which antibodies bind to specific proteins is described in Chapter 15).

Place the mixture of cells on solid medium containing the antibiotic. Cells containing the plasmid grow to form colonies.

Place a special filter on top of the solid medium and transfer some cells to the filter.

Add the antibody specific for the desired protein.

Wash away the unbound antibody and add special chemicals that bind to the antibodies. Find the colony of interest.

colony of interest

probes for locating genes on chromosomes, but are less useful in generating working genes. One potential problem is that the endonuclease may well have found its target sequence within a particular gene, and so cut the gene itself. Repeating the process with a different endonuclease may yield fragments with a gene intact. Rigorous screening is often required to isolate the bacteria with plasmids that have incorporated complete copies of the desired genes (Fig. 10.13). Worse yet, if the foreign DNA is from a eucaryote, as is often the case in recombinant DNA research, it will contain introns, which bacteria are unable to remove before translation. Thus the proteins produced by the bacteria include the introns, which renders the products inactive. To avoid these and other problems, many researchers use yeast, a simple eucaryote. Other researchers rely on more selective techniques to isolate and incorporate desirable genes into bacteria. One of these techniques is cDNA-based cloning.

■ HOW SPECIFIC GENES ARE CLONED

Recombinant techniques The crucial step in one method of mass-producing clones of a particular eucaryotic gene is to find cells that specialize in manufacturing that gene's product—pancreatic cells, for instance, if the desired product is insulin. The cytoplasm of such cells will have a high concentration of mRNA molecules coding for their special product, and these mRNAs will already have undergone intron removal in the nucleus. Many techniques exist to separate out the particular kinds of mRNA on the basis of physical characteristics like weight, so that the mRNA found in unusual abundance in the specialist cells can be isolated. A variety of other tricks are available to identify the mRNA of genes that are only rarely transcribed. (Another method of targeted gene cloning, the polymerase chain reaction, is described in the *Exploring Further* box on page 262).

Once the appropriate mRNA has been isolated, the next step is to produce from it the corresponding single-stranded DNA, using reverse transcriptase from a retrovirus. DNA polymerases are then used to replicate the DNA strand by complementary base pairing, a process that supplies the second strand for the double-helical structure of the transcript. The procedures already discussed for inserting foreign DNA into plasmids are then brought into play: a restriction endonuclease cuts open the plasmid, creating a pair of sticky ends; the cloned DNA with a complementary set of sticky ends is added, and the plasmid DNA and cDNA anneal; a ligase restores the bonds in the DNA backbone; and the plasmid is inserted into a bacterial host.

This procedure has two immediate advantages: the expensive process of sorting for bacteria bearing the desired gene is eliminated, since only the appropriate cDNA (in this case, the cloned DNA) is used; and introns, which cannot be removed by procaryotes, have already been eliminated in the production of the mRNA. The transformed bacterial cells grow and divide rapidly, creating limitless numbers of bacteria that may synthesize the desired product, particularly if the host plasmid has signals that enhance transcription of the cDNA. This technique is used to produce other hormones besides insulin that are difficult to synthesize, notably growth hormones. Recombinant bovine growth hormone is widely used to boost milk production by 10–40%, and synthetic human growth hormone is employed to prevent stunted growth in hormone-deficient children. A naturally produced but poorly understood agent called interferon, which sensitizes the immune system and so holds great promise in the treatment of various diseases, is also now widely available for research and medical applications as a result of recombinant DNA technology.

Recombinant techniques have also made possible several new approaches to vaccination. The one that seems most promising for finally defeating the common cold takes advantage of the need of viruses to locate and bind to the exposed portion of a particular membrane protein in their host's cells. Researchers can clone the portion of the gene that encodes the extracellular part of the target protein, and then mass-produce this fragment. When the body of a potential victim is flooded with these molecular decoys, the viruses attach themselves to the free-floating target segments, and so render themselves harmless (Fig. 10.14). Years of testing, however, will be needed before such treatments become widely available.

Gene cloning makes it possible not only to use the host cells as chemical factories that produce substances of medical or commercial importance, but also to study the sequencing and activity of genes from eucaryotic cells. Indeed, one of the many spinoffs from recombinant DNA technology is a method of mapping genes with enormous precision. A piece of single-stranded DNA is transcribed from an mRNA of known function in a medium containing ra-

10.14 Inactivating a cold virus with recombinant proteins

Rhinoviruses, which are responsible for about half of the cases of colds, gain entry to the cells they infect by first attaching to cell-adhesion molecules (CAMs) mounted on the membranes of many types of cells in the body. (The role of CAMs in development is described in Chapter 14.) By flooding the bloodstream with the small part of the CAM to which the virus binds (red), the pathogen is rendered harmless.

EXPLORING FURTHER

THE POLYMERASE CHAIN REACTION

Another way of making a vast number of copies of a gene depends on knowing the base sequence flanking each end of the region of interest. Many copies of short DNA primers complementary to the two known flanking sequences are then synthesized, one for the 3′ end of the region on each strand. The gene's chromosome is cut into fragments with an endonuclease and heated to denature the strands, and the two primer sequences are added. Because the primers greatly outnumber the chromosomes, they are far more likely to bind to the 3′ flanking target sequence on a strand than is the complementary strand of the chromosome.

After the mixture has cooled slightly, a DNA polymerase

from a species of bacterium adapted to life in the near-boiling waters of hot springs is introduced. The polymerases find the primers and begin synthesizing a complementary copy of each fragment. When enough time has passed for the polymerases to finish their work, the mixture is again heated to denature the two strands of each fragment. Once more, free primers bind to the flanking sequences, now present on four strands (the two original strands of the fragment with the gene the researchers want, plus two copies), and the mixture is again cooled to allow the polymerases to do their job. With each cycle of heating and cooling, the number of copies of the original fragment doubles, generating a million replicas in as

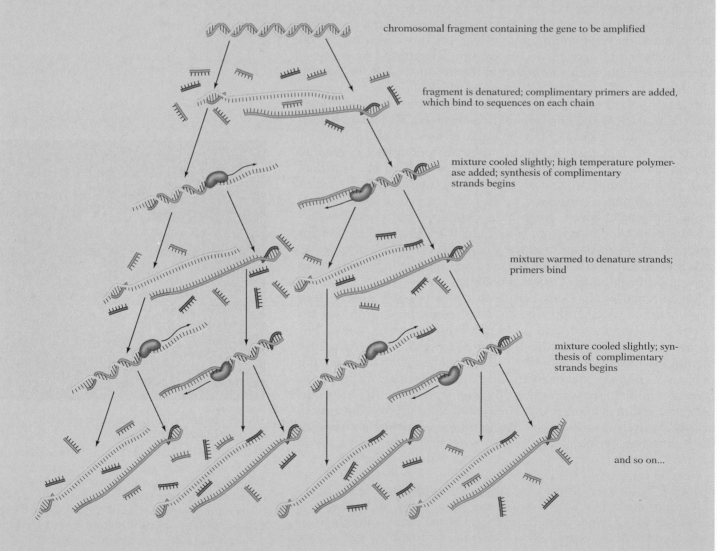

chromosomal fragment containing the gene to be amplified

fragment is denatured; complimentary primers are added, which bind to sequences on each chain

mixture cooled slightly; high temperature polymerase added; synthesis of complimentary strands begins

mixture warmed to denature strands; primers bind

mixture cooled slightly; synthesis of complimentary strands begins

and so on...

little as 8 hours.

This polymerase-chain-reaction (PCR) technique allows unlimited amplification and analysis of genes even if there is only a single copy present at the outset. In addition to the many obvious applications of the procedure to the study of genetic mechanisms, PCR has been used for a number of unusual tasks, ranging from the identification of criminals based on the DNA in a single hair, spot of blood, or drop of semen, to studying the evolutionary relationships of extinct organisms using nothing more than a minute quantity of surviving mitochondrial DNA in museum specimens or fossils. More recently, it was used to analyze whale meat sold in Japan; analysis revealed that a substantial proportion of the meat came from protected species, including fin, minke, and humpback whales—animals that had been killed and marketed illegally (and, until the invention of the PCR technique, with little risk of detection).

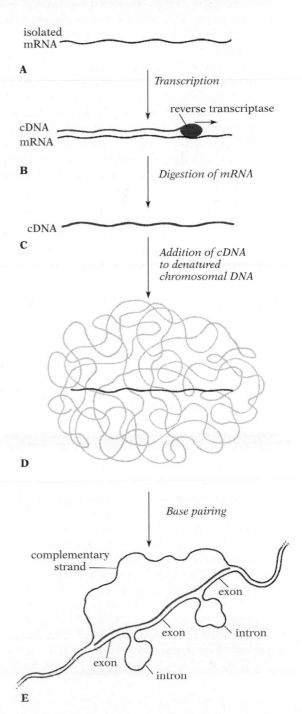

dioactive phosphorous (^{32}P). This cDNA is then mixed with a set of chromosomes treated to separate the double helix into single strands. When base pairing is made possible in the resulting mixture, the cDNA frequently pairs with the corresponding gene on the chromosome (Fig. 10.15). Because the cDNA carries a radioactive label, the exact location on the chromosome of the gene to which it binds is revealed (see Fig. 11.6, p. 284).

Prospects for the future Scientists have begun the complete mapping and sequencing of the human genome's 50,000 genes. (See *Exploring Further* box: DNA Sequencing.) This may make possible many kinds of therapeutic intervention, and cast considerable light on chromosomal organization and evolution. One of many uses of a complete map of the human genome may be to create DNA sequences complementary to a portion of the mRNA of a gene whose activity needs to be brought under control—a gene involved in cancer, for example (described in detail in Chapter 11). This cDNA—also known as antisense DNA—binds to the corresponding mRNA and blocks its translation (Fig. 10.16).

Other spinoffs of recombinant DNA research include techniques for (1) identifying nucleotide sequences in DNA quickly and precisely, (2) rearranging genes and their components to study gene interaction and gene control, and (3) creating specific mutations at specific sites in genes. This third procedure makes possible precise investigation of the roles of signal sequences in DNA and mRNA, and also enables the researcher to alter proteins at will to explore the bases of enzyme function.

Looking into the future, some researchers have suggested using cDNA technology to transfer genes for the fixation of atmospheric nitrogen to crop plants. This would be a formidable undertaking, but would eliminate the need to apply nitrogenous fertilizers. Researchers are also working to introduce the genes for proteins rich in essential amino acids into grains or beans. The result would

10.15 Locating a gene with a cDNA probe

(A) A particular kind of mRNA is isolated. (B) Reverse transcriptase is used to make a cDNA transcript incorporating a radioactive label, (C) then RNAse is added to digest the mRNA. (D) The cDNA is next mixed with denatured chromosomal DNA. (E) During base pairing, the cDNA binds to the gene coding for the mRNA, thus showing its location on the chromosome. Note that the intron segments loop out; they do not match up with the cDNA because they were absent from the mRNA.

EXPLORING FURTHER

DNA SEQUENCING

One of the cornerstones of recombinant DNA technology is sequencing—determining the order of bases in a gene. To obtain enough copies of the gene to be sequenced, a segment of the chromosome containing the gene is usually inserted into a plasmid and allowed to replicate many times. A widely used method for sequencing begins by breaking these multiple copies of the chromosome segment into pieces of manageable length by introducing an endonuclease. Because of the specificity of this enzyme, each copy of the chromosome segment is cut into an identical set of pieces. These pieces are then separated by any of a number of means (most often by molecular weight). All of the copies are tagged at one end with a radioactive marker. Next each collection of identical fragments is divided into four equal parts, each of which is treated with a low concentration of an enzyme that cuts out a specific nucleotide in the DNA; one part is exposed to a cytosine-specific enzyme, another to a guanine-specific chemical, and so on (Fig. A). The fragments are then denatured to separate the two strands; only one strand (the one labeled at the 5' end) is used in the subsequent analysis.

Though the DNA in each part is cut next to a specific base, different copies of the DNA are severed at different places along the DNA. (Some, by chance, are even cut two or three times, but every break point is adjacent to a base of the kind for which the cutting enzyme is specific.) The result is a collection of labeled fragments of varying length; each labeled fragment extends from the labeled end to one of the bases in question. These fragments are next separated by weight (length); the weight of each labeled fragment appears as a radioactive band on a gel, and so each band specifies the position (that is, the distance from the labeled end) of one copy of the base. The fragments in the other three parts provide similar information about the location of the other three bases. When the resulting bands from each sample are aligned, the sequence of bases can be read out directly (photo).

The same procedure must be repeated for each sample in order to sequence the entire set of chromosome pieces. At this point, however, there is still no way to put the sequences into the proper order with respect to each other. This is done by treating another collection of fragments of the same chromosome with a different endonuclease, and then sequencing those fragments (Fig. B). Because the second endonuclease will cut the DNA at different spots, the break points of the segments from the first analysis will lie within the sequences worked out in the second. Thus, by looking for the beginnings and ends of the sequences from the first digestion *within* the sequences from the second digestion, the relative order of the fragments can be established.

Nucleotide sequences produced by a sequencing technique: Each sample generates a group of four vertical bands on a gel.

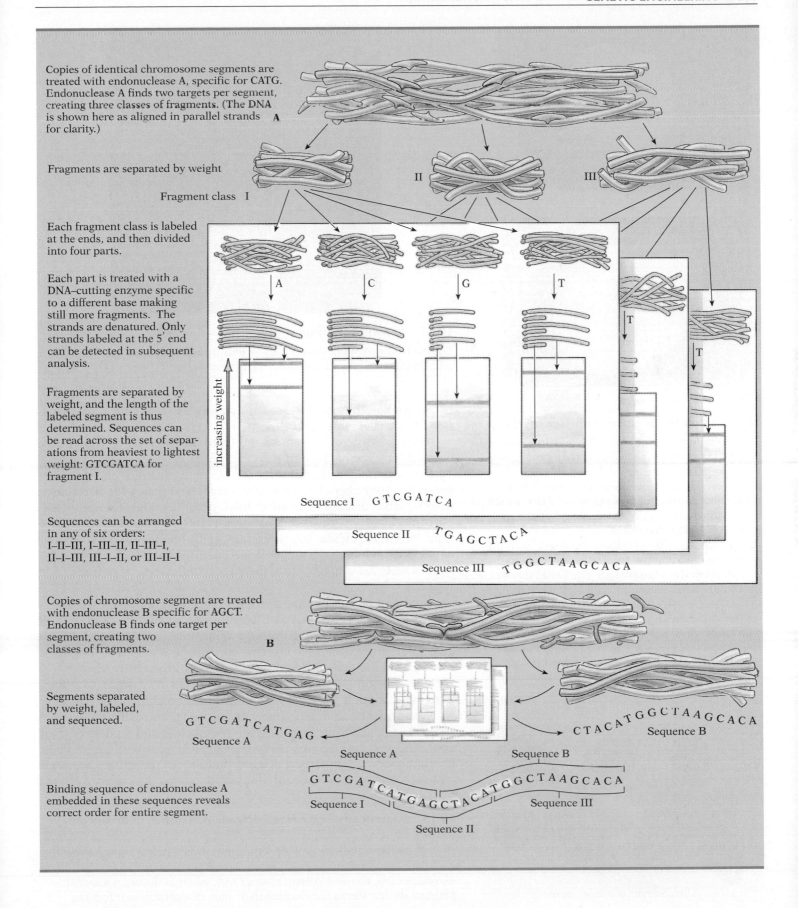

Copies of identical chromosome segments are treated with endonuclease A, specific for CATG. Endonuclease A finds two targets per segment, creating three classes of fragments. (The DNA is shown here as aligned in parallel strands for clarity.)

Fragments are separated by weight

Fragment class   I

Each fragment class is labeled at the ends, and then divided into four parts.

Each part is treated with a DNA–cutting enzyme specific to a different base making still more fragments. The strands are denatured. Only strands labeled at the 5′ end can be detected in subsequent analysis.

Fragments are separated by weight, and the length of the labeled segment is thus determined. Sequences can be read across the set of separations from heaviest to lightest weight: GTCGATCA for fragment I.

Sequences can be arranged in any of six orders: I–II–III, I–III–II, II–III–I, II–I–III, III–I–II, or III–II–I

Copies of chromosome segment are treated with endonuclease B specific for AGCT. Endonuclease B finds one target per segment, creating two classes of fragments.

Segments separated by weight, labeled, and sequenced.

Binding sequence of endonuclease A embedded in these sequences reveals correct order for entire segment.

Sequence I   GTCGATCA
Sequence II   TGAGCTACA
Sequence III   TGGCTAAGCACA

GTCGATCATGAG
Sequence A

CTACATGGCTAAGCACA
Sequence B

Sequence A   Sequence B
GTCGATCATGAGCTACATGGCTAAGCACA
Sequence I   Sequence II   Sequence III

A

B

10.16 Therapeutic use of antisense DNA

DNA complementary to the mRNA from a gene involved in stimulating cell growth (a gene also involved in some cancers) can be used to block dangerous "healing" inside arteries after therapeutic procedures like balloon angioplasty, which is used to widen clogged arteries. (When the unwanted healing response is triggered, the resulting proliferation of cells in the artery wall can narrow and harden the vessel, thereby undoing the effects of the angioplasty). The vessel shown in B was treated externally with a gel containing the antisense DNA; a newer technique uses viruses delivered during the angioplasty itself, which introduce a gene that makes cells in the artery wall susceptible to a chemical that blocks cell division. Part A shows an untreated vessel.

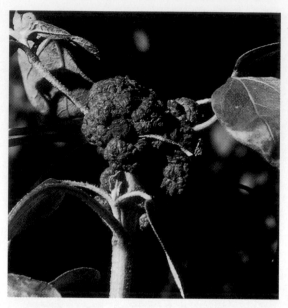

10.17 Crown gall tumor

The bacterium *Agrobacterium tumefaciens* inserts a plasmid into its host's cells, leading the plant to produce the tumor.

be a complete human food, as nutritious as meat and dairy products, but far less fatty and expensive.

Genes for disease resistance might be transferred to susceptible crops. Using the bacterium responsible for crown gall disease (Fig. 10.17), a pathogen that inserts its 10-gene plasmid into the host genome, researchers have already transferred genes for herbicide-, insect-, and disease-resistance into petunias and tobacco (Fig. 10.18). Plants currently being engineered for disease resistance (and often herbicide tolerance as well) include alfalfa, apples, canola, canteloupes, corn, cotton, cucumbers, papaya, potatoes, rice, soybeans, squash, strawberries, sunflowers, tomatoes, and walnuts.

Vaccines against the many viruses for which conventional vaccination techniques fail can also be mass-produced. For example, if a virus cannot be cultured in large quantities, a gene coding for a coat protein capable of triggering an immune response might be cloned, and the gene product used instead of the entire virus.

Gene therapy—which adds normal copies of a gene to organisms with mutant alleles—has already succeeded in tests in *Drosophila*, rodents, and humans. For example, retroviruses or liposomes (which can fuse with the cell membrane and so deliver their contents) can be used to carry therapeutic genes to target cells. Cystic fibrosis, an inherited disease that blocks digestive and sweat glands, causing painful and dangerous cysts, has already been

10.18 Genetically engineered disease resistance in tobacco plants

Both plants have been inoculated with a microbial disease at two spots on each leaf. The plant on the right, with resistance genes introduced by *Agrobacterium*, shows no disease symptoms, whereas the unengineered plant on the left shows the characteristic halo of a spreading infection.

firing pin

gunpowder cartridge

macroprojectile

DNA coated pellets

vent

stopping plate

target cells

10.19   A DNA particle gun

DNA to be inserted into target cells is coated onto metal pellets, which are then loaded into a projectile in the particle gun. An explosive charge accelerates the projectile, which is stopped by a constriction at the end of the barrel. Inertia carries the coated pellets into the cells, where some of the DNA is incorporated into the chromosomes.

cured in cultured laboratory cells using this delivery system, opening the door to a dramatic new way of treating otherwise incurable diseases. More recently, the genes to be inserted have been coated onto microscopic gold pellets, which are then literally shot into cells (Fig. 10.19). Though developed for plants, this approach has been used successfully to insert genes that promote tissue repair into cells near wounds, leading to a dramatic reduction of healing times. Another delivery technique involves encapsulating engineered cells in a polymer shell, and then inserting the capsules into specific tissues: the foreign cells, safe from the host's immune system, secrete the desired product, which diffuses out of the capsule, while host nutrients diffuse into the capsule and maintain the caged cells.

Gene transfers to domesticated animals are clearly possible. The germ line is altered by effecting the change in the egg, either with a retrovirus, a pellet gun, or by microinjection (Fig. 10.20). Extra

10.20   Microinjection of DNA

The very fine needle (barely visible at left) is being used to inject DNA into a male mouse's pronucleus (the nucleus of a sperm cell after it has entered an egg cell but before it has fused with the egg nucleus). Microinjection can also be performed after fertilization (that is, when the two gamete nuclei have fused), but the success rate is lower.

HOW WE KNOW:

DNA FINGERPRINTING

The techniques of genetic engineering have provided a new tool for identifying individuals based on samples of blood, skin, hair, or semen. These techniques are already being used for determining paternity, both in civil cases and in studying animal societies. The basic tool is the polymerase chain reaction, which generates enough copies of the DNA under study to permit quantitative analysis. But the key to individual identification lies in a particular class of genetic variation found in virtually all eucaryotes. As described in detail in Chapter 11, a large proportion of the human genome consists of untranscribed "junk" DNA, which includes hundreds of thousands of partial copies of transposons and still more copies of short nucleotide repeats—C-A-T, for instance. The repeats, which are concentrated near the centromeres and near the ends of the chromosomes, vary enormously in number and distribution from one individual to another (identical twins excluded). One person may have one copy of a particular eight-base repeat at a specific location on a given chromosome, while a sibling may have 20 copies of the same repeat at that spot.

To compare two samples to see if they are from the same individual, the DNA is cut with specific endonucleases, and the length and charge of the resulting fragments (as judged by the speed with which they move through a gel) are compared. This separation is very similar to the procedure used in gene sequencing (described in the box on p. 264). The position on the gel of fragments containing the so-called hypervariable regions of the genome differ greatly from one individual to another. In theory, the patterns are so specific that the chance

of two persons having the same fragment lengths is less than 1 in 700 trillion—far more specific than a set of conventional fingerprints.

The DNA fingerprinting technique was developed by the British geneticist Alec Jeffreys in the early 1980s. The first use in the United States involved an accused rapist in Orlando, Florida, in 1987, but the technique is used widely in murder and assault cases as well. DNA from the defendant ("D" in the figure), the victim (V), and the blood found on the suspect's clothes (amplified with the PCR technique to yield enough material; labeled "jeans" and "shirt" in the figure) are treated with the endonuclease and run on a gel, along with some calibration samples ("TS,", "1kb," and "γ"). A careful comparison of the columns on the gel show clearly that the blood on the suspect's clothes closely matches that of the victim, and is unrelated to the defendant's own blood. The defendant in this case was convicted on the DNA evidence.

DNA fingerprinting, like all forensic evidence, is only as reliable as the honesty of the investigators and the competence of the laboratory performing the analysis. In addition, DNA fingerprinting is not nearly as specific in practice as theory suggests. Compare the victim's blood in the figure with the blood found on the shirt in the "8μg" and "4μg" columns: the pattern is very similar, but the bands do not line up precisely. If the shirt sample were not available and you had only the very faint "jeans" column to compare with the victim's and defendant's blood, would you consider any of them a match? Clearly there is some degree of variation within a single gel, which amplifies a major problem with small sam-

DNA fragments from a murder victim's blood (V), the defendant's blood (D), and the blood (amplified through the PCR technique) from the suspect's clothes were treated with endonuclease and separated on a gel. (The three columns at the left and the two at the right contain control fragments for calibration.) The fragments are made visible by applying a radioactive stain and then placing the gel (or a replica created by pressing a mesh onto the gel) on a photographic emulsion; radioactive decays create the dark areas visible here.

ples; each limits the certainty of the test. If the blood on the clothing had begun to degrade from exposure to light, air, chemicals, or simply through the passage of time, numerous random fragments of unpredictable lengths might have been created before the endonuclease was applied. This factor also reduces the potential resolution of any comparison.

A more difficult problem with DNA fingerprinting is that some of the original calculations of a chance match did not take into account the fact that the more related two individuals are, the more similar their basic pattern of fragment-length variability. Thus while the patterns of victim and suspect from two different ethnic groups will be very dissimilar, members of the same group, innocent or guilty, will be more alike—a similarity that ought to be greater if they are from the same interbreeding subgroup (the same town, for instance). Moreover, some ethnic groups simply have less variability, particularly when the founder population from which the group is descended was small. To take an extreme example, using DNA fingerprinting to solve a violent crime within an Amish colony in the United States—a remote possi-

bility given the ethos of this sect—would be very difficult because the range of DNA variation is highly limited.

At present the exact probabilities in any given case, whether it entails animal research or human legal proceedings, are difficult to compute, and next to impossible for a jury to understand. In general, a worst-case scenario of maximum genetic similarity is used, which generates wildly conservative probabilities close to 1 in a million for obvious matches. How willing will juries be to convict on 1-in-a-million cases versus 1-in-a-trillion? Though this approach essentially eliminates the chance of false convictions, it raises the odds of mistaken acquittals. Increasingly, investigators are subdividing samples and analyzing them with additional endonucleases, generating extra patterns to compare. Each new set of digestion fragments reduces the probability of a chance match. Despite the lingering statistical controversy, and the resulting highly conservative interpretation of pattern matches, the DNA fingerprinting technique promises to revolutionize criminal investigation, and to increase the certainty with which juries can render judgments.

genes for growth hormone, paired with control regions that encourage high rates of transcription, have been added in this way, leading to strains of rodents, swine, cattle, and fish that grow larger and store less fat (Fig. 10.21). This shift in metabolic priorities overburdens some of the internal organs of these creatures, however, indicating that much remains to be learned and done to realize the full promise of the technique.

Recognition of the risk of undesirable side effects of recombinant DNA technology has led to strict regulation of how the technology is used. Laboratories must be equipped with facilities for sterile handling and for physical containment similar to those long used in the medical microbiology laboratories where dangerous pathogens are handled. In addition, there is a consensus that the microorganisms used in recombinant DNA research should be incapable of becoming pathogenic, or should be derived from strains with crippling mutations that make them incapable of surviving outside the laboratory.

10.21 Effects of microinjection of growth hormone genes

The eggs of coho salmon were microinjected shortly after fertilization with a gene for growth hormone and a control region that promoted its transcription. The five 14-month-old fish at the bottom are untreated; the five at the top were successfully injected with the gene and are on average 11 times heavier at the same age.

CHAPTER SUMMARY

THE DISCOVERY OF BACTERIAL PLASMIDS

DO BACTERIA MATE? Some kinds of bacteria can undergo conjugation, in which one bacterium donates genetic material to another. Usually the sex factor that allows a bacterium to initiate conjugation is found in a plasmid, which is all that is exchanged; when the plasmid has been incorporated into the chromosome, however, large sections of the donor's genome can be transferred. Donated chromosomal DNA can recombine with the recipient's DNA. (p. 248)

AUTONOMOUS PLASMIDS Most gene-carrying plasmids lack a sex factor; they reproduce autonomously and are passed to daughter cells. (p. 249)

TRANSFORMATION Certain bacteria can take up DNA from the surrounding medium; the DNA fragment may then recombine with the host chromosome. (p. 250)

HOW VIRUSES MOVE GENES

HOW VIRUSES REPRODUCE Some viruses simply inject DNA into hosts; the viral genes direct the replication of the viral chromosome and, via transcription and translation, the production of necessary enzymes and structural products. Lytic viruses reproduce and then lyse the host cell. In temperate viruses the DNA can be incorporated into the host chromosome and reproduced when the chromosome is replicated; later this lysogenic state can change into a lytic phase as the viral DNA becomes active and directs the production of new viruses. Retroviruses have RNA genes that are copied into DNA by reverse transcriptase; this copy is then incorporated into the host chromosome. (p. 250)

VIRAL TRANSFER OF NONVIRAL GENES Host genes sometimes become incorporated into new viruses, and can thus be carried to new hosts; this process is called transduction. (p. 253)

HOW GENES ARE MOVED WITHIN THE GENOME

THE BEHAVIOR OF TRANSPOSONS Transposons are genetic units that can move within the genome. Some move themselves, while others spread by making and dispatching RNA or DNA copies of themselves. (p. 253)

OTHER WAYS TO MOVE GENES In addition to breakage and fusion events, most other changes occur when cDNA copies of mRNA or other RNAs are made by reverse transcriptase and incorporated into the genome. (p. 255)

GENETIC ENGINEERING

ARTIFICIAL TRANSFORMATION Recombinant DNA techniques are used to incorporate new genes, or new versions of existing genes, into target genomes. Most methods utilize plasmids as gene carriers. (p. 256)

HOW SPECIFIC GENES ARE CLONED A variety of techniques are used to make copies of genes, including the polymerase chain reaction and the generation of cDNA copies of particular mRNAs. (p. 260)

STUDY QUESTIONS

1 Why can't cDNA genes—the coding sequence for insulin, for instance—be introduced directly into bacteria, thus dispensing with the plasmid and the several extra steps involved in using it to carry the gene? (pp. 256–269)

2 Construct a plausible scenario for the evolution of a functional RNA plasmid. How does it manage to maintain itself generation after generation? (pp. 249–256)

3 Now play the role of natural selection. How would you turn the plasmid into a working RNA virus, either single- or double-stranded?

4 What sort of evidence could convince you that plasmids evolved from viruses, not the reverse? (pp. 249–256)

5 Based on the evidence in Chapter 5 and in Chapters 8 and 9, summarize the case for the endosymbiotic hypothesis. If mitochondria and chloroplasts are really tame bacteria, why do you suppose that most but not all of their genes now reside in the nucleus (including many for products used only in organelles)? Surely it would have been simpler to have left them "on site," rather than to have to transport so many special products there. Given that most of their enzymes are being taken to the organelle anyway, why leave any genes behind? (pp. 149–150, 218–219, 239–241)

SUGGESTED READING

AHARONWITZ, Y., AND G. COHEN, 1981. Microbial production of pharmaceuticals, *Scientific American* 245 (3). *On how recombinant DNA techniques are used to make microbes produce antibiotics, hormones, and other drugs. Includes an explanation of how antibiotics work to destroy bacteria, which suggests how plasmid genes may confer resistance.*

AMÁBILE-CUEVAS, C.F., AND M.E. CHICUREL, 1993. Horizontal gene transfer, *American Scientist* 81 (4). *On gene movement between species.*

ANDERSON, W. FRENCH, 1995. Gene therapy, *Scientific American* 273 (3). *Concise review of gene therapy techniques and tests, with a look at future prospects.*

CAPECCHI, M.R., 1994. Targeted gene replacement, *Scientific American* 270 (3). *On the powerful technique that explores gene function by knocking out particular genes selectively.*

CHILTON, M.D., 1983. A vector for introducing new genes into plants, *Scientific American* 248 (6). *On bacteria that transduce host cells.*

COHEN, J. S., AND M. E. HOGAN, 1994. The new genetic medicines, *Scientific American* 271 (6). *On antisense DNA and related strategies for combating disease.*

FEDOROFF, N. V., 1984. Transposable genetic elements in maize, *Scientific American* 250 (6). *A modern interpretation of the transposition discovered by Barbara McClintock.*

GASSER, C.S., AND R.T. FRALEY, 1992. Transgenic crops, *Scientific American* 266 (6). *On the improvement of crops through genetic engineering.*

HOPWOOD, D. A., 1981. The genetic programming of industrial microorganisms, *Scientific American* 245 (3). *An excellent summary of how basic recombinant DNA techniques work.*

JOYCE, G. F., 1992. Directed molecular evolution, *Scientific American* 267 (6). *On how induced mutation and the polymerase chain reaction technique can be combined to select for novel forms of genes that encode products with more desirable properties.*

LASIC, D., 1992. Liposomes, *American Scientist* 80 (1). *On how liposomes can be used to deliver recombinant molecules to specific targets in the body.*

VARMUS, H., 1987. Reverse transcription, *Scientific American* 257 (3). *A detailed description of the process that reverses the usual direction of information flow.*

VERMA, I. M., 1990. Gene therapy, *Scientific American*, 263 (5). *On attempts to correct genetic defects.*

CHAPTER 11

CONTROL OF GENE EXPRESSION

Every cell in the body of a multicellular organism receives the same set of genetic instructions—the organism's full evolutionary endowment. Yet individual cells may look entirely different from other cells, and may behave in entirely different ways. In fact, only a particular subset of all the genes a cell contains will ever generate proteins in that particular cell, and only a small percentage of this DNA is active at any one time. The selective activation and control of the degree of expression of each of a cell's genes is critical to life. In this chapter we will examine the logic and mechanisms of gene control, a process that involves chemicals that bind directly or indirectly to DNA or mRNA.

ORCHESTRATING BACTERIAL GENES

Because bacteria have only about 3000 genes (compared with the 50,000 of humans, for instance) and can be grown rapidly in huge numbers, they have been especially useful for study of the control of gene transcription. The first models of gene control emerged from research on *Escherichia coli*. They were based on the assumption (now known to be largely correct) that the availability of a reactant in a cell will turn on transcription of the genes encoding the enzymes that process the reactant, while excessive amounts of product will turn off transcription of the genes encoding the enzymes of the biochemical pathway that produce the product.

■ HOW GENES ARE INDUCED

In the course of an extended investigation of enzyme synthesis in *E. coli*, beginning in the late 1940s, the French biochemists François Jacob and Jacques Monod (Fig. 11.1) formulated a powerful model of gene regulation in bacterial cells. They concentrated on the enzyme β-galactosidase, which catalyzes the breakdown of lactose into glucose and galactose.

The operon Lactose is not continuously available to *E. coli*, and so the gene for β-galactosidase is normally transcribed at a very low rate. Jacob and Monod found that further production of this digestive enzyme is triggered by the presence of an *inducer*, in this instance allolactose, a derivative of lactose automatically produced

A

B

11.1 François Jacob (A) and Jacques Monod (B)

273

in the cell when lactose is present. Normally, then, β-galactosidase is an *inducible enzyme*.

Four genes are involved in the production of β-galactosidase and the two other enzymes involved in lactose breakdown: three *structural genes*, each specifying the amino acid sequence of one of the three enzymes, and a *regulator gene*, which controls the activity of the structural genes. The regulator gene, which is located at some distance from the structural genes, normally directs the synthesis of a *repressor* protein that inhibits transcription of the structural genes.

Jacob and Monod also discovered that a special region of DNA contiguous to the structural genes for β-galactosidase determines whether transcription of the structural genes will be initiated. This special region is the *operator*, which itself is not a gene since it doesn't code for a specific product. The combination of an operator and its associated structural genes is an *operon*. This particular operon, since it is responsible for the synthesis of enzymes involved in the breakdown of lactose, is called the *lac* operon.

The operator is located between the promoter—the region to which RNA polymerase binds—and the structural genes. Hence, when the repressor binds to the operator, RNA polymerase cannot bind to the promoter, and transcription is blocked (Fig. 11.2A). If an inducer is present, it will bind to the repressor, thus causing a conformational change in the repressor that forces it to dissociate from the operator; in short, the inducer deactivates the repressor (Fig. 11.2B). RNA polymerase is now free to bind to the promoter and can initiate transcription of the structural genes and the production of mRNA (Fig. 11.2C). The mRNA carries the instructions for all three structural genes. This messenger binds to ribosomes in the cytoplasm, where its information is translated and the three enzymes necessary for lactose metabolism are synthesized. The number of β-galactosidase molecules rises to about 5000 per cell when the operon is not repressed; during repression, there are only about 10 copies of that enzyme.

Summary of the Jacob-Monod model The condition of the operator region is one key to whether or not there will be activation of the *lac* operon. If repressor protein is bound to the operator, there will be no transcription. If no repressor is bound to the operator (because the repressor has been inactivated by inducer), transcription can proceed freely. Only a few molecules of inducer are required to bind all the repressor molecules in a cell.

The three jointly controlled structural genes of the *lac* operon specify enzymes with closely related functions. It is typical for the structural genes of an operon to determine the enzymes of a single biochemical pathway; thus the whole pathway can be regulated as a unit.

■ **HOW GENES ARE REPRESSED**

Not all operons are regulated in the same way as the *lac* operon, which is an inducible operon—that is, one that is inactive until it

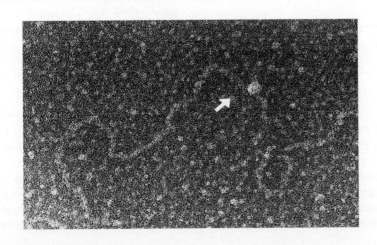

11.2 An inducible operon: the *lac* operon

(A) The operon consists of a promoter-operator region and three structural genes (Z, Y, and A). The operator sequence overlaps the beginning of the structural gene. The regulator gene codes for mRNA, which is translated on the ribosomes and determines synthesis of the repressor protein. When the repressor protein binds to the operator, it blocks one of the promoter's binding sites for RNA polymerase and thus prevents the initiation of transcription of the structural genes.
(B) Binding of inducer to repressor deactivates the repressor, and the RNA polymerase can then bind to the promoter regions. (C) Polymerases initiate transcription of the structural genes, which are transcribed as a unit, producing an mRNA coding for three gene products. The mRNA then binds to ribosomes and is translated into three enzymes. Enzyme I is β-galactosidase; enzyme II is a permease, which helps transport lactose into the cell; and enzyme III is a transacetylase, whose role in lactose utilization is not understood. The transcribed strand of the structural genes has been emphasized. The function of the activator sequence will be discussed presently. (D) Electron micrograph of the *lac* repressor protein bound to the operator region.

EXPLORING FURTHER

HOW CONTROL SUBSTANCES BIND TO DNA

For the repressor to bind to a particular operator, it must bear active sites that match the DNA substrate exactly. Such matching involves polar amino acids on the repressor and complementary polar groups exposed on the *sides* of the base pairs of the DNA.

The deoxyribose backbones of DNA are attached to the bases slightly off center. As detailed in Chapter 8, the two chains have opposite polarities, with one running from 3′ to 5′ and the other from 5′ to 3′. The result is that one of the exposed sides, or grooves, of the helix is somewhat wider

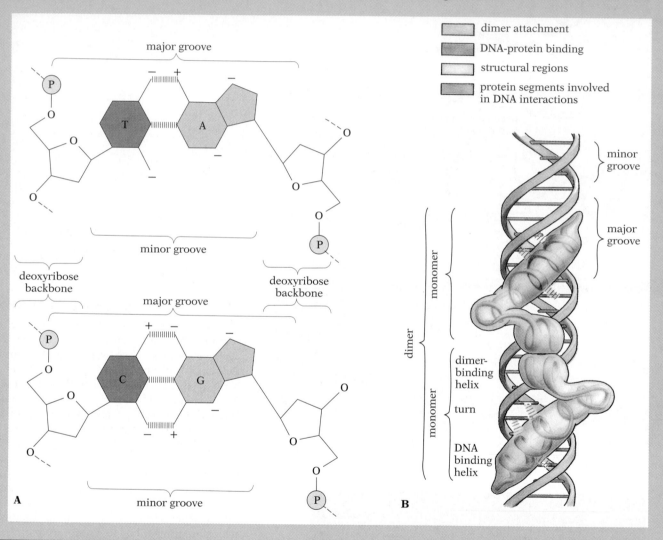

is turned on by an inducer substance. Many operons are, instead, continuously active unless turned off by a ***corepressor*** substance. One example is the operon whose five structural genes code for the enzymes necessary to synthesize the amino acid tryptophan. This operon is normally turned on, but when *E. coli* is grown in a medium containing tryptophan, it switches off. Enzymes encoded by genes that are usually active but can be repressed are called ***repressible enzymes***. Here the repressor protein encoded by the reg-than the other (Fig. A). The wider opening—the **major groove**—contains more sequence-specific information, and is therefore thought to be more important in binding repressors. For example, the major groove of a T–A pair has three polar groups—O, NH, and N, reading from T to A—their respective polarities being −, +, and −. The major groove of the C–G pair has the polar groups NH, O, and N, with polarities of +, −, and −. The **minor groove** is less useful: for T–A, the polar "code" is −, X, − (where "X" means no charge), while for C–G the sequence is −, +, −. Each particular repressor has a shape corresponding to the differing physical widths of the operator's major and minor grooves, and polarity patterns complementary to those of the operator's base pairs.

Several molecular morphologies are typical of DNA control substances. The "helix-turn-helix" motif (Fig. B) involves two α helices: one binds in the major groove, while the other attaches itself to a second copy of the binding chemical to create a dimer. The "leucine zipper" (Fig. B) is also a protein dimer with two helices. The shorter helix has a leucine at every seventh position, so that four of the leucine side groups project out on the same side of the short helix; these leucines interlock in a zipperlike fashion with the aligned leucine side groups of the other protein's short helix (Fig. C). The two long helices bind in the major groove of the DNA in what is called a "scissors grip." A third family of binding proteins also uses a dimer with a helix devoted to binding in the major groove (Fig. D). The "zinc-finger" proteins are stabilized by zinc ions, and (again) the dimers bind to each other as well as to the DNA. Other morphologies exist as well.

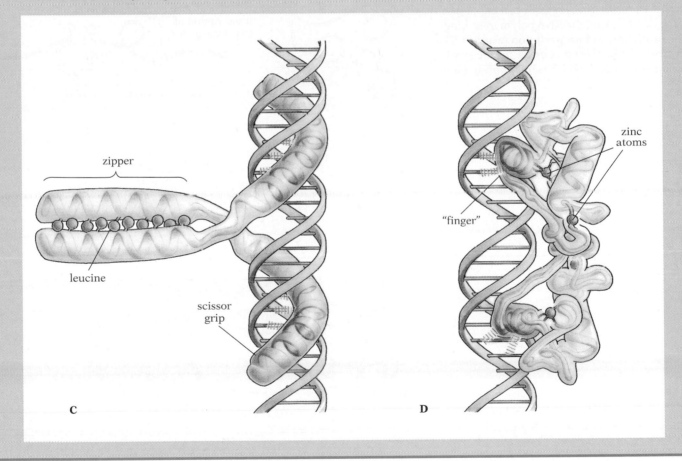

zipper

leucine

scissor grip

C

zinc atoms

"finger"

D

ulator gene is inactive when first produced. Only if a corepressor substance binds to and activates it can it bind to the operator and block RNA polymerase binding (Fig. 11.3). Unlike inducible enzymes, which are synthesized only if their operon is turned on by an inducer, repressible enzymes are automatically synthesized unless their operon is turned off by a corepressor. In tryptophan synthesis, the tryptophan itself activates the repressor protein, enabling it to bind to the operator.

11.3 A repressible operon

The repressor protein encoded by the regulator gene is initially inactive (A). As a result, polymerases can bind to the promoter and transcribe the structural genes (B). These genes frequently encode enzymes that lead to the synthesis of an end product such as an amino acid (C). When the inactive regulator protein combines with a specific corepressor molecule (often the end product of the biochemical pathway served by the enzymes encoded by the operon), it can bind to the operator and block transcription of the structural genes (D). After the operon has been repressed, the concentration of the corepressor falls as it is used in cellular metabolism, but no more is produced. When the corepressor becomes scarce, the repressor tends to lose it to metabolic enzymes. As a result, the repressor can no longer bind to the operator, the RNA polymerase binds to the promoter, and transcription resumes. The operon shown here is responsible for the synthesis of an amino acid, which is incorporated into new proteins.

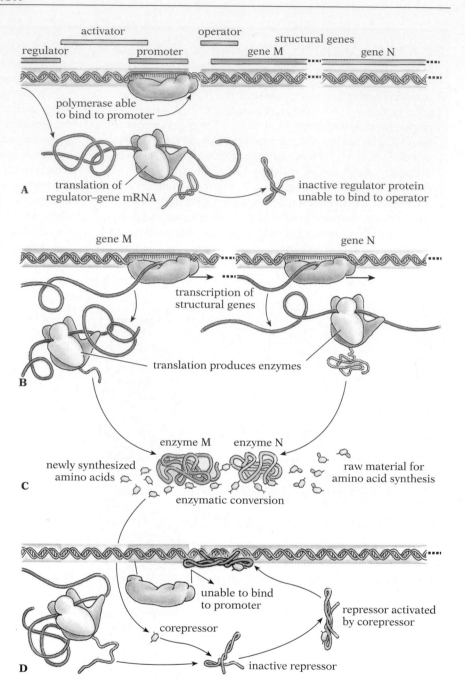

An inducer is often the first substrate in the biochemical pathway being regulated. Similarly, a corepressor is usually the end product of the biochemical pathway being regulated. In both substrate induction and end-product corepression, then, gene transcription is generally regulated by the cellular substances most affected by the transcription.

■ POSITIVE VERSUS NEGATIVE CONTROL

Both of the cases we have discussed are examples of **negative control**: a repressor bound to the operator turns off transcription.

Table 11.1 Summary of transcriptional control strategies in procaryotes

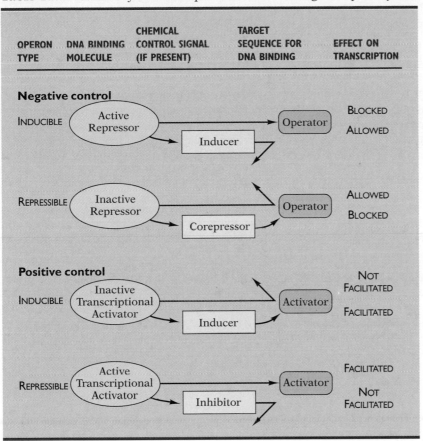

OPERON TYPE	DNA BINDING MOLECULE	CHEMICAL CONTROL SIGNAL (IF PRESENT)	TARGET SEQUENCE FOR DNA BINDING	EFFECT ON TRANSCRIPTION

Negative control

INDUCIBLE — Active Repressor → Inducer → Operator — BLOCKED / ALLOWED

REPRESSIBLE — Inactive Repressor → Corepressor → Operator — ALLOWED / BLOCKED

Positive control

INDUCIBLE — Inactive Transcriptional Activator → Inducer → Activator — NOT FACILITATED / FACILITATED

REPRESSIBLE — Active Transcriptional Activator → Inhibitor → Activator — FACILITATED / NOT FACILITATED

Control is effected in one case by deactivating the repressor with an inducer, and in the other case by activating the repressor with a corepressor (Table 11.1). Though negative control is the most common way of regulating gene expression in procaryotes, some systems are regulated by ***positive control***. In many of these cases a control protein binds directly to the DNA to activate the operon.

Positive control takes advantage of the variation in promoter sequences: RNA polymerase binds to the promoter sequence adjacent to the operator, but few genes have promoters with exactly the same nucleotide sequence. Genes with promoters that differ significantly from the optimal one are rarely transcribed, because the RNA polymerase will bind to them only weakly. In positive control of such genes, a control protein called a ***transcriptional activator (TA)*** binds to a region a little upstream from the promoter (Fig. 11.4), and then helps the polymerase bind. The presence of the bound activator protein compensates for the poor promoter sequence.

Two versions of positive control are known. In one, a chemical inducer binds to the TA, causing a conformational change that activates the TA, enabling it to bind to the DNA and facilitate transcription. A TA that works this way is the ***catabolic gene-activator protein (CAP)*** of *E. coli*, which helps control the transcription of

EXPLORING FURTHER

THE LAMBDA SWITCH AND CONTROL OF THE LYTIC CYCLE

The expression of genes is often regulated by systems that involve the interaction of three or more control substances, two or more separate operators, and sometimes other DNA-control regions in addition to the regulator genes, operators, and promoters already discussed. The most thoroughly investigated example of these more complex interactions involves the switching of the temperate virus lambda

from its lysogenic to its lytic cycle; similar control systems are used by the host itself.

Temperate viruses are able to insert their DNA into a specific section of the host chromosome and remain there without lysing the bacterial cell. Under normal conditions (that is, steady growth of the bacterial colony), this passive kind of infection rarely occurs, but once the virus does enter the host chromosome, it can remain there, in the lysogenic cycle, indefinitely. Exposure of an infected bacterium to ultraviolet radiation, however, may incite the viral DNA to leave the host chromosome and begin reproduction. In the lambda virus, which parasitizes *E. coli*, two interacting operons on complementary strands of the DNA control this switch from a lysogenic to a lytic cycle.

Excision of lambda DNA from the host chromosome and initiation of the lytic cycle are controlled by the *cro* gene and several other genes downstream from the operon. These genes are normally repressed: the *cI* gene on the complementary strand produces a repressor protein that binds to the operator region of the *cro* gene, and thereby blocks transcription (Fig. A). Repression of the *cro* and associated genes is stable until the host bacterium begins to synthesize unusually large quantities of DNA repair enzymes, often in response to UV damage (Fig. B). While fixing the DNA, one of the repair enzymes digests part of the repressor synthesized by the *cI* gene, freeing the *cro* operator to begin transcription of the normally repressed *cro* and associated viral genes. The *cro* gene codes for a second repressor, which binds to the *cI* operator and turns off transcription of *cI* (Fig. C). Transcription of lytic-phase genes can now proceed (Fig. D), leading to excision of the lambda genome from the host chromosome (Fig. E), the production of replicates of the viral DNA, synthesis of capsule proteins, viral assembly, and finally host lysis.

The lambda switch is irreversible: even the disappearance of the bacterial DNA repair enzyme does not stop production of the second repressor. The lambda virus reacts to the first sign of trouble (UV radiation, in this case) by initiating the synthesis of new phage, which will abandon the damaged bacterium and seek out other hosts.

transcription and translation of lytic–phase genes, leading to replication, production of coat protein, virus assembly, and lysis

genes encoding enzymes that digest unusual nutrients. When glucose, the bacterium's usual food, becomes rare, a messenger substance—cyclic AMP—is produced. The cAMP binds to and activates CAP, greatly increasing transcription of the alternative nutrient-metabolism operons (Fig. 11.4; Table 11.1). CAP, then, can affect the activity of many operons at once.

The other positive-control strategy involves a normally active

A

polymerase binds weakly to promoter

cAMP

inactive CAP transcription activator

binding of activated CAP induces 90° bend
in activator sequence and twists promoter open

B polymerase can now bind
strongly to open promoter

transcription proceeds
at 20 times normal rate

11.4   Induction of transcription by CAP

(A) Many genes have promoters that bind polymerase at low rates. (B) When a CAP transcriptional activator protein, itself activated when bound by a molecule of cAMP, binds to a sequence adjacent to the promoter, it induces a 90° bend in the DNA; the resulting twist opens the promoter, and increases the polymerase-binding rate 20-fold.

control protein that is turned off by the binding of a small molecule. Once switched off, the protein cannot help the RNA polymerase bind to the poor promoter sequence, so transcription is minimized (Table 11.1). The cAMP signal that, via CAP, activates the operons for alternative nutrient-digestion genes, simultaneously *inactivates* the TA proteins that help the polymerases bind to and transcribe the genes for transporting glucose into the cell. The metabolic systems for alternative food sources are activated, and the now-useless glucose system is temporarily shut down.

The *lac* system is one of the most complex procaryotic operons; it involves both negative control (a repressor, which can be inactivated by an inducer when lactose is present) and positive control (the CAP transcription factor, which can be induced to bind when glucose is rare). As a result, the *lac* operon is most active when lactose is present and glucose is absent (Table 11.2).

Table 11.2   Summary of *lac* operon control

ENVIRONMENTAL CONDITIONS		OPERON STATUS		TRANSCRIPTIONAL ACTIVITY OF LACTOSE-METABOLIZING GENES
GLUCOSE PRESENT	LACTOSE PRESENT	cAMP-CAP TA COMPLEX BOUND TO ACTIVATOR	REPRESSOR BOUND TO OPERATOR	
yes	no	no	yes	none
yes	YES[a]	no	NO[a]	LOW
NO[a]	no	YES	yes	none
NO	YES	YES	NO	HIGH

[a] Uppercase entries indicate factors that help enhance lactose digestion.

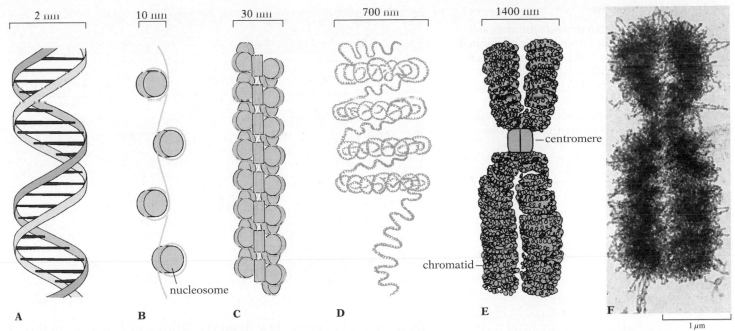

2 nm	10 nm	30 nm	700 nm	1400 nm

nucleosome

centromere

chromatid

A   B   C   D   E   F

1 μm

11.5 Packaging of DNA in a eucaryotic nuclear chromosome

The double-helical strand of DNA (A) is wound around nucleosomes about 10 nm in diameter (B). This chain of spools is itself coiled in some way to produce a thick strand about 30 nm in diameter (C). The arrangement shown here is hypothetical. Early in cell division the thick strands are collected into a long series of loops, which are wound into a helix (D). The result is the ragged appearance of the chromosome, as seen in drawing (E) and in the electron micrograph of a human chromosome (F). Some of the loops are tightly condensed while others are not, and can be seen extending out from the chromatid cores.

HOW EUCARYOTIC CONTROL IS DIFFERENT

The control of gene expression in eucaryotes is more complex, and it needs to be: even the simplest eucaryote has many more genes to regulate than any procaryote. In addition to functional complexity, there are also structural complications: as described in Chapter 5, eucaryotic DNA is wrapped on nucleosomes, wound into tightly packed coils, organized into loops, and condensed to form visible chromosomes (Fig. 11.5). Prior to transcription, a eucaryotic gene must be "unpacked."

■ THE ODD ORGANIZATION OF EUCARYOTIC CHROMOSOMES

Chromosomal proteins Eucaryotic chromosomes consist of DNA and protein, an association often called *chromatin*. Chromosomal proteins include *histones*, most of which are essential components of nucleosomes, and *nonhistone proteins*. Histones may be involved in gene expression, but their role seems to be passive: when the nucleosome cores are in place, transcription is not possible, since RNA polymerase cannot gain access to the DNA.

Nonhistone proteins are much more important in gene regula-

11.6 Locating a gene

In this version of the DNA probe technique, salivary-gland chromosomes from the much-studied fruit fly *Drosophila* have been treated to weaken base pairing; a radioactively labeled sample of previously identified DNA has been added and has bound to the complementary region. The dark dots reveal where radioactive decay is taking place, and therefore where this particular gene is located.

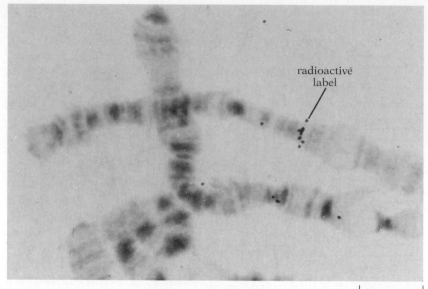

radioactive label

10 μm

tion. Some are bound directly to the DNA, while others are linked to the nucleosome cores. They exhibit a rich diversity, not only from organism to organism, but also from tissue to tissue within a given organism, and even within a single cell at various times, depending on its developmental stage and its current functional condition. Hence, nonhistone proteins may have the specificity necessary for control elements. Their role, at least in some cases, is to bind to specific control regions in the DNA to cause a chromosomal loop to decondense (uncoil from its nucleosomes). They also play a role in the second step of eucaryotic gene control. Since most loops are about 100,000 base pairs long, each usually contains several genes, which are often selectively activated.[1] This level of control probably includes removing one or more nucleosomes from the promoter regions of the genes destined to be switched on. As we will see, activation rather than inhibition of transcription appears to be the rule in eucaryotes.

Highly repetitive DNA The organization of eucaryotic chromosomes has been revealed in large part through the technique of **_DNA hybridization_**. The first step is to heat the chromosomal DNA under controlled conditions; this causes the two strands of each helix to separate. In one type of hybridization the next step is to add to the chromosomal DNA a labeled "probe" of a particular DNA sequence of interest, or an mRNA (or a cDNA made from it by reverse transcriptase). The mixture is then cooled very slowly. The RNA or DNA will frequently bind to the complementary region of the chromosomal DNA, and because of the labeling, the gene or sequence for it can then be located on the chromosome (Fig. 11.6). Clever variations on this basic method have produced a series of surprises for researchers, making it clear that the evolutionary histories of procaryote and eucaryote chromosomes have been very different. Perhaps the most consistent contrast is that eucaryotic

[1] The length of the average eucaryotic gene is 2000 nucleotides, of which only 1200 are in exons.

genomes are filled with sequences that are never transcribed, many of which exist in multiple copies. For instance, about 10% of most eucaryotic DNA contains base sequences that are found not once but thousands of times in the genome, constituting what is called **highly repetitive DNA**.

There are at least four classes of highly repetitive DNA. The first consists of a vast number of copies of a few kinds of short sequences located at the **centromere** (the specialized region that holds the two copies of a replicated chromosome together until cell division) and in large blocks in chromosome arms, especially at the ends, where they are called **telomeres** (Fig. 11.7A). The function of this DNA, which is never transcribed, is to facilitate particular steps in the process of cell division (described in Chapter 12), and to maintain stability in the chromosome.

Another class of highly repetitive DNA consists of long units repeated in tandem. These areas contain genes that code for the smallest of the four ribosomal RNAs, called 5S RNA (Fig. 11.7B). (The name refers to the rate at which this RNA sediments through a sucrose solution; the other three rRNAs—6S, 18S, and 28S—are all larger and heavier by this measure.) In one well-studied frog species, each 5S gene is associated with a nonfunctional region known as a **pseudogene** (so called because its sequence is almost identical to that of a functional gene) and a very long "spacer" region. Neither the pseudogene nor the spacer is ever transcribed; most of this region of the chromosome appears to be functionless. Though the gene for 5S rRNA makes up only a small fraction of the region, there is a clear reason for the large number of copies of the 5S gene. Developing eggs need enormous numbers of ribosomes (perhaps 10^{12}) to handle all the protein synthesis necessary for rapid growth. Repeated transcription of a single copy of this gene would be inadequate to meet the needs of the cell for rRNA; consequently most eucaryotes have approximately 25,000 copies of this sequence.

A third class of highly repetitive DNA, known as "long interspersed elements," consists of sequences that are several thousand base pairs long and are found tens of thousands of times in the chromosomes of many eucaryotes (Fig. 11.7C). Their function, if any, is unknown.

The fourth class consists of relatively short (300 base-pair) segments scattered throughout the genome. Humans have about 500,000 copies of one such sequence, which is simply a transposon (derived from an RNA polymerase promoter) that reproduced and spread rampantly at some stage in our evolution; it appears to have no function.

Moderately repetitive DNA About 20% of the typical eucaryotic genome consists of **moderately repetitive DNA** whose sequences are repeated hundreds rather than thousands of times. Moderately repetitive DNA comes in one of two varieties. The first is a tandem repeat of certain genes. In particular, genes for the other three kinds of rRNA are repeated tandemly as a group: 18S, 6S, 28S, untranscribed long spacer; 18S, 6S, 28S, long spacer; and so on (Fig. 11.7D). Since every ribosome must have one of each of the four kinds of rRNA, how do cells manage to get by with only 100–1000

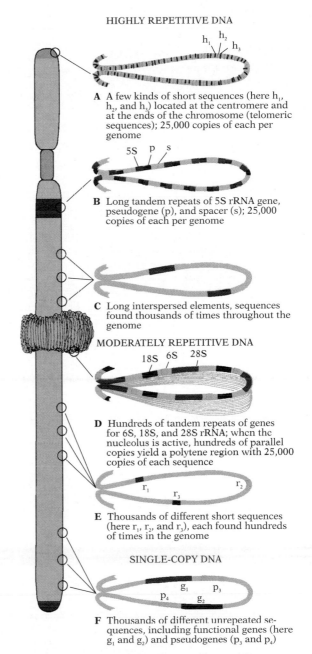

HIGHLY REPETITIVE DNA

A A few kinds of short sequences (here h_1, h_2, and h_3) located at the centromere and at the ends of the chromosome (telomeric sequences); 25,000 copies of each per genome

B Long tandem repeats of 5S rRNA gene, pseudogene (p), and spacer (s); 25,000 copies of each per genome

C Long interspersed elements, sequences found thousands of times throughout the genome

MODERATELY REPETITIVE DNA

D Hundreds of tandem repeats of genes for 6S, 18S, and 28S rRNA; when the nucleolus is active, hundreds of parallel copies yield a polytene region with 25,000 copies of each sequence

E Thousands of different short sequences (here r_1, r_2, and r_3), each found hundreds of times in the genome

SINGLE-COPY DNA

F Thousands of different unrepeated sequences, including functional genes (here g_1 and g_2) and pseudogenes (p_3 and p_4)

11.7 Organization of eucaryotic chromosomes

The short highly repetitive DNA (A), long tandem repeats (B), and polytene tandem repeats (D) are confined to different specialized regions of the chromosome. The long interspersed elements (C), moderately repetitive sequences (E), and single-copy DNA (F) are actually intermixed; for clarity, however, they are shown here in isolation on separate loops.

copies of these genes while there are many thousands of copies of the 5S gene? The answer is that when new ribosomes are needed, the 18S–6S–28S–spacer tandem repeat can be *replicated repeatedly, independent of the rest of the chromosome*. The result is a **polytene** region consisting of 25–250 parallel copies of the repeat region. Replicating the polytene region in turn produces up to 25,000 copies of each gene in active cells. The region of the chromosome with the many replicated repeating segments forms the nucleolus (described in Chapter 5).

The other class of moderately repetitive DNA is more mysterious, and varies widely in size and frequency between species. These sequences, of which there may be as many as 5000 different types, are only about 300–3000 bases long—as little as one-tenth the length of the average functional gene. Each type is usually scattered between functional genes throughout the chromosome in 30–500 different locations (Fig. 11.7E). Some of this DNA is clearly derived from mRNA containing the promoter sequence for reverse transcriptase; presumably some of the DNA copies were synthesized from mRNA by reverse transcriptase as an artifact of retroviral infection, and made their way into the host chromosome. One of the cell's tRNAs appears to be particularly susceptible to this process. Other cases, however, do not fit this model.

Single-copy DNA Though 70% of a typical eucaryotic genome consists of **single-copy** sequences (Fig. 11.7F), the majority of this DNA is *never* transcribed. In mammals, much of it exists as pseudogenes—nearly identical but untranscribed copies of functional genes. Only about 1% of eucaryotic DNA codes for mRNA that is subsequently translated. This is dramatically different from procaryotes, in which well over 90% of the genome is translated at one time or another. The high density of information in procaryote genomes—haploid and almost devoid of introns, spacers, and pseudogenes—is probably a result of intense selection for reproductive speed. In contrast, few eucaryotes find any reproductive advantage in high-frequency cellular doubling. This fact alone, however, seems insufficient to account for the eucaryotic cell's tolerance of countless introns, pseudogenes, and spacers, as well as transposons, retroinsertions, and other genetic debris. Some researchers believe that eucaryote genomes simply lack efficient ways of expunging genetic "junk," which therefore accumulates endlessly; others suspect that much of this vast proportion of untranscribed DNA is actively tolerated as a laboratory for evolutionary experiments in exon shuffling and mutation, or that it plays an undetected role in gene regulation. The logic of eucaryotic gene organization remains one of the most interesting and elusive questions in modern biology.

■ HOW CHANGES IN GENE TRANSCRIPTION CAN BE SEEN

Profound as the differences are in the organization of the procaryotic and eucaryotic genomes, it is not surprising that their systems of gene control differ as well. The eucaryotic genome, with its need to decondense tightly packed loops of DNA before transcribing them, must have an added level of control that procaryotes do not

require. In addition, eucaryotes must orchestrate multiple constellations of genes, each specific for different tissues and different
stages in their complex development from a single cell.

Heterochromatin versus euchromatin The internal structure of
a chromosome is not uniform; much of its variation reflects gene
activity. For example, some regions of chromosomes stain only
faintly when treated with basic dyes, whereas other regions stain
intensely. The nonstaining regions are called *euchromatin* and the
staining ones *heterochromatin*. Chromosomal mapping over the
past few years has shown that euchromatic regions contain active
genes whereas heterochromatic regions are inactive.

Some heterochromatic regions are simply devoid of genes, for
instance, the large region of heterochromatin with highly repetitive
base sequences located around the centromere. However, most heterochromatic regions do not lack genes; instead, the genes (often
long series of genes) are simply inactive, and may be euchromatic
at other stages of development.

Even within a region of euchromatin only a few widely spaced
genes may be active; the clumping of a small set of related genes
into an operon that is so characteristic of bacteria is rare in eucaryotes. Indeed, in many instances in which there appears to be
simultaneous control of the synthesis of functionally related enzymes, the enzymes are encoded by widely separated genes. The
genes that specify two polypeptide chains of a single protein (the α
and β chains of hemoglobin, for example) are often on different
chromosomes.

Lampbrush chromosomes and chromosomal puffs The developing egg cells of many vertebrates synthesize large amounts of
mRNA for later use. The mRNA is then ready during the early
stages of development, after fertilization, when the chromosomes
are so busy with DNA replication in support of rapid cell division
that they are largely unavailable for transcription. The parts of the
chromosome bearing genes that are being repeatedly transcribed—
the euchromatic regions—loop out laterally from the main chromosomal axis, while the parts bearing repressed genes remain
tightly compacted; chromosomes with many looped-out regions
are called *lampbrush chromosomes* (Fig. 11.8).

The giant chromosomes of the salivary glands of larval flies are
lateral arrays of replicated chromosomes (polytene chromosomes)
that have remained stuck together. Polytene chromosomes result
from the same processes that, on a much smaller scale, create the
polytene region called the nucleolus. About 10 cycles of replication
are involved in the creation of a polytene chromosome, yielding
1024 (2^{10}) parallel copies of each gene. The result of this unselective gene amplification is a capacity for rapid synthesis of large

**11.8   Lampbrush chromosome from a
developing egg cell of the spotted newt
(*Triturus viridescens*)**

The many feathery projections from the chromosome are regions bearing genes that are being
repeatedly transcribed.

20 μm

11.9 Chromosomal puffs

Puffs in polytene chromosomes consist of hundreds of parallel copies of the chromosomal DNA looped out to expose the maximum surface for synthesis of RNA. The interpretive drawing shows only nine of the loops, flanked by tightly coiled regions. The series of photographs indicates how the location of puffs in a *Drosophila* chromosome changes dramatically but predictably over time; we can clearly see the varying activation of four different genes or sets of genes on chromosome 3 (each identified by a series of connecting lines) over a period of 22 hours (left to right) in larval development.

amounts of RNA. When certain regions of a giant chromosome are especially active, all of its parallel DNA strands form brushlike loops (***chromosomal puffs***) in those regions (Fig. 11.9). The locations of puffs are different on chromosomes in different tissues. They are also different in the same tissue at different stages of development (Fig. 11.9), though at any given time all the cells of any one type in any given tissue show the same pattern of chromosomal puffing.

The puffs indicate the location of active genes, and thus the sites of rapid RNA synthesis. As such, the puffs provide a way of determining visually whether or not changes in the extranuclear environment—the cytoplasm or extracellular chemicals—can alter the pattern of gene activity. For example, if ecdysone, the hormone that causes molting in insects, is injected into a fly larva, the chromosomes rapidly undergo a shift in their puffing pattern, taking on the pattern characteristically found at the time of molting. If chromosomes from one type of cell are exposed to the cytoplasm of another type of cell or of the same type of cell at a different developmental stage, they quickly lose the puffs characteristic of their original cells and develop puffs characteristic of the type and stage of cells providing the new cytoplasm.

■ IMPRINTING: PRESERVING PATTERNS OF GENE ACTIVITY

Rapidly dividing cells in a developing eucaryote must have some way of preserving the patterns of loop decondensation and gene activity that give them the appropriate chemical "personality" for the tissue the cell is a part of. In fact, the patterns of euchromatin and heterochromatin can be passed directly from a growing cell to the two daughter cells it produces through a process called **imprinting.** One component of this imprinting system in vertebrates involves methylation of the cytosines in certain C–G sequences in inactive genes. Methylation can transmit activity patterns to daughter cells: Because only C–G sequences can be methylated by the enzyme involved, and since C–G is always paired with G–C on the other strand, the same pattern of methylation exists on both strands of the DNA:

$$
\begin{array}{c}
\text{m} \\
| \\
:\ \ :\ \ :A:T:C:G:T:C:A:\ \ :\ \ : \\
:\ \ :\ \ :T:A:G:C:A:G:T:\ \ :\ \ : \\
| \\
\text{m}
\end{array}
$$

(The colored *m* indicates the cytosine has been methylated by the addition of a CH_3 group.) Replication creates a hybrid structure wherever the parental DNA had been methylated:

$$
\begin{array}{ll}
\begin{array}{c}\text{m}\\ |\end{array} & \\
:\ \ :\ \ :A:T:C:G:T:C:A:\ \ :\ \ : & \text{parental strand} \\
:\ \ :\ \ :T:A:G:C:A:G:T:\ \ :\ \ : & \text{new strand}
\end{array}
$$

A special enzyme, maintenance methylase, scans the DNA for methylated cytosines in C–G sequences and methylates the corresponding cytosine on the new strand. Thus the pattern of methylation—gene inactivation—is passed on intact. If, however, both methyl groups at a site are lost by chance, cells are produced with altered gene activity. At least some of the problems associated with aging probably arise from the progressive loss of methylation over time, and hence the activation of inappropriate genes in cells. Demethylation in gametes can even pass mistaken gene-activation patterns on to offspring, and several diseases are now ascribed to such imprinting errors.

■ EUCARYOTIC CONTROL: INDUCERS AND ENHANCERS

Little is known about the selective decondensation of chromosomal loops—the necessary precursor to transcription—that creates polytene puffs and lampbrush projections. Certain master control chemicals like ecdysone can trigger major changes in looping, but how they act is a mystery. Despite this, knowledge about the basis of individual gene regulation on decondensed loops is growing rapidly.

Since eucaryotes rely mainly on positive control, transcription does not occur without active aid from a **transcription factor (TF)**, which helps the polymerase bind to the promoter. In addition, eucaryotes control TF binding in gene-specific ways from two other locations on the chromosome.

The first control locus is an inducer region just upstream of the TF sequence (Fig. 11.10A). It consists of one or (usually) more individual sites that specific control proteins can bind to. Some of these control molecules in turn help the TF protein bind (Fig.

loop folds back to allow enhancer molecules to interact with inducers, TF, and polymerase

enhancer region

gene to be transcribed

A

inducer region   TF   polymerase binding site
                  site (promoter)

B

inducer   TF

C

initiation factor

D

E

mRNA

11.10   Eucaryotic gene control

(A) For this hypothetical gene there are four inducer sites adjacent to the transcription-factor (TF) binding sequence, and four enhancer sites upstream. (B) In this simplified positive-control model an activating inducer binds to one site and loads the TF. (C) This in turn helps the polymerase to bind. An activating enhancer binds upstream. (D) Once the loop folds into position, the enhancer loads transcription-enhancing factor onto the waiting polymerase. (E) This action initiates transcription. As described in the text, there are other ways inducers and enhancers can participate in starting or blocking transcription.

11.10B–C), while others inhibit it. For instance, the β-globin gene (which encodes one of the two kinds of protein chains that combine to form hemoglobin) has seven of these control sites in its inducer region.

The other locus of action is a few thousand nucleotides away on the same decondensed loop. It consists of a cluster of sites collectively called the **enhancer region.** These sites exert their effects when the loop folds back on itself to bring the entire region into contact with the promoter and inducer sites (Fig. 11.10). Some enhancers help load the TF protein, which then aids in polymerase binding; others directly help the polymerase to bind. Still others add a protein to the polymerase complex that can greatly enhance its efficiency in transcription. Other regulatory substances that can bind to the enhancer region inhibit one or another of these steps.

This action-at-a-distance system is nearly ubiquitous in eucaryotes, and a few bacterial genes have been found to be controlled in this way as well. Why do most eucaryotic genes have so many alternative ways to regulate transcription? Eucaryotes have much more complex life histories than procaryotes. Eucaryotes are usually multicellular, and so must grow in a controlled and directed way to achieve a species-specific morphology. Moreover, eucaryotes usually have distinct tissues and organs whose cells must develop particular attributes to fulfill their specialized roles. In short, gene activity depends not only on immediate metabolic needs, but on a cell's location and developmental age as well.

Studies of specific clusters of enhancer and inducer sites support this idea. Some sequentially active enhancer and inducer sites are found in many genes scattered throughout the chromosomes, indicating that those genes are expressed in synchrony. There are four versions of the β-globin gene, for instance: two encode hemoglobin subunits whose affinity for oxygen is appropriate early in development, before the fetal circulatory system matures; another version of the gene takes over for the interval during which oxygen must be obtained from the maternal placenta; and the fourth version leads to the production of a subunit adapted to the high-oxygen conditions in the lungs of children and adults. Each has a regulatory sequence shared with the many other genes active at similar ages.

Other sequences are less common, and seem to promote gene transcription only in particular tissues; the activity of β-globin genes, for instance, in addition to being restricted to particular developmental stages, is further limited to cells in the bone marrow, along with all the other specialized genes needed to produce red blood corpuscles and to maintain the marrow itself. It is only in the marrow cells that the necessary enhancers for these genes are found; the enhancers, in turn, are produced in response to earlier chemical signals from neighboring cells (signals that identify tissue type), and internal signals encoding developmental age.

Finally, there are control sequences in these same marrow cells that are unique, or are shared by only a few genes at most; at least some of these appear to adjust the concentration of hemoglobin to ensure an adequate supply. An active gene in a eucaryote, then, would probably need the cooperation of at least three control substances bound to the inducer and enhancer regions, as well as the

Table 11.3 Comparison of procaryotic and eucaryotic gene control

	PROCARYOTES	EUCARYOTES
Control of DNA	Usually negative, blocked by a repressor	Usually positive, possible only with aid of transcription factor (TF)
Direct control of TA/TF binding: inducer sites	Allosteric change after binding of transcriptional activator to activating (inducer) site	Inducer must bind to DNA before helping TF bind; inhibitors can bind to DNA and block TF or inducers
Indirect control: enhancer sites	Very rare; exerts negative effect	Very common: can exert positive or negative effect
Levels of control	Usually one; rarely two	Usually at least three; very often four or more

absence of any inhibitors, in order to be transcribed. The comparison of the main features of procaryotic and eucaryotic gene regulation are summarized in Table 11.3.

■ GENE AMPLIFICATION

Ordinarily a single copy of each gene per chromosome is sufficient to meet the needs of a cell for the substance coded by the gene. The rate of transcription of a single gene is such that in 4 days it may produce 100,000 mRNA molecules, which can lead to the synthesis of as many as 10 billion (10^{10}) protein molecules. However, in some cases the demand for a product is so great that a single gene cannot meet it. We've already seen that so much rRNA is required by a eucaryotic cell for construction of ribosomes that the 5S rRNA gene is repeated in tandem 25,000 times (see Fig. 11.7B). The other three rRNA genes are repeated only 100–1,000 times in tandem, but when large numbers of ribosomes are needed, many parallel copies of this region of the chromosome are synthesized (Fig. 11.7D). This polytene region usually loops out from the main axis of one of the chromosomes to form the nucleolus, and rRNA is produced there at a high rate (Fig. 11.11). The production of these extra copies of the rRNA genes from moderately repetitive DNA is one type of **gene amplification.** Inappropriate gene amplification is common in many kinds of cancer.

■ CONTROL AFTER TRANSCRIPTION

Cellular control mechanisms are not limited to regulating mRNA synthesis. For example, some primary transcripts can be spliced (processed) in more than one way, yielding slightly different products, depending on the cell's needs. In *Drosophila,* for example, the way one gene's mRNA is spliced controls the expression of a constellation of other genes that help orchestrate the development of either a male or female fly (Fig. 11.12). The critical event occurs when the splicing complex prepares to join the end of exon 3 to the beginning of the next exon. If a particular gene (*transformer,* or *tra*) is fully expressed, it causes exon 4 to be added, but blocks further

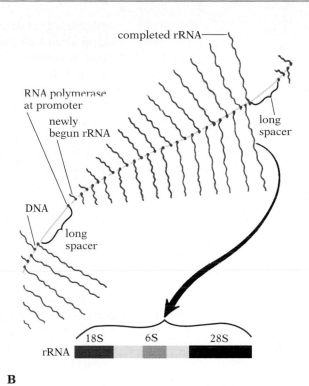

A

11.11 Transcription of rRNA on DNA in the nucleolus

(A) Electron micrograph of a portion of the nucleolus in an egg cell of the spotted newt. This continuous strand of DNA bears multiple copies of the genes for three RNAs. Strands of rRNA at progressive stages of synthesis feather out from each set of genes. Because the successive gene sets are separated by regions of intergenic DNA spacers where no rRNA is being synthesized, we can see more or less exactly where on the DNA

B

molecule each set of genes begins and ends. (B) Diagram of rRNA synthesis on one gene set. Many molecules consisting of the three kinds of rRNA joined together in a single strand (red) are being synthesized, as RNA polymerase molecules (red circles) specialized for transcribing rRNA genes move along the DNA. Strands of rRNA attached to the DNA near the beginning of the gene set are still short, their synthesis having just begun.

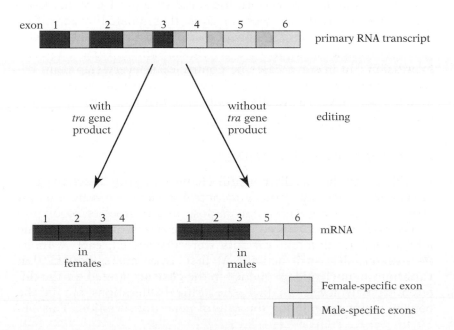

11.12 Alternative splicing in *Drosophila*

The primary transcript of the "double-sex" gene in *Drosophila* can be spliced in two ways, depending on whether the *tra* gene product is present to intervene in the process. If the exon sequence is 1-2-3-4, the protein translated from the mRNA leads to the expression of female-specific genes, presumably by binding to the appropriate places in the DNA. If the sequence is 1-2-3-5-6, the mRNA encodes a product that leads to the development of a male fly.

splicing; if active *tra* gene product is not available, exon 4 is skipped and all the remaining exons are spliced normally. (The *tra* protein mimics the portion of the splicing complex that binds to the leading end of exons, but how it alters splicing is not known.) The *tra* product itself is generated in response to signals in the nucleus that reflect whether the developing organism is a male or a female fly.

Even after splicing, almost half of the mature mRNA produced in the nuclei of most cells never reaches the cytoplasm for translation. There must be a system that can selectively block the export of specific mRNA molecules, probably in response to chemical signals that reflect the cell's needs. Even when mRNA reaches the cytoplasm, it may not be translated. Inhibitor substances can bind to specific messengers in the cytoplasm and block either ribosomal binding or complete translation. This strategy can reduce the cell's response lag to changing conditions by maintaining a pool of mRNAs ready to begin or complete translation when an appropriate chemical signal binds to the inhibitor and causes it to disassociate from the messenger. In most cases, the translation inhibitor is a protein, but in a few instances a segment of RNA with a sequence complementary to the mRNA binds though base pairing. This binding of **antisense RNA** is the mechanism by which herpes viruses are able to remain dormant in cells for extended periods. How antisense inhibitors come to be removed is not yet understood. Other control chemicals facilitate rather than inhibit translation; these bind to specific mRNAs and help them to bind ribosomes.

Finally, the expression of mRNA can be regulated by the rate at which the messenger is broken down: some types of mRNA have long life spans, while others are destroyed within minutes. An mRNA's degree of resistance to digestion by RNAse is encoded near the trailing end of the messenger. Control can also be exerted after translation: for example, protein longevity (that is, resistance to proteases) is written into the sequence of subunits, and ranges from a few minutes to days or, for some structural proteins, even years. The activity of enzymes is frequently regulated by activators or inhibitors. Even the assembly of many structural proteins (collagen, for example) is regulated by chemical signals. Hence, a gene's expression can in some cases be controlled at every step from before transcription until after translation. Failure of these control systems can have disastrous results, producing a variety of diseases, most notably cancer.

■ THE IMPACT OF MUTATIONS

In addition to the usually harmful changes in protein activity created by mutations, alterations in control sequences can have major effects: levels of regulation can be lost, and degrees of regulation can be reduced or enhanced. Consider the starch-digesting enzyme amylase. Many different variants were discovered, each with its own unique degree of activity. At first, researchers thought that variations in nucleotide sequence in the gene accounted for the different activity levels. In fact, the critical alterations are in the control regions regulating the rate of gene transcription. This observation has had a major impact on the way scientists now think about gene evolution.

Alterations of regulatory regions can also be important to human health. Consider, for example, the two steps involved in detoxifying alcohol in humans:

$$\text{ethyl alcohol} \xrightarrow{\substack{\text{alcohol} \\ \text{dehydrogenase}}} \text{acetyl aldehyde} \xrightarrow{\substack{\text{aldehyde} \\ \text{dehydrogenase}}} \text{acetic acid}$$

In many individuals, the transcriptional activity of the aldehyde dehydrogenase gene is unusually low, leading to a buildup of the toxic intermediate acetyl aldehyde; the result is a low tolerance for alcohol. Most individuals of Asian and Native American descent have characteristically low aldehyde dehydrogenase levels. Alcoholics, on the other hand, almost invariably have elevated levels of aldehyde dehydrogenase transcription; presumably, one of their control sequences regulating transcription of this gene is unusually active. Such patterns suggest a genetic basis for susceptibility to at least some diseases or behavior patterns. The most dramatic cases of altered gene regulation, however, come from studies of the many forms of vertebrate cancer.

CANCER: WHEN GENE CONTROL FAILS

Biologists hope to learn more about how normal cellular controls operate by studying how they fail—by investigating how the normal controls can become so deranged as to permit cancerous growth. The most distinctive feature of cancer cells is their unrestrained proliferation, which results in the formation of malignant tumors. These tumors grow, consuming the body's resources while also crushing nearby structures (the lungs, for instance), rendering them useless. Worse yet, the tumor cells often spread from the original site of growth to many other parts of the body, a process known as **metastasis.**

■ TISSUE CULTURE: A WAY OF STUDYING CANCER CELLS

One of the most profitable ways of studying the properties of specific kinds of cells is to grow them in tissue culture in the laboratory. Embryonic cells and tumor cells grow best. Among nontumor cells, embryonic fibroblasts outgrow all other cells in culture. Fibroblasts (literally, "fibrous germinating cells") serve many regenerative functions such as producing scar tissue. The usual lab procedure is to seed a large number of cells onto (or into) a sterile culture medium to which a variety of nutrients and growth-stimulating factors have been added. The nutrients and other factors are frequently supplied by serum—the liquid portion of blood. Serum contains, for example, the platelet-derived growth factor that signals fibroblasts to initiate the growth of scar tissue. In the body, the growth factor escapes from the circulatory system and binds to receptors on fibroblasts whenever a wound causes a break in the blood vessels.

20 μm

11.13   Cultured neuroblastoma cells from a mouse
These cells, from a malignant tumor of the nervous tissue (neuroblastoma), are growing on a solid surface.

A cell culture, no matter how elaborately set up and controlled, is far from the normal environment of the cells. Nevertheless, some of the cells may survive and engage in many of their usual developmental and functional activities, including cell division (Fig. 11.13). Most cultured lineages of animal cells stop dividing when they become crowded or, if crowding is prevented, after a certain number of cellular generations. The number of generations tends to be specific for the species and tissue of origin.

These two mechanisms limiting the growth of cells in culture are found in organisms as well. The first one—the cell's reaction to the effect of crowding—is called ***contact inhibition.*** When receptors in the cell membrane recognize markers in the cell coats (glycocalyces) of adjacent cells of the same type, the receptors signal the nucleus, and further cell division is inhibited. This inhibition is not absolute, however: adding high concentrations of growth factors, for example, can often stimulate a further round of cell division even in a crowded culture. The second mechanism—the automatic cessation of cell division after a set number of divisions—is an independent system for limiting cell proliferation. In most tissues it comes into play only if contact inhibition fails to work normally.

Cultured cells that continue to divide indefinitely fall into one of two categories: those that have lost the fixed-number-of-divisions control, and will go on growing as long as they are not crowded, and those that have lost both kinds of control. The cells of each category are potentially "immortal" in culture, but the former resemble ***benign*** (that is, self-limiting) tumors, while the latter resemble cancers: they will grow indefinitely regardless of crowding to create an ever-larger mass of tissue.

■ HOW ARE CANCEROUS CELLS DIFFERENT?

The loss of normal control systems can produce a variety of changes in a cell's anatomy, chemistry, and general behavior.

2 μm

11.14 A typical cancer cell

This cell, from the HeLa line, is growing in tissue culture. The cell is rather spherical, and is covered with many small "blisters," called blebs, whose significance is not understood.

Studies of these changes tell us quite a bit about how cancer cells differ from normal ones, and sometimes suggest potential avenues of treatment.

One property of cultured cancer cells is that they almost always have an abnormal set of chromosomes. For example, the cells of the human cancer line called HeLa (Fig. 11.14), by far the most widely studied line of cultured human cells, typically have 70–80 chromosomes instead of the normal 46. Interestingly enough, noncancerous cell lines that become potentially immortal in culture also have extra chromosomes; hence it seems likely that possession of extra chromosomes somehow frees cultured cells from some of the normal constraints on proliferation.

In cancerous tissue in organisms, however, such an increase in chromosome number is not typical. Instead, there are frequently numerous self-perpetuating chromosomal segments called minute chromosomes, or specific rearrangements of the normal chromosomes. For example, in 90% of patients with Burkitt's lymphoma (a cancer of cells in the immune system), the tip of chromosome 8 has been moved to the end of chromosome 14 (see Fig. 12.4, p. 311). The insertion point in chromosome 14 is next to a gene coding for an immune-system polypeptide while the translocated region contains a gene that is frequently involved in cancers. The other 10% of patients with Burkitt's lymphoma show other translocations, from different sites on chromosome 8 to the same target site on chromosome 14. The implication is that the critical site in chromosome 8 has a control region that can contribute to cancer by stimulating the transcription of specific genes. Predictable translocations are also found in the cultured cells of several other sorts of cancer.

Besides differences within the nucleus, cancer cells and normal

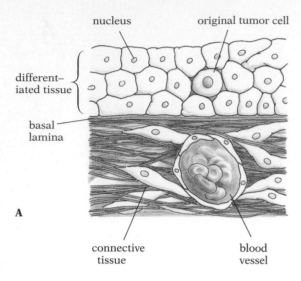

nucleus   original tumor cell

different–
iated tissue

basal
lamina

A

connective   blood
tissue   vessel

11.15   Development of a malignant tumor

(A) Normal tissue has differentiated cells with small, quiescent nuclei, separated from the circulatory system and loosely packed connective tissue by a tough basal lamina (the basement membrane). (B) When a cell incurs enough changes to begin proliferating, its nucleus enlarges, its shape changes, and it begins to replicate its DNA. As the cell reproduces, the clone it forms begins to push other cells back. If it lacks contact inhibition, it will grow regardless of crowding, and if it is not limited to a fixed number of divisions, the clone will grow as long as nutrients (blood supply) permit, but may still be contained by the lamina. (C) If a cell in the clone develops the ability to penetrate the basal lamina, its descendants will move across and proliferate. The tumor will metastasize widely when it penetrates the lamina protecting blood vessels.

growing
tumor
contained
(benign)
until one
tumor cell
develops
ability to
penetrate
lamina

B

tumor
metastasizes
as
malignancy
spreads
through the
circulatory
system

C

cells display a host of significant differences related to cell shape and to the nature of the cell surface. Cultured cancer cells, for instance, tend to have a spherical shape, one seen in normal cells only during a short period immediately following cell division. This peculiarity of shape probably results from their having an abnormally small number of functional structure-stabilizing microfilaments, a factor that also makes these cells more mobile than normal cells. This characteristic is also related to a striking feature of cultured cancer cells known as anchorage-independence: most cells must "cling" to a solid surface in order to grow—a useful constraint, since they will require support later in order to function properly. Cancer cells are not constrained in this way, and so they can grow in liquids or on soft surfaces.

In order to metastasize in an individual, and so escape from its tissue or organ of origin, a cancerous cell must usually be able to produce and transport to its own membrane a receptor called laminin, which enables cancer cells to bind to the basal membrane (a tough layer, or lamina, that underlies or surrounds many tissues and organs, including the vessels of the circulatory system). It must then be able to secrete collagenase, which digests the lamina and allows the cell to cross the barriers that otherwise contain the growth of tumors (Fig. 11.15). In particular, the capacity to break through the lamina surrounding capillaries allows cancer cells to spread throughout the body. Since most normal cells can neither produce laminin nor secrete collagenase, the majority of tumors are contained at their site of origin, and so are benign.

Cancer cells, with their abnormal surfaces, appear to lack the ability to react appropriately to environmental changes. Cancer cells typically have fewer and different glycolipids and glycoproteins in their cell coats; these differences are probably correlated both with the absence of normal contact inhibition during cell division, and with the apparent inability of cancer cells to recognize other cells of their own tissue type. When mixed in culture, normal cells from two different tissues (such as liver and kidney) tend to sort themselves out and reaggregate according to tissue type; cancer cells, however, don't do so. This absence of normal cellular affinities is probably one of the reasons why malignant cells of many cancers can spread.

■ THE STEPS THAT LEAD TO CANCER

The conversion of a normal cell into a cancer cell involves several changes: loss of fixed-number-of-division control, loss or reduction of contact inhibition, loss of anchorage-dependence and, sometimes, tissue-specific cell-surface changes. But in many tissues even these changes result only in benign tumors that grow slowly and then stop, apparently because the cells fail to produce the chemical signal or signals that cause vascularization (the establishment of a capillary system that supplies oxygen and nutrients to growing tissue). Even when it can recruit capillaries, the tumor cannot metastasize without being able to bind to and destroy the basement membrane. Do all these changes take place as a result of a single genetic event, or is each change the consequence of a separate event?

Health statistics suggested the answer to this question before the actual mechanisms began to be elucidated. If just a single genetic event were required, the probability of an individual's having contracted a particular type of cancer should increase proportionally with age. For example, if the chance of contracting skin cancer were 1% per year, then the chance of having contracted it would be 2% for a two-year-old, 3% for a three-year-old, and so on. In fact, however, cancer is normally a disease of old age (Fig. 11.16). To account for the exponential rise of the incidence of cancer, several independent genetic events are required. The curve would be the product of the individual probabilities of the separate events.

Some cancers require only two or three genetic events; examples of this group are leukemia and immune-system cancers because they involve tissues in which tissue-type affinity, vascularization, crossing of a lamina, and a solid surface for growth are not necessary for the normal cells. Other kinds of cancer require four to seven genetic events.

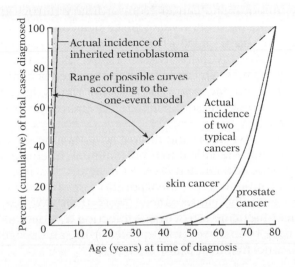

11.16   Incidence of representative cancers

If each type of cancer were caused by a single transforming event, the probability of developing a particular cancer would increase linearly with age; the shaded area indicates the range of possible curves. Actually, the incidence of most cancers increases exponentially, indicating that several contributing events are usually involved. The incidence curve for skin cancer closely approximates a three-event curve, while the curve for prostate cancer approximates a five-event curve. By contrast, the curve for inherited retinoblastoma is linear, approximating a single-event curve; this is because the other transforming event required for this form of cancer is already incorporated in a mutant gene carried by these individuals.

Additional evidence for a multistep model comes from studies of individuals who inherit a tendency to get certain cancers. Individuals with a genetic susceptibility to retinoblastoma (a cancer of the eye) usually develop it when young. The cumulative-incidence curve for this cancer shows a linear rise with age, rather than the more common exponential one. This suggests that in such individuals only a single transforming event is necessary to trigger the disease. (All other necessary changes must already be present in the form of mutant genes.) Because inheritance of genetic susceptibility in this kind of cancer behaves as a single-gene characteristic, developing retinoblastoma must require a total of *two* transforming events among individuals who are not genetically predisposed to it; genetically susceptible individuals require only one, because the other event is already built into the genome.

Statistical studies cannot tell us what constitutes a genetic event leading to cancer—among the possibilities are single-base changes, deletions, insertions, or chromosomal rearrangements. Nor do the studies tell us the precise effects of the events. Most steps are probably linked directly to the loss of a particular level of control. The best data on this come from studies of oncogenes.

■ THE ROLE OF ONCOGENES

The discovery of **oncogenes**—genes that cause cancer—has a complex history. In 1910 Peyton Rous found that what is now known as Rous sarcoma virus could cause cancer in chickens; how it did so was then a mystery. We know now that the causative agent is a retrovirus, which, like other retroviruses, produces a cDNA copy of its RNA that is incorporated into the host chromosome (see p. 250). Some strains of Rous sarcoma virus cause cancer quickly, and at a high rate, because the cDNA brings with it an oncogene (known as *v-src*—the *v* identifies it as a viral gene while *src* specifies that it causes a sarcoma). The expression of this gene in the host chromosome, which does not depend on the site of incorporation, makes the cell cancerous.

Ordinary retroviruses—those not containing oncogenes—can also cause cancer, but only at a very low rate, and even then only after a long latency. Incorporation of cDNA into the host genome is apparently random, and cancer from ordinary retroviruses occurs only when the incorporation takes place near one of a few particular genes in the host, called **proto-oncogenes.** Presumably an active control region in the cDNA alters the expression of one or more of these genes. Oncogenes, then, are genes that inevitably cause cancer; proto-oncogenes, on the other hand, have the potential to cause cancer, but some change is required to convert them into oncogenes.

Perhaps the most surprising discovery about the oncogene *v-src* is that it is virtually identical to a normal gene in chickens; indeed, this normal gene (designated *c-src* to indicate that it is cellular in origin) is a proto-oncogene: incorporation of ordinary cDNA near *c-src* can (eventually) cause cancer. More than two dozen rapidly acting, cancer-inducing retroviruses infecting a variety of birds and mammals have been discovered, each carrying an oncogene almost identical to a normal host proto-oncogene.

Table 11.4   Events that can create oncogenes

I. Altered gene, resulting in altered product
  1. Incorporation of retroviral oncogene
  2. Mutation of normal proto-oncogene to create an oncogene

II. Altered gene expression, in which a gene is transcribed at the wrong time or at too high a rate
  1. Insertion next to a proto-oncogene of retroviral cDNA containing an active control region
  2. Translocation of a proto-oncogene to a position next to an active control region
  3. Translocation of an active control region to a position next to a proto-oncogene
  4. Mutation of a control region next to a proto-oncogene

III. Loss of anti-oncogene activity

The viral cancers are convenient to study, but as yet few forms of cancer in humans are definitely viral in origin. However, studies of cultured human and nonviral animal tumors reveal strong similarities to the observed behavior of retroviruses: some of the genetic changes that are responsible for making the cells in these cultures cancerous involve the same oncogenes also found in cancer-inducing retroviruses. Some of the nonviral oncogenes appear to have arisen through a mutation in a proto-oncogene that altered the base sequence through a base change, insertion, or deletion. Others appear to be proto-oncogenes that have been moved (translocated) from their usual locations, and perhaps released from their normal controls, to become oncogenes.

Conversely, either mutation or translocation involving a control region next to a proto-oncogene can convert a proto-oncogene into an oncogene. Indeed, when grown in culture, cells of several leukemias, one ovarian cancer, and several other kinds of cancer display predictable chromosomal translocations or losses. The translocations seem to involve moving a structural gene and its control region, which is responsible for its transcription in the tissue at a high rate, to a position next to a proto-oncogene, or moving the proto-oncogene next to a structural gene and its control region. Finally, there are ***antioncogenes***—genes that encode proteins that deactivate oncogene products. Failure of such an inhibitor gene is equivalent to activating a proto-oncogene. The gene involved in retinoblastoma is an antioncogene.

The various mechanisms that may result in the presence of an oncogene in the genome are summarized in Table 11.4.

Most oncogenes encode products that fall into one of four general categories (Fig. 11.17):

1 Growth factors (extracellular signal molecules that stimulate cell proliferation)

2 Receptors (for growth factors, contact inhibition, or surface adhesion)

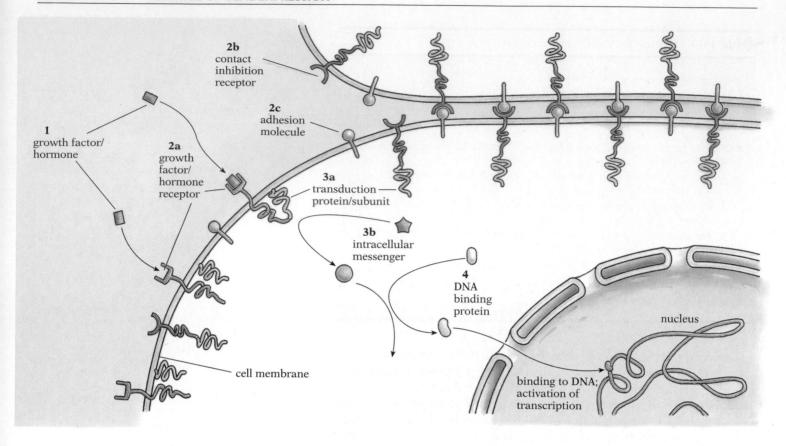

1
growth factor/
hormone

2a
growth
factor/
hormone
receptor

2b
contact
inhibition
receptor

2c
adhesion
molecule

3a
transduction
protein/subunit

3b
intracellular
messenger

4
DNA
binding
protein

cell membrane

nucleus

binding to DNA;
activation of
transcription

11.17 General sites of oncogene action

Oncogenes can exert their effect at any point in the flow of information from the extracellular environment to the chromosomes. Some act to create high levels of growth factor (1) or growth factors with unusual properties. Others alter the sensitivity of membrane receptors for growth factor or other relevant signals, like those involved in adhesion or contact inhibition (2). Still others alter the communication between the receptors and the intracellular messenger chemicals (3); some receptors incorporate domains that alter intracellular messengers, while others interact with an intermediate enzyme that makes the appropriate change. Finally, some oncogenes alter the DNA-binding proteins that are activated or repressed by the intracellular messengers (4).

3 Intracellular signalling systems (which communicate information from receptors to intracellular enzymes or binding proteins)

4 DNA binding molecules (which regulate transcription of specific genes or replication of the entire genome)

These categories mirror most of the array of normal cellular control strategies: cell division, contact inhibition, anchorage-dependence, tissue-type affinities, vascularization, and the expression of the genes needed to breach the basement membrane.

The first evidence concerning the actual operation of oncogenes came from work on the *src* oncogene by Raymond Erikson and Marc Collett. They found that the enzyme encoded by *src* is a type of intracellular signal molecule called a protein kinase—one of a class of enzymes that phosphorylate a particular protein or a particular component of proteins. Phosphorylation is frequently involved in controlling chemical pathways, usually serving to

Table 11.5 A partial list of classes of cancer-causing and cancer-suppressing genes

NAME	TYPE
Extracellular signals	
sis	Mutant growth factor (GF)
Receptors	
erbB, neu	Mutant GF receptor
ros, erbA	Mutant hormone receptor
C-myc	Mutant adhesion molecule
Transducers	
mos, raf	Mutant serine protein kinase (PK)
src, met	Mutant tyrosine PK
ras	Mutant guanine binding protein
Second messengers	
(none known)	
DNA binding	
fos, jun	Mutant transcription factor
myc, myb	Mutant inducer or enhancer
RB	Antioncogene; blocks binding at some stage in the control pathway

activate an enzyme; the *src* enzyme phosphorylates the amino acid tyrosine. In cells with *src* oncogenes, tyrosine phosphorylation increases to 10 or more times the normal level. Several other oncogenes are now known to code for tyrosine kinases, while a few others encode serine kinases.

Some oncogenes seem to encode products that bridge two steps in the pathway from extracellular signals to DNA binding. The epidermal growth factor (EGF) receptor, for instance, has an extracellular component that binds to EGF, and an intracellular section that acts as a kinase. At least one oncogene appears to code for an enzyme that is very similar to the kinase part of the activated EGF receptor, but lacks the extracellular portion that binds to EGF. This oncogene product may act by signalling the cell to divide continuously, whether or not EGF is present.

Understanding cancer in the near future may be a realistic hope: there appear to be only a limited number of ways cancer is triggered, only a limited number of proto-oncogenes in any cell (fewer than a hundred in humans), and an even smaller number of functions for oncogene products (Table 11.5). Understanding oncogenes will probably prove the most important step in finding effective treatments for cancer.

■ ENVIRONMENTAL CAUSES OF CANCER

Mutations, translocations, and retroviruses can each play a role in triggering cancer. Though many, if not all, mutations and translo-

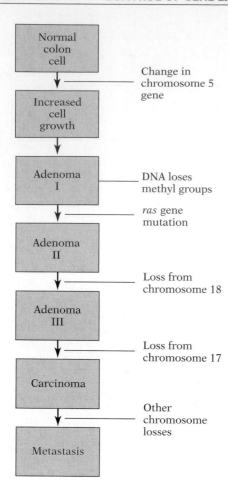

Normal colon cell

↓ ——— Change in chromosome 5 gene

Increased cell growth

↓

Adenoma I

——— DNA loses methyl groups

↓ ——— *ras* gene mutation

Adenoma II

↓ ——— Loss from chromosome 18

Adenoma III

↓ ——— Loss from chromosome 17

Carcinoma

↓ ——— Other chromosome losses

Metastasis

11.18 Chromosomal changes leading to colon cancer

Colon cancer usually begins with small growths called adenomas, which are categorized according to morphology. The cancer is unusual because some of the chromosomal changes involved are cytologically visible and occur more or less in a particular order. These peculiarities have made it possible to outline the steps involved. The *ras* oncogene encodes an intracellular signalling enzyme, which couples the response of growth-factor receptors to DNA-binding protein, which in turn regulates part of the cell-division control system.

cations are caused by external agents—radiation and mutagenic chemicals—linking particular chemicals or radiation sources to cancer in humans and evaluating the relative danger of each source is difficult. Not every mutation occurs in a cancer-causing location, and even a mutation that is in such a location will trigger only one of the several steps necessary for development of a cancer. Thus the final step may occur years after the first significant exposure to a mutagenic chemical or radiation. Only in the case of colon cancer are researchers even close to understanding the steps involved (Fig. 11.18).

The absence of strict cause-and-effect relationships between particular mutagenic agents and particular cancers has made the identification and evaluation of causative agents difficult, even in the most obvious cases. For example, though health statistics have shown that cigarette smoking is probably responsible for about 150,000 fatal cancers annually (and for 25% of fatal heart disease), the tobacco industry denies the link. While granting that smokers develop lung cancer and heart disease far more often than non-smokers, they point out that many smokers never develop lung cancer, and that some nonsmokers do. An informed understanding of how oncogenes work and of the probabilistic nature of their development, however, reveals the weaknesses in these arguments: The genetic events necessary for inducing lung cancer must be greatly facilitated by the highly mutagenic tar in cigarette smoke. (Tar is a deadly combination of chemicals that on average kills two-pack-per-day smokers 8 years early.) However, the changes must occur in the correct locations (including a specific deletion on chromosome 3) in at least one cell. Thus, even with regular exposure to mutagenic smoke, some smokers will escape unscathed. At the same time, other sources of mutation, such as radiation, are also present, and though less potent, they will sometimes cause the changes necessary to trigger lung cancer even in nonsmokers.

Another way to evaluate the carcinogenic potential of chemicals and radiation is to expose animals—usually mice—to measured doses of these agents. This procedure is expensive and time-consuming, and in addition it assumes that what causes cancer in mice is equally carcinogenic in humans. Another assay is the Ames test, described on page 243, which involves exposing bacteria to potential carcinogens to see if mutations result. The assumption is that a mutagen for *E. coli* is a carcinogen for humans. The Ames test has identified many substances as potential carcinogens. Others having no initial effect are converted into mutagens when liver homogenate is added to the bacterial culture, and are presumably also thus converted in the liver of an organism. A few of the substances identified by the Ames test as possible carcinogens are hexachlorophene soaps, the coal-tar components of many hair dyes and cosmetics, the seared protein of grilled meat, the smoke from wood fires, and several chemicals in certain vegetables and spices. When tested on animals each of these substances has proven to be carcinogenic. It seems only sensible to assume that all mutagens are carcinogenic, but given the ubiquity of environmental mutagens, avoiding them all is impossible. The only sensible course, then, is to eliminate our exposure to the most potent ones—especially cigarette smoke and coal tars—and, weighing the costs and benefits (just as we do each time we decide to take the risk of driving a car), to minimize contact with the rest.

CHAPTER SUMMARY

ORCHESTRATING BACTERIAL GENES

HOW GENES ARE INDUCED The genes for inducible enzymes are normally not transcribed. An inducer binds to and deactivates a gene's repressor protein, which then can no longer bind to the gene's operator region; a deactivated repressor can no longer prevent RNA polymerase from binding to the promotor region and transcribing the gene. (p. 273)

HOW GENES ARE REPRESSED Genes that are normally transcribed are shut down when a corepressor substance binds to and activates the gene's repressor. (p. 274)

POSITIVE VERSUS NEGATIVE CONTROL Genes that use repressors are under negative control. Positive control usually depends on transcription factors binding to activator regions; this binding changes the ability of RNA polymerase to bind to the gene's promoter. (p. 278)

HOW EUCARYOTIC CONTROL IS DIFFERENT

THE ODD ORGANIZATION OF EUCARYOTIC CHROMOSOMES The nucleosomes onto which eucaryotic DNA is wound must be temporarily removed to allow transcription. Eucaryotic chromosomes have regions of highly repetitive DNA: in the centromeres and telomeres it has a structural role or aids in cell division; other repetitive regions encode rRNA; yet others have no known function. Some moderately repetitive DNA encodes rRNA, while the function of other sorts is unknown. Most single-copy DNA is never transcribed, and its function is not known. (p. 283)

HOW CHANGES IN GENE TRANSCRIPTION CAN BE SEEN DNA transcription requires decondensation of a chromosomal loop; this decondensation is sometimes visible as chromosomal puffs in polytene chromosomes and as a "lampbrush" pattern in certain chromosomes. (p. 286)

IMPRINTING: PRESERVING PATTERNS OF GENE ACTIVITY The activity of some genes is controlled by methylation of the bases. Special enzymes scan the parental strand after replication and methylate the daughter strand to match. Gametes containing methylated genes carry with them these methylation patterns. (p. 288)

EUCARYOTIC CONTROL: INDUCERS AND ENHANCERS Most eucaryotic genes utilize positive control, but the binding of transcription factors to a gene's inducer region is controlled by a multitude of other molecules—especially those that bind at its distant enhancer region. The large number of control elements seems to reflect the need for complex developmental control. (p. 289)

GENE AMPLIFICATION Genes are said to be amplified when multiple copies of a gene are created, thus generally increasing the transcription rate of the gene. (p. 292)

CONTROL AFTER TRANSCRIPTION Gene activity can be controlled after transcription through alternative ways of splicing its exons, which include sequences that can block or facilitate export of an mRNA to the cytoplasm, alter its translation rate, and determine its degree of resistance to proteases in the cytoplasm. (p. 292)

THE IMPACT OF MUTATIONS Mutations can alter the sequence of amino acids in the protein a gene encodes and the activity of control elements, including promoters, operators, inducers, and enhancers. (p. 294)

CANCER: WHEN GENE CONTROL FAILS

TISSUE CULTURE: A WAY OF STUDYING CANCER CELLS Most cells fail to become cancerous because they undergo only a fixed number of divisions, are inhibited from division by crowding (contact inhibition), cannot recruit additional capillaries to support extra tissue, and cannot bind to and penetrate the basement membrane that confines tissues. (p. 295)

HOW ARE CANCEROUS CELLS DIFFERENT? Cancerous cells have lost levels of control that keep normal cells from multiplying in their tissue type. Many cancer cells have extra chromosomes or translocations, which alter patterns of gene expression. (p. 296)

THE STEPS THAT LEAD TO CANCER Most cancers develop after several independent events—events that presumably remove each level of control that normally prevents unrestrained growth. (p. 299)

THE ROLE OF ONCOGENES Oncogenes are genes that cause cancer; proto-oncogenes are genes that can turn into oncogenes when their sequence or control regions are altered. Most oncogenes encode growth factors, cell-surface receptors, intracellular signalling components, and gene-control molecules. (p. 300)

ENVIRONMENTAL CAUSES OF CANCER Cancers are caused by mutations (either of genes or control regions), translocations, gene amplification, or the insertion of genes by retroviruses. Each change removes one of the defenses against uncontrolled growth of the cell. Most mutations and translocations are caused by ionizing radiation or chemicals in the cell. The chemicals may be normal, highly reactive by-products of cellular chemistry, or unusual molecules obtained through the diet. (p. 303)

STUDY QUESTIONS

1 Are there good reasons why eucaryotes should rely much more than procaryotes on positive gene control? (pp. 283–295)

2 Why do you suppose that skin cancer is so often controllable? (pp. 295–299)

3 The eucaryotic genome has 50,000 genes, each managed by one or more control genes, which in turn are orchestrated by thousands of middle-level executive genes, which themselves respond to hundreds of upper-management sequences, and so on. Why is the genome not strangled by this bureaucracy? (pp. 289–292)

4 Why bother keeping inactive regions of chromosomes condensed?

5 What sorts of mutations could affect the *lac* operon, and what effects would they have? What would happen if a cell had two of these mutations at once? (pp. 273–275, 278–282)

SUGGESTED READING

BISHOP, J. M., 1982. Oncogenes, *Scientific American* 246 (3). *An illuminating look at the relationship between cancer genes carried by viruses and the similar noncancerous genes in normal cells.*

CAVENEE, W. K., AND R. L. WHITE, 1995. The genetic basis of cancer. *Scientific American* 272 (3). *A clear presentation of the multiple steps that lead to cancer.*

CROCE, C. M., AND G. KLEIN, 1985. Chromosome transloca-tions and human cancer, *Scientific American* 252 (3). *A clear discussion of the translocations involved in Burkitt's lymphoma.*

FELDMAN, M., AND L. EISENBACH, 1988. What makes a tumor cell metastatic? *Scientific American* 259 (5). *About the oncogenes that permit tumor cells to stop adhering to other cells or structures, and so spread to other parts of the body.*

FELSENFELD, G., 1985. DNA, *Scientific American* 253 (4). *The role of DNA structure in the regulation of gene expression.*

MCKNIGHT, S. L., 1991. Molecular zippers in gene regulation, *Scientific American* 264 (4). *On the operation of the leucine zipper.*

PTASHNE, M., 1989. How gene activators work, *Scientific American* 260 (1). *An excellent summary of how promoters work in bacteria and yeast.*

SAPIENZA, C., 1990. Parental imprinting of genes, *Scientific American* 263 (4). *On the inheritance of gene switches bound to the DNA of gametes.*

TIOLLAS, P., AND M. A. BUENDIA, 1991. Hepatitis B virus, *Scientific American* 264 (4). *On the life history of a cancer-promoting virus.*

TIJAN, R., 1995. Molecular machines that control genes, *Scientific American* 272 (2). *On the elaborate organization of transcription complexes.*

WEINBERG, R. A., 1988. Finding the anti-oncogene, *Scientific American* 259 (3). *On the genes that restrain cell growth, focusing on retinoblastoma.*

WEINTRAUB, H., 1990. Antisense RNA and DNA, *Scientific American* 262 (1). *On the use of antisense RNA to regulate mRNA activity.*

CHAPTER 12

CELLULAR REPRODUCTION

The previous four chapters trace the flow of genetic information within individual cells. This chapter describes how cells themselves reproduce, passing their genetic endowment from cell to cell and from parent to offspring. Chapters 13 and 14 describe how the genome organizes the development and cellular specializations typical of complex, multicellular organisms.

FISSION VERSUS DIVISION

Before a cell can divide to produce two new cells, it must first replicate all the genetic information in its nucleus and then make sure that a full set of information is given to each daughter cell. The process of DNA replication is detailed in Chapter 8; the emphasis here is on the larger-scale organizational task of duplicating the entire genome and accurately segregating the copies into two progeny cells.

■ PROCARYOTIC FISSION

Cell division is less complex in procaryotes than in eucaryotes. As the procaryotic cell elongates sufficiently to form two independent daughter cells, the single circular chromosome replicates and the resulting second chromosome attaches to a different point from the first one on the expanding plasma membrane (Fig. 12.1). Next, new plasma membrane and wall material form near the midpoint of the parental cell and grow slowly inward, cutting through both

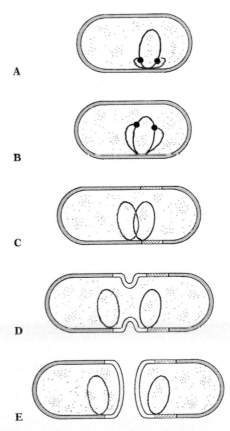

A

B

C

D

E

12.1 Binary fission of a procaryotic cell

(A) The circular chromosome of a cell is attached to the plasma membrane near one end; it has already begun replication (with the partially formed second chromosome shown in red). (B) Replication is about 80% complete. (C) Chromosomal replication is finished, and the second chromosome now has an independent point of attachment to the membrane. During replication additional membrane and wall (stippled) has formed. (D) More new membrane and wall (tan) has formed between the points of attachment of the two chromosomes. Part of this growth forms invaginations that will give rise to a septum cutting the cell in two. (E) Fission is complete, and two daughter cells have formed. (For clarity, the chromosomes are depicted here as small open circles; actually they are so long that they must be looped and tangled to fit into a cell; at the scale of this drawing, the true circumference of each chromosome would be about 30 m.)

12.2   Electron micrograph of a dividing bacterial cell

New wall is growing inward, cutting this *E. coli* in two. The nucleoid (white areas inside cells) has already divided.

0.5 µm

the cytoplasm and the nucleoid (Fig. 12.2). Each new daughter cell receives a complete chromosome. This process is known as transverse fission, or ***binary fission***.

■ **WHY EUCARYOTIC CHROMOSOMES POSE PROBLEMS**

As described in Chapter 11, the chromosomes of eucaryotes differ from those of procaryotes in several ways. First, except in mitochondria and other organelles, eucaryotic chromosomes are linear rather than circular. Second, eucaryotic chromosomal DNA in the nucleus is wound on nucleosome cores. Third, each chromosome is organized into loops, all of which remain in a highly condensed state during cell division, probably to prevent tangling (Fig. 12.3).

The final major difference between eucaryotic and procaryotic chromosomes arises because most eucaryotes receive chromosomes from two parental cells—from a sperm cell and an egg cell, for instance—rather than from just one. Most eucaryotes are diploid, and their chromosomes thus occur in ***homologous pairs***, each consisting of one chromosome from each parent bearing basically the same genes in the same order (Fig. 12.4). Procaryotes are haploid: their chromosome is unpaired. Though chromosomal number may vary enormously between species—cats, for instance, have 19 pairs, fruit flies 4 pairs, onions 8 pairs, and humans 23 pairs—the chromosomal number within a species is normally uniform.

When eucaryotic chromosomes condense and become visible early in cell division, each chromosome has already been replicated to produce compact twins, connected by a centromere (Fig. 12.5). Each member of this bound pair is referred to as a ***chromatid***.

There are two kinds of cell division in eucaryotes: division as part of the growth of an organism, also known as mitosis or somatic cell division, and division to produce the gametes that give rise to new individuals, also known as meiosis. They differ in important ways.

0.1 µm

12.3   Condensed chromosome showing highly condensed loops

12.4   Chromosomes of a human male

At left, the chromosomes have been arranged as homologous pairs and numbered according to accepted convention. A human somatic cell (any cell except the egg or sperm cells) contains 23 pairs of chromosomes, including a pair of sex chromosomes (for a male, X and Y).

HOW SOMATIC CELLS DIVIDE

Cell division in eucaryotic cells involves division of the nucleus and division of the cytoplasm. The process by which the nucleus divides to produce two new nuclei, each with the same number of chromosomes as the parental nucleus, is called **mitosis** (from the Greek *mitos*, "thread"). The process of division of the cytoplasm is called **cytokinesis**.

■ NUCLEAR DIVISION: MITOSIS

For convenience, each mitotic cycle, from one cell division to the next, is divided into a series of stages, each designated by a special name. Though each stage will be discussed separately here, the entire process is continuous rather than a series of discrete occurrences.

Interphase and the cell cycle   The nondividing cell is in the interphase state. The nucleus is clearly visible as a distinct membrane-bounded organelle, and one or more nucleoli are usually prominent. Chromosomes are not visible in the nucleus; there are none of the distinct rodlike bodies that microscopes and sophisticated staining techniques reveal in a dividing cell (Fig. 12.6). Interphase chromosomes are so thin and tangled that they cannot

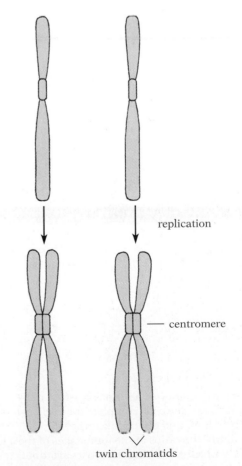

replication

centromere

twin chromatids

Homologous chromosomes

12.5   Production of chromatids through replication

At some time prior to cell division the genetic material of each chromosome duplicates itself. As a result, twin chromatids take form upon condensation, joined to each other by a centromere. For clarity, the unduplicated chromosomes are shown above in condensed form, even though condensation normally takes place only after duplication.

12.6   Chromosomes in a dividing cell of the African blood lily

The separating chromosomes for each of the new nuclei are easily distinguishable.

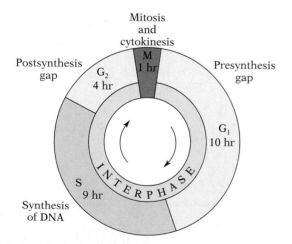

12.7   The cell cycle

This particular cycle assumes a period of 24 hours, but some cells complete the cycle in less than an hour and others take many days. Similarly, the ratios of the four stages of the cycle vary, with G_1 exhibiting the most variation.

be recognized as separate entities. They appear only as an irregular granular-looking mass of chromatin.

In interphase animal cells, there is a special region of cytoplasm just outside the nucleus that contains two small cylindrical bodies oriented at right angles to each other. These are the **centrioles** (see Figs. 5.34 and 5.35B, p. 142), which will duplicate themselves, move apart, and become associated with the poles of the mitotic apparatus of the dividing cell. Because they are needed for producing flagella and cilia, the cells' centrioles must be distributed to daughter cells as reliably and accurately as the chromosomes themselves. In many animal cells this separation of the centrioles occurs just before the onset of mitosis, but in some cells it occurs during interphase, long before mitosis begins. There are no centrioles in the cells of most seed plants, which is not surprising since these species lack cilia and flagella; centrioles do occur in some algae, fungi, bryophytes (mosses, liverworts, and so forth), and ferns that produce motile sperm. Centrioles are not essential for cell division even in animal cells; they remain close to the spindle-organizing centers during mitosis, but do not participate in spindle formation.

Interphase cells carry out all the innumerable activities of a living, functioning cell—respiration, protein synthesis, growth, differentiation, and so forth. In addition, during interphase the genetic material is replicated and all the cellular machinery is duplicated in preparation for the next division sequence.

Replication of the genetic material does not begin immediately after completion of the last division sequence. There is a gap, designated the G_1 stage, before genetic replication; this is when ribosomes and organelles begin to be duplicated. Next comes the S stage, during which the synthesis of new DNA takes place, along with the further duplication of organelles. Another period, G_2, occurs after replication, during which time the cell prepares for mitosis. Then cytokinesis and mitosis proper (together referred to as the M stage) occurs. The three subdivisions of interphase, along with mitosis and cytokinesis, constitute the **cell cycle** (Fig. 12.7).

The duration of the complete cell cycle can vary greatly. Though usually lasting 10–30 hours in plants and 18–24 hours in animals, it may be as short as 20 min in some organisms or as long as several days or even weeks. All the stages can vary in duration to some degree, but by far the greatest variation occurs in the G_1 stage. At one extreme very rapidly dividing embryonic cells may pass so quickly through the G_1 stage that it can hardly be said to exist at all; at the other extreme some cell types become arrested in the G_1 stage. Differentiated skeletal-muscle cells and nerve cells, for example, are suspended in the G_1 stage and normally never divide again. There are a few cases in which cells are arrested in the G_2 stage: they have replicated their DNA but do not divide.

Arrest in either the G_1 or G_2 stages results from a failure to produce an essential control chemical. If the nucleus from a cell in G_1 arrest is transplanted into a cell that is just entering the S stage, the transplanted nucleus will promptly enter the S stage, apparently because it has been stimulated by a control substance present in the cytoplasm of the host cell. Similarly, when a cell in G_2 arrest is fused with a mitotic cell, its chromosomes soon begin to condense and the cell enters mitosis.

The chemicals that control the cell cycle are called *cyclins*. S-cyclin is involved in stimulating replication (DNA synthesis), while M-cyclin helps trigger mitosis. Each can bind to the same **cell-division-cycle protein (cdc),** which in turn can activate a cellular messenger specific for either replication or mitosis.

To see how the system works, let's follow a complete cycle beginning with the G_1 phase (Fig. 12.8). As G_1 begins, the cdc protein is

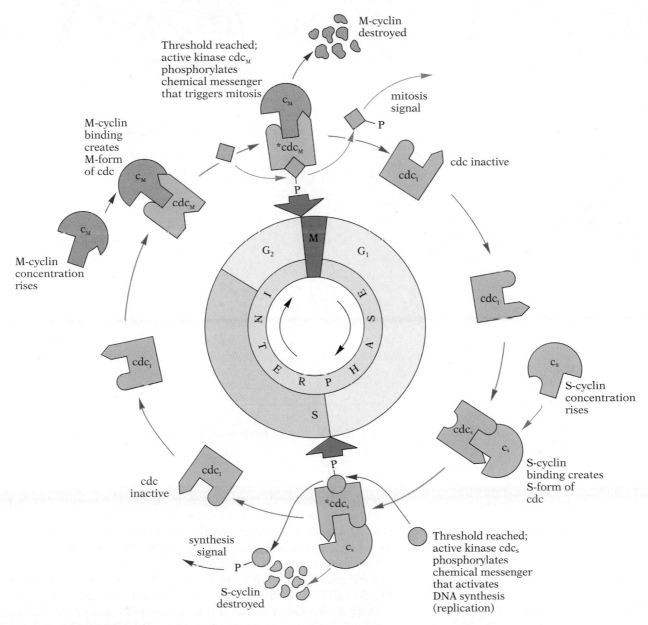

12.8   Control of cell division

A single protein kinase, cdc, is responsible for signalling the cell to replicate or undergo mitosis. It adopts the replication-promoting form as the concentration of S-cyclin rises in G_1 (lower right), becoming activated at a certain threshold (activation indicated here by an asterisk). It then phosphorylates a messenger that triggers replica-

tion and the destruction of S-cyclin. Similarly, as the M-cyclin concentration rises in G_2 (upper left), it binds to cdc, which adopts a mitosis-promoting form. Once the threshold has been reached, the cdc becomes activated and catalyzes the phosphorylation of a messenger that triggers mitosis and the destruction of M-cyclin.

2. EARLY PROPHASE
Centrioles begin to move apart.
Chromosomes appear as long thin threads.
Nucleolus becomes less distinct.

3. MIDDLE PROPHASE
Centrioles move farther apart.
Asters begin to form.
Twin chromatids become visible.

aster

nucleolus

centrioles

1/9. INTERPHASE
Cytokinesis is complete.
Nuclear membranes are complete.
Nucleolus is visible in each cell.
Chromosomes are not seen as distinct
    structures.
Replication of genetic material occurs
    before the end of this phase.
Centrioles are replicated.

8. TELOPHASE
New nuclear membranes begin to form.
Chromosomes become longer, thinner,
    and less distinct.
Nucleolus reappears.
Cytokinesis is nearly complete.

in its inactive form, cdc_i. If the cell is to divide again, S-cyclin is steadily synthesized; the faster it is made, the sooner the S phase will begin. As the S-cyclin concentration rises, it starts to bind to cdc_i, converting it into its synthesis-promoting form, cdc_s. When a threshold is reached, cdc_s becomes activated and begins phosphorylating a messenger compound (Fig. 12.8); thus activated, the messenger triggers replication, and other enzymes quickly destroy the S-cyclin, restoring cdc_s to its inactive cdc_i form.

During G_2, the mitosis-inducing cyclin—M-cyclin—is steadily synthesized. As its concentration rises, it binds to cdc_i and converts it into a mitosis-promoting form, cdc_m. Again, when a threshold is passed, cdc_m is activated and begins phosphorylating a different messenger compound; this new messenger triggers mitosis and the digestion of M-cyclin, thus turning cdc_m into cdc_i (Fig. 12.8).

The potential links between control of the cell cycle and the control failures that underlie cancer are becoming clear. Excess cyclin (leading to uncontrolled cell division) could be produced, either

4. LATE PROPHASE
Centrioles nearly reach opposite sides
of nucleus
Spindle begins to form and kinetochore
microtubules project from centromeres
toward spindle poles.
Nuclear membrane is disappearing.
Nucleolus is no longer visible.

5. METAPHASE
Nuclear membrane has disappeared.
Kinetochore microtubules move each
twin-chromatid chromosome to the
midline; other spindle microtubules
interact with spindle tubules from
opposite pole.

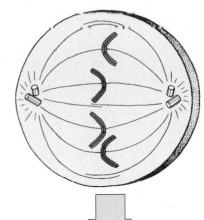

polar microtubule

kinetochore microtubule

spindle

7. LATE ANAPHASE
The two sets of new single-chromatid
chromosomes are nearing their
respective poles.
The poles are being pushed apart.
Cytokinesis begins.

6. EARLY ANAPHASE
Centromeres have split and begun moving
toward opposite poles of spindle.
Spindle microtubules from opposite poles
force the poles apart.

through alteration of cyclin transcriptional control, or (as in the case of several oncogenes mentioned in Chapter 11) through overstimulation of growth-factor receptors. Alternatively, a mutant form of cdc could remain constantly active, or be impossible to inactivate; the *RB* proto-oncogene of retinoblastoma, for instance, is thought to be involved in keeping cdc turned off. With the same effect, the messenger compound itself may be made permanently active. Finally, the DNA target sequence of a messenger could be altered, throwing the gene it regulates into a transcriptionally active form. Studies of the control of replication and mitosis may lead to techniques for restoring quiescence to tumor cells.

Let's assume now that the cell we are examining has passed through the G_1, S, and G_2 stages of interphase and is entering mitosis proper. Because mitosis is such a complex process, it is customarily divided into four stages: prophase, metaphase, anaphase, and telophase (Fig. 12.9).

12.9 Mitosis and cytokinesis in an animal cell

In these drawings, purple and green are used to distinguish the two chromosomes inherited from each parent. For clarity, only a few microtubules are shown. For simplicity, only single kinetochore microtubules are shown pulling the chromatids toward the poles; in reality, each centromere is attached to a bundle of microtubules.

Prophase During prophase, the cell readies the nucleus for the crucial separation of two complete sets of chromosomes into two daughter nuclei. As the two centrioles of an animal cell move toward opposite sides of the nucleus, the chromosomes begin to condense into visible threads. (During early prophase, they appear as long, thin, intertwined filaments, but by late prophase the individual chromosomes will become much shorter, thicker rodlike structures.) As the chromosomes become more distinct, the two nucleoli become less distinct, often disappearing altogether by the end of prophase (Fig. 12.9).

The shortening of the chromosomes during cell division has one obvious advantage. In their shorter form, they can be moved about freely without becoming hopelessly tangled. Staying in the shorter form permanently, however, would not be advantageous, since only in their long uncoiled form can they participate in the critical task of replication during interphase.

Each chromosome from a late-prophase nucleus consists of separate twin chromatids (Fig. 12.5). The replication that occurred during interphase creates two identical copies of the DNA molecules of the original chromosome. Hence, the two chromatids of a prophase chromosome are genetically identical.

The means by which chromosomes are separated during cell division, the mitotic *spindle*, also becomes visible in late prophase. As the centrioles of an animal cell begin to move apart, a system of microtubules appears near each pair of centrioles (Fig. 12.9, stages 2–3). These series of blind-ended microtubules are called the *asters* (Fig. 12.10C–D). As a pair of centrioles approaches a pole, some microtubules attach to microtubules from the opposite pair of centrioles forming *polar microtubules*; asters and polar microtubules together form the basketlike spindle (Fig. 12.9:4). In late prophase, the nuclear membrane gradually disappears and some aster microtubules bind to protein plates called *kinetochores*, which form on the centromere of each chromatid; the kinetochore microtubules thus connect the centromeres to the poles.

Metaphase The stage of metaphase proper is preceded by a short period known as the prometaphase. The chromosomes begin to move toward the equator, or middle, of the spindle. This movement occurs because the two bundles of kinetochore microtubules that anchor each chromosome are able to alter their lengths relative to each other: the shorter microtubule bundle adds tubulin subunits at its aster-pole end, while the longer one digests (removes) them. In this way the position of each chromosome continues to shift until, by the end of prometaphase, it is positioned on the midline.

During the brief stage of metaphase proper, the chromosomes are arranged on the equatorial plane of the spindle, and in side view appear to form a line across the middle of the spindle (Figs. 12.9:5, 12.10D). Metaphase ends when the centromeres of each pair of twin chromatids split apart. Each chromatid then becomes an independent chromosome with its own centromere. By the end of metaphase the total amount of genetic material is unchanged though the number of independent chromosomes in the nucleus has doubled.

Anaphase Next comes the separation of the two complete sets of chromosomes. At the beginning of anaphase the centromeres that

A

B

C

D

E

F

10 μm

12.10 Mitosis in a cell from the endothelium of a frog's heart

(A) Interphase: a diffuse network of microtubules (stained red) is visible. (B) Early prophase: the chromosomes begin to condense; the centrioles have not yet separated. (C) Late prophase: prophase ends as the nuclear envelope disappears; the centrioles have moved apart, and the two asters are prominent. (D) Metaphase: the chromosomes are lined up along the equator of the spindle, midway between the two asters. (E) Middle anaphase: the two groups of chromosomes have begun to move apart from each other, toward their respective poles of the spindle. (F) Early telophase: the two groups of chromosomes are being organized into new nuclei; cytokinesis has begun, and the line dividing the two daughter cells is clearly visible.

hitherto held the twin chromatids together have just broken apart. The two new sets of single-chromatid chromosomes now begin to move away from each other, one going toward one pole of the spindle and the other going toward the opposite pole. This movement toward the respective poles is accomplished in two ways. A process similar to the one that lined up the chromosomes at the midline during prometaphase now pulls the centromeres, with their attached chromosomes, to the poles. (The only difference is that now the tubulin subunits are digested at *both* poles, and no tubulin is added.) Meanwhile, the polar microtubules from opposite ends of the dividing cell form cross bridges between themselves; by adding tubulin subunits to their ends, they increase their lengths, thus pushing the poles apart (Figs. 12.9:6, 12.10E). By late anaphase the cell contains two groups of chromosomes that are widely separated, the two clusters having almost reached their respective poles of the spindle (Fig. 12.9:7). Cytokinesis often begins during late anaphase.

Telophase Telophase (Figs. 12.9:8, 12.10F) is essentially a reversal of prophase. The two sets of chromosomes, having reached their respective poles, become enclosed in new nuclear membranes as the spindle disappears. Then the chromosomes begin to uncoil and to resume their interphase form, while the nucleoli slowly reappear. Cytokinesis is often completed during telophase. Telophase ends when the new nuclei have fully assumed the characteristics of interphase, thus bringing to a close the complete mitotic process. What was a single nucleus containing one set of twin-chromatid chromosomes in prophase is now two nuclei, each with one set of single-chromatid chromosomes.

■ THE EVOLUTION OF NUCLEAR DIVISION

The preceding description of mitotic division fits most but not all eucaryotes. Most of the deviations from the pattern illustrated in Figure 12.9 are seen in primitive eucaryotes, and they may indicate how the typical eucaryotic pattern of nuclear division evolved from the sequence of cell division in procaryotes. Some primitive eucaryotic dinoflagellates, for example, even though they have numerous chromosomes packaged in a membrane-bounded nucleus, still divide by a process remarkably like the binary fission of procaryotes. The chromosomes, which are attached to the inner surface of the nuclear membrane by very short kinetochore microtubules running from their centromeres, are first replicated. Then they are separated into two groups by growth of the membrane between the points of attachment of the original chromosomes and their replicates. As a result, two new nuclei are organized.

These dinoflagellates rely on microtubules in their cytoplasm to determine the direction of nuclear division. Bundles of these tubules run in parallel from each end of the cell toward the other, meeting in "tunnels" in the nucleus (Fig. 12.11A); the two new nuclei move away from each other along this scaffolding.

Advanced dinoflagellates and some fungi and protozoans illustrate a possible further evolutionary progression in their method of cell division, while still relying on cytoplasmic microtubules to orient the process. As in the primitive dinoflagellates, their chromosomes are attached to the nuclear membrane by very short kinetochore microtubules from the centromeres, but there is just a single tunnel. Additional spindle microtubules, originating at the same poles as the transnuclear bundles, run to the nuclear membrane and physically pull the two newly formed nuclei apart (Figs. 12.11B, 12.12).

As we have seen, in higher eucaryotes the nuclear membrane disappears before division, and there is no apparent bundling of microtubules. Instead of securing the chromosomes to the nuclear envelope, as in some dinoflagellates, kinetochore microtubules

12.11   Evolution of nuclear division

(A) In primitive dinoflagellates, parallel bundles of microtubules run from each end toward the other, meeting and interacting in tunnels in the nucleus. The chromosomes are attached to the nuclear membrane and separate as the nucleus divides, and the daughter nuclei slide along the microtubules. (B) In advanced dinoflagellates there is only a single microtubule tunnel through the nucleus. Additional microtubules run from the poles to the nuclear membrane and pull the daughter nuclei apart. (C) In higher eucaryotes the nuclear membrane disappears during division, and certain of the polar microtubules interact with the kinetochore microtubules to pull the individual chromosomes toward the poles.

connect to polar microtubules or to the poles themselves. These tubule complexes exert a direct pull on the chromosomes, rather than on the daughter nuclei (Fig. 12.11C).

■ CELL DIVISION: CYTOKINESIS

Division of the cytoplasm usually accompanies division of the nucleus, but mitosis without cytokinesis is common in some algae and fungi, producing *coenocytic* plant bodies (bodies with many nuclei but few, if any, cellular partitions). It regularly occurs during certain phases of reproduction in seed plants and certain other vascular plants. It is also common in a few lower invertebrate animals with coenocytic bodies. During the early development of insect eggs, mitosis without cytokinesis produces hundreds of nuclei in a limited amount of cytoplasm; later, cytokinesis cuts up this cytoplasm to produce many new cells in a very short time.

How animal cells divide Division of an animal cell normally begins with the formation of a *cleavage furrow* running around the cell (Fig. 12.13). When cytokinesis occurs during mitosis, the location of the furrow is ordinarily determined by the orientation of

12.12 Persistent nuclear membrane during mitosis in the fungus *Catenaria*

In this fungus, part of the spindle runs through the nucleus. Here two new telophase nuclei, at opposite ends of the photograph, are being pinched off by constrictions of the nuclear envelope. The middle part of the envelope, between the new nuclei, will eventually disintegrate.

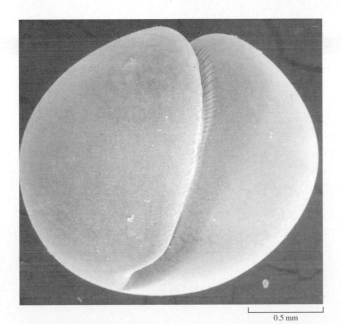

0.5 mm

12.13 A dividing frog egg

The cleavage furrow is not yet complete (see bottom of cell). Note the puckered stress lines in the furrow visible in this scanning electron micrograph.

Animal cell   Algal cell   Higher–plant cell

12.14   Three mechanisms of cytokinesis

Cytokinesis in animal cells typically occurs by a pinching-in of the plasma membrane. In many algal cells cytokinesis occurs by an inward growth of new wall and membrane. In higher plants cytokinesis typically begins in the middle and proceeds toward the periphery, as membranous vesicles fuse to form the cell plate.

the spindle, in whose equatorial region the furrow forms (see Fig. 12.9:7). The furrow becomes progressively deeper until it cuts completely through the cell (and its spindle), producing two new cells.

Since the location of the furrow is usually related to the position of the centrioles and spindle, a dense belt of actin and myosin microfilaments at the site of the cleavage furrow is probably responsible. Drugs like cytochalasin, which block the activity of microfilaments, stop cytokinesis. Since agents that bind actin and myosin also stop the cleavage process, the sort of actin-myosin interaction known to be involved in cell movement, as discussed in Chapter 5, is probably responsible for cytokinesis as well.

How plant cells divide The relatively rigid cell walls of plants, essential for their support, required the evolution of a different kind of cytokinesis. In many fungi and algae, new plasma membrane and cell wall grow inward around the wall midline until the growing edges meet and completely separate the daughter cells (Fig. 12.14B). In higher plants a special membrane, called the ***cell plate***, forms halfway between the two nuclei at the equator of the spindle if cytokinesis accompanies mitosis (Figs. 12.14C, 12.15D). The cell plate begins to form in the center of the cytoplasm and slowly becomes larger until its edges reach the outer surface of the cell and the cell's contents are cut in two. Thus cytokinesis of higher-plant cells progresses from the middle to the periphery, whereas cytokinesis in animal cells progresses from the periphery to the middle.

The cell plate forms from membranous vesicles that are carried

12.15   Cell division in a plant, the African blood lily

(A) Prophase: the chromosomes have condensed; microtubules (red) are visible. (B) Metaphase. (C) Anaphase: the two groups of chromosomes are moving to opposite poles of the cell. (D) Telophase: a cell plate has begun to form.

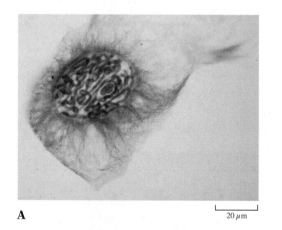

A   20 μm   **B**   20 μm

C   20 μm   **D**   20 μm

1 μm

12.16   A late-telophase cell in corn root, showing formation of the cell plate

Mitosis has been completed, and the two new nuclei (N) are being formed; the chromosomes (dark areas in the nuclei) are no longer visible as distinct structures, but the nuclear membranes are not yet complete. A cell plate (CP) is being assembled from numerous small vesicular structures. At the lower end of the nucleus on the right, a length of endoplasmic reticulum (ER) can be seen in this electron micrograph that appears to run from the nuclear envelope to the cell wall, through the wall, and into the next cell.

to the site of plate formation by the microtubules that remain after mitosis. There they first line up and then unite (Fig. 12.16). The vesicles are derived mainly from the Golgi and, to a lesser extent, from the ER. As the vesicle membranes fuse with one another and, peripherally, with the old plasma membrane, they constitute the partitioning membranes of the two newly formed daughter cells. The contents of the vesicles, trapped between the daughter cells, give rise to the middle lamella and to the beginnings of primary cell walls.

MAKING GAMETES: MEIOSIS

Mitosis maintains a constant number of chromosomes in somatic cells. What would be its effect in reproductive cells? If two gametes (an egg and a sperm, for example) each carried the normal number of mitotic chromosomes, the *zygote* (the first cell produced by their union) would have double that number. At each successive generation the number would again double. Gamete production requires

1. EARLY PROPHASE
Chromosomes become visible as
long, well-separated filaments;
replication has already occurred.

2. MIDDLE PROPHASE I
Homologous chromosomes become
shorter and thicker, and synapse;
crossing over takes place.

3. LATE PROPHASE I
The tetrad structure of the synapsed
chromosomes, and the chiasmata created
by crossing over, become visible.
Nuclear membrane begins to disappear.
Kinetochore microtubules connect
chromosomes to the poles the centromeres.

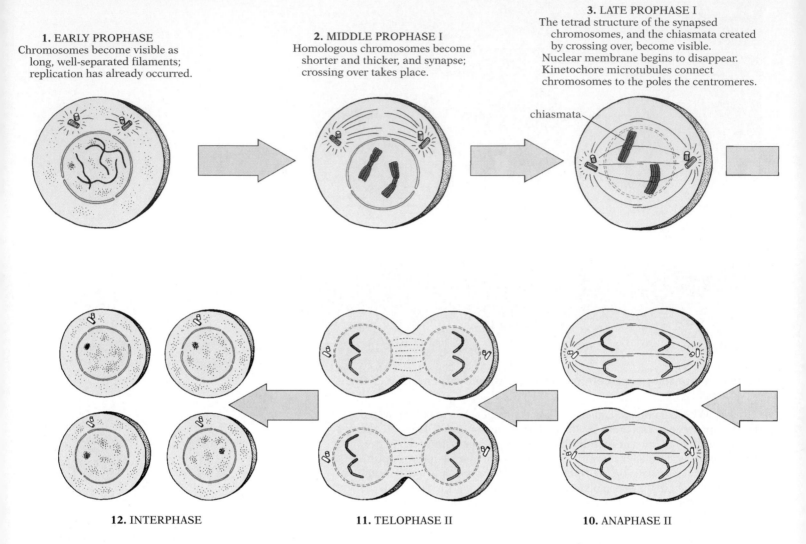

chiasmata

12. INTERPHASE

11. TELOPHASE II

10. ANAPHASE II

a division strategy that reduces the number of chromosomes to
half, so that when the egg and sperm unite in fertilization the normal diploid number is restored. This special process of reduction
division is ***meiosis*** (from the Greek word for "diminution"). In all
multicellular animals meiosis occurs at the time of gamete production. Consequently each gamete possesses only half the species-typ-
ical number of chromosomes.

Note that in the reduction division of meiosis, the chromosomes
of the parental cell are not simply separated randomly into two
halves. The diploid nucleus contains two of each type of chromosome, and meiosis partitions these chromosome pairs so that each
gamete contains one of the two homologues. Such a cell, with only
one of each type of chromosome, is therefore haploid. When two
haploid gametes unite in fertilization, the resulting zygote is
diploid, having received one of each chromosome type from the
sperm of the male parent and one of each type from the egg of the
female parent.

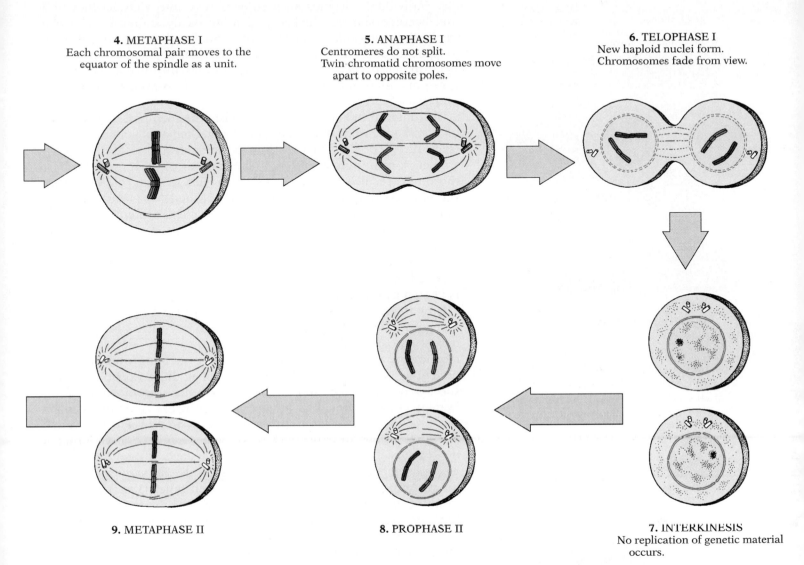

4. METAPHASE I
Each chromosomal pair moves to the equator of the spindle as a unit.

5. ANAPHASE I
Centromeres do not split. Twin chromatid chromosomes move apart to opposite poles.

6. TELOPHASE I
New haploid nuclei form. Chromosomes fade from view.

9. METAPHASE II

8. PROPHASE II

7. INTERKINESIS
No replication of genetic material occurs.

12.17 Meiosis in an animal cell

The chromosomes inherited from each parent are shown in different colors. As a result, each member of the homologous pair is a different color, which aids in visualizing the results of crossing over.

■ THE MEIOTIC CYCLE

Production of two haploid gametes from a diploid cell could be accomplished by a single division; instead, complete meiosis (Fig. 12.17) involves two division sequences, which result in four new haploid cells. As in mitosis, meiosis is preceded by replication, which creates chromatid pairs. This step seems unnecessary, and may exist solely to make possible the process called crossing over (to be discussed presently). The result is that the gamete-producing cell now has 4 times as much DNA as the gametes require, and meiosis proceeds to correct this excess. The first division sequence reduces the number of chromosomes; the second separates chromatids. The same four stages found in mitosis—prophase, metaphase, anaphase, and telophase—are recognized in each division sequence in meiosis.

Prophase I: crossing over Many of the events in prophase I of meiosis superficially resemble those in the prophase of mitosis.

DNA of one pair of twin chromatids   Axial protein   DNA of the homologous pair of twin chromatids

0.2 μm

12.18   Formation of the chromosomal synapse

Axial proteins gather the replicated DNA of each chromosome—that is, the twin chromatids—into a long series of paired loops, as seen in this longitudinal section through a synaptonemal complex from a meiotic cell of the ascomycete *Neottiella*.

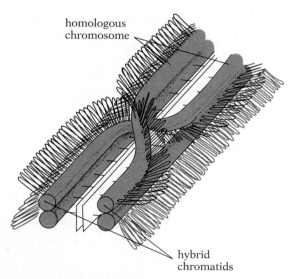

homologous chromosome

hybrid chromatids

12.19   Crossing over

When two homologous chromosomes have synapsed, chromatids from one chromosome exchange fragments with chromatids from the other chromosome to create hybrid chromatids. One such exchange is shown here.

The individual chromosomes come slowly into view as they coil and become shorter, thicker, and more easily stainable. The two nucleoli slowly fade from view and, finally, the nuclear envelope disappears and the spindle is organized. The replication of the genetic material occurs during interphase, as in mitosis. There are, however, important differences between the prophase of meiosis and that of mitosis.

The chief difference is that in meiosis the members of each pair of homologous chromosomes move together and lie side by side (Fig. 12.17:2). Also, in addition to being fastened at their centromeres, as in mitosis, the twin chromatids of each meiotic chromosome are also held together by a special pair of long, thin protein axes that run their entire length. The DNA is gathered into loops as part of the condensation process. The protein axes of the two homologous chromosomes now join by means of protein cross bridges to form an intricate compound structure known as a *synaptonemal complex*, which lines up the four chromatids (often referred to as a tetrad) in perfect register (Fig. 12.18). This process is known as *synapsis*.

Now the remarkable process of *crossing over* begins. Large protein complexes called recombination nodules appear along the ladderlike cross bridges. These nodules probably determine the locations at which genetic material will be exchanged between homologous chromosomes. The number of nodules depends on the species of organism and the length of the chromosome; each human chromosome has an average of three. In many species there is also a sexual difference in crossing over, with fewer crossover events in male cells. Next, each nodule begins a process by which two of the chromatids, one from each of the homologous chromosomes, are clipped open at precisely the same place, and the resulting fragments are spliced to each other (Fig. 12.19). The result of this splicing is that the chromatids of the tetrad no longer form two sets of twins; the recombined chromatids are now hybrids, containing genetic material descended from both the mother's and the father's homologous chromosomes.

When the synaptonemal complex begins to break up in late prophase, the points at which crossing over has taken place become visible. The hybrid chromatids produced by crossing over link the two homologous chromosomes at these points, called *chiasmata* (Fig. 12.20). Each crossover event can involve a different pair of chromatids (Fig. 12.21). Crossing over is not rare or accidental: it is a frequent and highly organized mechanism.

As prophase ends, most chromosomes now consist of two hybrid chromatids rather than twin chromatids, since genetic material has been exchanged between the chromatids of homologous chromosomes. Spindle microtubules appear, radiating from the two poles of the cell, and kinetochore microtubules attach to the centromeres. There is a distinct difference between the activity of the centromeric microtubules in mitosis and in meiosis. In meiosis, microtubules from one centromere of each homologous pair attach to a pole so the two pairs are pulled toward opposite poles (Fig. 12.22).

Metaphase I In mitosis each chromosome consists of a pair of twin chromatids, and moves independently to the midline of the

chiasma

chiasma

chromatid

centromere

2 μm

12.20 Two chiasmata between homologous chromosomes

Each chromosome is clearly recognizable as a pair of chromatids. The centromeres, too, are clearly visible. Crossing over has taken place at two points—the chiasmata. Note that the crossing over involves only one chromatid from each chromosome. These chromosomes of a Costa Rican salamander are from a spermatocyte in prometaphase I.

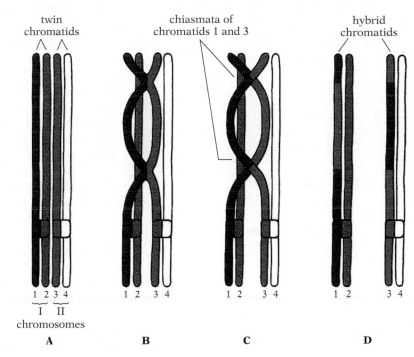

twin chromatids

chiasmata of chromatids 1 and 3

hybrid chromatids

1 2 3 4   1 2 3 4   1 2 3 4   1 2 3 4

I   II

chromosomes

A   B   C   D

12.21 Schematic summary of synapsis and crossing over

(A) Crossing over begins as homologous chromosomes (I and II), each consisting of twin chromatids (1,2 and 3,4), are brought into register and synapse. (The synaptonemal complex has been omitted for clarity.) (B) Parts of separate chromatids are spliced together, while the protein axis (not shown) keeps each segment firmly attached to its twin. (C) When the synaptonemal complex breaks up, the homologous chromosomes begin to drift apart, but the chiasmata of spliced hybrid chromatids prevent them from separating. (D) The homologous chromosomes separate fully during anaphase.

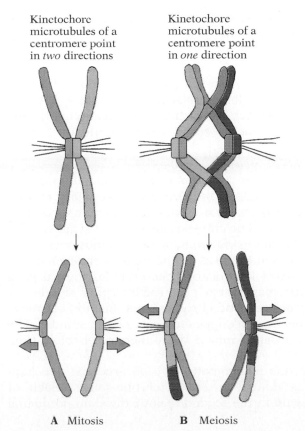

Kinetochore microtubules of a centromere point in *two* directions

Kinetochore microtubules of a centromere point in *one* direction

A   Mitosis   B   Meiosis

12.22 Kinetochore microtubules in mitosis and meiosis

(A) In mitosis, each centromere is attached by kinetochore microtubules to both poles. As a consequence, the two chromatids are pulled apart and wind up at opposite ends of the cell during anaphase. (B) In meiosis I, however, one centromere in each pair is attached to one or the other of the two poles. The result is that in anaphase I the chromatids remain joined, and homologous two-chromatid chromosomes wind up in separate cells.

cell. In meiosis each chromosome typically consists of two hybrid chromatids, and homologous chromosomes move as a unit to the midline in preparation for anaphase. As a result, in meiosis the number of independent units waiting at the midline for cell division is only half the number of mitosis (Fig. 12.17:4).

Anaphase I In mitosis, metaphase ends and anaphase begins when the centromere of each twin chromatid splits, and the two independent single-chromatid chromosomes thus formed move away from each other toward opposite poles of the spindle. However, in the first division of meiosis, the separation occurs instead between the two chromosomes that have been joined since the middle of prophase (Fig. 12.22). Since each chromosome has its own centromere, there is no splitting of centromeres. With their respective centromeres attached by microtubules to only one pole, the homologous chromosomes move away from each other toward opposite poles during anaphase. Thus in a four-chromosome organism, only two chromosomes, each with a hybrid chromatid, move to each pole (Fig. 12.17:5), in contrast to the four single-chromatid chromosomes that move to each pole in mitosis. Because the synaptic pairing was not random, but involved the two homologous chromosomes of each type, the two daughter nuclei get not just any two chromosomes, but rather one of each type.

Telophase I Telophase of mitosis and meiosis are essentially the same, except that in mitosis each of the two new nuclei has the same number of chromosomes as the parental nucleus, whereas each of the new nuclei formed in meiosis has half the chromosomes present in the parental nucleus (Fig. 12.17:6). At the end of telophase in mitosis, the chromosomes are single; at the end of telophase I of meiosis, the chromosomes are composed of two chromatids each.

Interkinesis Following telophase I of meiosis, there is a short period called interkinesis, which is similar to an interphase between two mitotic division sequences except that no replication of the genetic material occurs and hence no new chromatids are formed (Fig. 12.17:7).

The second set of meiotic divisions The second division sequence of meiosis, which follows interkinesis, is essentially mitotic from the standpoint of mechanics, though the functional result is different (Fig. 12.17:8–12). The chromosomes do not synapse; they cannot, since the cell contains no homologous chromosomes. Each two-chromatid chromosome moves to the midline independently, and its centromere sends kinetochore microtubules toward each pole. At the end of metaphase II each centromere splits, and during anaphase II the single-chromatid chromosomes thus formed move away from each other toward opposite poles of the spindle. The new nuclei formed during telophase II are therefore haploid (Fig. 12.23).

In summary, then, the first meiotic division produces two haploid cells containing double-chromatid chromosomes. Each of these cells divides again in the second meiotic division, producing

12.23 Meiosis in a cell from the grasshopper _Mongolotetix japonicus_

(A) Early prophase I: the chromosomes are seen as long filaments. (B) Middle prophase I: synapsing of homologous chromosomes takes place. (C) Metaphase I: chromosomal pairs line up at the equator. (D) Anaphase I: homologous chromosomes have separated and are moving to opposite poles. (E) Telophase I: division into two haploid nuclei has begun. (F) Prophase II: early in this stage, separate chromosomes are barely distinguishable. (G) Metaphase II: chromosomes in each cell are at the equator, and the spindles are evident. (H) Telophase II: four new haploid nuclei can be seen.

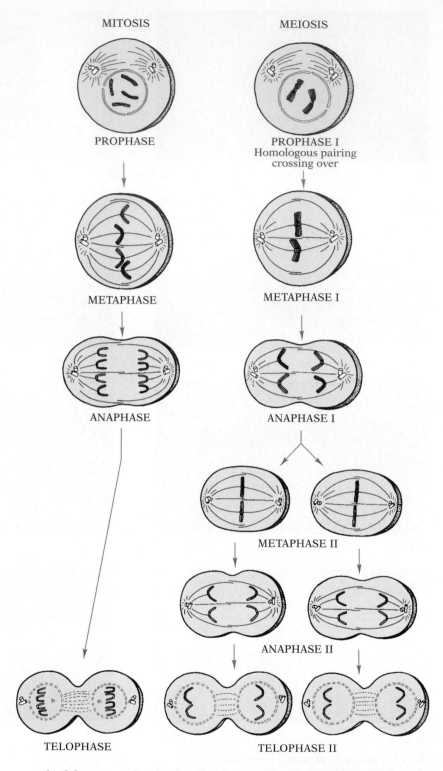

MITOSIS   MEIOSIS

PROPHASE   PROPHASE I
                                 Homologous pairing
                                 crossing over

METAPHASE   METAPHASE I

ANAPHASE   ANAPHASE I

                                 METAPHASE II

                                 ANAPHASE II

TELOPHASE   TELOPHASE II

12.24   Comparison of mitosis and meiosis

The first step, interphase (during which replication occurs), is omitted, as are the stages between anaphase I and metaphase II of meiosis.

a total of four new haploid cells containing single-chromatid chromosomes. Mitosis and meiosis are compared in Figure 12.24.

■ WHAT IS THE PURPOSE OF SEX AND CROSSING OVER?

Life depends on an optimal balance between stability and change—maintaining the fidelity of the genetic message while at the same time providing enough variation to permit selection for improved genes, particularly in the face of unpredictable changes in competition, predation, habitat, and climate. Sexual organisms go to great lengths to produce haploid gametes for mating (**sexual recombination**): asexual cloning of diploid offspring is metabolically less expensive, and gamete-producing cells are put to considerable trouble to make crossing over possible. Most researchers believe these two processes serve the goal of generating change.

The most important consequence of sexual reproduction is that it creates offspring with new combinations of characteristics. Each member of a homologous pair of chromosomes comes from a different parent, and contains genes coding for the same kinds of RNA, structural proteins, and enzymes, but the exact base sequences in two homologous genes are not necessarily identical. Instead, the copy of a gene inherited from an organism's male parent is often at least slightly different from the copy inherited from the female parent. Different versions of the same gene are called **alleles**, and the enzymes or structural proteins they produce are likely to have slightly (or even very) different activities. For example, blue eyes develop in people when both copies of their eye-pigment gene code for a defective (colorless) screening pigment; people with brown eyes have at least one copy of the allele that codes for a functional pigment. Since an individual organism is a mix of alleles from two different parents, its chromosomes are almost certain to contain an ensemble of alleles different from that of either parent, and many aspects of its morphology, physiology, and behavior will be correspondingly different.

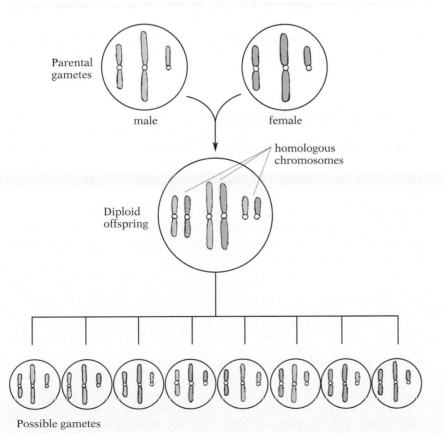

12.25 Variation in gametes

Even without crossing over, a diploid organism produces 2^n different kinds of gametes, where n is the number of pairs of homologous chromosomes. In this hypothetical organism, $n = 3$, so there are eight (2^3) possible kinds of gametes.

To understand why variation arises in sexual recombination, let's consider a hypothetical six-chromosome organism (Fig. 12.25). Three chromosomes came originally from the gamete of the male parent (the product of meiosis), and three from the gamete of the female parent, to form the three pairs of homologous chromosomes in this diploid organism. Each time meiosis occurs in an individual producing gametes for reproduction, all of that organism's maternal chromosomes *might* go into one gamete and all the paternal ones into another, but many other combinations are equally likely. If an organism has three pairs of chromosomes, eight different combinations may be found in the gametes. In the zygote formed by two gametes in such a species, 64 (8 × 8) different combinations of chromosomes are then possible. For humans, with our unusually large number of chromosomes, the number of different chromosome combinations possible in the offspring of the same two parents is about 7×10^{13}.

The highly ordered process of crossing over increases the variety of possible gametes astronomically, since crossover points are essentially random, and virtually every hybrid chromatid is spliced together at a unique set of points. To take a specific example, assume that the father contributes a gene coding for black fur (which we can call *F*) and a gene coding for a large body (*S*), while the alleles from the mother code for gray fur (*f*) and a small body (*s*). Though the *F* and *S* of the chromosome contributed by the father remain together (or **linked**) throughout the *mitotic* life of a cell, as do the *f* and *s* on the maternal chromosome, crossing over in *meiosis* can rearrange things (Fig. 12.26). As a result, the chromosome in the offspring's gamete is likely to have a new combination—*F* and *s*, or *f* and *S*—rather than one of the parental combinations. In fact, crossing over can occur even within a gene, so if the copies of a particular gene are different alleles, two entirely novel alleles can also be created.

■ WHEN DOES MEIOSIS OCCUR?

Meiosis in the life cycle of plants Meiosis in plants usually produces haploid reproductive cells called *spores*, which often divide mitotically to develop into haploid multicellular plant bodies. At one extreme in plant reproductive patterns are some primitive plants like algae (Fig. 12.27). The haploid spore cells (stage 1) divide mitotically and develop into a rapidly growing, poorly differentiated, haploid, multicellular stage (stage 2) in which the

Parental gametes

male   female

Diploid offspring

Meiosis I

crossing over

Meiosis II

Possible gametes

12.26   Variation in gametes from crossing over

In this example, the parents contribute chromosomes with two different alleles of genes for fur color (*F* and *f*) and body size (*S* and *s*); these chromosomes form a homologous pair in the diploid offspring. At meiosis, crossing over breaks up the paternal and maternal combinations, so some of the resulting gametes have hybrid chromosomes, unlike those of either parent.

A PRIMITIVE PLANT

B INTERMEDIATE PLANT

C ADVANCED PLANT

D ANIMAL

■ Haploid stages
■ Diploid stages

12.27 Four types of life cycles

(A) In some very primitive plants the diploid phase is represented only by the zygote, which quickly divides by meiosis to produce haploid spores, which may in turn divide mitotically to produce a multicellular haploid plant. (B) In most multicellular plants there are two multicellular stages, one haploid and one diploid (stages 2 and 5). The relative importance of these two stages varies greatly from one plant group to another; the cycle shown here is an intermediate one, in which stages 2 and 5 are nearly equal. (C) In flowering plants the multicellular diploid stage (stage 5) is the major one, and the multicellular haploid stage (stage 2) is much reduced, being represented by a tiny organism with very few cells. (D) Animals and a very few plants have a life cycle in which meiosis produces gametes directly—the spore stage and the multicellular haploid stage (stages 1 and 2) being absent.

organism passes most of its life. This multicellular plant eventually produces, by mitosis, cells specialized as gametes (stage 3), which unite to form the diploid zygote (stage 4); the zygote promptly undergoes meiosis to produce four haploid spores, thus beginning the cycle again. In such an organism, then, the haploid phase of the cycle (particularly stage 2) is dominant, and the only diploid stage—the zygote—is brief.

Most plants devote much of their life cycle to the diploid state. Many ferns, for instance, divide their time about evenly between haploid and diploid phases (Fig. 12.27). While the fern is in the diploid portion of its life cycle, certain cells in its reproductive organs divide by meiosis to produce haploid spores (stage 1). These spores divide mitotically and develop into haploid multicellular plants (stage 2). The haploid multicellular plant eventually produces cells specialized as gametes (stage 3). As with algae, the gametes are produced by mitosis, not meiosis, because the cells that divide to produce the gametes are already haploid. Two of these gametes unite in fertilization to form the diploid zygote (stage 4), which divides mitotically and develops into a diploid multicellular plant (stage 5). In time, this plant produces spores and the cycle starts over again.

As described in Chapter 22, the several groups of plants vary greatly in the relative importance of the diploid and haploid phases in their life cycles. A very few plants have cycles almost like the animal life cycle shown in Figure 12.27D; stages 1 and 2 are absent, and the haploid phase of the cycle is represented only by the gametes. In the flowering plants (Fig. 12.27C), stages 1 and 2 have not been abandoned altogether, but stage 2 has been reduced to a tiny three-to-eight-cell entity that is not free-living, and the plant spends most of its life as a multicellular diploid organism (stage 5). In general, haploid-dominated plants tend to be short-lived, to reproduce rapidly and prolifically, and have little differentiated tissue; diploid-dominated plants are at the other end of the scale, and more like higher animals in each regard.

Meiosis in the life cycle of animals With rare exceptions, higher animals exist as diploid multicellular organisms through most of their life cycle. At the time of reproduction, meiosis produces haploid gametes, which, when their nuclei unite in fertilization, give rise to the diploid zygote. The zygote then divides mitotically to produce the new diploid multicellular individual. The gametes—sperm and egg cells—are thus the only haploid stage in the animal life cycle (see Fig. 12.27D).

In male animals, sperm cells (spermatozoa) are produced by the germinal epithelium lining the seminiferous tubules of the testes (Fig. 12.28). When one of the epithelial cells undergoes meiosis, the four haploid cells that result are all quite small, but approximately equal in size (Fig. 12.29). All four differentiate into sperm cells with long flagella, but with very little cytoplasm in the head,

12.28 Cross section of rat seminiferous tubule, showing spermatogenesis

The dark-stained outermost cells in the wall of the tubule are spermatogonia (Sg), which divide mitotically, producing new cells that move inward. These cells enlarge and differentiate into primary spermatocytes (Sc), which divide meiotically to produce secondary spermatocytes and then spermatids (St). The spermatids differentiate into mature sperm cells, or spermatozoa (Sp), whose long flagella can be seen in the lumen of the tubule in this photograph.

0.1 μm

12.29   Schematic illustration of spermatogenesis and oogenesis in an animal

In some animals the first polar body does not divide.

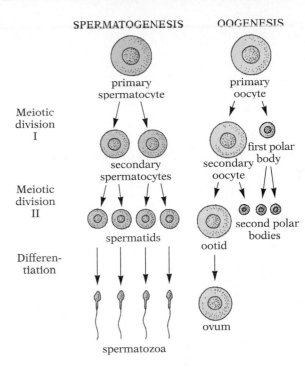

which consists primarily of the nucleus. This process of sperm production is called ***spermatogenesis***. Meiosis in human males begins at puberty.

In female animals the egg cells are produced within the follicles of the ovaries through ***oogenesis***. When a cell in the ovary undergoes meiosis, the haploid cells that result are very unequal in size. The first meiotic division produces one relatively large cell and a tiny one called a first ***polar body***. The second meiotic division of the larger of these two cells (secondary oocyte) produces a tiny second polar body and a large cell that soon differentiates into the egg cell (or ovum). The first polar body may or may not go through the second meiotic division. If it does redivide, there are three polar bodies altogether (Fig. 12.30). Thus, when a diploid cell in the ovary undergoes complete meiosis, only one mature ovum is produced (Fig. 12.29). In human females, the oocytes complete the first meiotic prophase in the fetal ovaries, and then complete meiosis upon ovulation (when a mature egg is released from an ovary). The interval between these two events can exceed 40 years.

The unequal cytokinesis of oogenesis provides an unusually large supply of cytoplasm and stored food to the ovum for use by the embryo that will develop from it. In fact, the ovum provides almost all the cytoplasm and initial food supply for the embryo. The tiny, highly motile sperm cell essentially contributes only its genetic material.

0.5 μm

12.30   Human egg cell with polar bodies

The polar bodies are the three small circular structures at the right.

CHAPTER SUMMARY

FISSION VERSUS DIVISION

PROCARYOTIC FISSION Bacteria divide by replicating their single circular chromosome, and then dividing into two daughter cells, each containing a copy of the chromosome. (p. 309)

WHY EUCARYOTIC CHROMOSOMES POSE PROBLEMS Eucaryotic chromosomes are generally linear, organized as homologous pairs, wound on nucleosomes, and organized into condensed loops. All these factors make cell division more complicated. (p. 310)

HOW SOMATIC CELLS DIVIDE

NUCLEAR DIVISION: MITOSIS Eucaryotic cells must distribute a copy of each chromosome and one pair of centrioles to each daughter cell. During interphase the centrioles and chromosomes are duplicated; the duplicated chromatids of each chromosome are joined at the centromere. Interphase is divided into an initial gap, G_1 (when ribosomes and organelles are first duplicated), the synthesis phase, S (when the chromosomes are duplicated), and a second gap, G_2 (when the cell prepares for division). The interactions between cyclins and the cell-division-cycle protein (cdc) regulate interphase. During prophase the centioles move to the poles, the chromosomes condense, the nuclear membrane is disassembled, and the spindle forms. During metaphase the spindle attaches to microtubules from the centromeres, the chromosomes are moved to a plane equidistant between the poles, and the centromeres break. In anaphase the single-chromatid chromosomes move to the poles. In telophase the nuclear membrane reforms around the two daughter nuclei, the chromosomes decondense, and physical division of the daughter cells—cytokinesis—is completed. (p. 311)

THE EVOLUTION OF NUCLEAR DIVISION In some primitive eucaryotes the nuclear membrane remains intact throughout division, and the nucleus fissions in the manner of organelles and bacteria. (p. 318)

CELL DIVISION: CYTOKINESIS The cleavage furrow that marks where animal cells will divide forms equidistant between the two poles. Plants form a membraneous cell plate midway between the poles, which grows until it divides the two daughter cells. (p. 319)

MAKING GAMETES: MEIOSIS

THE MEIOTIC CYCLE Meiosis reduces the number of chromosomes by half, so that gametes have only one chromosome from each homologous pair. Meiosis proceeds in two steps: all chromosomes are duplicated and two diploid daughter cells are produced; then another round of division produces four haploid gametes. The first step differs from mitosis in that after duplication the homologous chromosomes pair up, and crossing over occurs; then one member of each homologous pair is moved into each daughter cell. Thus, the centromeres that connect the chromatids do not break until the second step of meiosis. (p. 323)

WHAT IS THE PURPOSE OF SEX AND CROSSING OVER? Sexual recombination occurs when haploid gametes combine to create a diploid zygote. Sexual recombination creates variation in three ways: crossing over recombines the homologous chromosomes inherited from an individual's two parents, creating novel hybrid chromosomes; a randomly selected member of each homologous pair is incorporated into each gamete, creating many different gametes; and gametes from different individuals combine to generate a unique diploid. (p. 328)

CHAPTER 13

WHEN DOES MEIOSIS OCCUR? Plants alternate between a diploid spore-producing phase and a haploid gamete-producing phase. In most plants the spore-producing form is the dominant phase; the spores are produced through meiosis, and grow into the form that generates the gametes that recombine. Animals have no such alternation. Most plants and all animals produce small male gametes (pollen and sperm) and large female gametes (eggs). Animal eggs are produced by unequal meiotic division that concentrates all the cytoplasm in just one of the four haploid daughter cells; sperm are produced by equal division of the precursor cell entering meiosis. Egg production in many animals begins long before reproduction and is then arrested until eggs are needed; males generally produce sperm as needed. (p. 330)

STUDY QUESTIONS

1 It has been said that the only possible function of the odd-process of meiosis is to make crossing over possible. Compare mitosis and meiosis with this suggestion in mind. (pp. 314–315, 322–323)

2 Crossing over is less frequent in males than in females in many species—male *Drosophila*, for instance, never undergo crossing over in meiosis. Even in hermaphroditic species, in which every organism can have both male and female reproductive organs, the same pattern holds. What might be the evolutionary logic behind this curious asymmetry? (pp. 328–330)

3 Interpret the following observations and formulate a master hypothesis about what sorts of chemical signals must appear in the cell at various points, and what effect they have:

a When a cell just entering S phase is fused with a G_1-phase cell arrested in that state, a process that mixes their cytoplasms but leaves the nuclei separate, both nuclei enter S phase.

b When a mid-S-phase cell is fused with a G_1-phase cell, the S nucleus remains frozen until the G_1 nucleus catches up.

c When a G_1-phase cell and a G_2-phase cell are fused, the G_2 nucleus remains frozen until the G_1 nucleus catches up.

d When an M-phase cell is fused with an S-phase or either sort of G-phase cell, the M nucleus is unaffected, but the other nucleus loses its nuclear envelope and its chromosomes condense. (pp. 312–315)

4 Do animals with larger numbers of chromosomes but similar crossover rates have more, less, or the same general amount of variation in their progeny? (pp. 328–330)

SUGGESTED READING

GOULD, J. L., AND C. G. GOULD, 1989. *Sexual Selection.* Scientific American Library, New York. *A wide-ranging, nontechnical account of the many theories that seek to account for the evolution of sex and gender.*

MCINTOSH, J. R., AND K. L. MCDONALD, 1989. The mitotic spindle, *Scientific American* 261 (4).

MURRAY, A. W., AND M. W. KIRCHNER, 1991. What controls the cell cycle, *Scientific American* 264 (3). *On the biochemical control of cell division, with emphasis on cyclin and cdc.*

THE COURSE OF ANIMAL DEVELOPMENT

Probably no aspect of biology is more amazing than the development of a complete new organism with perhaps more than 10^{12} cooperating cells arising from one fertilized egg. The process is so precisely controlled that the entire intricate organization of cells, tissues, organs, and organ systems that will characterize the functioning adult comes into being, each element in just the right place with respect to the others.

Development requires a precisely programmed and coordinated series of changes in gene expression, cell-surface proteins, cell shape, and cell motility, as cells interact to create a characteristic species morphology and to give rise to the lines of specialist cells that constitute the various tissues and organs. This chapter will describe the major morphological and physiological events of development in animals, from fertilization through embryonic development, birth, and postnatal development. The next chapter will examine the molecular and biochemical mechanisms that control and coordinate development.

We will concentrate in these chapters on the patterns and mechanisms of animal development; for a number of reasons plant development is treated separately (in Chapter 32). For one thing, the vast majority of plants are autotrophic, and are adapted to remain rooted during a life cycle to a specific locale, and to collect sunlight. Since plants are not adapted for locomotion, they can enjoy the structural advantages of more rigid cell walls—a specialization that fixes each cell in place, making impossible the dynamic aspect of animal development that is examined in the next chapter: the migration of cells during the growth of the embryo.

Furthermore, because few plants feed on other organisms, they do not need nervous systems and muscles to guide and power the capture or harvesting of food; nor do they need mouths, stomachs, or digestive systems, or the more than 300 specialized cell types that make up kidneys, bladders, rectums, and other specialized tissues and organs of animals. Because plants compete with each other by growing taller or broader to gather more light, and by extending their roots to obtain more water and minerals, they produce new organs like leaves and flowers whenever and wherever they are needed throughout their life cycle. Animals usually generate their entire array of tissues and organs early in development, and devote later energies to their maintenance and growth.

0.02 mm

13.1 Human sperm

(A) Three sections are easily distinguishable: the head, containing the nucleus; the midpiece, tightly packed with mitochondria, which produce the ATP that fuels the sperm's swimming; and the long flagellum. The chromosomes in the nucleus are supercondensed into an almost crystalline, genetically inactive form. In the apex of the head, immediately in front of the nucleus, is the acrosome, a membrane-bounded vesicle derived from the Golgi apparatus. There is only a tiny amount of cytosol in the cell. (B) The photograph shows human sperm in a tube leading to the egg.

THE EVENTS OF FERTILIZATION

In the egg-producing organs, or ***ovaries***, certain cells are set aside early as egg primordia. These cells grow to an unusually large size, and when they then undergo meiosis, the divisions are unequal; almost all the cytoplasm is retained in the ripe ovum. The sperm, on the other hand, are unusually small cells with very little cytoplasm. The ovum thus furnishes most of the initial cytoplasm for the ***embryo***—the entity that develops from the fertilized egg and orchestrates its own conversion into a new individual. Moreover, even in eggs with relatively little stored food, or ***yolk***, the yolk granules are more abundant in some parts of the eggs than others. Often one side of the egg is richer in yolk; this side is then called the vegetal pole. As described in the next chapter, this uneven distribution of yolk reflects an underlying polarization that helps to provide order and direction in early development.

The fusion of sperm and egg cell creates the ***zygote***, which then immediately starts to develop into an embryo. However, it is the contact between the two cell membranes—not the fusion of the sperm nucleus with egg nucleus to form a diploid nucleus—that causes development to begin. Apparently ***fertilization***—the joining of two haploid sets of chromosomes—is not necessary to induce embryonic development in many animals. For example, it is easy to induce unfertilized frog eggs to begin development in the laboratory by pricking them with a fine needle dipped in blood. A few such eggs will develop into viable, apparently normal tadpoles.[1] Adult rabbits have been produced from unfertilized eggs by similar procedures. Unfertilized eggs can even be stimulated to begin development through a mild electric shock, a change in the salt concentration in the surrounding fluid, or a physical jolt. In most species, however, the development aborts after only a few divisions.

In many mammalian species the egg is initially enclosed in a thin protective layer of cells called the ***follicle***. This barrier may be loosened by enzymes released by the ***acrosome***, a vesicle located at the apex of the sperm in many species (Fig. 13.1). After the follicle-cell barrier has been loosened, the sperm encounters a thick jelly-like coat, called in mammals the ***zona pellucida*** (Fig. 13.2A). Receptors in the head of the sperm then bind to a long, filamentous glycoprotein embedded in the coat. Species-specific features of the glycoprotein prevent the sperm of other species from attaching.

When the acrosomal vesicle releases its enzymes, it also everts to form a long filament that penetrates the jelly coat (Fig. 13.2B–C). The membrane of the filament fuses with the membrane of the microvilli of the egg, thus fully fusing the two gametes (Fig. 13.2D–E). Once this fusion has occurred, the electrical potential of the egg-cell membrane of many species changes dramatically (thus making fusion with the other sperm difficult for the next minute or so), and the sperm nucleus moves into the cytoplasm of the egg. At the

[1] Most frog embryos developed from unfertilized eggs are haploids and die. The embryos that reach adulthood are the few that undergo spontaneous chromosomal doubling to become diploid (though the two sets of chromosomes are identical).

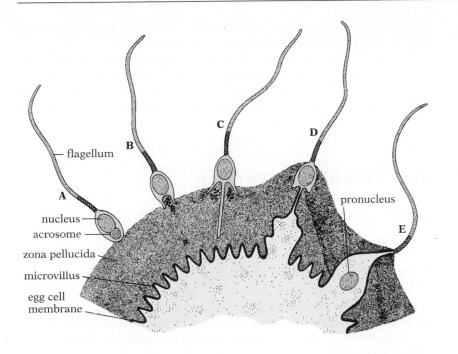

13.2 Fertilization process

(A) After the follicle-cell coat (not shown) has been broken, a sperm cell comes into contact with the zona pellucida surrounding the egg cell membrane. (B) If the recognition of species-specific markers on the two gametes is successful, the membrane of the acrosome fuses with the plasma membrane of the sperm and ruptures, releasing the acrosomal contents, which include enzymes that act on the jelly coat and on the membrane of the egg cell. (C) A tubular filament derived from the acrosome membrane pushes through the zona pellucida. (D) The tube fuses with an enlarged microvillus of the egg cell; there is now no membranous barrier between the contents of the sperm cell and the egg cell. (E) The sperm pronucleus moves into the egg cell.

same time, the structure of the egg-coat glycoprotein responsible for sperm binding is altered to prevent the attachment of additional sperm even after the membrane potential returns to normal.

As the sperm nucleus, now known as a pronucleus, moves into the egg, many changes begin. First, calcium-containing vesicles in the egg release their contents, triggering a wave of calcium release from vesicles throughout the egg, initiating the earliest steps in postfertilization development (Fig. 13.3). (The techniques for inducing development in unfertilized eggs, mentioned earlier, probably succeed by triggering calcium release.) Among the many other

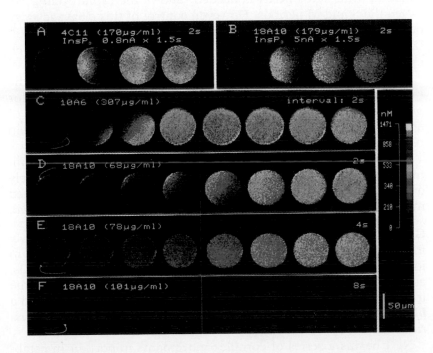

13.3 Calcium release spreading across an egg cell

The release of calcium (orange) across this hamster egg from the site of sperm fusion occurs over a period of just 2 sec.

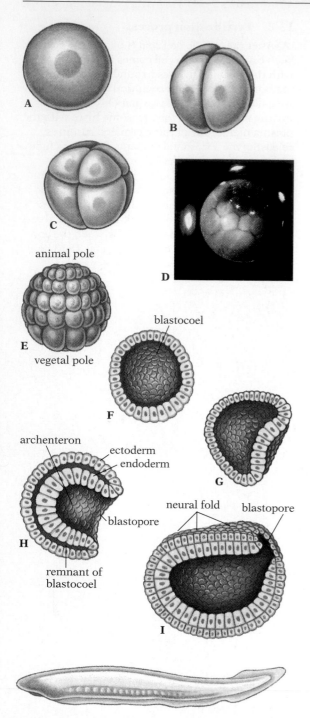

13.4 Early embryology of amphioxus

(A) Zygote. (B–D) Early cleavage stages, culminating in formation of a blastula (E). (F) Longitudinal section through a blastula, showing the blastocoel. (G–H) Longitudinal sections through an early gastrula and a late one. Notice that the invagination is at the vegetal pole of the embryo, where the cells are largest. (I) The blastopore becomes the anus of the gastrula, and a neural fold begins to form. (Subsequent steps are traced in Fig. 13.5.) An adult amphioxus is also shown at the bottom.

changes regularly seen in the egg after fusion is a striking alteration in the permeability of the plasma membrane.

True fertilization—the union of the two gamete nuclei—depends on some attraction of the sperm pronucleus by the mature egg pronucleus. The activated egg also induces the decondensation of the sperm chromosomes, and the preparations for DNA replication prior to the first mitotic division of embryonic development.

HOW EMBRYOS DEVELOP

■ EARLY CLEAVAGE AND MORPHOGENESIS

In normal development the zygote begins a rapid series of mitotic divisions soon after fertilization occurs. In most animals these early cleavages are not accompanied by cytoplasmic growth; they produce a cluster of cells that is no larger than the single egg cell from which it is derived (Fig. 13.4C). In some animals, however—notably reptiles and birds—some cytoplasmic growth does occur, as nutrients from the yolk (stored food) are consumed and transformed into cellular material.

Subdividing the ovum During this early cleavage stage of development in most eggs, the nuclei cycle very rapidly between chromosomal replication (the S period of the cell cycle, when DNA is synthesized) and mitosis (the M period); the G_1 and G_2 (gap) periods are practically absent. Such rapid cycling, in which G_1 and G_2 are skipped, is possible because the ovum already contains huge quantities of the DNA polymerase necessary for catalyzing repeated chromosomal replication, as well as most of the mRNA required for synthesis of proteins during early cleavage.

Perhaps the brevity of the interval taken up by transcription (because so little new mRNA is needed) allows the rapid cycling between the S and M periods. However, control of this cleavage stage depends largely on mRNA synthesized in the oocyte prior to fertilization, so the paternal genes have little input until later in development.

Ova are remarkably large cells. They are so large, in fact, that the ratio of nuclear to cytoplasmic material would be too low for proper control of ordinary cellular activities. The early cleavages of embryonic development, with their minimal cell growth, thus help restore a more normal ratio of nuclear to cytoplasmic material.

The one exception to the pattern outlined here is seen in mammals: the eggs are small, have little yolk, and divide slowly: they rely on RNA transcription from the outset. They probably do not need much yolk because the placenta will provide nutrients to the embryo.

Giving the embryo shape As cleavage continues, the newly formed cells (blastomeres) of many species begin to pump sodium ions into the intercellular spaces in the center of the mass of cells; as a result, water diffuses into this space and the blastomeres come to be arranged in a sphere surrounding a fluid-filled cavity called a **blastocoel** (Fig. 13.4F). An embryo at this stage is termed a **blastula**.

Next begins a series of complex movements important in estab-

13.5 Mechanism of some morphogenetic movements in cells

Contraction of microfilament-microtubular complexes (red), asymmetrically positioned in the cells, may change the shapes of the cells and produce evaginations (A), invaginations (B), or other alterations of the arrangement of cells in a developing organ. The contractile interactions are of the same type as those involved in amoeboid movement and cytokinesis.

lishing the definitive shape and pattern of the developing embryo. The establishment of shape and pattern in all organisms is called *morphogenesis* (meaning "the genesis of form"). Morphogenesis in animals always involves movements of cells in large masses during the early stages. Often it also involves changes in cell shapes, probably effected by interactions between actin microfilaments and myosin. These changes may be relatively small (Fig. 13.5), or they may be extensive. Some cells actively move. Other movements involve changes in the adhesive affinities of the cells for neighboring cells or, in the case of epithelial (surface-covering) cells, for the basement membrane on which these cells sit.

Since the pattern of cleavages and cell movement is greatly influenced by the amount of yolk in the egg, we will examine separately the patterns in animals whose eggs have small, medium, and large amounts of yolk.

Development of zygotes with little yolk In amphioxus (Fig. 13.4), a tiny marine animal whose egg has very little yolk, the movements that occur after formation of the blastula (when it is composed of about 500 cells) convert it into a two-layered structure called a *gastrula*. The process of *gastrulation* begins when a broad depression, or invagination, starts to form at a point on the surface of the blastula where the cells are somewhat larger than those on the opposite side (Fig. 13.4F). The smaller cells make up the *animal hemisphere* of the embryo, and the larger cells make up the *vegetal hemisphere* (Fig. 13.4E). It is at the pole of the vegetal hemisphere that the invagination of gastrulation occurs in amphioxus.

As gastrulation proceeds, the invaginated layer bends farther and farther inward, until eventually it comes to lie against the inside of the outer layer, nearly obliterating the old blastocoel (Fig. 13.4H). In some animals, pseudopods extend from the inner membranes of the cells where invagination began into the blastocoel. The pseudopods may adhere to the membranes of other blastula cells lining the blastocoel. In at least some cases contractions of these pseudopods assist invagination.

The resulting gastrula is a two-layered cup, with a new cavity that opens to the outside via the *blastopore*. The new cavity, called the *archenteron*, will become the cavity of the digestive tract, and the blastopore will become the anus. (Amphioxus is a chordate. In all animals except echinoderms, chordates, and two minor phyla, however, the blastopore becomes the mouth rather than the anus).

Gastrulation, as it occurs in amphioxus, first produces an embryo with two primary cell layers: an outer *ectoderm* and an inner layer. The inner layer subsequently separates in turn into two layers: the *endoderm* and the *mesoderm*; the mesoderm lies dorsally (toward the top side) between the ectoderm and the endoderm. In

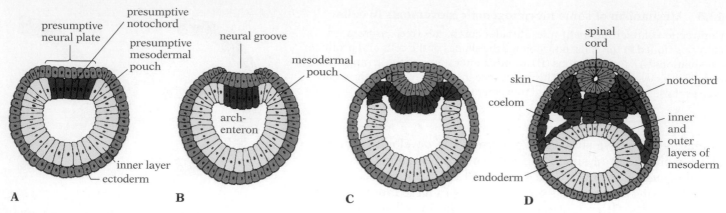

A — presumptive neural plate / presumptive notochord / presumptive mesodermal pouch / inner layer / ectoderm

B — neural groove / mesodermal pouch / arch-enteron

C — (unlabeled)

D — spinal cord / skin / coelom / endoderm / notochord / inner and outer layers of mesoderm

13.6   Neurulation in amphioxus

Cross sections through amphioxus embryos show the progressive formation of mesoderm and the neural tube. (A) When gastrulation is complete, the dorsal part of the inner layer has already been segregated as presumptive mesoderm—presumptive notochord and mesodermal pouches. (Presumptive tissue is tissue which, though not yet differentiated, is destined for a given developmental fate.) Similarly, part of the ectoderm has differentiated as presumptive neural plate. (B–C) The notochord and mesodermal pouches form as evaginations from the inner layer, and the neural plate invaginates from the ectoderm as it begins to form the spinal cord. (D) In this later embryo, called the neurula, both the spinal cord and the mesoderm are taking their definitive form. Notice that there is a cavity (the coelom) in the mesoderm.

amphioxus the mesoderm originates as pouches flanking the **notochord**, a supportive central rod found in all chordates (at least in the embryo); these pouches all pinch off from the inner layer (Fig. 13.6). The remaining part, the endoderm, pinches together to form a tube that becomes the digestive tract. Figure 13.6 shows the further development of the embryo to the **neurula** stage, so called because it incorporates the beginnings of the nervous system (discussed in a later section). All these changes are orchestrated by well-timed changes in gene activity, described in the next chapter.

In the amphioxus egg, where the distinction between animal and vegetal hemispheres is only slight (owing to the small amount of yolk in the vegetal hemisphere), the early cleavages are nearly equal. The new cells are thus of nearly the same size, and gastrulation occurs in a direct and uncomplicated manner. The eggs of many organisms, however, have far more yolk in their vegetal hemisphere, and this deposit of stored food imposes complications and limitations on cleavage and gastrulation. Generally, the more yolk an egg contains, the more cleavage tends to be restricted to the animal hemisphere and the more gastrulation departs from the pattern in amphioxus.

Development when there is more yolk   The frog egg contains far more yolk than that of amphioxus but much less than that of a bird, say. The first two cleavages, which are perpendicular to each other, cut through both the animal and vegetal poles, producing cells of roughly the same size (Fig. 13.7B). The next cleavage is equatorial (parallel to the egg's equator) and located decidedly nearer the animal pole (Fig. 13.7C); hence the four cells produced at the animal end of the egg are considerably smaller than the four at the vegetal end. From this stage onward, more cleavages occur in the animal hemisphere of the embryo than in the vegetal hemisphere as the blastula develops. As in amphioxus, there is no increase in total mass during these early cleavage stages (Fig. 13.8).

After the blastula has been formed, the frog embryo begins gastrulation. Simple invagination of the vegetal hemisphere is not mechanically feasible, because of the large mass of inert yolk. Instead, portions of the cell layer of the animal hemisphere apparently expand and move down around the yolk mass and then turn in at the edge of the yolk. This involution begins at what will be the dorsal side of the yolk mass, forming initially a crescent-shaped blasto-

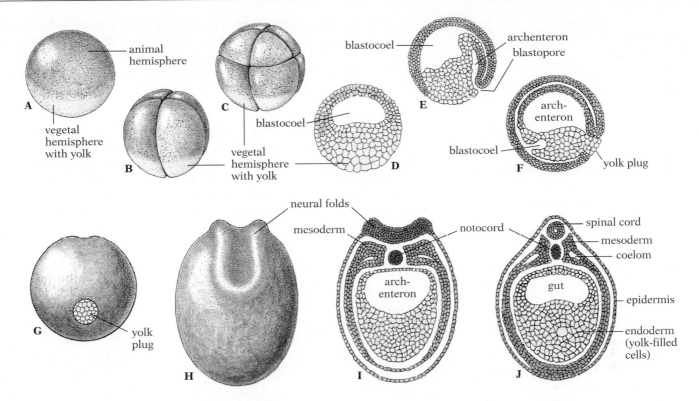

13.7 Early embryology of a frog

The large amount of yolk in the frog egg causes its pattern of gastrulation to differ from that in amphioxus. (A) Zygote. (B–C) Early cleavage stages. Note that because the first horizontal cleavage is nearer the animal pole, the cells at the vegetal pole are much larger. (D) Longitudinal section of a blastula. (E–F) Longitudinal sections of two late gastrula stages. (G) End view of late gastrula with first hint of developing neural fold at top. (H) End view of an early neurula, showing the neural folds and neural groove. (I) Cross section of a neurula after formation of mesoderm. (J) Cross section of a later embryo, showing definitive spinal cord; this stage is reached about 24 hours after fertilization.

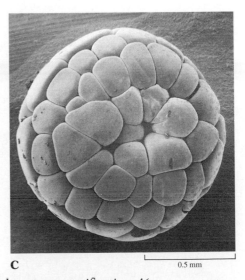

13.8 Frog egg and two of its early cleavage stages

(A) Unfertilized egg. (B) eight-cell stage. (C) 32–64-cell stage. All three scanning electron micrographs are at the same magnification: 46x. Note that there has been no overall growth in size during these cleavage stages—the 32–64-cell embryo is no larger than the egg cell.

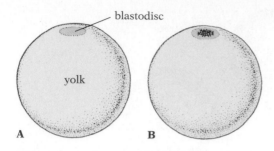

13.9   Egg and early-cleavage embryo of a chick

(A) The zygote. A small cytoplasmic disc—the blastodisc—lies on the surface of a massive yolk. (B) Early cleavage. There is no cleavage of the yolk.

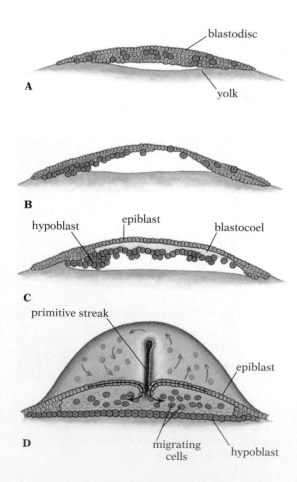

pore at the edge of the yolk. The infolding of surface cells slowly spreads to all sides of the yolk, so that the crescent blastopore is converted into a circle. Movement of the other cells around the yolk eventually encloses this material almost completely within the cavity of the archenteron.

Development with a massive yolk supply Birds' eggs contain so much yolk that the small disk of cytoplasm on the surface is dwarfed by comparison. No cleavage of the massive yolk is possible, and all cell division is restricted to the small cytoplasmic disk, or **blastodisc** (Fig. 13.9). (Note that the yolk and the small lighter-colored disk on its surface constitute the true egg cell. The albumen, or "white," of the egg lies outside the cell.) The gastrulation process is of necessity greatly modified in such eggs. Neither invagination of the vegetal hemisphere (as in amphioxus) nor involution along the edge of the yolk mass (as in a frog) can occur. Instead, the disk floating on the yolk becomes multilayered as some of the surface cells move inward into the fluid-filled space between the cellular layer and the underlying cavity. Eventually two layers are produced, an outer **epiblast** and an inner **hypoblast** (Fig. 13.10A–C). In the posterior portion of the disk, the cells of the epiblast converge toward the midline, giving rise to a clearly visible line or depression on the epiblast (Fig. 13.10D); this line, called the primitive streak, is in effect a very elongate closed blastopore. Individual cells move downward from this region. Some of these cells stay between the epiblast and the hypoblast and give rise to the mesoderm, while others intrude into the hypoblast and help form the endoderm.

Developmental fates The fates of cells in different parts of the three primary layers of vertebrates have been traced by staining them with dyes of different colors, or by marking them with radioactive carbon or other isotopes, and then following their movements. The ectoderm eventually gives rise to the outermost layer of the body—the top layer of the skin, called the **epidermis**—and to structures derived from the epidermis, such as hair, nails, eye lens, pituitary gland, and epithelium of the nasal cavity, mouth, and anal canal. The endoderm gives rise to the innermost layer of the body: the epithelial lining of the digestive tract and of other structures derived from the digestive tract, such as respiratory passages and lungs, liver, pancreas, thyroid, and bladder. The mesoderm gives rise to most of the tissues in between, such as muscle, connective tissue (including blood and bone), and the notochord. The origins

13.10   Gastrulation in the chick embryo

(A) Longitudinal section through a blastula. Larger yolk-laden cells are intermixed with smaller cells. (B) The larger cells begin to accumulate on the lower surface of the cell mass. (C) The layer of larger cells separates from the layer of smaller cells to become the hypoblast; the cavity between the two layers is the blastocoel. (D) Surface view of a gastrula. The inward movement of cells along the midline of the embryo during gastrulation (arrows) produces a clearly visible primitive streak, which is essentially a very elongate blastopore. Some epiblast cells of the primitive streak move downward to form the mesoderm; others invade the hypoblast to form the endoderm.

of various body organs and tissues are summarized in Table 13.1.

Oddly enough, nervous tissue is derived from the ectoderm. Soon after gastrulation, the ectoderm becomes divided into two components, the epidermis and the neural plate. A sheet of ectodermal cells lying along the midline of the embryo dorsal to the newly formed digestive tract and developing notochord bends inward in a process called *neurulation*, and forms a long groove extending most of the length of the embryo (Figs. 13.6A–C and 13.7H–I). The dorsal folds that border this groove then move toward each other and fuse, converting the groove into a long tube lying beneath the surface of the back. This neural tube becomes detached from the epidermis dorsal to it, and in time differentiates into the spinal cord and brain (Figs. 13.6D, 13.7J).

■ LATER EMBRYONIC DEVELOPMENT

Gastrulation and neurulation provide the organization that shapes the development of the early embryo. The late embryo must be converted from this promising beginning into a fully developed young animal ready for birth: the individual tissues and organs must be formed, and an efficient circulatory system must quickly come into being (Fig. 13.11). In a vertebrate the four limbs must develop, the elaborate system of nervous control must be established, and so on. The complexity and precision characterizing these developmental changes are staggering to contemplate: in each human arm and hand, for example, approximately 43 muscles, 29 bones, and many hundreds of nervous pathways must form. To function properly, all these components must be precisely correlated. Each muscle must have exactly the right attachments; each bone must be jointed to the next bone beyond it in a certain way; each nerve fiber must have all the proper connections with the central nervous system (CNS) and muscle cells.

Surprisingly enough, the developmental processes that produce these remarkably precise later embryonic changes are the same ones at work in the early embryo—cell division, cell growth, cell *differentiation* (as cells take on increasingly specialized roles), and morphogenetic movements. Bursts of mitotic activity in some areas and cessation of cell division in other areas alter the balance

Table 13.1 Origins of certain organs and tissues

PRIMARY LAYER	TISSUE OR ORGAN
Ectoderm	Epidermis
	Hair
	Nails
	Eye Lens
	Nervous System
	Lining of nose and mouth
Mesoderm	Muscle
	Bone
	Blood
	Notochord
	Kidneys
	Gonads
Endoderm	Lungs
	Liver
	Pancreas
	Lining of digestive tract
	Bladder

13.11 Chick embryo after 4 days of incubation

The tiny embryo lies on the surface of the yolk. It has a functional circulatory system, including a beating heart, even at this early stage of its development. Note the long branching blood vessels that run out of the embryo into the yolk; they transport nutrients to the embryo.

13.12 Human embryos at successive stages of development

(A) The 5-week embryo, 1 cm long, shows the beginnings of eyes but no distinctive face; note its mittenlike hands and feet, with no separation between the digits. (B) By contrast, the 7-week embryo, 2 cm long, has a distinct face, and there is separation between its fingers. (C) At 13 weeks the fetus is over 7 cm long and weighs about 30 g, 15 times more than it did at 7 weeks. (D) By 17 weeks the fetus is over 15 cm long and 7 times heavier than it was a month earlier; all the internal organs have formed.

A

B

C

D

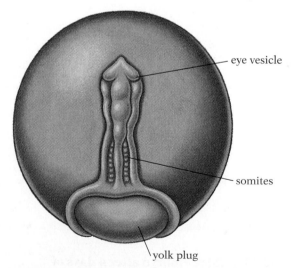

13.13 Formation of somites in the chick embryo

After the primitive streak forms, it develops into a neural groove, beginning at the anterior end and proceeding posteriorly. Somites begin to appear, each of which organizes the development of a separate segment of the body.

between the parts. Special patterns of cell growth produce important changes in size and shape. Through differentiation cells may lose particular capacities, but become more efficient at performing other functions. Folds and pouches establish the primordia of lungs and glands, of eyes and bladder. Even cell death plays an important role in the normal development of the living animal: fingers and toes, for example, become separated by the death of the cells between them (Fig. 13.12B).

One simplification in the organization of development is the repetitive organization of older embryos. Once the notochord or primitive streak is fully formed (which occurs about 1 day after fertilization in birds), regularly spaced clumps of cells called *somites* begin to appear along the dorsal midline (Fig. 13.13). In vertebrates each pair of somites gives rise to a vertebra, and helps to organize the development of the nerves, muscles, bones, and other nearby structures. Hence, the developmental program of most animals in compartmentalized. As detailed in the next chapter, specialized characteristics of segments (like those of somites that produce limbs) arise as these organizing centers "read" their anterior-posterior position in the embryo and then activate the genes appropriate for that position along the axis.

13.14 Vertebrate embryos compared at three stages of development

(A) All the embryos—whether fish, amphibian, reptile, bird, or mammal—strongly resemble one another. (B) At a later stage the fish and salamander are noticeably different, but the other embryos are still very similar; note the pharyngeal pouches in the neck region and the prominent tail. (C) By this stage, each embryo has taken on many of the features distinctive of its own species.

Fish   Salamander   Tortoise   Chick   Rabbit   Human

One fact of embryology that pushed Darwin toward the idea of evolution is that the early embryos of most vertebrates closely resemble one another. The early human embryo, with its well-developed tail and a series of pouches in the pharyngeal region (where the neck will appear), looks very much like an early fish embryo; it looks even more like an early rabbit embryo (Fig. 13.14). Only as development proceeds to later stages do the distinctive traits of each kind of vertebrate become apparent (Fig. 13.14B–C). About 100 years ago, Ernst Haeckel proposed that the development of each organism retraces in detail its evolutionary history—that "ontogeny recapitulates phylogeny." According to this hypothesis, early human embryos resemble fish because mammals are the evolutionary descendants of fish. It is certainly true that the suites of genes working together to control developmental patterns are conserved by natural selection, changing little over time as compared with superficial morphology. Nevertheless, we now know that the general developmental pattern of a species can skip the steps of ancestors, or create new structures *de novo*. There is even evidence to suggest that novel groups of organisms can arise through major and relatively rapid changes in developmental programs.

DEVELOPMENT AFTER BIRTH

The extent to which an animal has developed by the time of birth varies greatly among different species. Some young animals are entirely self-sufficient from the time they are born, and neither need

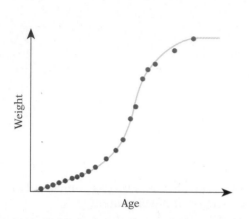

13.15 Typical S-shaped growth curve

nor receive parental care. Others are born while still at an early stage of development, and are nearly helpless and totally dependent on parental care. Newly hatched robins, for instance, are blind, almost devoid of feathers, and unable to stand.

The extent of development at birth is often (though not always) a reflection of the length of the embryonic period, which is usually correlated in animals that lay eggs with the amount of yolk the eggs contain. Among birds in particular, species that have a short incubation period for the eggs characteristically have altricial (incompletely developed) young (see Fig. 25.30, p. 724), while species that have a longer incubation period characteristically have precocial (well-developed) young. For example, robins (which are altricial) incubate their eggs for only 13 days, while precocial chickens have a 21-day incubation period. Regardless of their state of development at birth, all animals have a complete circulatory system, gastrointestinal tract, and respiratory system; nevertheless, even the most precocial young continue to undergo major developmental changes during their postembryonic life.

■ PATTERNS OF GROWTH

Though development after birth seldom involves any major morphogenetic movements, there is some cell multiplication and cell differentiation. But the most obvious characteristic of postembryonic development by far in many animals is growth in size. Usually growth begins slowly, becomes more rapid for a time, and then slows down again or stops. This pattern yields a characteristic S-shaped growth curve (Fig. 13.15). The slope of the curve is different for different species, depending on whether they grow very rapidly for a shorter time or more slowly for a longer time. The shape of the curve is seldom as smooth as it appears in a generalized growth curve because many factors can affect the rate of growth. For example, in most mammals growth slows down for a while immediately after weaning, and it often varies greatly during puberty (Fig. 13.16). An especially marked departure from the smooth generalized curve is seen in the growth of arthropods, a group that includes the insects (Fig. 13.17). These animals can undergo only limited growth between molts, because the hard ex-

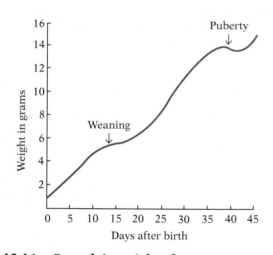

13.16 Growth in weight of a mouse

The rate of growth is slower at weaning and at puberty.

13.17 Growth in weight of an insect

The growth spurts of the water boatman, an aquatic insect, occur at the time of molt, when the old exoskeleton has been shed and the new one has not yet fully hardened.

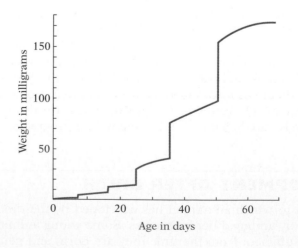

oskeleton that encases their bodies can be stretched only slightly. At each molt, however, there is a sharp burst of growth during the short period after the old exoskeleton has been shed and before the new one has hardened.

Growth does not occur at the same rate or at the same time in all parts of the body. It is obvious to anyone that the differences between a newborn baby and an adult human are differences not only in overall size but also in body proportions. The head is far larger in relation to the rest of its body in a young child than in an adult, while a child's legs are disproportionately short. Normal adult proportions arise because the various parts of the body grow at quite different rates or stop growing at different times (Fig. 13.18).

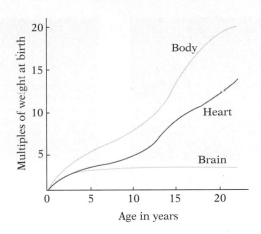

13.18   Differences in relative growth of body, heart, and brain of a human

■ HOW LARVAE DEVELOP

Growth in size is not always the principal mechanism of postembryonic development. Many aquatic animals, particularly those leading sessile (nonmobile) lives as adults, go through **larval** stages that bear little resemblance to the adult (Fig. 13.19). The series of sometimes drastic developmental changes that convert an immature animal into the adult form is called **metamorphosis**. It often involves extensive cell division and differentiation, and sometimes even morphogenetic movement.

In many aquatic animals dispersal of the species depends on the larval stage; the tiny larvae either swim or are passively carried by currents to new locations, where they settle down and undergo metamorphosis into sedentary adults. In other species where the adult is not sedentary, the adaptive significance of the larval stage seems less a matter of dispersal than of exploiting alternative food sources. For example, frog larvae (tadpoles) feed primarily on microscopic plant material, while adults are carnivorous and take fairly large prey.

Though a larval stage occurs in the life history of many aquatic animals, probably the most familiar larvae are those of terrestrial insects, including butterflies and moths. The butterfly or moth larva is called a caterpillar. In the course of their larval lives, these insects molt several times and grow much larger, but this growth does not bring them any closer to adult appearance; they simply become larger larvae (Fig. 13.20B–D). Finally, after they have completed their larval development, they enter an inactive stage called the **pupa**, during which they are usually enclosed in a case or co-

ciliated band — mouth — anus

Ciliated larva   Bipinnaria larva   Brachiolaria larva   Adult

13.19   Three larval stages and adult of the sea star *Asterias vulgaris*

The gastrula develops into the ciliated larva, which changes into the bipinnaria larva, which changes into the brachiolaria larva, which metamorphoses into the characteristically shaped sea star with five arms.

A   Monarch butterfly egg

B   First-stage larva eating its own egg shell

C   Later-stage larva eating its own molted skin

D   Full-grown larva going into pupation

E   Early chrysalis

F   Late-stage chrysalis

G   Inflation of wings by newly emerged butterfly

13.20   Developmental stages of a monarch butterfly, an insect with complete metamorphosis

(A–C) From the egg hatches a small larva which goes through a series of molts, growing larger at each. (D–E) Eventually the full-grown larva attaches to a plant and goes into pupation, forming a case around itself; in butterflies the encased pupa is called a chrysalis. (F) During pupation most of the larval tissues are broken down and new adult tissues are formed; in the monarch some of the adult structures can be seen through the case of the late chrysalis. (G) The newly formed adult emerges from the pupal case and rests while it inflates its wings. (H) It then flies off to feed at flowers before mating and laying eggs (almost always on milkweed) to start the cycle over again.

H   Adult butterfly at flower

coon (Fig. 13.20E). During the pupal stage most of the old larval tissues are destroyed, and new tissues and organs develop from small groups of cells called **imaginal discs,** which were present in the larva but never underwent terminal development. The adult that emerges from the pupa (Fig. 13.20G–H) is therefore radically different from the larva; it is the product of an entirely different developmental program—one could almost consider it a new organism built from the raw materials of the larval body.

Insects with a pupal stage and the type of development just described are said to undergo **complete metamorphosis**. The sharp distinction in form between the larval and adult stages in such insects has meant evolution in two markedly different directions; in general, the larva is more specialized for feeding and growth, while the adult is more specialized for active dispersal and reproduction.

Many insects, such as grasshoppers, instead undergo **gradual metamorphosis** (Fig. 13.21). Young larvae of such insects resemble the adults, except that their body proportions are different (the wings and reproductive organs, especially, are poorly developed). They go through a series of molts during which their form gradually changes, becoming more and more like that of the adult, largely as a result of a differential growth of the various body parts. They have no pupal stage and experience no wholesale destruction of the immature tissues.

■ ARE AGING AND DEATH PROGRAMMED?

The adult organism is not a static entity; it continues to change, and hence to develop, until death brings the developmental process to an end. The term *aging* is applied to the complex of developmental changes that lead, with the passage of time, to the deterioration of the mature organism (Table 13.2) and ultimately to its death. Modern scientific progress and improved medical techniques have

13.21 Gradual metamorphosis of a grasshopper

The insect that emerges from the egg (top) goes through several nymphal stages that bring it gradually closer to the adult form (bottom). The increase in size is more dramatic than it appears: note that the scale in this drawing changes gradually by 10-fold from egg to adult.

Table 13.2 Average decline in a human male from ages 30 to 75

CHARACTERISTIC	PERCENT DECLINE
Weight of brain	44
Number of nerve cells in spinal cord	37
Velocity of nerve impulse	10
Number of taste buds	64
Blood supply to brain	20
Output of heart at rest	30
Speed of return to normal pH of blood after displacement	83
Number of filtering subunits in kidney	44
Filtration rate of kidney	31
Capacity of lungs	44
Maximum O_2 uptake during exercise	60

greatly increased our ability to protect ourselves against disease, starvation, and the destructive forces of the physical environment. More people are living to an advanced age. As the life expectancy increases and the proportion of the population in the upper age brackets rises, the changes associated with aging become more obvious and more important to all of us.

The process of aging seems to be correlated with specialization of cells for one or a few highly specific functions. Cells that remain relatively unspecialized and continue to divide do not age as rapidly (if at all) as cells that have lost the capacity to divide. Cancer cells divide continually, and so can live indefinitely. Bacteria and some other unicellular organisms do not age, for any cell that is not destroyed eventually divides to produce two young cells; division is thus a process of rejuvenation. Animals that continue to grow throughout their lives (many fish and reptiles, for example) show fewer obvious symptoms of aging than, say, mammals and birds, which cease growing soon after they reach maturity. Within a single species, individuals whose period of growth and development is slowed and extended by a very limited diet are usually older than normal before they begin to show signs of aging.

Clearly, then, the aging and death of individual cells and the aging and death of the multicellular organism as a whole are two different things. Paradoxical as it may seem at first, cell death is an essential part of life: the death of individual cells, as noted previously, plays an essential role in the development of the animal embryo and in the complete metamorphosis of some insects. Early death of individual red blood corpuscles and epidermal cells is entirely normal even in a young healthy mammal. Aging of the whole organism, therefore, is a matter not simply of the death of its cells, but of the deterioration and death of those cells and tissues that cannot be replaced.

Only a few of the factors that correlate with aging are understood:

1 When cells in most types of tissue die as a result of disease or injury, no new cells are formed to replace the ones lost. Wound healing in such cases involves the growth of connective (scar) tissue, which serves as a patching material but cannot function like the original cells. As more and more irreplaceable cells die, the increased burden placed on the remaining cells of that type may contribute to their aging.

2 The changing hormonal balance, such as that caused by a drop in the level of sex hormones, may disturb the function of a variety of tissues.

3 As cells become older, they tend to accumulate some metabolic wastes, and these wastes—particularly highly reactive products like hydrogen peroxide—may contribute to the eventual deterioration of the cells. At the same time, aging cells produce ever smaller amounts of the antioxidant molecules needed to detoxify these dangerous chemicals.

These factors are not really explanations but symptoms. The real question is why these changes occur. Perhaps somatic cells slowly cease to function and eventually die as a result of damage by radiation (particularly X rays and cosmic rays). In the laboratory, however, radiation damage is greatest in actively dividing cells—the cells that age most slowly. Furthermore, the amount of radiation damage should be proportional to the chronological age of the cells, but aging is a function of physiological age—a 5-year-old rat is physiologically very old indeed, and its tissues show pronounced symptoms of aging, whereas 5-year-old tissues in a human being are not yet even mature.

The changes characteristic of aging may be programmed in the genes just like the earlier developmental changes. Though extrinsic environmental factors doubtless influence aging, they do so only by speeding up or slowing down processes that would occur anyway. These processes may involve a decline in the production of important enzymes or an altered chemical balance or physical structure, with an ensuing loss of ability to perform certain functions. They could also result in development of autoimmune reactions (allergies against parts of the organism's own body) that result in destruction of essential tissues; or they may involve increased rupture of lysosomes and release of destructive hydrolytic enzymes within cells.

A particularly interesting proposal is a variation of two hypotheses already mentioned: that aging is a developmentally programmed termination of life that results from an organism's failure to repair somatic-cell mutations or other damage. According to this hypothesis, the different rates of aging in various species reflect different inherited capacities for DNA repair and the production of antioxidant molecules. Long-lived species are the ones with high levels of DNA repair enzymes, and this pattern holds true for the concentration of antioxidants as well (Fig 13.22). Conversely, species with short life spans and early aging invest little in either DNA repair or cellular detoxification. This does not appear to be a chance relationship: genetically engineered *Drosophila* having 50% higher concentrations of the enzymes that detoxify hydrogen peroxide and oxygen radicals live far longer than their unengineered

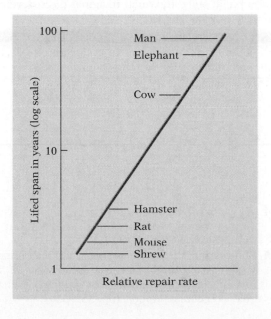

13.22 Activity of DNA repair systems in various animals

Longer-lived animals invest more energy and material in repair enzymes.

counterparts. If, as seems indicated by such data, aging is a genetically programmed trait (and a critical gene from carp has been isolated that significantly extends the life span of mice), it must have originated and been maintained because of a selective advantage over a longer average life span. As yet, however, there is no general explanation of why allowing itself to age and die should be to an individual's advantage. The opportunity to synthesize concepts and data from molecular biology and evolution is one of the most exciting challenges of modern biology.

CHAPTER SUMMARY

THE EVENTS OF FERTILIZATION

Fusion of egg and sperm creates a zygote. Fusion triggers calcium release in the egg, initiating the first steps of embryonic development. (p. 338)

HOW EMBRYOS DEVELOP

EARLY CLEAVAGE AND MORPHOGENESIS The zygote is divided into many smaller cells by cell division: it then forms a hollow blastula. Morphogenic movements cause part of the blastula to fold into itself, forming a two-layer gastrula. The inner layer becomes endoderm, which lines the gastrointestinal and respiratory tracts; the outer layer becomes ectoderm, which forms the skin, nervous system, and other structures; a new third layer, the mesoderm, forms from ingrowths of the endoderm, and develops into muscle and bone. (p. 340)

LATER EMBRYONIC DEVELOPMENT Organs and other structures are formed by progressive differentiation and morphogenesis. (p. 345)

DEVELOPMENT AFTER BIRTH

PATTERNS OF GROWTH Development after birth involves mostly differential growth of organs and structures. (p. 348)

HOW LARVAE DEVELOP Many animals have a larval stage that may be very different from the adult. Many insects go through a series of larval molts followed a pupal stage in which imaginal discs develop to produce the adult (complete metamorphosis). (p. 349)

ARE AGING AND DEATH PROGRAMMED? Aging is species-specific, and correlates with the organism's investment in repair and its capacity for continued growth and differentiation; thus aging seems to be an inborn property. (p. 351)

STUDY QUESTIONS

1 What evolutionary pressures and aspects of life history are likely to favor eggs with small versus large quantities of yolk? (p. 340)

2 What would be the consequence of having several sperm fertilize an egg?

3 Why do you suppose sperm are small and streamlined? Why should selection not have favored male gametes that bring large quantities of cytoplasmic resources for the zy-gote to use?

4 Why might the nervous system have evolved from ectodermal tissue? What does this suggest about the course of that evolution? (pp. 344–345)

5 Summarize the case for programmed aging and death. Explain what might be the adaptive value of such a system in a species whose life history you know well enough to be able to speculate about. (pp. 351–354)

SUGGESTED READING

PARTRIDGE, L., AND N. H. BARTON 1993. Optimality, mutation, and the evolution of aging, *Nature* 362, 305–311. *An evolutionary analysis of aging.*

SHARON, N., AND H. LIS, 1993. Carbohydrates in cell reognition, *Scientific American* 268 (1). *On the cell-surface markers that allow cell-to-cell recognition during development.*

WASSARMAN, P. M., 1988. Fertilization in mammals, *Scientific American* 259 (6). *On how eggs manage to be fertilized by only one sperm.*

CHAPTER 14

MECHANISMS OF ANIMAL DEVELOPMENT

A single fertilized egg cell contains all the genetic information necessary to orchestrate the precise series of changes from a blastula to a gastrula, then a neurula, and finally to a mature multicellular organism. The intricacies of eucaryotic gene control, discussed in Chapter 11, can account for the many changes in gene expression that successful development requires. Throughout development, cells need to know their location in the organism, what their immediate neighbors are doing, and the organism's developmental age. The major sources of all this information are chemical signals that modify each cell's pattern of gene expression.

This chapter will examine how genes control development: the differentiation of cells into ever more specific classes, the movement of cells at particular times, the formation of morphological patterns to create organs and limbs, and the repeating cycles of growth. Much remains to be learned about development, but the essential molecular mechanisms are becoming clear. In particular, the process of *induction*, by which chemical signals alter the expression of a cell's DNA to specify or determine its developmental fate, is yielding to analysis. We will see in broad terms how, in the process of cell differentiation, each cell type follows a different avenue of determination leading to a distinctive cellular biochemistry and morphology.

HOW POLARITY ORGANIZES DEVELOPMENT

The early steps in the development of an embryo establish the basic body plan, a kind of coordinate system by which the blastula will orient during gastrulation and further development will proceed. The egg, as seen in Chapter 13, is often polarized into animal and vegetal hemispheres by the concentration of yolk (stored food) at the vegetal end. There are also crucial differences in the concentration of various proteins and messenger RNAs (mRNAs) between different parts of eggs; this molecular polarization may underlie, at least in part, the organization of the developing embryo. In frogs, for instance, the animal pole becomes the anterior part of the organism (the head), while the vegetal pole becomes the posterior.

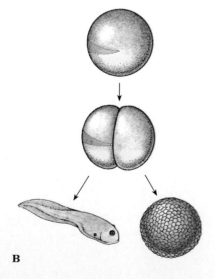

14.1 Polarity of the fertilized frog egg

After a sperm fuses with the egg, a gray crescent develops opposite the point of fusion (A). The crescent, in turn, helps define the anterior-posterior and dorsal-ventral axes of the tadpole (B). The first cleavage divides the embryo's left side from the right. Not all parts of the tadpole arise from corresponding sites on the egg: the anterior half of the egg contributes all of the ectoderm; cells developing from the posterior half fold back into the anterior and give rise to mesoderm.

The site of sperm fusion can also provide a polarizing cue: as soon as the egg cell of a leopard frog fuses with a sperm cell, some of the contents of the egg shift position, and a crescent-shaped grayish area appears on the egg roughly opposite the point where the sperm entered (Fig. 14.1); these two points define the embryo's dorsal-ventral axis. This **gray crescent** is the site where gastrulation will later begin. The first cleavage of the frog zygote normally passes through the gray crescent, so that each daughter cell receives half (Fig. 14.2A). If these two cells are separated, each will develop into a normal tadpole, since each cell retains the information specifying both axes. If the plane of the first cleavage is experimentally made to shift to the side of the gray crescent, however, the result of separating the daughter cells will be very different; the cell that contains the gray crescent will develop into a normal tadpole, but the other cell will form only an unorganized mass of cells (Fig. 14.2B). In other words, the way in which the material of the egg, especially the material near the gray crescent, is distributed is critical to the development of the embryo.

14.2 Importance of the gray crescent in the early development of a frog embryo

(A) If the two cells produced by a normal first cleavage, which passes through the gray crescent, are separated, each develops into a normal tadpole. (B) If the first cleavage is experimentally oriented so that it does not pass through the gray crescent, and the two daughter cells are separated, the cell with the crescent develops normally, but the other cell develops into an unorganized cellular mass.

14.3 Cleavage of the fertilized egg of the sea snail *Ilyanassa*

(A) At one pole of the zygote is a region of clear cytoplasm (yellow). (B) Just before the first cleavage, this polar cytoplasm moves into a large protuberance, the polar lobe. (C) The first cleavage partitions the zygote in such a way that the entire polar lobe goes to one of the daughter cells. (D) The lobe recedes during interphase, but it will form again prior to the next division sequence. Only the cell that receives the polar-lobe material can give rise to the specific set of external structures seen in normal larvae.

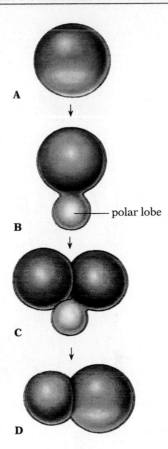

After the first cleavage of the frog zygote, both of the new cells are **totipotent**: they have the full developmental potential of the original zygote (since the polarized cytoplasmic substances are equally distributed in each). In this respect their development is characteristic of sea stars and many of their relatives, and of most vertebrates, including humans. Mammalian blastomeres generally remain totipotent until at least the eight-cell stage. In fact, eight-cell embryos of two different strains of mice can be mixed to create a 16-cell **chimera** (after the Chimera, a monster in Greek mythology composed of parts from several different animals) that develops normally. This technique enables researchers to study how cells from one strain interact with those of another during development. Only animals with this kind of development can give birth to identical twins.[1]

In some other groups of animals, such as molluscs and segmented worms, the normal first cleavage partitions critical cytoplasmic constituents asymmetrically; as a result, when the daughter cells are separated, they do not have equivalent developmental potential. For example, in some molluscs, a protuberance called the polar lobe develops on the fertilized egg cell just before the first cleavage occurs. The plane of cleavage is oriented in such a way that one of the two daughter cells receives the entire polar lobe (Fig. 14.3). If the two daughter cells are separated, the one with the lobe material (which was drawn back into the main body of the cell soon after division was accomplished) will form a normal embryo possessing two prominent structures—a so-called apical organ and the posttrochal bristles—that are seen on normal larvae; the one with no lobe material forms an aberrant embryo lacking these structures. Something in the polar-lobe material must be essential for formation of the apical organ and the bristles.

In sea urchins (relatives of sea stars), cytoplasmic determinants promoting various kinds of differentiation are distributed along the animal-vegetal axis of the embryo. If the animal and vegetal halves of the embryo are separated (along the plane of the third cleavage), the animal half will give rise to an abnormal blastulalike larva with overdeveloped cilia, and the vegetal half will often give rise to a different type of abnormal larva with overdeveloped digestive organs (Fig. 14.4B). By contrast, if the eight-cell embryo is di-

---

[1] Identical twins develop from the same zygote, usually as a result of a double gastrulation event. Nonidentical (fraternal) twins develop from separate zygotes when two egg cells are released from the ovaries at the same time and are fertilized by different sperm cells.

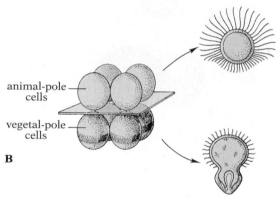

animal-pole cells

vegetal-pole cells

B

14.4 Experimental separation of cells after the third cleavage in the embryo of a sea urchin

(A) If the embryo is partitioned meridionally, so that each half receives both animal-pole and vegetal-pole cells, each half develops into a normal larva. (B) If the embryo is partitioned equatorially, so that one half receives only animal-pole cells and the other half only vegetal-pole cells, the animal half develops into an abnormal blastulalike larva with overdeveloped cilia, and the vegetal half develops into an abnormal larva with an overdeveloped digestive cavity.

vided along its animal-vegetal axis, each half will develop into a small but normal larva (Fig. 14.4A). In an intact embryo each of these halves would have developed into only half of the normal larva; evidently, as in the frog, interactions between the different regions of the embryo normally act to modify the course of differentiation of neighboring cells, integrating them into a single properly structured and proportioned larva.

Clearly, then, some unevenly distributed cytoplasmic substances must play a prominent role during early embryonic development. Some substances function by activating some genes and repressing others. Others are maternal mRNAs produced by genes expressed during oogenesis, as well as the products resulting from translation of such RNAs. The unequal distribution of these products of previous gene activity along the animal-vegetal or another axis leads to the expression of different traits in different parts of the embryo. Some polarization is present at oogenesis; additional polarization appears early in embryonic development (as early as the first cleavage in some organisms).

INDUCING DEVELOPMENTAL EVENTS

■ HOW GASTRULATION AND NEURULATION ARE INDUCED

The polarity of the embryo is first revealed when gastrulation takes place. The site of gastrulation (the blastopore of vertebrates) becomes the anus in vertebrates and echinoderms; in all other animals the gastrulation site becomes the mouth. In addition to knowing where to initiate gastrulation, the embryo also needs to know *when* the appropriate moment for gastrulation has arrived. This is not simply a matter of counting cell divisions until the right number is reached. In the frog embryo these developmental events are cued by the concentration of specific molecules, a kind of molecular developmental clock. Researchers have inferred these cues by studying a variety of so-called heterochronic mutations, which cause developing cells to misread the molecular clock, so that structures are produced at the wrong times.

Once the time has arrived for gastrulation in a frog embryo, cells in the region of the gray crescent begin the process of involution. In the early gastrula the cells derived from the gray-crescent portion of the egg become the **dorsal lip** of the blastopore. These cells move inward and form parts of the roof of the mouth and the

14.5 Spemann's experimental transplantation of the gray crescent

(A) When Spemann and Mangold transplanted dorsal-lip material from a light-colored salamander embryo to a dark-colored one, the blastula with two dorsal lips proceeded to form an embryo with two gastrulation zones—one at the original dorsal lip of the blastopore, and one at the transplanted dorsal lip. A double larva of mostly dark-colored tissue was the result. (B) A similar transplantation experiment in the frog *Xenopus laevis* produced an embryo with two body axes, including two heads and a second spinal cord.

chordamesoderm, which is at first located in the roof of the newly forming archenteron (see Fig. 13.7E, p. 343). The chordamesoderm soon detaches from the roof of the archenteron to form the notochord and other mesodermal structures (see Fig. 13.7I–J). The chordamesoderm also induces the ectodermal tissue lying over it to fold inward during neurulation, forming the neural tube, which differentiates into the brain and spinal cord.

In 1924 Hans Spemann, who had previously demonstrated the importance of the gray crescent in early cleavage, and Hilde Mangold discovered another critical developmental structure. They transplanted the dorsal lip from its normal position on a light-colored salamander embryo to the belly region of a darker-colored embryo. After the operation, gastrulation occurred in two places on the recipient embryo—at the site of its own blastoporal lip and at the site of the implanted lip. Eventually two nervous systems were formed, and sometimes even two nearly complete embryos developed, joined together ventrally (Fig. 14.5). Most of the tissue in both embryos was dark-colored, an indication that the transplanted blastoporal lip had altered the course of development of cells derived from the host. Transplanting other tissues had little effect.

The dorsal lip of the blastopore plays a crucial role even after it takes the lead in gastrulation. Signals from those prospective chordamesodermal cells, as they make their way anteriorly under the dorsal ectoderm, induce the formation of the neural plate, which in turn is important in establishing the longitudinal axis of the embryo and in inducing formation of other structures.

■ HOW ORGANS ARE INDUCED

Some of the most definitive studies on embryonic induction were begun in 1905 by Warren H. Lewis. Lewis worked on the development of the eye lens in frogs. In normal development the eyes form as lateral outpockets (optic vesicles) from developing brain tissue. When one of these outpockets comes into contact with the epidermis on the side of the head, the contacted epidermal cells form a thick plate of cells that sinks inward, becomes detached from the epidermis, and eventually differentiates into the eye lens (Figs. 14.6, 14.7). Lewis cut the connection between one of the optic vesicles and the brain before the vesicle came into contact with the epidermis. He then moved the vesicle posteriorly into the trunk of

A

B

A

B

C

0.1 mm

14.6   Development of optic vesicles and their induction of lenses in the vertebrate eye

(A–B) The proximity of the optic vesicle helps induce the nearby cells of the epidermis to fold inward and form a lens. (C–D) The infolding presumptive lens tissue, in turn, helps mold the optic vesicle into a two-layered structure called the optic cup. The cup cells then differentiate, the layer adjacent to the lens forming visual receptor cells and nerve cells.

14.7   Development of a mammalian eye

Photographs of mammalian eye development over a 4-day period. (A) The contact with the tip of the optic vesicle causes the epidermis to invaginate. (B) The epidermal region differentiates to form the lens vesicle, while the optic vesicle develops into the optic cup. (C) The optic cup ultimately becomes the retina—the tissue at the rear of the eye that contains light receptors.

the embryo. Despite its lack of connection to the brain, the optic vesicle continued to develop, and when it came into contact with the epidermis of the trunk, that epidermis differentiated into a crude lens. The epidermis on the head that would normally have formed a lens usually failed to do so. Clearly, the differentiation of epidermal tissue into lens tissue depends on some inductive stimulus from the underlying optic vesicle. Later work showed that the ability to form a lens is triggered by a substance released from cells in the ectoderm dorsal to the usual site of lens formation. This chemical spreads over a limited area of the ectoderm, priming these cells for a brief period to form a lens at the site specified by the arrival of the developing optic vesicle.

The regulation is not all one-way; the lens, once it begins to form, also influences the further development of the optic vesicle. If epidermis from a species with normally small eyes is transplanted to the sides of the head of a species with large eyes, the eyes that are formed do not have a large optic cup (formed from the optic vesicle) and a small lens. Instead, both the optic cup and the lens are intermediate in size and correctly proportioned to each other. Obviously, each influences the other as they develop together.

Of the many chemicals that can act as intercellular inducers in embryonic development, some play an instructive role, others play a more permissive role. The ***instructive inducers*** restrict the developmental potential of the target cell and thus help determine the course of differentiation. The ***permissive inducers*** amplify potentialities already expressed. For example, a rudimentary organ like the pancreas may form fully committed cells during embryonic development, but it will not complete its development and become functional until acted on by a permissive inducer.

Instructive inducers do not give their target cells specific instructions about the design of the tissues or organs they are to form; they instruct only in the sense that, through repression of some gene expression and induction of expression in others, they tell the cells what part of their genetic endowment they are to use. For example, when Spemann and Oscar E. Schotté transplanted ectoderm from the flank of a frog embryo to the mouth region of a salamander embryo, they found that, though induced by salamander endoderm, the transplanted frog tissue formed the typical horny jaws of a frog. The constellation of genes for producing a jaw was activated by inducers from the adjacent tissue in the salamander, but the induced genes, being from frog ectoderm, encoded the morphology of a frog jaw.

■ HORMONES AS INDUCERS

The body uses many internal chemical signals, called ***hormones***, to pass messages over relatively long distances, often through the circulatory system; testosterone (a male sex hormone) and adrenalin (an alerting hormone) are examples. Hormone action is discussed in Chapter 33.

In vertebrates, hormones play a predominantly permissive role in development, helping differentiated tissue assume its full character. The essential interplay between instructive inducers and permissive hormones is clear in the gonads of amphibians. The ***gonads*** are the primary sex organs—the sperm-producing ***testes*** of the male, and the egg-producing ***ovaries*** of the female. Two distinct kinds of cells are set aside early to become gonads in amphibian embryos: peripherally located cortical cells and centrally located medullary cells (Fig. 14.8). The cortical cells have the potential for forming ovarian tissue, and the medullary cells for form-

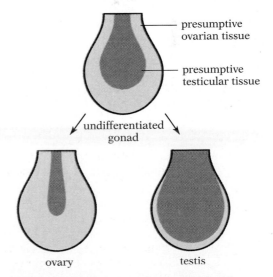

presumptive ovarian tissue

presumptive testicular tissue

undifferentiated gonad

ovary

testis

14.8 Development of ovary or testis from the undifferentiated gonad of an amphibian
Depending on the hormonal condition of the embryo, one or the other of the two types of tissue in the undifferentiated gonad gains developmental ascendancy and the other type is repressed.

ing testicular tissue. Only when sex hormone acts on the sexually uncommitted embryonic gonad does one or the other set of cells begin to express its potential.

Similarly, the same embryonic primordia give rise to the accessory sexual organs of both sexes in humans (Fig. 14.9). Whether these primordia form male or female structures depends on whether or not male sex hormone is present in the embryo at the critical stages of embryonic development. In birds, the situation is reversed; the sexual organs will follow a masculine developmental course unless female hormone is present at the critical embryonic stages.

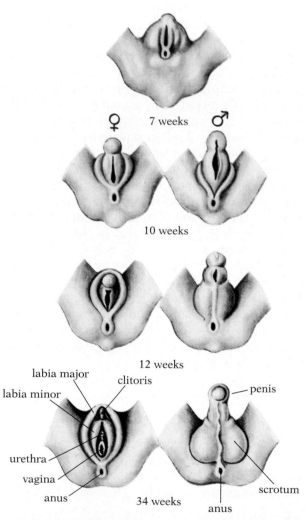

14.9 Development of the external genitalia of human beings

At 7 weeks the genitalia of male and female fetuses are virtually identical. At 10 weeks the penis of the male is slightly larger than the clitoris and labia minor, which form from the same primordium in the female. At 12 weeks these differences are more pronounced, and the male scrotum has formed from the tissue that becomes the labia major in the female. At 34 weeks the distinctive features of the genitalia of the two sexes are fully apparent. It is largely the concentration of male sex hormones like testosterone that determines which of these developmental pathways will be followed.

■ HOW MIGRATING CELLS KNOW WHERE TO GO

One of the most remarkable phenomena of development is the organized movement of cells and the folding of tissues into new shapes (morphogenesis). Dorsal-lip cells of amphibian gastrulas crawl in the right direction and stop at the correct spot; cells of the optic vesicle "know" when to begin their evagination outward toward the ectoderm, when they have reached their target, and then, how to create a suitable cup. What molecules provide directional information to migrating cells, and how are the coordinates read?

At least some orientation is to gradients of diffusible chemicals—molecular beacons—produced by landmark cells. Tissue-dissociation experiments also suggest another strategy: if cells from two different developing organs—the liver and spleen, say—are separated from one another and the two groups of cells mixed together, they slowly but accurately sort themselves out into two clumps. Each cell extends several pseudopods, touching and adhering to neighboring cells, and then pulls itself toward the most similar cells.

Each cell membrane is richly supplied with one or (usually) several kinds of **cell-adhesion molecules** (CAMs). Unlike the typical binder-receptor systems we have dealt with up to now, in which two different molecules—glucose and the glucose receptor, for instance—bind to one another, some kinds of CAMs attach themselves exclusively to other molecules of the same sort of CAM (Fig. 14.10), a phenomenon called **homophilic binding**. The proportion of the several different classes of CAMs on the membranes of various liver cells is fairly consistent and is different from the corresponding ratio on spleen cells. As a result, liver cells stick only slightly to spleen cells, but more firmly to their own kind. It is possible that when a migrating liver cell attempts to withdraw its pseudopods from a spleen cell, the pseudopod loses its hold, but when it tries to withdraw from another liver cell, it remains stuck and instead drags itself toward the other liver cell. This differential affinity would permit the two classes of organ cells to sort themselves out.

The same thing may occur when the dorsal-lip cells begin to move. Their developmental clock tells them it is time for gastrulation, which may cause a new set of CAMs to be mounted in the membrane. General chemical gradients may provide some initial orientation, and the presence of other cells or structures to attach to must constrain the choice of routes available to migrating cells. However, many researchers believe CAMs to be the decisive factor in determining which among a cell's alternative routes to follow once the cell begins to grow. Pseudopods begin to form and seek attachments, and would find the strongest interactions in the anterior-dorsal direction. Pulled along by the discovery of even better

14.10 Homophilic binding of cell-adhesion molecules

The complementary structure of the arms of CAMs allows each type to bind to others of the same class. Several dozen kinds of CAM are thought to exist, though fewer than 10 have yet been characterized. Heterophilic CAMs (which bind to CAMs not of the same class) also exist.

14.11 A model of cell migration

The CAM-based migration hypothesis: (A) A cell induced to begin searching for a better CAM match extends pseudopods and attaches more or less strongly to other cells. (B) When the pseudopods are periodically retracted, the cell is pulled in the direction of the best match. (C) After many such steps, the cell reaches a point from which there is no available improvement in CAM matching.

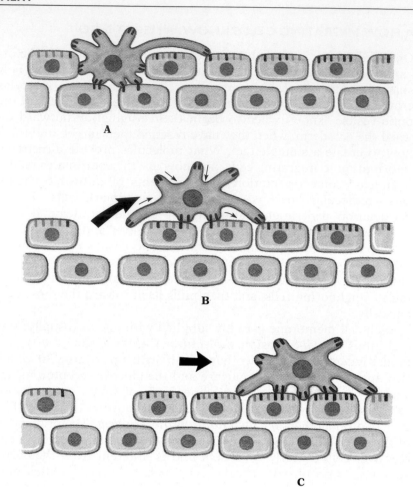

matches farther forward, the dorsal lip cells would drag themselves (and the posterior surface of the blastula) forward until their CAMs found the best matches available; when no direction offers the prospect of any stronger attachments, migration would stop (Fig. 14.11).

Analyses of the structure of CAMs have yielded a major surprise: they are closely related to specialized proteins of the immune system. Since CAMs are also found in organisms that lack immune systems such as the fruit fly *Drosophila* and lower vertebrates, this developmental recognition system probably provided the basis for the evolution of the molecules of the immune system of mammals (described in Chapter 15).

PATTERN FORMATION

Once the embryo has undergone gastrulation and neurulation, different organs begin to appear up and down its length. The strategy of further steps in development, from insects to humans, is one of subdivision of the embryo into a series of domains, followed by largely independent development of each domain.

14.12 Somite formation

(A) The dorsal portion of embryonic mesoderm differentiates into a series of cell blocks, called somites, which give rise to vertebrae and dermis, as well as (depending on the somite) ribs and muscles of the limbs. (B) A photograph of an amphibian larva in frontal section with somites clearly visible.

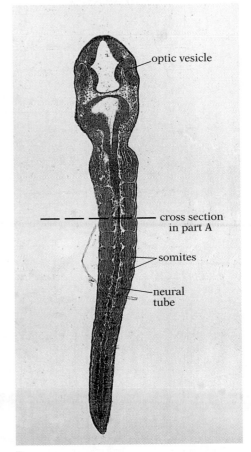

■ WHY EMBRYOS ARE SEGMENTED

Vertebrate somites Once the spinal cord has fully formed in a vertebrate embryo, the most dorsal region of mesoderm begins to differentiate into blocks of tissue called somites (Fig. 14.12). This reorganization of the dorsal mesoderm may be based both on an anterior-posterior chemical gradient, which triggers formation of the somites, and on local interactions between somites, which assure proper grouping and segregation.

Each somite goes on to produce a vertebra, the pair of ribs (if any) associated with it, the muscles unique to that vertebra, and the dermis (the layer of cells just below the epidermis). Each somite "knows" which set of bones, nerves, and muscles to construct on the basis of its anterior-posterior location, and becomes *determined*—that is, committed to that fate. Some embryologists believe that positions along axes in vertebrates are interpreted by the genes of somites and other units by measuring the concentration of one or more chemical *morphogens*, which are secreted from a specific point in the embryo and diffuse away.

Morphogens and segmentation in *Drosophila* Morphogens are no mere hypothesis in invertebrates. The genetic and molecular bases of subdivision, and the use of morphogens for position finding, are well understood in *Drosophila*, where the segmentation in both the larva and adult is clearly visible externally. (In the adult, there are three segments in the head, three in the thorax—the middle section, to which the legs and wings are attached—and eight in the abdomen.) Two chemical axes are established in the blastoderm—one anterior-posterior, the other dorsal-ventral—which determine the overall organization of the larva and adult. The dorsal-ventral axis is established, and locations on it are interpreted, by a group of about 20 genes; another 30 or so genes are involved in the construction and reading of the anterior-posterior axis that generates segmentation.

Segmentation depends on gradients of two protein morphogens already being produced in the egg; one diffuses from the front toward the rear, while the other (or perhaps a pair) emanates from the posterior end forward. The result is two overlapping concentration gradients (Fig. 14.13A). The genes of blastula cells apparently respond to the ratio of these two chemicals. Even before segmentation begins, if cytoplasm is removed from the anterior end of the egg and replaced with posterior cytoplasm, the larva develops

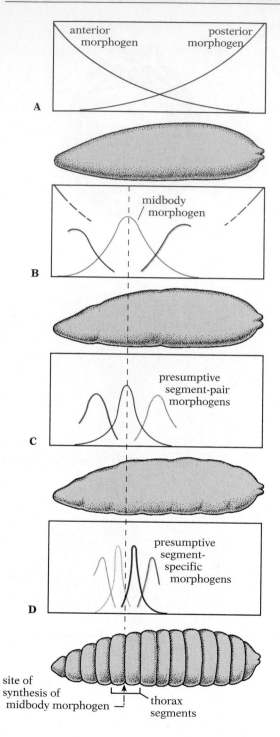

14.13 Model of morphogen concentration gradients responsible for segmentation in *Drosophila*

(A) Two morphogens important in establishing segmentation are produced in *Drosophila* eggs. One (blue curve), encoded by the gene *bicoid*, is synthesized near the anterior end and diffuses toward the rear; the other (red curve), the product of the gene *nanos,* is generated at the posterior end and diffuses toward the front. (A second posterior morphogen may also exist.) The result is a pair of overlapping concentration gradients. (The gradient pictured for the posterior morphogen may not be this regular; there is some evidence that it is actively transported anteriorly before beginning to diffuse.) (B) Additional gradients are created later in development. One, based on a morphogen encoded by the gene *Krüppel*, helps organize the middle of the larva. (C) These more local morphogens allow the larva to divide itself into seven segments. (D) These give rise to the final complement of 14, each of which then organizes itself internally.

without a head but with an abdomen at each end. Mutants lacking one of these polarizing morphogens can be "rescued" through an injection of the absent chemical into the appropriate end of the egg.

Though the overlapping concentration gradients inform cells of their approximate location in the embryo, this information is not precise enough to direct accurate development. At least one intermediate landmark provides a closer point of reference for cells in the long midregion of the larva. A tiny band of cells about a third of the way back along the larva, stimulated by the appropriate ratio of anterior-to-posterior morphogens, begins synthesizing and releasing a different morphogen that helps organize the middle of the embryo (Fig. 14.13B); other morphogens are produced by bands elsewhere along the axis.

The next step in organizing the embryo requires the genes in the various cells to interpret the anterior and posterior morphogen gradients and the local morphogen to determine which of seven segment pairs the cell is located in (Fig. 14.13C). Presumably this triggers the release of further chemicals, which enable cells to localize themselves to the anterior or posterior member of their segment pair (Fig. 14.13D). Finally, a set of 10 or so genes sets to work polarizing each segment individually. The sequential and hierarchical organization of whole-body morphogens, followed by ever-more-local morphogens, produces a gradual, step-by-step determination of cells toward increasingly specific fates.

The roles of homeotic genes At this point, as segment identity is fixed and the polarizing coordinates established, particular members of a family of about 20 **homeotic genes** (so called because the mutant alleles of these genes alter a segment or structure so that it resembles another) are activated according to which segment they are in. The products of homeotic genes, like the morphogens that induced them, are often control substances that bind to DNA and orchestrate the operation of many other genes; in this way, they cause each segment to express its unique character. Like the other

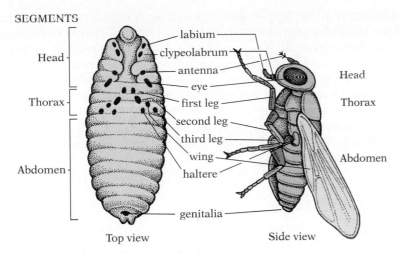

SEGMENTS

Head

Thorax

Abdomen

labium
clypeolabrum
antenna
eye
first leg
second leg
third leg
wing
haltere
genitalia

Top view

Head

Thorax

Abdomen

Side view

14.14   Segmentation of a *Drosophila* larva

Induced by morphogens, homeotic genes in each segment give rise to imaginal discs (shown as black spots) in the larva that control the development of particular structures in the adult.

control genes, homeotic genes make permanent changes in a cell's genome that are passed on to its daughters when it divides, thus inducing and stabilizing cellular determination.

The most dramatic consequence of the action of homeotic genes in insects is the creation of imaginal discs—the platelike clumps of cells mentioned in Chapter 13 that differentiate during the pupal stage to form adult structures (Fig. 14.14). Although differentiation occurs in the pupal form, determination took place much earlier, in the larval stage, soon after segmentation, when homeotic genes first became active: if a presumptive leg disc is exchanged with an antenna disc, the resulting adult will have a leg on its head and an antenna on its thorax.

Most of the homeotic genes are organized on the chromosome in two tight groups, the bithorax and antennapedia complexes. A gene's position within a complex correlates with its site of action on the embryo: beginning with those in the antennapedia complex that operate at the extreme anterior end of the embryo, genes farther along the chromosome operate on ever-more-posterior structures, and the pattern holds for the bithorax group as well. The logic of this organization is not yet understood.

Another intriguing discovery is the 180-nucleotide sequence called the **homeobox**, which is found in each homeotic gene as well as in many other developmental-control genes. Although the exact sequence of the homeobox varies a little from gene to gene, a strong similarity is clear. Homeobox gene products are transcription factors, which suggests that the variation in basic homeobox sequence accounts for the specificity of homeotic genes in activating the particular constellation of genes needed in each segment. One particular mutation, for example, causes a gene normally transcribed only in the thorax to become active in the head, leading to the production of legs in place of antennae (Fig. 14.15); this mimicking of the effects of disc transplants described earlier is almost certainly the consequence of a mutation in a control element that regulates the activity of a homeotic gene.

extra wings

A

misplaced legs

B

14.15   Two homeobox mutations in *Drosophila*

(A) Bithorax. (B) Antennapedia.

Studies of other species have turned up homeoboxes in all kinds of multicellular organisms, from plants to roundworms to humans. An understanding of the workings of the segmentation genes in *Drosophila*, therefore, will probably contribute to our understanding of how our own embryos go about organizing themselves. The evolutionary origin of homeobox genes is equally intriguing. They seem to be derived from genes controlling very simple developmental events and polarities: one is involved in mating-type determination in unicellular yeast and others in anterior-posterior differentiation in hydra.

■ HOW LIMBS FORM

Once a particular somite or imaginal disc is told to help produce a limb or an organ, there comes the major developmental task of orchestrating its construction. One popular model to account for vertebrate limb growth and patternings is based on investigations into wing development in chicks. The model assumes a bicoordinate system: a proximo-distal axis (from the body to the end of the extremity) and an anterio-posterior (front-to-rear) axis.

The hypothetical proximal-distal coordinate is linked with the ***progress zone*** of the wing, the area in which new cells are produced as the bud grows. The progress zone is associated with an ectodermal ridge running across the tip of the limb bud (Fig. 14.16A). The new cells produced in the progress zone are left behind as the area is pushed farther and farther away from the body. A cell's proximal-distal positional values might be determined by the time it spent in this progress zone (Fig. 14.16B). Cells left behind by the progress zone very early in development of the wing would have a low positional value, which might cause them to develop into the basal part of the wing (the humerus portion; see Fig. 1.10). Cells left behind a bit later would have an intermediate posi-

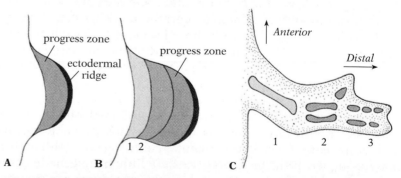

14.16 Pattern formation in the development of the chick wing

Left: Scanning EM of the ectodermal ridge of a chick wing bud. Right: Three stages of development of the wing bud, according to the progress-zone model. (A) Just behind the ectodermal ridge is a progress zone, where new cells are produced. (B–C) The first band of cells derived from the progress zone will become the humerus section of the wing; the second band will become the section containing the radius and ulna; and a third band, derived from the progress zone late in the development of the wing bud, will become the distal part of the wing.

14.17 Experiment with a grafted progress zone of a chick wing bud

(A) A normal wing developed from an intact wing bud. The tan area in the bud is the progress zone. (B) The wing that developed when an early wing bud (lighter tan) was grafted onto the original bud (darker tan) after the basal part of the wing had already begun to develop. The wing has extra humerus and radius-ulna sections.

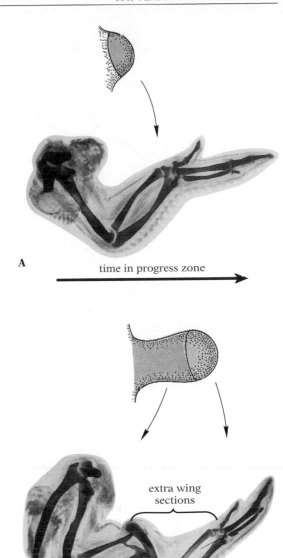

A　　　time in progress zone

B　　　extra wing sections

tional value, appropriate to formation of the middle part of the wing (the radius and ulna portion). Cells left behind late in the development, having spent a long time in the progress zone, would have a high positional value, appropriate to formation of the distal part of the wing. This information must somehow be stored in the cells as they are left behind, enabling them to regenerate new limbs from any point of amputation.

For the anterior-posterior coordinate, the model imagines that a diffusion gradient of a morphogen (which may be retinoic acid or triggered by it) is secreted by a small group of cells at the rear margin of the wing bud and polarizes the bud. With these two coordinates, the cells could "read" their position with sufficient accuracy to ensure their differentiation into an appropriate structure.

One test of the model involved grafting the progress zone (the ectodermal ridge and the associated area of actively dividing cells) from an early wing bud onto the end of an older bud whose own progress zone had been removed. The result was development of a wing with two humerus sections and two radius-ulna sections (Fig. 14.17). The cells of the graft had no way of telling they were so far out on the wing that they should form only its distal parts. Because they had spent so little time in the progress zone, some of these cells were left behind, programmed to read their position as being near the wing base, and developed into structures appropriate to such a position. The converse experiment of grafting the progress zone from an older bud onto an early bud produced a wing with only the distal parts, the phalanges: the transplanted tissue had no indication that the humerus, radius, and ulna had not yet developed. Having spent a long time in the progress zone, it developed structures appropriate to the wing tip.

To test the anterior-posterior gradient part of the hypothesis, part of the supposed polarizing region of the bud from the rear (posterior) edge was transplanted to the front (Fig. 14.18A–C). The results were dramatic: a partial set of mirror-image phalanges developed on the front edge. Their orientation to the transplanted part of the polarizing region was identical to the orientation of the normal set of bones to the polarizing tissue at the rear of the wing. Evidently this is because the morphogen from the transplanted region polarized the front half of the bud, diffusing from front to rear, just as the morphogen from the intact polarizing region on the rear edge polarized the rear half of the bud, diffusing from rear to front. Implants of retinoic acid mimic these transplants; moreover, retinoic acid receptors in wing-bud cells, once bound, are known to move to the nucleus and bind to the DNA, just as we

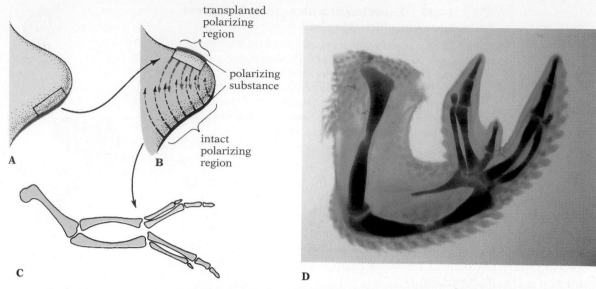

14.18   Experiment with a transplanted polarizing region

Tissue from the rear margin of one wing bud (A) is transplanted to the leading edge of another (B). There the morphogen diffusing from the transplant causes cells to differentiate into a second set of phalanges (C). The same effect can be achieved by treating the leading edge of a limb bud with retinoic acid (D).

would expect from a gene-control substance (Fig. 14.18D). In addition, which homeotic gene is expressed correlates with the concentration of retinoic acid. Whether retinoic acid is actually the primary morphogen, however, is controversial.

REVERSING DEVELOPMENT

The usual sequence of development, as we have seen, is for a morphogen or other chemical (often from an adjacent cell) to act as an inducer, altering the pattern of gene expression of a cell. This responding cell has now become to some degree determined. Subsequent experience with additional morphogens or other control substances further focuses a cell's fate, until it may become fully determined. At the same time, a cell's morphology and chemistry are responding to the activity of the genes in its nucleus, so that the cell differentiates (though often after considerable delay). Determination and differentiation, therefore, are usually gradual and proceed in that order.

Though differentiation usually follows determination, certain cells become differentiated *before* determination fully fixes their fate; they are capable of some degree of **dedifferentiation**—that is, they can regress to less differentiated states, and then begin differentiating into a new type of tissue. In mammals and birds, dedifferentiation is usually restricted to embryos. In amphibians, however, this ability to maintain incomplete determination in at least some cells into adulthood leads to remarkable powers of re-

14.19   Regeneration of a salamander arm

generation. For example, if the leg of a salamander is amputated, cells near the wound begin to dedifferentiate under the epidermis at the tip of the stump. Gradually, as mitotic activity and cellular redifferentiation take place within this area, the mound comes to look more and more like the normal limb bud of an embryonic salamander (Fig. 14.19). It slowly elongates, and after several weeks a distinct elbow and digits appear, complete with muscle, tendon, bone, connective tissue, etc.

Dedifferentiation is most common in animals that grow throughout their life cycles; many fish, reptiles, and amphibians have no typical adult size, but instead simply grow larger until they die. With the need to be able to grow indefinitely, the developmental options of at least some cells may need to be kept open.

Although dedifferentiation is possible for some cells in some species, full determination is seldom reversed. Cancerous cells are one exception: they may suffer mutations that remove some of the determining constraints, begin dividing again, and produce cells with a different identity (most often that of an earlier, less differentiated cell type). Obviously this requires profound changes in patterns of gene expression.

HOW NERVOUS SYSTEMS DEVELOP

Most aspects of development are illustrated with particular clarity in the nervous system. Nerve cells, or **neurons**, are "born," migrate to their proper places, send out fibers called axons and dendrites to specific target locations, and so come to form a highly integrated functional network that is more complex than any other organ system in the body.

■ NERVE-CELL MIGRATION

The first step in the life of a newly formed presumptive neuron, like that of many other cells in a developing organism, is usually movement from where it was formed to where it is destined to be. The cells that give rise to the retina, for instance, must grow out from the developing brain to where the eyes are to be located, and then form the optic vesicle, while the cells of the outer layer of the brain, the cerebral cortex, must move through layers of older cells to get from the core of the brain, where they were engendered, to the outside layer of the cortex, where they end up.

The movement of neuronal cells during development is quite

14.20 Migration of nerve cells

Before nerve cells join together in a network, many must move from their place of origin to their final location. (A) Cells from the neural crest move down and along glycoprotein filaments to positions at the future sites of spinal ganglia and the adrenal glands. Other neural-crest cells encounter another set of filaments, leading to positions just under the ectoderm; cells following this pathway become pigment cells in the skin. (B) New cells of the cerebral cortex are generated by a basal layer of stem cells lining each fluid-filled cavity, or ventricle, and migrate out along filaments to form a layer on the outside of the cortex. Cells produced later will move through this layer to positions still farther out.

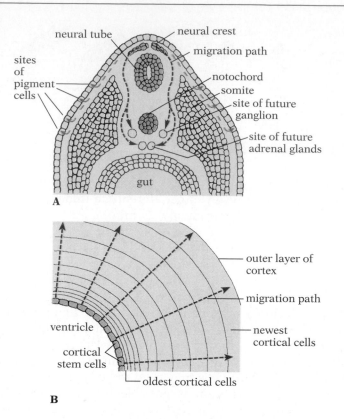

well oriented. Consider the cells that give rise to the adrenal glands of the kidney and to nearby clumps of nerve cells (ganglia). They originate in the neural crest—the region just above the developing spinal cord—and migrate down to specific spots near the notochord (Fig. 14.20A). They move in the proper general direction from the outset, follow highly predictable pathways, and stop at precise spots. Similarly, cells of the cerebral cortex are generated in a basal layer of tissue and then migrate outward through other, older cortical cells until they reach the outer layer of the cortex (Fig. 14.20B). The three stages the neural-crest and cortical cells pass through—determining the initial direction, following a path, and determining the stopping point—are characteristic of the migration of developing cells. At least three mechanisms are utilized: diffusing chemicals, CAMs, and tactile cues.

The gradients of the diffusing chemicals help guide neuronal cells by causing them to move in an amoeboid fashion toward (or, in some cases, away from) the source of the chemical, leading neural-crest cells "down," cortical cells "outward," and other cells in an anterior or posterior, or dorsal or ventral, direction. One current model proposes that most path finding also involves the other two mechanisms: CAMs and tactile cues. Neural-crest cells (guided by filaments of glycoproteins that lead around the spinal cord and past the notochord) and cortical cells (following a radial array of filaments up through the cortex) apparently recognize their respective pathways by means of the particular ratios of CAMs they encounter on the other cells along the pathway. They partially envelop the guide cells, and then move along them, maintaining intimate tactile contact, finding ever-better matches between self- and substrate-borne CAMs.

According to this model, when a migrating cell encounters the optimum CAM correspondence on the cells it has touched, it stops moving and proceeds to form the cell-to-cell attachments that will anchor it in place. The molecular specificity of nerve-cell surface markers is so precise that each class of neurons probably has distinctive molecules in its membrane.

■ AXON GROWTH AND TARGETING

Once a neuronal cell has reached its permanent place in the nervous system, it must send axons (the long, thin processes specialized for transmitting information) and, in the case of some sensory cells, dendrites to specific target cells. Here again, both chemical and tactile information seem to play a role. The leading edge of the developing axon displays an unusual type of structure known as a *growth cone* (Fig. 14.21). The growth cone continually extends and retracts spikelike pseudopods called filopodia, which probably sample the environment for specific chemicals and for the actual presence of certain guide cells. If the cone encounters the chemical or tactile stimulus of such a guide cell (usually the axon of another nerve), it partially envelops it and grows along it. Other chemicals can repel growing axons, and thus help prevent mistargeting.

In vertebrates, as many as 1 million axons may project information from the body to one of several large arrays of target cells like those found in the visual, auditory, and tactile areas of the cortex. These connections of axons to target cells are spatially organized and genetically predetermined. It seems unlikely that 10^6 specific molecular labels exist in the visual system to assure that all axons find their correct target cells. Nor does it seem plausible to imagine that a simple bicoordinate morphogen gradient could be sufficiently precise. In fact, the initial wiring is not very exact; the final pattern is determined dynamically through neural competition.

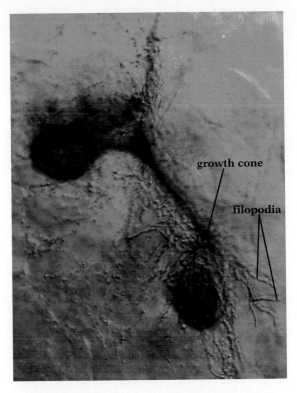

14.21 Growth cone of a developing axon in a grasshopper

The filopodia have made contact with another neuron (lower right).

■ DEATH AND COMPETITION IN DEVELOPMENT

In most animal nervous systems, many more cells are born than are actually needed. In vertebrates, for example, identical ganglia containing vast numbers of cells develop next to each of the vertebrae. And yet only the ganglia serving the many muscles and sensory receptors of the arms and legs require so many cells; in the other ganglia, the extra cells die. Apparently it is easier or more efficient for the developmental program to build all segments alike initially, and then to allow functionless cells to die.

In nematodes (roundworms) a similar pattern is evident. Their simple nervous system consists of a long ventral nerve cord containing 12 ganglia. Each ganglion arises during development from one of 12 precursor cells, each a part of a clump of embryonic cells. The ganglia mature to control rhythmic swimming movements and mating. The pattern of cell division is the same in all 12 ganglia even though the eventual organization of each ganglion may be very different. Cells that become neurons in only some (or even just one) of the ganglia are nevertheless formed in all 12, and

14.22 Cell death during the formation of nematode ganglion 12

This family tree of the cells in the last ganglion of a nematode suggests that the developmental program for the neurons involves a fixed series of divisions regardless of which neurons are needed by the ganglion. For example, the ganglion at the tip of the tail (illustrated here) does not need cells b and S_2, which in other ganglia connect to the next ganglion to the rear. Cell c is also unnecessary except in the middle ganglia of males, where it controls some of the animal's reproductive behavior. Nevertheless, a cell can apparently be produced only if cells b and c are produced as well.

those not needed are then allowed to atrophy and die (Fig. 14.22).

An analogous phenomenon occurs in many animals at the level of neuron-to-neuron connections: many more of the axonal endings specialized to communicate neural activity from one cell to another—structures called synapses—are formed than survive. When cells wired to a particular target are prevented from responding to external sensory stimuli (either by experimental manipulation or because the cells are misplaced or defective), their synapses seem to get crowded out on the target neuron by synapses from other cells that do fire normally. In animals as diverse as crickets and mammals, cells that lose in this competition seem to atrophy and disappear. This process serves to fine-tune the connections in large spatial arrays.

CHAPTER SUMMARY

HOW POLARITY ORGANIZES DEVELOPMENT

Egg cells are polarized by virtue of having different concentrations of critical molecules in different parts of the egg. This polarity, the site of sperm fusion, and the plane of an early division can orient later development. (p. 357)

INDUCING DEVELOPMENTAL EVENTS

HOW GASTRULATION AND NEURULATION ARE INDUCED Developmental timers tell the embryo when to gastrulate. In frogs the cells derived from the gray crescent organize gastrulation and much of early development. (p. 360)

HOW ORGANS ARE INDUCED Many organs are formed in response to inducer chemicals produced by cells near where the organ is destined to lie. (p. 361)

HORMONES AS INDUCERS Some hormones—particularly the sex hormones—act as organ inducers. (p. 363)

HOW MIGRATING CELLS KNOW WHERE TO GO The migration of cells from their point of origin to their ultimate location may be guided by cell-adhesion molecules (CAMs). Moving cells appear to seek the best CAM match with the other cells they encounter until they reach a location from which no better matches are within reach. (p. 365)

PATTERN FORMATION

WHY EMBRYOS ARE SEGMENTED Most embryos develop as a series of segments that are initially very similar, but specialize later based on their anterior-posterior position—information that they read from concentrations of morphogens. Misreading the correct location creates homeotic changes that transform one region into another. (p. 367)

HOW LIMBS FORM Limb formation requires organization along two axes: a distal-proximal one (which seems to depend on developmental age) and an anterior-posterior axis (which relies on a morphogen released from the rear edge of the developing limb). (p. 370)

REVERSING DEVELOPMENT

Some cells that appear to be fully differentiated are nevertheless able to dedifferentiate, divide, and take up new roles under special circumstances. Some cancerous cells mimic this pattern. (p. 372)

HOW NERVOUS SYSTEMS DEVELOP

NERVE-CELL MIGRATION Many nerve cells migrate from their site of birth to a ganglion along specific routes guided by chemical gradients; CAMs and tactile cues may also be involved. (p. 373)

AXON GROWTH AND TARGETING Axon growth shows the same pattern as nerve-cell migration, except that the distances are often far longer and the precision much greater. (p. 375)

DEATH AND COMPETITION IN DEVELOPMENT Many more nervous-system connections are made than ultimately survive. Competition between connections based on the degree of neural activity seems to fine-tune the connections. Each ganglion has many more neurons than it eventually needs; these excess cells, which play a role in other ganglia with different specializations, atrophy and die. (p. 375)

STUDY QUESTIONS

1 When the anterior half of a frog egg was separated from the posterior half (Fig. 14.2B), one segment became a normal tadpole while the other turned into a disorganized mass of cells. What might happen if you were to perform the same experiment on a *Drosophila* egg or blastula? (pp. 367–369)

2 How does the overlapping two-gradient strategy of *Drosophila* allow for greater accuracy in reading anterior-posterior location than the use of a single morphogen? Is there any evidence that *Drosophila* needs this degree of precision? (pp. 367–370)

3 Can you think of any reason why it might make sense to arrange the homeotic genes in anatomical order?

4 Think of five different ways unnatural chemicals in the diet of a pregnant mammal might disrupt the development of an embryo. (pp. 361–372)

5 Why might incomplete determination be essential to organisms with indefinite growth? Why might incomplete determination be a disadvantage to species with fixed growth—species that usually live longer and grow larger? (pp. 372–373)

SUGGESTED READING

COOKE, J., 1988. The early embryo and the formation of body pattern, *American Scientist* 76 (1). *A wide-ranging review looking for common mechanisms.*

EDELMAN, G. M., 1984. Cell-adhesion molecules: A molecular basis for animal form, *Scientific American* 250 (4). *On the likely molecular basis of cell-to-cell adhesion and changes in adhesion during embryonic development.*

GILBERT, S. F., 1994. *Developmental Biology, 4th ed.* Sinauer, Sunderland, Mass. *Excellent, highly detailed text.*

GOODMAN, C. S., AND M. J. BASTIANI, 1984. How embryonic nerve cells recognize one another, *Scientific American* 251 (6). *An excellent description of how axons of invertebrates employ the stepping-stone strategy, following first gradients and then one preexisting axon after another to reach their targets.*

MCGINNIS, W., AND M. KUZIORA, 1994. The molecular architects of body design, *Scientific American* 270 (2). *On how the homeobox genes orchestrate development in both insects and vertebrates.*

CHAPTER 15

IMMUNOLOGY

O nce we have had measles or mumps, we cannot contract them again. This immunity develops because the cells that fight the disease while it is present are able to recognize and destroy the disease-causing **pathogen** far more rapidly in the future. The number of foreign cells and unfamiliar chemicals the immune system can learn to recognize is essentially infinite, yet only a few genes are involved. How does this remarkable defense system work? The answer, as we will see in this chapter, involves the cell-adhesion and cell-surface molecules discussed in Chapter 14, as well as powerful mechanisms of genetic variation.

HOW THE IMMUNE RESPONSE WORKS

Nearly all animals have phagocytic cells that ingest bacteria and dead cells. These **macrophages** are attracted by chemicals released by damaged tissue and many foreign cells; they create a localized inflammation where they are active, as well as a liquid mass of dead cells and other debris (commonly called pus). This slow and unselective line of defense has been elaborated in vertebrates into a highly specific immune system adapted to larger body size and longer life.

The best-known receptor molecules of the immune system are the **antibodies**, which bind to molecules that are foreign to the organism. Foreign substances, collectively called **antigens** (short for "antibody generating"), are almost always large molecules (usually proteins or polysaccharides). Antigens may be free in solution, as are the toxins secreted by some pathogenic microorganisms. Antigens may also be built into the outer surfaces of the pathogens themselves (as with some viruses and bacteria), or they may be part of the coating of otherwise innocuous entities like grains of pollen. The antigens stimulate certain cells in the immune system to produce highly specific antibodies—proteins that bind to these antigens exactly as enzymes bind to reactants. The binding of antibodies to toxins inactivates them, often by simply covering the toxin's active site, and therefore altering their toxicity. Antibodies inactivate viruses by binding to the receptors by which they recognize their hosts. When antibodies bind to microorganisms, however, they target these pathogens for subsequent destruction by one of several mechanisms described in the following sections.

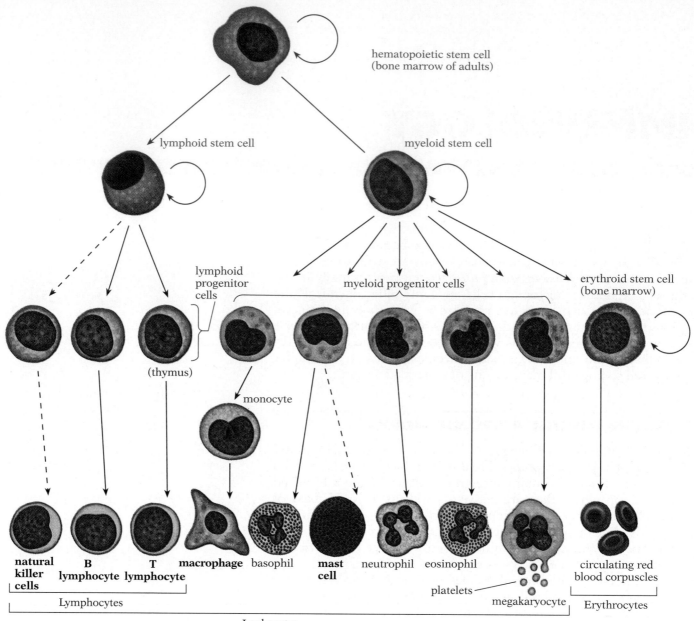

hematopoietic stem cell
(bone marrow of adults)

lymphoid stem cell

myeloid stem cell

lymphoid
progenitor
cells

myeloid progenitor cells

erythroid stem cell
(bone marrow)

(thymus)

monocyte

**natural
killer
cells**

**B
lymphocyte**

**T
lymphocyte**

macrophage

basophil

**mast
cell**

neutrophil

eosinophil

platelets

megakaryocyte

circulating red
blood corpuscles

Lymphocytes

Erythrocytes

Leukocytes

15.1   Hematopoiesis: the origin of blood cells

Hematopoietic stem cells give rise to at least two kinds of more highly determined stem cells: lymphoid and myeloid stem cells. Both kinds of stem cells have the ability to regenerate themselves. The more highly determined stem cells in turn give rise to more specialized cells. Lymphoid stem cells give rise to three still more highly determined, progenitor cell types, which in turn produce the natural killer cells, B lymphocytes, and T lymphocytes that we will discuss in this chapter. Myeloid stem cells produce at least six classes of more highly determined cells, which themselves go on to generate the more specia-

ized monocytes (which mature into the cell-scavenging macrophage discussed later), as well as red blood corpuscles, basophils and (perhaps) mast cells. (Myeloid stem cells also give rise to neutrophils and eosinophils, which attack bacteria and larger parasites, as well as megakaryocytes, which produce platelets—substances crucial to clotting, discussed in Chapter 30.) The presentation cells (which we will encounter later in this chapter) are derived from hematopoietic stem cells along an unknown pathway. The text discussion focuses on the interactions of the best-understood immunocytes, designated here with boldface type.

■ THE CAST: CELLS AND ORGANS OF THE IMMUNE SYSTEM

Where immune-system cells come from In most vertebrates, all *immunocytes*—cells with immunological function—derive from precursors that form in the yolk sac of the early embryo. These hematopoietic stem cells (so named because they have become developmentally determined as progenitors of blood cells) migrate to specific tissues and organs, where they give rise to the red and white blood cells (Fig. 15.1). Red cells carry oxygen in the blood and have no immunological role. A class of white cells that comes from the thymus gland and from bone marrow gives rise to *lymphocytes*; it is these cells that respond to the presence of foreign antigens or kill microorganisms tagged by antibodies. Lymphocytes derive their name from the high proportion of time they spend in the lymphatic system, where dead cells and debris tend to collect, and into which toxins and infectious organisms usually find their way.

The role of the lymphatic system Blood consists of a solid portion (cells) and a liquid portion (plasma). As blood flows through the many fine capillary beds in the body, a portion of the plasma leaks out between capillary cell walls into surrounding tissue. This fluid, supplemented by liquid transported endocytotically or lost osmotically, brings nutrients to tissue cells outside the bloodstream, and picks up waste products. As blood moves from the capillaries into larger vessels, most but not all of the lost plasma and metabolic wastes of tissues is reabsorbed. The rest is collected as *lymph* into a parallel system of lymphatic veins and funneled into a large thoracic duct in the chest, from which it rejoins the blood (Fig. 15.2). Fluid and unanchored cells (generally dead or foreign) from body tissues can also enter the lymphatic system.

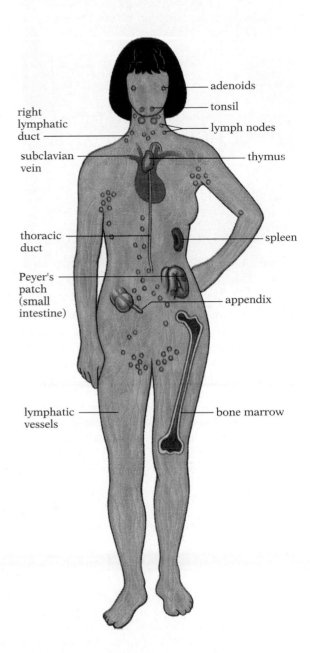

15.2 Human lymphatic system

Above: The primary organs of the lymphatic system—the thymus and bone marrow (only one bone is shown)—produce lymphocytes that circulate from extremely fine blood vessels, called capillaries, into the lymphatic system. Other tissues and organs with connections to the lymphatic system are shown. Left: The source of lymph is fluid forced from the blood capillaries, which is picked up by lymph capillaries; these merge to form lymph ducts, which in turn empty into the lymph vessels shown above.

15.3   Lymph node seen from the inside

Reticular cells (RC) and their extensions create a meshwork that acts as a crude filter and provides a support surface for lymphocytes (L) and macrophages (M). The wall of the node (W) is also visible. (From *Tissues and Organs: A Text-Atlas of Scanning Electron Microscopy* by Richard G. Kessel and Randy H. Kardon. Copyright ©1979 W.H. Freeman and Company. Reprinted with permission.)

The transport of lymph is passive: body movements squeeze lymph past one-way check valves on its slow journey to the chest.

Lymphatic veins from capillary beds in the chest contain one or more **lymph nodes;** these are regions with a fine tissue mesh that filters dead cells and other large fragments from the lymph (Fig. 15.3). The nodes are home to many phagocytic cells (which consume the material trapped in the mesh) and lymphocytes. Like the spleen, through which all of the body's blood is frequently filtered, the nodes provide a convenient place for immunocytes to monitor the plasma for foreign antigens, and immune responses are frequently localized in the nodes, particularly those in the tonsils.

The diversity of immunocytes   The two main types of lymphocytes are B cells (which mature in the bone marrow) and T cells (which mature in the thymus). B cells manufacture and secrete antibodies, and are responsible for the **humoral immune response**— the response triggered by antibodies circulating in the body fluids, blood, and lymph. Because their antibodies can recognize and bind to cell-surface markers, B cells are particularly effective against bacteria, fungi, parasitic protozoans, and viruses, as well as toxins free in the blood and plasma. The T cells, on the other hand, are responsible for the **cell-mediated response**, which kills those of the organism's own cells that have become infected. T cells can also bind to and modulate the activity of B cells.

The major phagocytic cells of vertebrates are the macrophages. The macrophages of vertebrates can be far more selective than those of invertebrates because they look specifically for cells to which antibodies have been bound, thus identifying them as foreign. Another kind of white cell, the natural killer (NK) cell, is also found in invertebrates, but like the macrophages, it has evolved in vertebrates to home in on and attack antibody-tagged cells. The discussion that follows will emphasize the best-understood actors in immune reactions: macrophages, B cells, T cells, NK cells, and mast cells. Other white blood cells with immunological functions include neutrophils (Fig. 15.4), eosinophils (which attack bacteria and larger parasites), megakaryocytes (which produce platelets crucial to clot formation), basophils (which, like mast

15.4   Neutrophil

Neutrophils home in on invading organisms (usually bacteria) by detecting some of their waste products. The neutrophil seen here crawls toward its target, releases toxic chemicals and enzymes onto it, and then envelopes it by phagocytosis.

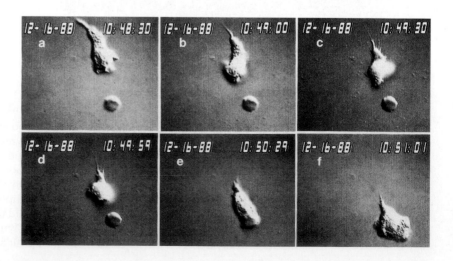

cells, release a chemical histamine that signals injury or infection, and thus create inflammation), and presentation cells. Basophils are specialized for circulating in the blood, while mast cells reside in tissues. Presentation cells are derived from hematopoietic stem cells along an unknown pathway.

■ WHAT THE HUMORAL RESPONSE ACCOMPLISHES

Most B-cell lymphocytes migrate back and forth from the blood to the lymph, taking up temporary residence in the spleen and lymph nodes where, using their membrane-mounted antibodies, they are well placed to monitor the body's fluids or dispense free antibodies. When needed, they can disperse to the circulation and tissues.

The B-cell antibody Each antibody molecule consists of four polypeptide chains—two identical "heavy" chains and two identical shorter "light" chains, linked by disulfide bonds (Fig. 15.5). There are five classes of heavy chains—A, D, E, G, and M—that differ in the amino acid sequence at the COOH (tail) ends. The tail regions play no role in antigen specificity; they determine which reaction of the humoral antibody response will take place. For example, after an antigen has been bound, heavy chains with a G tail undergo an allosteric change in conformation that allows them to be recognized by macrophages, which can then ingest the antibody along with whatever the bound antigen is attached to—a virus, for instance.

Antibodies with other heavy chains are used to activate other parts of the immunological reaction: antibodies with E tails become mounted on the membranes of mast cells, signalling these early-warning cells to begin releasing histamine when an antigen is encountered. Others are specialized for tasks to be described presently, like conferring maternal immunity on newborns, or activating a series of enzyme reactions known as the complement system. Together, all the antibodies of all classes are called *immunoglobulins (Ig)*, and the individual classes are often referred to with added letters, such as IgG, IgE, and so on.

Regardless of which class of heavy or light chain is incorporated into an antibody, most of each chain has a constant amino acid sequence and structure; the variability crucial to antigen specificity lies mostly at the free amino ends. The binding sites for antigens (two identical sites on each antibody molecule) are at the ends of the variable portions. Each binding site is a pocket bounded partly by the heavy chain and partly by the light chain (Fig. 15.5). This region can bind to roughly six amino acids or carbohydrate units of an antigen.

How the humoral response develops Before an organism has encountered a particular antigen, its *B lymphocytes* which can bind that antigen are small, metabolically quiescent *virgin cells* that move freely between the blood and the lymphatic tissues. Each of the millions of virgin cells generated during embryonic development has thousands of identical antibodies mounted in the membrane, but no two virgin cells are likely to display antibodies with the same antigenic specificity. When a cell's antibodies begin

A

B

15.5 B-cell antibody molecule

(A) B-cell antibodies consist of two identical pairs of polypeptides; each pair has a heavy chain and a light chain, which is readily seen in this schematic representation. The sections shown in gray have relatively constant sequences, while the colored portions vary greatly from one B cell to another. Antigens bind in the cleft between the heavy and light chain of each pair. B-cell antibodies can exist as freely circulating molecules (as shown here) or mounted in the membrane of B lymphocytes or mast cells. (B) A space-filling model showing the three-dimensional shape of antibodies. Each sphere represents one amino acid. The antibody shown here is IgG.

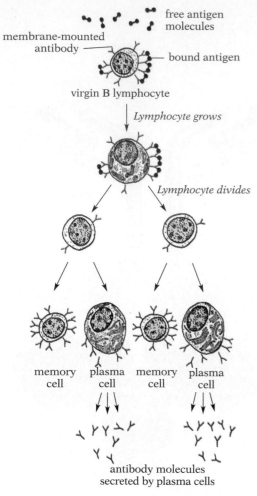

15.6   Stimulation of B lymphocyte by antigen

When the membrane-mounted antibodies of a virgin B cell are able to bind a particular antigen, the lymphocyte first grows larger and then begins a series of cell divisions (only two being shown here). Some of the cells produced by this proliferation are memory cells, which resemble the original lymphocyte; others become specialized as plasma cells, which secrete antibodies. The antigen in this example is a toxin.

15.7   Agglutination by antibodies bound to antigens

Each antibody molecule can bind to two antigen molecules; hence the microorganisms or viruses bearing the antigens can be held together in large clumps. Here the antigens are surface proteins or carbohydrates on invading viruses. (For clarity, the antibodies and antigens have been enlarged; they are in fact much smaller than viruses.) This agglutination aids in the destruction of antibody-bearing pathogens by macrophages, killer cells, or complement-system proteins.

binding antigen, these small lymphocytes start to grow larger and divide (Fig. 15.6).

Cell division of a stimulated B lymphocyte gives rise over a period of several days to numerous **plasma cells**, and it is primarily these cells that secrete antibody molecules. A stimulated B lymphocyte also gives rise to other lymphocytes like itself; these serve as **memory cells**, which will make possible a more rapid response in the future should that particular antigen be encountered again. (How this is accomplished will be described later in the chapter.) It is the rapidity of this subsequent response that confers immunity.

Each of the millions of kinds of antibodies has a different set of active sites, so each binds to one or more different antigens or antigen regions. Some antibodies are so well matched to an antigen that this bonding is rapid and strong; other antibodies bind with a lower affinity. Because each antibody molecule can bind to two antigen molecules, the antibodies can agglutinate (lump together) any microorganisms or viruses bearing the antigen (Fig. 15.7).

This agglutination in turn can trigger three reactions. First, large phagocytic macrophages (see Fig. 4.25, p. 108) in the lymph recognize antigen-bound antibodies, and engulf them and their targets (Fig. 15.8A). This reaction is effective against toxins, viruses, and most bacteria. Second, **natural killer (NK) lymphocytes** recognize the bound antibodies, bind to them, and destroy any foreign eucaryotic cells they have marked (Fig. 15.8B). The mechanism by which NK lymphocytes kill is not yet fully understood, but it involves making holes in the membrane of the target cell; osmotic entry of extracellular water then kills the cell.

Finally, the bound antibodies can trigger the **complement system**, a cascade reaction involving more than 20 plasma proteins (mostly zymogens—enzymes that exist in a nonfunctional conformation until activated). In the cascade reaction, each protein in turn catalyzes the activation of another. The complement-system proteins assemble to create membrane channels in the invading cell; the channels allow water to enter by osmosis into the microor-

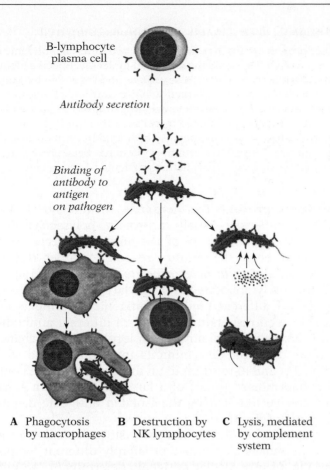

B-lymphocyte
plasma cell

Antibody secretion

*Binding of
antibody to
antigen
on pathogen*

A Phagocytosis
by macrophages

B Destruction by
NK lymphocytes

C Lysis, mediated
by complement
system

15.8 How humoral antibodies facilitate the destruction of pathogens

Once bound to antigens, the antibodies secreted by B lymphocytes can trigger three sorts of reactions. (A) Bound antibodies, and the pathogen to which they are attached, are ingested by phagocytic macrophages; this is also the primary mechanism by which agglutinated toxins and viruses are eliminated. (B) Bound antibodies are recognized by NK lymphocytes, which destroy the marked pathogen by creating holes in the target's membrane. (C) Bound antibodies trigger a cascade reaction by which zymogens of the complement system are activated and catalyze the construction of a membrane channel in the invading microorganism; water moves into the invader through this channel by osmosis, causing the cell to swell and burst. (The action of circulating antibodies on mast cells is not shown in this drawing.)

ganism, which swells and bursts (Fig. 15.8C). Some membrane-enclosed viruses are neutralized in this way.[1] This system operates almost exactly like the hole-punching strategy of NK cells. Having detected an invader, the immune system clearly takes no chances; several redundant systems work simultaneously to attack anything foreign.

In addition to binding directly to antigens and thus targeting them for destruction, some circulating antibodies can become attached by their bases to **mast cells**. The antibodies thus mounted on the cell membrane of the mast cell have their antigen-binding region free and exposed to the surrounding medium. When an antigen is bound by a mast-cell-mounted antibody, the mast cell is induced to release histamine and other chemicals. Histamine in turn causes nearby blood vessels to dilate and become "leaky," allowing the plasma, rich in antibodies and complement-system proteins, to reach the site of the histamine release. Lymphocytes and macrophages are also attracted. The mast cell–antibody system, then, acts as a sort of cellular burglar alarm to recruit the other elements of the immune system to concentrations of antigens. It is particularly important in the response to parasites embedded in tissue rather than moving through the blood or lymph.

---

[1] Some viruses seem to have evolved defenses against the complement system. Herpes and Epstein-Barr viruses, for instance, bear receptors that bind and inactivate the third protein in the cascade, bringing the reaction sequence to a halt; vaccinia virus (one of the pox viruses) protects itself by binding the fourth protein.

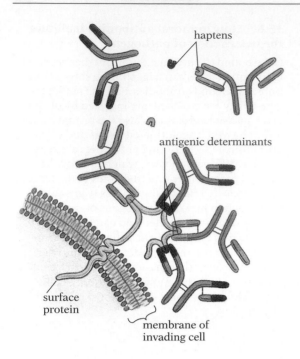

haptens

antigenic determinants

surface protein

membrane of invading cell

15.9 Antigenic determinants and haptens compared

The surface protein on this invading cell has two different antigenic determinants, each of which occurs twice in each molecule. Specific membrane-mounted antibodies can bind to these regions and trigger an immune response, which includes the production of free antibodies (shown). The surface protein and its antigenic determinants may be unique to this kind of invading cell, or they may be shared by cells of other kinds. Isolated antigenic determinants, or haptens, can also be bound by antibodies and stimulate an immune response if the immune system has been previously exposed to a large molecule bearing that particular determinant.

How antigens provoke the response An antigen is almost always a large molecule—usually a protein, polysaccharide, glycoprotein, or glycolipid. Not all of the antigen molecule stimulates lymphocytes to begin the immune response. Rather, restricted regions serve as ***antigenic determinants***—sites of interaction with the receptors on lymphocytes. A single large antigen molecule may include several different kinds of antigenic determinants, which are bound by a corresponding number of different antibodies (Fig. 15.9). Conversely, different antigen molecules may, by chance, have one or more antigenic determinants in common, and so "share" antibodies. The initial reaction to an antigen is triggered only if the antigenic determinant is part of a large molecule, but subsequent reactions can be initiated by the isolated antigenic determinant, called a ***hapten*** (Fig. 15.9).

Proper functioning of the immune system depends on the availability of an enormous number of slightly different lymphocytes, each specific for a particular antigenic determinant. An individual may have 10 billion (10^{10}) or more antibody types. Each antigen reacts only with those very few lymphocytes that have antibodies capable of binding to part of it, and this binding is necessary to induce proliferation of the appropriate lymphocyte types (Fig. 15.10).

Each stimulated lymphocyte gives rise to a clone of cells (a group of genetically identical cells descended from a common virgin cell). Hence the proliferation of the particular lymphocytes that react with a specific antigen is called ***clonal selection***. Each of the plasma cells to which a given B lymphocyte gives rise when stimulated may transcribe as many as 20,000 mRNA molecules from its genes for antibody, enabling each of the plasma cells to secrete 5 million identical antibody molecules per hour.[2] But the immune system is careful not to trigger this massive response by accident.

[2] A partial but rapid immune response can also be triggered in individuals directly inoculated with the appropriate antibodies. Though there are no memory cells to begin supplying new antibodies, the antibody molecules introduced from outside can deactivate pathogens, or mark them for phagocytosis or complement-mediated destruction. Breast feeding is one means of creating this "passive immunity": maternal antibodies specialized to cross the intestinal wall (by virtue of a special heavy-chain tail) and enter the fetal bloodstream provide a significant degree of temporary protection until the infant's own immune system gains experience. Another source of antibodies is an injection of gamma globulin, a protein extract from blood containing antibodies from individuals immune to a particular disease, though the extract contains many other kinds of antibodies as well. Gamma globulin treatment may reduce the severity of symptoms in someone who has been exposed to certain diseases, and can be useful in the absence of an effective vaccine, or when there is no time for an immunization to "take."

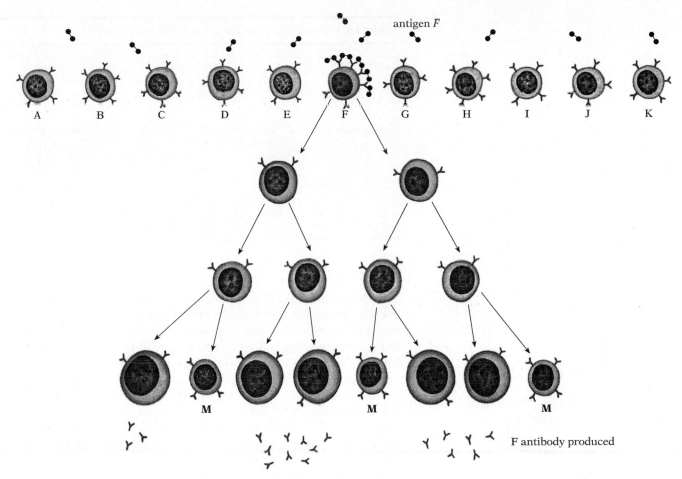

antigen F

F antibody produced

15.10 Clonal selection

Antigens are able to react with the membrane-mounted antibodies of only one or a few very specific lymphocytes from among the billions of kinds of lymphocytes in the organism's body. In this example antigen *F* can be bound only by the B lymphocyte of type F; it does not affect the other lymphocytes (top row). Lymphocyte F, stimulated by the binding of the antigen, proliferates to form a clone of genetically identical cells. Some cells of the clone are memory cells (M); others are plasma cells that actively secrete F antibody. The cross-linking of the antibodies on lymphocyte F greatly facilitates the proliferation of the F clone.

The cross-linking of receptors (which, by requiring at least two simultaneous antigen-binding events, serves to reduce false alarms, and probably explains why haptens cannot stimulate virgin cells) appears to be the event that induces the lymphocytes to begin dividing (Fig. 15.10).

■ WHAT THE CELL-MEDIATED RESPONSE ACCOMPLISHES

Another set of immune reactions is mediated by *T lymphocytes*. One kind—the *cytotoxic T cell*—kills infected cells of an individual's own body before the pathogen (usually a virus) can spread further. Cytotoxic cells, sometimes called killer T cells, are often confused with NK cells; the distinction between the two is made clear in Table 15.1. However, the B-cell and T-cell systems are not

Table 15.1 Simplified outline of immune-system cells

FUNCTION	B-CELL SYSTEM (HUMORAL RESPONSE)	T-CELL SYSTEM (CELL-MEDIATED RESPONSE)
Early-warning	Mast cells	Presentation cells
Diversity generation	Virgin B cell	Virgin T cells
Modulation		Helper T cells
Effector systems	Complement cascade	Cytotoxic T cells
	Natural killer cells	
	Macrophages	

15.11 T- cell receptor

T-cell receptors consist of two peptide chains, each with its own sequence. Each chain has a relatively constant section (gray) and a highly variable region (purple) to which an antigen can bind. Another section in each chain (white) binds to a portion of complementary MHC molecules (described later in the text).

as independent as Table 15.1 suggests. For example, other sorts of T cells modulate the activity of both humoral and cell-mediated systems, accelerating the response at the outset and later preventing it from getting out of hand. Controlling the immune reaction involves tuning a lymphocyte's activity and is antigen-specific. For example, a particular clone of T cells will manage the activity of the clone of B cells that reacts to the antigen in question. Each of these jobs requires T cells to recognize specific antigens, but to interact only with other cells of the body; free antigens or antigenic markers on pathogens are ignored. The mechanisms by which this dual recognition is accomplished will become clear when we look at the structure of the T-cell receptor.

The T-cell receptor The receptor molecules of T lymphocytes are not secreted, but remain instead firmly attached by their tails to the lymphocyte membrane (Fig. 15.11). Like the B-cell antibody, each polypeptide arm of the T-cell receptor has a constant region at its base and a variable section farther out. The cleft lies between the two variable-region arms, which cooperate to bind the antigen. Like B cells, each T cell produces receptors specific to only one antigenic determinant, and nearly every T cell has a unique specificity.

T-cell receptors differ from B-cell antibodies in that they can bind only one antigen at a time. In addition, each arm has a region that is used to bind to cell-surface markers on other cells within the organism's body. It is this binding that keeps T cells from duplicating the work of B cells. The "self" markers involved in this recognition are membrane-mounted proteins produced by the genes of the *major histocompatibility complex (MHC)*. As we will see, the MHC molecules are active participants in the T-cell response.

Self-protection: the MHC system There are two general types of MHC molecules: MHC-II proteins are found on the membranes of B cells and certain immune-system cells which are specialized for antigen presentation and are located in the tissues; MHC-I proteins are found on all other cells in the body. This dichotomy reflects the dual role of T cells as modulators of B-cell activity and assassins of disease-infected cells: cells bearing MHC-II proteins participate in regulation of the immune system, while those with MHC-I molecules can be killed by cytotoxic T cells.

MHC molecules, like the immune-system antibodies and receptors we have already discussed, are composed of two chains, and involve both constant and variable regions (Fig. 15.12A–B); T-cell receptors probably evolved from MHCs. MHC molecules bind antigens and then "present" them on the surface of the cell for binding

15.12 Structure and function of MHC molecules

(A) MHC-II molecules are similar to T-cell receptors: they have a pair of distinct chains, each with a constant region (gray), a variable section that binds an antigen (color), and a portion (white) that is complementary to a corresponding T-cell receptor (C). (B) MHC-I molecules are functionally almost identical to MHC-II, though structurally they are dissimilar. They too bind antigen (though the two variable domains are on the same chain) and a matching T-cell receptor (D). (C–D)When either MHC molecule binds to a T-cell receptor, the antigen-binding portions of the T-cell receptor and MHC protein each interact with different parts of the antigen they hold in common. The binding is stabilized by a cluster determinant protein (CD4 or CD8).

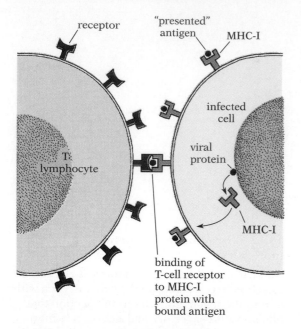

15.13 Antigen display by MHC-I molecules

In this model, MHC-I molecules in the cytoplasm of a cell bind antigens and carry them to the membrane for display. A T cell with a specific affinity for that antigen will bind to it and the MHC-I simultaneously, and become activated.

by appropriate T-cell receptors (Fig. 15.12C–D). This binding is then stabilized by a specific class of glycoproteins called cluster determinants: CD8 for MHC-I, and CD4 for MHC-II.

How the cell-mediated response develops Like their humoral counterparts, T lymphocytes begin as antigen-specific virgin cells. The first step in their activation occurs when MHC-I molecules in an infected cell begin presenting pathogenic antigens. The antigens are usually viral-coat proteins synthesized by the infecting organism in preparation for its further spread in the host. MHC-I molecules probably intercept these foreign compounds in the cytoplasm or on the ER, and then carry them to the cell surface for display (Fig. 15.13). If the antigen is too large to fit in the MHC pocket, it is "processed" (digested) into smaller pieces functionally equivalent to haptens. Though MHC molecules have a variable region, it is not very specific: a given MHC can bind 10–20% of the antigens it encounters.

Once the MHC-antigen complexes are mounted on the surface of an infected cell, a matching virgin T cell will eventually bind to both the antigen and the MHC-I protein. Thus stimulated, a virgin T cell grows and divides to produce the lymphocytes involved in the immediate response, as well as the memory cells that make future reactions more rapid.

The simplest variety of activated T lymphocyte is the cytotoxic T cell mentioned earlier. It destroys cells bearing the MHC-I–antigen complex to which its receptor can bind by perforating the infected cell (Fig. 15.14), which allows water to enter and burst it. Another class of specialized T lymphocytes—the helper cells—react to

15.14 Cell-mediated immune response

When a virgin T lymphocyte binds to an infected antigen-bearing body cell, it becomes activated, grows, and begins dividing to generate memory cells (which permit faster responses when the antigen is encountered again), cytotoxic T cells (which kill infected cells), and helper T cells (which regulate the responses of B lymphocytes, cytotoxic T cells, macrophages, and other elements of the immune system to the same antigen).

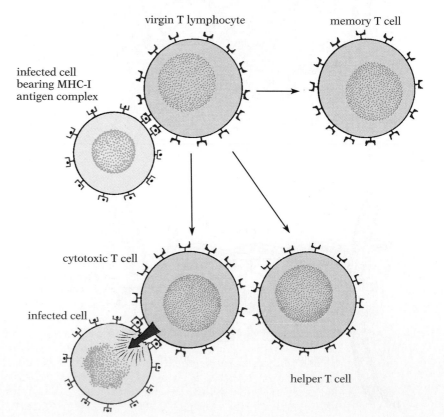

MHC-II–antigen complexes, and are involved in modulating the immune response.

■ HOW T CELLS FINE-TUNE THE RESPONSE

So far we have seen several strategies for capturing antigens: cells with MHC-I complexes collect antigens found in the cytoplasm; the immune-system cells with MHC-II proteins actively import antigens, process them if necessary, and then display them on the MHC-II complex; B cells use their membrane-mounted antibodies to collect specific antigens (Fig. 15.15A); and T cells use the T-cell receptor. *Presentation cells*, whose structure we have not yet explored, have neither receptors nor antibodies to capture antigens. These curious cells disperse from the bone marrow to the tissues, differentiate in tissue-specific ways, and take up their role as antigen presenters. They continually and unselectively endocytose material from the surrounding tissue fluid, process it, and then display the fragments on their MHC-II molecules.

Helper T cells begin their modulation of the immune response by binding to MHC-II proteins displaying the appropriate processed antigens on the cell surface of other immune-system cells (Fig. 15.15B). However, helper T cells do not attack the antigen or its source directly; instead, once bound to B cells or presentation

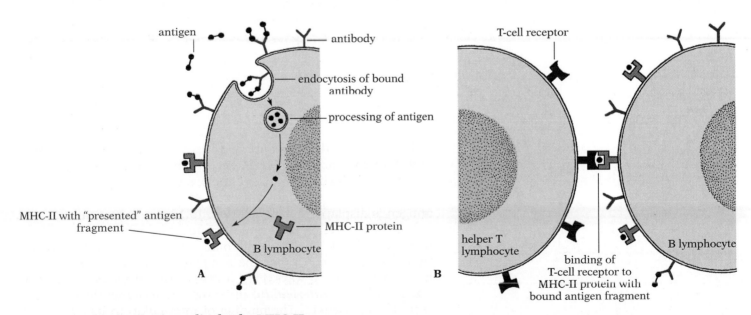

15.15   Antigen display by MHC-II molecules

MHC-II molecules are found in B lymphocytes and presentation cells. (A) When the membrane-mounted antibodies of a B cell (shown here) or the receptors of a cytotoxic T cell bind their particular antigen, the antibodies are taken in through endocytosis. Antigens are removed in special proteolytic (protein-digesting) vesicles and, if necessary, broken into convenient lengths; these fragments are then bound by MHC-II molecules in the cytoplasm and displayed on the cell surface. (B) When a helper T cell specific for that antigen is encountered, the two lymphocytes bind and the T cell becomes active. (Activation by a presentation cell, not shown, involves nonselective endocytosis of intercellular material, processing, and display on its MHC-II molecule.) This system of indirect activation is essential if the immune reaction is to be accurately modulated.

15.16 Modulatory action of T lymphocytes

(A) Helper cells regulate the humoral immune response by binding to B lymphocytes that display a unique antigen. (B) The bound helper secretes various interleukins, one of which causes the helper to multiply; another form of interleukin induces the B cell to secrete antibodies, which bind the antigens on the invading cell (A). (C) In the cell-mediated response (which normally would be active simultaneously only if the invading microorganisms can infect host cells), a third kind of interleukin induces nearby bound cytotoxic T lymphocytes to kill their targets. (D) A fourth interleukin activates nearby macrophages to ingest antibody-marked targets.

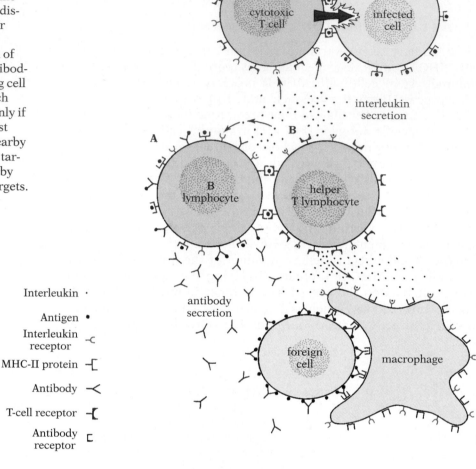

Interleukin	·
Antigen	•
Interleukin receptor	
MHC-II protein	
Antibody	
T-cell receptor	
Antibody receptor	

cells displaying the appropriate antigenic determinant, they help orchestrate the activation of these elements of the system.

Helper cells become aware of the presence of an antigen when it is exposed in the binding cleft of an MHC-II protein. Whether the antigen is displayed by a B cell (Fig. 15.16A) or a presentation cell, only a helper cell specific for the antigenic determinant *and* the MHC-II molecule holding it can bind and become activated. An activated helper can advance the campaign against the antigen and its source in several ways. First it installs receptors for a chemical-signal molecule, ***interleukin***, in its own membrane; then it begins secreting interleukin. The binding of interleukin to its own receptors causes the helper cell to proliferate (Fig. 15.16B). The secreted interleukin also induces multiplication of any activated cytotoxic T cells nearby that have recently encountered their specific antigens; more often than not, the helper cell and its cytotoxic neighbor will be responding to the same pathogen (Fig. 15.16C).

When bound to stimulated B lymphocytes, helper T cells produce an interleukin that encourages the B cells to secrete antibodies (Fig. 15.16A). Yet another kind of interleukin energizes nearby macrophages (Fig. 15.16D).

This two-step activation mechanism—in which B lymphocytes and cytotoxic T cells must both bind an antigen *and* be induced by a helper T cell specific for the same antigen—is probably a safety feature designed to prevent the potent cell-destruction capacity of the immune system from making mistakes. In particular, this double-checking strategy prevents the immune system from erroneously attacking the organism's own proteins, which can lead to a slow and sometimes fatal process of self-digestion. We will look at the problem of recognizing and ignoring friendly molecules in the next section.

One consequence of over stimulation by helper T cells is an excess of antigen-specific mast cells, and therefore a hypersensitivity to the antigen in question (Fig. 15.17). Even a slight exposure to the antigen under these conditions can lead to a massive release of histamine, triggering excessive loss of fluid from the blood. The result is an allergic reaction that, if extreme, can cause anaphylaxis: loss of consciousness as blood pressure falls, and even asphyxiation as fluid-induced swelling in the throat flattens the trachea (windpipe).

15.17 Mast cell
Massive exocytosis of histamine by a mast cell from a rat.

■ INITIAL CALIBRATION: RECOGNITION OF SELF

While the immune system of an organism is able to recognize and destroy almost any foreign antigen or antigen-bearing cell, it must not confuse its self antigens—the cell-surface proteins on its own cells—with foreign antigens. This recognition of self, or self-tolerance, is learned early in the embryonic development of the immune system: if a piece of tissue from one organism is transplanted into or grafted onto another adult animal, the recipient's immune system will almost always reject the tissues by mounting an immune reaction against the donor's cell-surface proteins. However, foreign tissue transplanted prenatally, while the recipient's immune system lacks experience, will be accepted; transplants from that same donor will be accepted later, even when the recipient is an adult.

Self antigens are generally membrane-anchored glycoproteins of the glycocalyx (see pp. 114–115). Some are specific to particular tissues, and seem to play a role in cell-type recognition and cell-to-cell adhesion. Others are encoded by the MHC, which devotes at least seven genes to the task, each of which can exist in as many as 100 versions. The diversity of MHC proteins is important if they are to bind and display the full range of processed antigens; it also has important consequences for cell recognition.

Inactivation of virgin cells Self-tolerance originates during fetal development when a process of selective lymphocyte inactivation takes place. Millions of virgin B cells are released into the prenatal blood at a time when no foreign antigens are present. Because any cells whose antibodies bind during this stage of development must be attaching themselves to normal proteins, the immune system is designed to inactivate B cells that bind prenatally. This still leaves vast numbers of virgin cells specific for nonself antigens capable of proliferating and mounting an immune response should the need arise.

The inactivation of T cells is more complicated. A T cell must be repressed not only if the receptor binds to a normal protein present during fetal development, but also if it *fails* to bind to one of the organism's MHC complexes. Some such T cells are actually killed, in a process known as **clonal deletion**. Inactivation or deletion is essential if T cells are to perform their normal role of killing infected cells and modulating the activity of other immune-system cells.

Inactivation is an ongoing process that depends on the continued presence of each self antigen. Even with continued exposure, however, inactivation may not be permanent: should control over inactivation be loosened, as by infection with certain bacteria or viruses, an **autoimmune disease** can result. One example is myasthenia gravis, a neurological disorder involving the loss of tolerance of the body's billions of receptors for acetylcholine, a chemical that is used for communicating between nerve cells and muscles. The result is a debilitating deterioration of muscle control. Multiple sclerosis, rheumatoid arthritis, and Type I diabetes are among the more familiar autoimmune diseases.

Why transplants fail When a foreign cell is introduced into an organism as part of an organ transplant, it bears several different MHC-I proteins on its surface. The diversity of MHC proteins plays an important role in cell recognition, as we have said. Most of the MHC variants on the cells of the transplant are novel to the host's immune system, and so B-cell antibodies will bind to them and trigger a humoral response. At the same time, the overwhelming odds are that the foreign cell and the host share at least one MHC-I variable region; as a result, at least one class of T-cell receptor is able to bind to one region of this one kind of MHC protein. Unfortunately, the rest of that foreign MHC molecule almost certainly contains novel regions, and thus is seen as an antigen by the T cell; as a result, a cell-mediated response is also triggered.

The resulting two-pronged immune reaction leads to transplant rejection. This attack on transplanted tissue can be suppressed with certain drugs (like cyclosporin, which blocks an enhancer of interleukin genes), but not without cost: the recipient, whose entire immune system is now depressed, is at risk even from common colds. A recipient's lymphocytes may, with the passage of time, come to learn to ignore the foreign antigens, and then immunosuppression therapy can be relaxed or discontinued.

■ SUBVERTING THE IMMUNE SYSTEM: AIDS

The adaptive value of the vertebrate immune system is graphically illustrated by **acquired immune deficiency syndrome (AIDS)**, a disease that destroys most of the immune response by eliminating the all-important helper T cells.

The AIDS virus The virus that causes AIDS (human immunodeficiency virus type 1, or HIV-1) is a complex retrovirus (Fig. 15.18). It consists of two copies of its RNA genome, each coated by a pair of proteins (called P7 and P9), and all enclosed by two very differ-

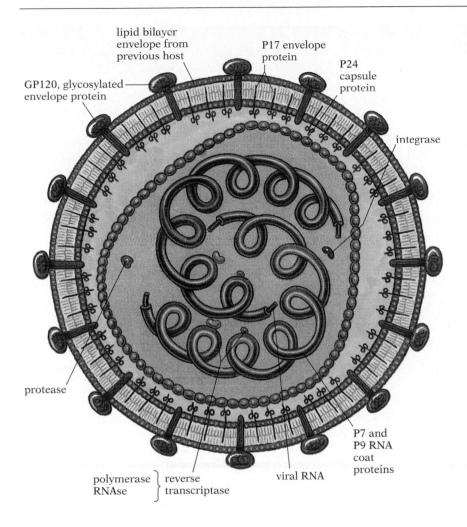

lipid bilayer envelope from previous host

P17 envelope protein

P24 capsule protein

GP120, glycosylated envelope protein

integrase

protease

P7 and P9 RNA coat proteins

polymerase RNAse } reverse transcriptase

viral RNA

15.18 Structure of HIV

The two copies of the viral RNA are protected by a protein coat and enclosed in a capsule. Also contained in the capsule are reverse transcriptase enzymes (which convert the single-stranded RNAs into double-stranded DNA copies), integrase (which inserts the DNA version into the host genome), and protease (which catalyzes the budding off of the virus). The capsule is itself enclosed in a bilayer membrane obtained from the previous host cell; in the membrane are mounted P17 (the remains of a protein involved in budding off) and GP120 (the glycoprotein that binds the helper T-cell receptor and so enables the virus to gain entry into its host).

ent layers. The inner layer is a self-assembling structure of one kind of protein (P24). The outer envelope is more complex: its primary constituent is a lipid bilayer obtained from the host cell as the virus buds off (Fig. 15.19), but embedded in this vesicle are one sort of protein facing inward (P17) and a glycoprotein (GP120) directed outward. HIV locates and enters its host by means of GP120, which binds specifically to the CD4 protein of helper T cells.

The HIV-1 retrovirus carries four enzymes. The first and second work together as the reverse transcriptase: a polymerase synthesizes a complementary DNA copy of the single-stranded RNA genome, and an RNAse separates the viral template from the complementary DNA allowing the polymerase to complete the job of making the HIV DNA double-stranded for insertion into the host genome. The third enzyme, an integrase, cuts the host DNA and inserts the parasite's own instructions into the genetic library. The fourth enzymatic passenger, a protease, activates certain viral proteins when needed.

The viral genome contains several regulatory sequences. The virus can exist in a low-impact, wholly lysogenic or semilytic phase, during which it produces a slow stream of progeny. As with

15.19 HIV viruses budding from a cultured lymphocyte

MONOCLONAL ANTIBODIES

A growing understanding of the workings of the immune system has enabled researchers to develop techniques for learning about cellular architecture—approaches that can provide new medical applications while they help reveal the structure of the cell. In one such technique, researchers select and clone a single type of lymphocyte—typically a B lymphocyte—whose antibodies bind to an antigenic region on a particular kind of cell, a specific structure in a cell, or some other particular substance. This **monoclonal antibody technique**, as it is called, can be used to locate all the tubulin in a cell, or the sodium-potassium pumps, the F_1 complexes of mitochondria, the RNA polymerases, or any of the thousands of enzymes whose distribution and function in a cell are not yet well understood.

To clone antibodies specific to a particular substance, researchers first inject that substance into mice (see figure). Soon there is a substantial increase in the B lymphocytes producing various antibodies specific to the many different antigens on the foreign cell or chemical. Most of these lymphocytes are producing antibodies to antigens common to many different foreign cells or chemicals; however, a few may, by chance, be producing antibodies specific to an antigen unique to the foreign material. If just these lymphocytes could be selected and cultured, they would produce a supply of the single antibody desired.

There are two major problems: the specific antibody-producing cells must be separated from the others, and they must be propagated. In the actual cloning procedure, the second of these problems is solved first. Normal cells have a set number of cell divisions, after which they age and die, while cancer cells are immortal. If the lymphocytes were cancerous, they would multiply indefinitely. Hence the next step is to force cells of the mixed collection of lymphocytes to fuse with cells from a cancerous line of B cells. Those that fuse become immortal, and are called hybridomas.

This still leaves the problem of selection. It is necessary to eliminate all but the hybridoma cells, and then to select those producing the desired antibody. The elimination is relatively easy. The nonhybrid lymphocytes from the mouse simply die off after completing their set number of cell divisions. The more difficult problem of sorting out the nonhybrid cancerous lymphocytes is accomplished by using a line of cancer cells lacking a crucial synthetic pathway. When these cancer cells are cultured in a medium without the substance they cannot make, they die. Only the hybridomas will survive, since they are able to synthesize the missing nutrient because of genes from the mouse lymphocytes, and are immortal by virtue of genes from the cancer cells.

The next step is to separate the individual hybridoma cells,

to culture each one separately, and then to test each culture by adding the antigen. If the cells of a particular group are producing the desired antibody, the test will result in antigen-antibody clumping. These cells can then be cultured further, to form a clone of immortal cells producing a continuous supply of a specific antibody. Researchers have even been able to

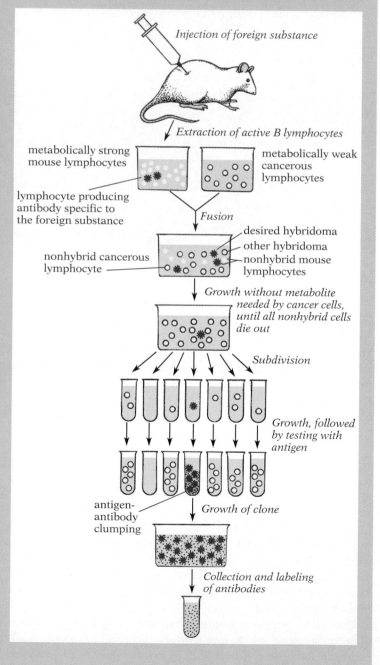

Injection of foreign substance

Extraction of active B lymphocytes

metabolically strong mouse lymphocytes

metabolically weak cancerous lymphocytes

lymphocyte producing antibody specific to the foreign substance

Fusion

desired hybridoma
other hybridoma
nonhybrid mouse lymphocytes

nonhybrid cancerous lymphocyte

Growth without metabolite needed by cancer cells, until all nonhybrid cells die out

Subdivision

Growth, followed by testing with antigen

antigen-antibody clumping

Growth of clone

Collection and labeling of antibodies

transplant the antibody genes from certain hybridoma cells into plants, and (via the phage lambda) into *E. coli*, where these immunoglobins are produced far more rapidly and cheaply.

Researchers use monoclonal antibodies to tag specific cells, cellular structures, or chemicals by labelling the antibodies with fluorescent dye, the electron-dense substance ferritin, or a radioactive marker. Once a labelled antibody has reached its target, the location of the antigen can be found with a light microscope, an electron microscope, or any of a variety of radioisotope analyses. A labelled antibody can even be used to mark certain kinds of cancer cells selectively for destruction; ovarian cancer cells, for example, can be bound by ferritin-linked antibodies and then destroyed by ferritin-specific T cells. The monoclonal antibody technique, then, has enormous potential both for basic research and for medical technology.

more conventional viruses, induction into the fully lytic state seems to require a shock or threat to the host cell, including, for example, its participation in an immune response. The virus reproduces both by replicating its entire genome and by producing mRNAs that subsequently direct synthesis of the envelope protein GP120, the "P" proteins, and the integrase and protease. The proteins and RNA aggregate in the host membrane, with only the GP120s facing out (Fig. 15.20). When a sufficient collection of membrane proteins accumulates, the membrane begins to bud, and the protease activates the enzymes destined to be packaged with the virus, as well as the proteins of the coat, or **capsid**. Once the capsid is formed, the virus buds off and is free to bind another host cell.

How AIDS attacks the immune system Initially, the immune system responds normally to HIV, with B cells releasing suitable antibodies, helper T cells amplifying the response, and macrophages consuming free viruses. But the virus lives on, hidden in helper T cells and, undigested, in macrophages. Though the total population of T cells remains normal for about a year (on average), infected cells slowly begin to be recruited into the lytic or semilytic phase. As GP120 antigens begin to appear on the surface of T cells, these hosts are destroyed by cytotoxic T cells and NK cells, but this is not of any real benefit. Not only is the body's arsenal of helper T cells being culled (dropping, on average, to 25% of normal 3 years after infection); GP120s are released into the blood and lymph, where they bind to other helper cells, which, though not actually infected, are then attacked and killed. Worse, an infected cell with GP120 on its membrane can bind to other T cells; just as the viral envelope fuses with the host, the membrane of the infected cell fuses with the bound cell. Dozens or hundreds of T cells are drawn into this fusion event, with the result that all die together, either at the hands of killer or cytotoxic cells, or because the resulting multinucleate cell becomes too large to function.

AIDS also affects the nervous system. Macrophages can cross from blood vessels into the nervous system where, as the viruses they harbor escape, HIV infects the glial cells that insulate neurons. As the insulating cells die, conduction becomes slower and less efficient, and the accuracy with which new neurons are routed to their targets declines.

Things go from bad to worse. Soon the number of helper cells is down to 5% of normal or below, and other infections that come along face only a minor immune response. AIDS patients generally

15.20 Assembly of a new HIV virus

Assembly occurs on the host-cell membrane where viral RNA, GP120, the P-protein precursor, and the long P-protein–enzyme precursors aggregate. When the collection of parts is sufficient, the piece of membrane begins to bud outward. Then a protease is freed and quickly cuts out the many enzymes and P-proteins, allowing coating of the RNA and capsid formation, steps that lead to complete budding off.

die 5–10 years after infection, often of diseases that pose no particular threat to humans with healthy immune systems.

The epidemiology of AIDS Studies of how AIDS spreads indicate that HIV is almost always passed from the blood or seminal fluid of one individual to the circulatory system of another. Hence the disease spreads readily through transfusion of infected blood, reuse of hypodermic needles, and anal intercourse. The virus is also transmitted, though at a lower rate, through vaginal intercourse.

The spread of AIDS (like any epidemic) depends on the average incubation period of the infective agent, the length of the infectious period, the rate of contact (most often, in this case, the number of sexual partners or shared needles), and the efficiency of transferring the infection with each encounter. Most of this information can now be estimated for the United States. It appears that the average time from infection to death is about 8 years (the mortality rate as of 1995 was essentially 100%), and carriers are infectious throughout this period, though more so early and late in the cycle.

Among American homosexuals and intravenous drug users, the infection rates are as high as 70% in urban areas like San Francisco and New York City, where contact rates are very high; on average, each infected individual communicates the disease to several others. In less urban regions, where the contact rate is lower and the susceptible pool of potential victims smaller, the spread is slower. Overall rates of transmission in the United States are declining primarily because a large portion of the high-risk group is already infected; several million Americans are thought to carry the virus.

Among heterosexuals, both transmission rates and contact rates are lower. It is not clear whether there will be an epidemic among heterosexuals in Europe or the Americas: the statistic that defines an epidemic is whether the typical carrier will infect, on average, more than one other individual. In Africa, HIV transmission is primarily heterosexual, and risk correlates with the number of sexual partners, which is larger on average than in most other parts of the world. The World Health Organization estimates that nearly 60% of all AIDS deaths, which totalled 3.6 million through 1995 (and will rise beyond 20 million by the end of the decade), have occurred in Africa, where millions more are infected. Clearly a heterosexual epidemic *is* possible.

Can AIDS be cured? Early attempts to combat AIDS suffered from a lack of knowledge about the mode of transmission of this disease, its long latency, and ignorance of its elaborate life history. Now that more is known, one thing is immediately obvious: though nearly all conventional vaccines use a surface antigen as the means for teaching the immune system to recognize a particular disease, this approach will not work with AIDS since a surface-antigen vaccine merely prepares the body to destroy its own immune-system cells. Instead, some step in the life cycle must be blocked. Preventing binding by masking the CD4 molecules is useless, since this would inactivate the immune response. Blocking the GP120 active site is a possibility — unless, of course, it turns out to be an analogue of the MHC molecule to which the T-cell receptors normally bind, which seems all too likely.

Preventing a step in viral duplication inside the infected cell is more promising, though that step must be unique to the virus or the treatment will be toxic to the host. A good example of this trade-off is seen with 3'-azido-2',3'-dideoxythymidine (commonly called azidothymidine, or AZT; Fig. 15.21). This thymidine analogue is actually preferred by HIV's DNA polymerase over thymidine, whereas host-cell polymerases prefer conventional thymidine. When AZT is incorporated into a growing DNA chain, replication stops because the host-cell polymerase is unable to add the next nucleotide. At suitably low doses of AZT, host cells replicate fairly normally but HIV reproduces less successfully. There are side effects, however; these include substantial attrition in the bone marrow (where rapid production of blood cells is always under way) and often anemia. Nevertheless, AZT can delay, for about a year, the death of patients with clear symptoms.

One big risk with AZT is that the HIV polymerase gene, which has an unusually high error rate (one mispairing per 2000 bases in a genome only about 10,000 nucleotides long) will mutate to discriminate against AZT. The search for a more practical and effective treatment is being intensively pursued. An antisense RNA complementary to part of the HIV genome, for instance, would be a better treatment because it would be completely specific and nontoxic. To make it work, however, means must be found to deliver it efficiently to target cells and make it resistant to RNAse. Injecting the portion of CD4 to which the virus binds holds promise as well. At the moment, though, prevention through behavioral changes among those at risk seems the most practical alternative.

15.21 AZT

AZT is a DNA thymidine mimic. The base (T) is normal, but the deoxyribose unit (R) differs from DNA's sugar in that the hydroxyl group to which the next nucleotide is attached is replaced by N_3, which cannot be used by DNA polymerase.

THE SOURCE OF ANTIBODY DIVERSITY

■ SHUFFLING EXONS

How is the vertebrate genome able to produce an almost infinite variety of antibodies and T-cell receptors without employing millions or billions of different genes, one for each potential antigen? The answer to this question involves a remarkable strategy of exon selection, which operates initially during the fetal development of the immune system, and leads to the vast diversity of lymphocytes available to the organism during its lifetime. In B cells, the process involves selection from a pool of exons in only three genes—one for the heavy chains and one for each of the two classes of light chains. The strategy used for T cells is identical in most details.

Let's look at the chromosomes of a developing human B lymphocyte in the bone marrow before the cell specializes to produce a single type of antibody. In the human genome, the part of the DNA coding for the constant regions of the heavy chain is composed of 22 exons arranged in five sets, one corresponding to each class of heavy chain—A, D, E, G, and M. Each set contains exons for re-

15.22   Antigen-binding site of an antibody molecule

The binding site is a pocket formed by the inter-action of a heavy chain and a light chain. The sectors labeled C_H3, C_H2, hinge, C_H1, J_H, D, V_H, C_L, J_L, and V_L on this schematic representation are encoded by separate exons in the genes. The five different classes of heavy chains have differ-ent C_H2 and C_H3 regions; class E and class G chains have a fourth C_H region. Only one of the antigenic determinants of the antigen (beige) is bound by this antibody. Completely different antibodies may bind to separate sites on the same kind of antigen. (The left half of the antibody beyond the hinge is omitted.)

15.23   Organization of immature and fully mature heavy-chain antibody gene, and corresponding mRNA

(A) The immature heavy-chain gene originally contains five sets of constant-region exons (one set for each class of heavy chain), four different joiner exons, approximately 12 different D-region exons, and about 200 different variable-region exons. (B) When the gene is fully mature, many of the exons have been removed, and the remaining exons and introns are transcribed. (C) How-ever, only the exon closest to the 5' end of the primary RNA transcript of each exon group survives processing and appears in the mature mRNA. The two steps of exon removal enable a B lymphocyte to produce a unique kind of antibody from millions of possible alternatives.

gions C_H1, hinge, C_H2, C_H3, and, in the sets that code for heavy chains of classes E and G, C_H4 (Fig. 15.22).

This arrangement of exons in the heavy-chain gene seems at first glance to be counterproductive, since so much has to be tran-scribed that is ultimately not used: thus, even if the virgin lympho-cyte later produces only antibodies with class A chains, exons for the other classes would be transcribed as well. In the portion of the gene encoding the variable region, the organization seems even more counterproductive: there are four different exons for the J_H (joiner) region, even though each heavy chain has only one J_H seg-ment. Still farther toward the 5' end, are approximately 12 differ-ent exons for the D region. Again, each antibody produced by a virgin cell has only one D segment. Still farther toward the 5' end lies a string of roughly 200 different V_H exons, only one of which is eventually translated.

How is it, then, that the roughly 240 exons in the original heavy-chain antibody gene give rise to the protein product encoded by only seven or eight exons? Exon removal seems to be accomplished in two stages. In the first step, which takes place during fetal devel-opment, large regions of the gene are cut out of the chromosome, and the remaining ends are spliced together. One such excision, for example, extends a random distance from the D region into the V_H exons (Fig. 15.23B). Any number of V_H exons from 0 to 199 may be removed. The first V_H exon not excised is the one ultimately translated and expressed. During transcription, RNA polymerase copies all the remaining V_H exons on the 5' side, but these superfluous transcribed exons are removed during RNA processing to yield an mRNA with only the first V_H exon that survived excision. As a result, only the V_H exon closest to the D-region exon will be translated (Fig. 15.23C). All but one transcribed joiner exon and D-region exon are removed by means of two similar steps. A corresponding process is seen in the gene for the light chain, which has one C_L exon, four alternative joiner exons, and about 300 alternative V_L exons.

The diversity made possible by an active but random assembly of exons like the one just described is vast. For instance, since the gene for the heavy chain has four alternative joiner exons and 12 alternative D-region exons, there are 48 (4 x 12) possible joiner–D-region combinations. Moreover, since there are roughly 200 V_H alternative exons, a lymphocyte can produce antibodies with one of 9600 (48 x 200) forms of the variable region of the heavy chain. Similarly, the light chain, with its four alternative joiner exons and 300 V_L exons, can be produced in 1200 different forms. Since each antibody consists of one kind of randomly selected heavy chain and one kind of light chain, these different forms give rise to about 12 million (1200 x 9600) different antibodies during the development of the immune system. Additional variation is generated by a process of active mutation in the joiner and D-region exons and in parts of the V_L and V_H exons. Since the mutations take place in developing B cells, which are somatic cells, they are unique to each cell and are not perpetuated in the germ line.

■ HYPERMUTATION

As an immune response gets under way, the affinity of the B-cell antibodies produced by some cells in the activated clone actually increases. This fine-tuning involves yet another level of immunological learning. After a virgin cell is stimulated by an antigen, the exons for the variable regions of the antibody genes are replicated, during which time mistakes are permitted at a million times the normal rate. This error rate means that roughly half of the daughter cells will have slightly modified versions of the original antibody. Because the further cloning of cells from this first generation of daughter cells depends on how well each lymphocyte's version of the antibody binds the antigen on the cell membrane, a mutant with a better match will outreproduce its clone mates. As a result, the cells producing the better-binding antibody come to dominate the response. Here is a kind of natural selection in miniature.

HOW THE IMMUNE SYSTEM EVOLVED

The molecules of the immune system are related to cell-adhesion molecules (CAMs) described in Chapter 14. Of course, differential cell adhesion is critical for immune-system cells: not only must they attach themselves to their targets; lymphocytes must also be able to move to and from the blood, lymph, and any tissue that is

EXPLORING FURTHER

SLEEPING SICKNESS: HOW THE TRYPANOSOME CHANGES ITS COAT

Sleeping sickness is one of the most debilitating diseases and remains virtually impossible to treat. Its cause is a protozoan known as a trypanosome (see figure), which bloodsucking tsetse flies carry from an infected host to a potential host. The trypanosomes multiply in the blood of the host and later spread into the nervous system. Symptoms begin with fever and fatigue, and progress to drowsiness during the day and insomnia at night. The victim ultimately becomes too sleepy to eat, then comatose, and finally dies.

Since the trypanosomes are in the bloodstream of the host for weeks before causing death, how is it that the immune system, with its millions of antibodies generated by exon shuffling, fails to react effectively? The answer is that during these weeks the trypanosomes are shuffling their genes as well.

Trypanosomes are unusual in having only one kind of protein exposed on their cell membranes. Within a few days of infection the immune system of the host mounts a full-scale immune response. By the fifth day, however, all the surviving trypanosomes have changed their surface proteins. The trypanosome genome has hundreds—perhaps thousands—of surface-protein genes, and a new one is transposed into an active site in the genome roughly every 5 days. The number of trypanosome antigens increases by the fifth day from the few present at the time of infection to a great many, a different one on nearly every surviving trypanosome. As a result, the existing immune reaction becomes irrelevant, and a new one, responsive to a larger set of antigens, must begin from scratch: a new, different set of lymphocytes must now bind to the parasites, to begin again the process in which virgin B and T lymphocytes are activated and divide, antibodies are produced, and so on.

Again, even if the battle is being won, the surface-protein genes are shuffled. New surface proteins appear, and the number of new antigens multiplies. Eventually, the body's immune system is overwhelmed. Transposition in trypanosomes, then, seems to be a precisely controlled event. Transposons may be important elements in many diseases.

4 μm

A trypanosome and a red blood corpuscle

releasing the chemical signals that draw these cells to the site of trouble. In fact, the molecules on which the lymphocytes rely evolved from proteins that animals have been using for hundreds of millions of years—long before the first immune systems appeared. For example, some vertebrate immunoglobulins are virtually identical to CAMs discovered in *Drosophila*, where they are involved in axonal guidance. As outlined in Chapter 17, the process involves repeated gene duplication, followed by the independent evolution of the different copies.

Let's look closer at a partial cast of these molecular actors (Fig. 15.24). One of the most telling similarities is between the light

15.24 Some molecules used by the immune system

Most of the proteins and protein complexes used by immunocytes for cell binding fall into three molecular "families," each of which is part of a larger family of developmental-control substances. The similarities are obvious at a glance: compare the structure of antibodies to MHC complexes, T-cell receptors, and CAMs. Other immune-system-specific members of developmental control families are shown in the lower half of the figure.

cell-adhesion molecules   T-cell receptors

CD4   CD8

T-cell accessory   MHC-I MHC-II Antibody
binding molecules

Integrin molecules

Integrin molecules   Selectin molecules

chains of antibody molecules, the α chain of the T-cell receptor, and the cell-adhesion molecule ICAM-2. This particular CAM is expressed on the membranes of cells in inflamed regions (in response to locally secreted chemicals like the interleukins) and thus encourages T-cell adhesion. ICAM-2 is also very similar to the MHC molecules, to CD2 (which helps T cells find antigen concentrations in lymph nodes), to LFA-3 (the molecule to which CD2 binds), and the β chain of the T-cell receptor. Except that they are missing a transmembrane anchor, the light chains of antibody molecules also strongly resemble the members of this ancient cell-adhesion family.

Three other components of the T-cell receptor (ε, γ, and δ, which communicate the fact of T-cell receptor binding to the interior of the T cell) are just truncated versions of ICAM-2, while CD8 is very like ICAM-2 with a long insertion. CD4, the heavy chains of antibody molecules, and two other immune-system CAMs are essentially ICAM-2s with a doubled outer segment.

Two other families of developmental CAMs are also involved in the immune response. The immune system's **integrins** help lymphocytes home in on infection sites, and modulate T-cell proliferation. **Selectins**, for their part, regulate lymphocyte binding to tissue at sites of inflammation. In short, lymphocytes have evolved to use the selective CAMs of morphogenesis to track down infections, where they bind both normal cell receptors and unfamiliar antigens prior to exerting their specific effects. In a way, many lymphocytes are like undifferentiated stem cells in early development, migrating to their proper location by means of CAMs and chemical gradients. The cells of the immune system, however, come equipped with dozens of CAMs and other morphogenic binding molecules, and respond to any of a large number of gradients; hence, they have the potential to go anywhere in the body in search of cells broadcasting chemical cries for help. Selection has acted to transform the existing genetic systems of organismal development into a self-guided internal security system.

CHAPTER SUMMARY

HOW THE IMMUNE RESPONSE WORKS

THE CAST: CELLS AND ORGANS OF THE IMMUNE SYSTEM Antibodies bind to foreign antigens, marking them for engulfment by macrophages or destruction by killer cells; binding to antibodies on mast cells causes the release of histamine and other signalling molecules. These cells are found in both the bloodstream (particularly the spleen) and lymph (particularly the nodes). (p. 381)

WHAT THE HUMORAL RESPONSE ACCOMPLISHES The humoral response involves the B cells, each of which produces a unique antibody. When part of an antigen (an antigenic determinant, called hapten when it is isolated from the rest of the antigen) binds to a surface-mounted antibody on a B cell, the cell begins dividing and releasing antibodies into the blood. Some cells produced by this clonal selection specialize in antibody secretion while others become memory cells, which allow a faster reaction when the antigen is next encountered. Some of the secreted antibodies become mounted on mast cells. Others bind to the antigen directly, and trigger various reactions, including engulfment by macrophages, perforation by complement enzymes, and attack by natural killer cells. (p. 383)

WHAT THE CELL-MEDIATED RESPONSE ACCOMPLISHES The cell-mediated response involves the T cells. Each T cell mounts in its membrane a receptor molecule with unique specificity. The version of the receptor found on cytotoxic T cells can bind to haptens presented by cell-surface markers (MHC-I molecules) on the body's cells; the MHC-I molecules in each body cell constantly scavenge molecular fragments from within the cell and carry them to the surface for display. Cytotoxic T cells can kill infected cells displaying MHC molecules with antigen haptens. Clonal selection leads to the proliferation of T cells that bind antigen haptens. (p. 387)

HOW T CELLS FINE-TUNE THE RESPONSE Immune-system cells use their surface-mounted antibodies (B cells) or receptors (T cells) to bind circulating antigens; these are imported into the cell, where they are processed, and then displayed on the cell surface by the MHC-II molecules unique to the cells of the immune system. Helper T cells stimulate B cells with the same antigen specificity—that is, with MHC molecules displaying haptens to which the T-cell receptor can bind. (p. 391)

INITIAL CALIBRATION: RECOGNITION OF SELF Early in development, B and T cells that bind molecules in the body are deactivated or repressed (clonal deletion) to prevent an autoimmune response; at the same time, T cells that fail to bind the body's MHC molecules are deactivated. (p. 393)

SUBVERTING THE IMMUNE SYSTEM: AIDS The AIDS virus infects helper T cells and also survives engulfment by macrophages. As the virus slowly reproduces, it spreads to more T cells, eventually disabling the ability of T cells to stimulate immune responses. (p. 394)

THE SOURCE OF ANTIBODY DIVERSITY

SHUFFLING EXONS The number of possible antibodies is nearly infinite. Much of the diversity is accomplished by reassembling the antibody-coding region in each B and T cell as it matures, using a randomly selected exon from among many alternatives for each of several segments of the binding region of the antibody. (p. 399)

HYPERMUTATION Additional diversity is generated when a stimulated B or T cell begins to multiply. Mutations are induced in the antibody-coding regions of the progeny cells. As a result, any of these that have a higher affinity for the antigen are favored in further cloning. (p. 401)

HOW THE IMMUNE SYSTEM EVOLVED

The immune system molecules are closely related to the molecules involved in cell-to-cell recognition during development. Apparently the high specificity of these molecules was used to create a system for recognizing foreign chemicals. (p. 401)

STUDY QUESTIONS

1 Compare and contrast clonal selection with clonal deletion. (pp. 386–387, 393–394)

2 What would be the physiological effect on an individual born with a defect in either histamine release or reception? (pp. 385, 393)

3 What happens when, by chance, the virgin B cell for an antigen is never generated (that is, the appropriate antibody gene is not created by recombination and hypermutation) but the helper T cells are present? (pp. 388–393)

4 What would be the consequences of having a B-cell system that produced hybrid antibodies—that is, every cell produced antibodies, one arm of which bound to one antigen while the other bound to another? (pp. 383–393)

5 What happens to a cell if it suffers a mutation in its MHC gene? (Consider both immunocytes and other cell types.) (pp. 388–394)

6 How could the MHC system be exploited by the body to avoid inbreeding (mating with close kin)? (pp. 388–394)

SUGGESTED READING

ANDERSON, R. M., AND R. M. MAY, 1992. Understanding the AIDS pandemic, *Scientific American* 266 (6). *On the epidemiology of the AIDS epidemic from an ecological and evolutionary point of view.*

GREY, H. M., A. SETTE, AND S. BUUS, 1989. How *T* cells see antigen, *Scientific American* 261 (5). *A clear discussion of how processed antigens are "presented" by MHC proteins, and recognized by T-cell antibodies.*

HASSELTINE, W. A., AND F. WONG-STAAL, 1988. The molecular biology of the AIDS virus, *Scientific American* 259 (4). *An excellent description of how HIV-1 works, emphasizing gene regulation. This same issue has nine related articles on various aspects of the AIDS epidemic.*

MARRACK, P., AND J. KAPPLER, 1986. The T cell and its receptor, *Scientific American* 254 (2). *An excellent summary of how T cells interact with the other elements of the immune system.*

MILLS, J., AND H. MASUR, 1990. AIDS-related infections, *Scientific American* 263 (2).

OLD, L. J., 1988. Tumor necrosis factor, *Scientific American* 258 (5). *On the many messenger chemicals that immune-system cells use to regulate each other's activity.*

YOUNG, J. D.-E., AND Z. A. COHN, 1988. How killer cells kill, *Scientific American* 258 (1). *Compares the action of the complement system with the analogous strategy of killer cells in perforating the membranes of target cells.*

CHAPTER 16

INHERITANCE

Many observable traits are inherited. In dogs, for example, in addition to such obvious characteristics as height and build, strong breed propensities, such as the drive toward herding, hunting, or retrieving, are the result of inheritance. Thousands of years before the discovery of DNA humans used this knowledge to develop better grains, vegetables, fruits, and domesticated animals. Without understanding inheritance, or why selective breeding sometimes worked, breeders knew that when plants produce runners (reproducing asexually by cloning), the progeny were exactly like the parents. When plants reproduce sexually, however, breeders have long known that the progeny can be very different, not only from the parents but from each other as well.

The most common hypothesis until the beginning of this century, that of *blending inheritance*, predicted that progeny would always embody a recognizable blend of parental traits. Nearly everyone knew this hypothesis to be wrong, or at least highly incomplete: in populations of European descent, two brown-eyed parents sometimes have a child with blue eyes. In any group, traits absent for generations can mysteriously recur, while altogether new ones occasionally appear out of the blue.

In this chapter we will see how the discovery of the laws of inheritance made sense of these seeming contradictions, clearing the way for the discovery of their genetic basis. This understanding of how inheritance works makes possible a deeper appreciation of how selection operates to create evolution, the topic of Part III.

HOW SINGLE-GENE TRAITS ARE INHERITED

The Austrian monk Gregor Mendel was the first to make some sense of the conflicting phenomena of inheritance. In 1856, he began breeding garden peas. The experiments he performed over the next 12 years laid the groundwork for what is now called the *Chromosomal Theory of Inheritance*.

■ WHAT MENDEL FOUND

Mendel began with several dozen strains of peas, mostly purchased from commercial sources. He raised each variety for several years to discover which strains had recognizable morphological varia-

Table 16.1 Mendel's results from crosses involving single character differences

P CHARACTERS	F_1	F_2	F_1 RATIO
1 Round × wrinkled seeds	All round	5474 round; 1850 wrinkled	2.96:1
2 Yellow × green seeds	All yellow	6022 yellow; 2001 green	3.01:1
3 Red × white flowers	All red	705 red: 224 white	3.15:1
4 Inflated × constricted pods	All inflated	882 inflated: 299 constricted	2.95:1
5 Green × yellow pods	All green	428 green: 152 yellow	2.82:1
6 Axial × terminal flowers	All axial	651 axial: 207 terminal	3.14:1
7 Long × short stems	All long	787 long: 277 short	2.84:1

tions that bred true (in other words, that reappeared consistently in each generation).

Dominant and recessive traits In the end, Mendel reported his work on seven of the characteristics he studied. Each of these seven traits occurred in two contrasting forms: the seeds were either round or wrinkled, the flowers were red or white, the pods were green or yellow, and so on (Table 16.1). When Mendel bred plants with contrasting forms of just one of these characteristics, he found that all the offspring (usually referred to as the F_1, or first filial, generation) were alike and resembled only one of the two parents (the *P*, or parental, generation). When these offspring were crossed among themselves, however, some of their offspring (the F_2, or second filial, generation) showed one of the original contrasting traits and some showed the other (Table 16.1). Thus, a trait that had been present in one of the parents, but not in any of their children, reappeared in the third generation, just as blue eyes can reappear after a generation or more of brown eyes.

Mendel's cross of red-flowered with white-flowered plants, for example, produced red flowers in all plants of the F_1 generation (Fig. 16.1). Apparently one form of each characteristic takes precedence over the other: red takes precedence over white in the flowers. Mendel referred to the traits that appear in the F_1 offspring of such crosses as ***dominant characters***, and the traits that are latent in the F_1 generation (white flowers, for example) as ***recessive characters***.

When Mendel allowed the F_1 peas from the cross involving flower color, all of which were red, to breed freely among themselves, their offspring (the F_2 generation) consisted of 705 plants with red flowers and 224 with white flowers. The recessive character had reappeared in approximately one-fourth of the F_2 plants. The same pattern held true for each of the other six characters he described crossing (Table 16.1); in each case the recessive character disappeared in the F_1 generation, but reappeared in approximately one-fourth of the plants in the F_2 generation. We can

P generation
Red flower   White flower

F₁ generation 100% red

F₂ generation 75% red, 25% white

16.1   Results of Mendel's cross of red-flowered and white-flowered garden peas

summarize the flower-color experiment as follows:

$$P \quad \text{red} \times \text{white}$$
$$\downarrow$$
$$F_1 \quad \text{all red}$$
$$\downarrow$$
$$F_2 \quad \text{3/4 red} \quad \text{1/4 white}$$

Segregation These experiments led Mendel to formulate and test a series of hypotheses. Finally, he drew the revolutionary conclusion that each pea plant possesses two hereditary "factors" for each character, and that when gametes are formed the two factors segregate and pass into separate gametes. Each gamete, then, possesses only one factor for each character. Each new plant thus receives one factor for each character from its male parent and one for each character from its female parent. That two contrasting parental traits, such as red and white flowers, can both appear in normal form in the F_2 offspring indicates that the hereditary factors must exist as separate entities in the cell; they do not blend or fuse with each other. Thus the cells of an F_1 pea plant from the cross involving flower color contains one factor for red color and one factor for white color, the factor for red being dominant. These two factors remain distinct, and segregate unchanged when new gametes are formed. Mendel referred to this pattern of inheritance as the **Principle of Segregation**.

Mendel's conclusions are consistent with what we now know about the chromosomes and their behavior in meiosis: the diploid nucleus contains two of each type of chromosome, one homologous chromosome from each parent. Each of the two chromosomes in any given pair bears genes for the same characters; hence the diploid cell contains two copies of each type of gene; these are the two hereditary factors for each character that Mendel described. Since homologous chromosomes segregate during meiosis, gametes contain only one chromosome of each type and hence only one copy of each gene, just as Mendel deduced. At the time, however, no one had even observed chromosomes.

Mendel may have doubted his own conclusions: some of his unpublished results could not be reconciled with his model, for, as we will see, not all genetic characters can be categorized as simply dominant or recessive, while others do not segregate independently. Mendel abandoned his efforts in 1868. In 1900, after cell division and chromosome segregation had been observed and described in detail, three biologists—Hugo De Vries in Holland, Carl Corren in Germany, and Erich von Tschermak-Seysenegg in Austria—independently rediscovered the phenomenon of segregation and, almost immediately, Mendel's original paper.

A modern interpretation of Mendel's results Let's consider Mendel's results on flower color from a modern perspective. In the cells of a pea plant, a gene for flower color exists at the same position, or **locus**, on two homologous chromosomes (Fig. 16.2). The gene can exist in many different forms, or **alleles**, but no individual can have more than two—one on each homologous chromosome containing that gene. Genes are designated by letters, using capital letters for dominant alleles and small letters for recessives. Flower color comes from the presence of pigment molecules, whose struc-

gene loci

homologous chromosomes

16.2 Anatomy of segregation
Somatic cells have pairs of homologous chromosomes, a pair consisting of one chromosome from each parent. Each gene contains two copies, one on each chromosome of the homologous pair at corresponding loci. When the copies at the corresponding loci are different, they are called alleles. In meiosis, each gamete receives a copy of only one chromosome from each homologous pair, and hence only one of the two alleles.

ture is determined by the genetic information in the chromosomes. Mendel's red allele (*C*) encodes a functional pigment that absorbs all wavelengths of light except the red that we see reflected, while the white allele (*c*) encodes a nonfunctional pigment molecule.

Since a diploid cell contains two copies of each gene, it may have two copies of the same allele or one copy of one allele and one copy of another allele (Fig. 16.2). Thus the cells of a pea plant may contain two copies of the allele for red flowers (*C/C*), or two copies of the allele for white flowers (*c/c*), or one red allele and one white allele (*C/c*). Cells with two copies of the same allele (*C/C* or *c/c*) are said to be **homozygous** for that trait. Those with two different alleles (*C/c*) are said to be **heterozygous**. (The slash between letters indicates that the alleles are on separate chromosomes.)

We cannot tell by visual inspection whether a given pea plant is homozygous dominant (*C/C*) or heterozygous (*C/c*), because each type of plant has red flowers. Where one allele is dominant over another, the dominant allele takes full precedence over the recessive allele, and heterozygous organisms exhibit the dominant trait; one copy of the dominant allele is as effective as two. For this reason, there is often no one-to-one correspondence between the different possible genetic combinations, or **genotypes**, and the possible appearances, or **phenotypes**, of the organisms. In the example of flower color in peas, there are three possible genotypes—*C/C*, *C/c*, and *c/c*—but only two possible phenotypes, red and white. The heterozygote (*C/c*) produces a red phenotype because the red allele is dominant.

Just as we often cannot tell whether an individual is heterozygous or homozygous for a dominant allele, natural selection cannot act on genotypes directly. Hidden recessive traits are immune to selection; it is an organism's phenotype—the sum of its expressed traits—that matters in the struggle to survive and reproduce. Only when recombination produces a homozygous recessive, and the trait it encodes is therefore expressed, can selection operate on that allele.

The functional-nonfunctional dichotomy is the most common basis of dominant and recessive phenotypes. Wrinkled peas, for instance, result from a gene encoding an enzyme that polymerizes sugar to make starch. The recessive allele encodes a defective version of the enzyme. In the homozygous condition, this leads to a buildup of sugar, which draws extra water in osmotically; the developing seed thus becomes abnormally large. Water is removed from seeds when they mature; because these swollen seeds have more water and wall area than normal, wrinkles develop when the water is removed.

Using the modern conception of gene function, we can rewrite Mendel's pea cross to show both the genotypes and phenotypes:

P		*C/C* red	×	*c/c* white
			↓	
F$_1$		*C/c* red	×	*C/c* red
			↓	
F$_2$	*C/C* red	*C/c* red	*c/C* red	*c/c* white

Mendel began with a cross in the parental generation between a plant with a homozygous dominant genotype (red phenotype) and a plant with a homozygous recessive genotype (white phenotype). All of the F₁ progeny had red phenotypes, because all of them were heterozygous, having received a dominant allele for red (*C*) from the homozygous dominant parent and a recessive allele for white (*c*) from the homozygous recessive parent. However, when the F₁ individuals were allowed to cross freely among themselves, the F₂ progeny were of three genotypes and two phenotypes; 1/4 were homozygous dominant with the red phenotype, 1/2 (2/4) were heterozygous and showed red phenotypes, and 1/4 were homozygous recessive and produced white phenotypes. The ratio of genotypes in F₂ is 1 : 2 : 1; the ratio of the phenotypes is 3 : 1.

Figuring out the possible genotypic combinations in F₂ is simple in such a **monohybrid cross** (a cross involving only one character). All individuals in the F₁ generation are heterozygous (*C/c*); that is, they have one of each of the two types of alleles. Each allele is located on a separate homologous chromosomes and thus the two alleles segregate in meiosis. This means that half the gametes produced by a heterozygous individual will contain the C allele and half the c allele. When two such individuals are crossed (Fig. 16.3), there are four possible combinations of their gametes:

C from female parent, *C* from male parent
C from female parent, *c* from male parent
c from female parent, *C* from male parent
c from female parent, *c* from male parent

Notice that the first combination produces homozygous dominant offspring (red), the second and third produce heterozygous offspring (also red), while the fourth produces homozygous recessive offspring (white). Since each of the four combinations is equally probable, we would expect, if a large number of F2 progeny are produced, a genotypic ratio close to 1 : 2 : 1 and a phenotypic ratio close to 3 : 1, just as Mendel found.

An easy way to figure out the possible genotypes produced in F₂ is to construct a **Punnett square** (named after the English geneticist R. C. Punnett). Along a horizontal line, write all the possible kinds of gametes the male parent can produce; in a vertical column to the left, write all the possible kinds of gametes the female parent can produce; then draw squares for each possible combination of these:

Next, write in each box the allele letters for the female and male gamete (in that order) separated by a slash. Each box then contains the symbols for the genotype of one possible zygote combina-

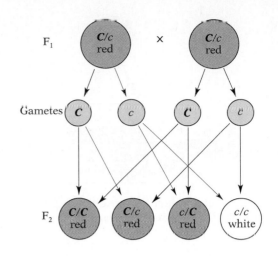

16.3 Gametes formed by F₁ individuals in Mendel's cross for flower color, and their possible combinations in F₂

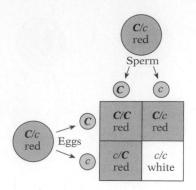

16.4   Punnett-square representation of the information shown in Figure 16.3

tion from the cross. The completed Punnett square in Figure 16.4 shows that the cross yields the expected 1 : 2 : 1 genotypic ratio and, since dominance is present, the expected 3 : 1 phenotypic ratio.

Whenever there is dominance, the expected phenotypic ratio is 3 : 1. If the samples are large and these ratios are not obtained, some complicating condition must be present. The failure of monohybrid crosses to yield the expected results has often led to the discovery of genetic phenomena that are critical to the understanding of evolution. Two of these phenomena—linkage and sex-specific effects—will be discussed presently. Others include strict maternal inheritance (which led to the discovery of organelle genes), genetic "imprinting" (which can also lead to early expression of only maternal or paternal alleles; see pp. 288–289), and alleles that eliminate any heterozygous allele on the homologous chromosome, thereby assuring their own transmission in all gametes.

■ PARTIALLY DOMINANT TRAITS

The seven characters of peas that Mendel reported on involved an allele, known as a full dominant, that showed complete dominance over the other. Some characteristics do not show this pattern, including some that Mendel himself studied but did not discuss. When a heterozygous individual clearly displays both alleles, the alleles are referred to as partial (or incomplete) dominants.

In many cases of partial dominance, heterozygous individuals have a phenotype that is intermediate between the phenotype of individuals homozygous for one allele and the phenotype of individuals homozygous for the other—a kind of blending inheritance. For example, crosses between homozygous red snapdragons or sweet peas and homozygous white varieties yield plants producing pink blossoms (Fig. 16.5). When these pink-flowering plants are crossed among themselves, they yield red, pink, and white offspring in a ratio of 1 : 2 : 1, as follows:

$$
\begin{array}{lcccc}
\text{P} & & R/R & \times & R'/R' \\
 & & \text{red} & & \text{white} \\
 & & & \downarrow & \\
\text{F}_1 & & R/R' & \times & R/R' \\
 & & \text{pink} & & \text{pink} \\
 & & & \downarrow & \\
\text{F}_2 & R/R & R/R' & R'/R & R'/R' \\
 & \text{red} & \text{pink} & \text{pink} & \text{white}
\end{array}
$$

When there is partial dominance, both alleles are designated by a capital letter, and one is distinguished from the other by a prime, as here, or by a superscript (for example, C^r for red, C^w for white). Since both alleles affect the phenotype, each is subject to selection.

The chemical events underlying the production of pink flowers are not fully understood. Heterozygous pea flowers, as we have seen, appear red. In pea flowers the red pigment is so intense that it probably masks the white pigment totally, while the less saturated red pigment of snapdragons does not. You can see how this is

16.5   Partial dominance in sweet peas

When plants homozygous for red flowers are crossed with plants homozygous for white blossoms, the heterozygous F_1 sweet-pea plants have pink flowers.

possible by noticing how little pure red pigment it takes to turn a gallon of white base paint bright red.

In other cases of partial dominance, the heterozygous phenotype is not necessarily intermediate between the two homozygous phenotypes. For example, in certain strains, a mating between a black chicken and a so-called splashed white chicken produces offspring all of which have a distinctive appearance called blue Andalusian. A cross between two Andalusians produces black, blue Andalusian, and splashed white offspring in a ratio of 1 : 2 : 1:

$$P \qquad black \quad \times \quad white$$
$$\downarrow$$
$$F_1 \qquad blue \quad \times \quad blue$$
$$\downarrow$$
$$F_2 \quad black \quad blue \quad blue \quad white$$

Here the gene products of two alleles, one for black and one for white, interact to produce a phenotype that is not intermediate between black and splashed white.

HOW MULTIPLE AND MULTIGENIC TRAITS ARE INHERITED

We have limited our discussion so far to monohybrid crosses, those in which a single phenotypic character is studied. Moreover, our examples have involved a single character controlled by a single gene. Such a character is called a *monogenic* trait. However, most characters are controlled by several different genes at once, and so involve *multigenic* inheritance. In addition, organisms have many different contrasting characters, and so all crosses are actually *multihybrid crosses*.

■ INHERITING TWO TRAITS AT ONCE

Mendel's experiments sometimes involved two or more of the characters listed in Table 16.1. For example, he crossed plants bearing round yellow seeds with plants producing wrinkled green seeds. The resulting F_1 plants all had round yellow seeds—all showed dominant phenotypes. When these plants were crossed among themselves, the F_2 progeny showed four different phenotypes:

315 had round yellow seeds
101 had wrinkled yellow seeds
108 had round green seeds
32 had wrinkled green seeds

These numbers represent a ratio of about 9 : 3 : 3 : 1 for the phenotypes.

This experiment produced a new and interesting result: even if full dominance is present, a dihybrid cross can produce new plants phenotypically different from either of the original parental plants; here the new phenotypes were wrinkled yellow and round green

P generation

Round
yellow

Wrinkled
green

F₁ generation 100% round yellow

F₂ generation four phenotypes in ratio 9 : 3 : 3 : 1

Round
yellow

Wrinkled
yellow

Round
green

Wrinkled
green

Novel phenotypes

16.6   Expression of novel phenotypes

Crosses between plants homozygous for round yellow seeds and plants homozygous for wrinkled green seeds produce two novel phenotypes in the F₂ generation: wrinkled yellow seeds and round green seeds.

(Fig. 16.6). This confirms that during meiosis the genes for seed color and the genes for seed form do not necessarily remain paired as they were in the parents, but can separate and reassemble in different allelic combinations in the gametes. This implies that the genes for seed color are on one homologous pair of chromosomes while the genes for seed form are on another pair. Consequently, the genes for the two characters segregate independently during meiosis, sometimes producing new phenotypes on which selection can operate.

The 9 : 3 : 3 : 1 phenotypic ratio is characteristic of the F₂ generation of a dihybrid cross (with dominance) in which the genes for the two characters are ***independent*** (located on nonhomologous chromosomes). Each independent gene behaves in a dihybrid cross exactly as in a monohybrid cross. For example, if we view Mendel's F₂ results as the product of a monohybrid cross for seed color (ignoring seed form), we find that there were 416 yellow seeds (315 + 101) and 140 green seeds (108 + 32), which closely approximates the 3 : 1 F₂ ratio expected in a monohybrid cross. Similarly, if we treat the experiment as a monohybrid cross for seed form and ignore seed color, the F₂ results also show a phenotypic ratio of approximately 3 : 1. The dihybrid F₂ ratio of 9 : 3 : 3 : 1 is simply the product of two separate and independent 3 : 1 ratios.

We can summarize this cross using the following symbols: (with *R* symbolizing the round-seed allele and *r* for the wrinkled-seed allele, *G* for the yellow-seed allele and *g* for the green-seed allele[1]; the dash means that it does not matter phenotypically whether the dominant or the recessive allele occurs in the spot indicated):

P		*R/R G/G*	×	*r/r g/g*	
		round yellow		wrinkled green	
			↓		
F₁		*R/r G/g*	×	*R/r G/g*	
		round yellow		round yellow	
			↓		
F₂	9 *R/– G/–*	3 *r/r G/–*		3 *R/– g/g*	1 *r/r g/g*
	round	wrinkled		round	wrinkled
	yellow	yellow		green	green

The round yellow parent could produce gametes of only one genotype, *RG*. The wrinkled green parent could produce only *rg* gametes. When *RG* gametes from the one parent united with *rg* gametes from the other parent in the process of fertilization, all the resulting F₁ offspring were heterozygous for both characters (*R/r G/g*) and showed the phenotype of the dominant parent (round yellow). Each of these F₁ individuals could produce four different types of gametes, *RG*, *Rg*, *rG*, and *rg*. When two such individuals were crossed, there were 16 possible combinations of gametes (4 × 4). These 16 combinations included nine genotypes, which determined four phenotypes in the ratio of 9 : 3 : 3 : 1 (Fig. 16.7).

One way to determine the genotypic and phenotypic ratios in a dihybrid cross is to construct a Punnett square and then count the

---

[1]Mendel used *G* and *g* to denote color factors because the typical color of peas is green; he was surprised to discover later that green is recessive.

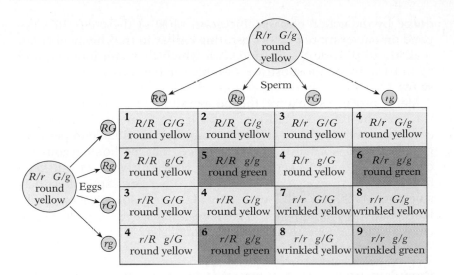

16.7 Punnett-square representation of an F₁ dihybrid cross

When two individuals heterozygous for both seed color and shape were crossed, for every plant that produced wrinkled green seeds, Mendel obtained roughly nine plants that produced round yellow seeds, three that produced round green seeds, and three that produced wrinkled yellow seeds. (Combinations of alleles that give rise to identical phenotypes have the same shading.) Underlying these results were nine different genotypes, as indicated by the numbers in the boxes. (Combinations of alleles that represent identical genotypes are shown with the same number.)

number of boxes representing each genotype and each phenotype. This method becomes prohibitively tedious in any cross involving more than two characters. An alternative procedure is much easier. It is based on the principle that *the chance that a number of independent events will occur together is equal to the product of the chances that each event will occur separately*, a principle known as the **Product Law**. It is best explained by an example.

Suppose we want to know how many of the 16 combinations in Mendel's cross will produce the wrinkled yellow phenotype. We know that wrinkled is recessive; hence it is expected in one-fourth of the F_2 individuals in a monohybrid cross. We know that yellow is dominant; hence it is expected in three-quarters of the F_2 individuals. Multiplying these two values (1/4 × 3/4) gives us 3/16; three of the 16 possible combinations will produce a wrinkled yellow phenotype. Similarly, if we want to know how many of the combinations will produce a round yellow phenotype, we multiply the expectancies for two characters (both of which are dominant) (so 3/4 × 3/4) to get 9/16.

The Product Law applies equally well to more complex examples. Suppose we want to find out, for a trihybrid cross involving the two seed characters and flower color, what fraction of the F_2 individuals (produced by allowing *C/c R/r G/g* individuals to cross among themselves) will exhibit a phenotype combining red flowers, wrinkled seeds, and yellow seeds. The separate probability for red flowers in a monohybrid cross is 3/4, that for wrinkled seeds is 1/4, and that for yellow seeds is 3/4. Multiplying these three values (3/4 × 1/4 × 3/4) gives us 9/64.

■ HOW GENES INTERACT

We have seen that different alleles of the same gene can interact when both are partially dominant. Separate genes can also interact. Indeed, most traits or characters (phenotypic characteristics), whether they are morphological, like flower color, or more subtle chemical characteristics, such as enzyme pathways, are controlled by several genes. Glycolysis, for instance, is a dozen-step process. Each step requires a different enzyme, and each enzyme is pro-

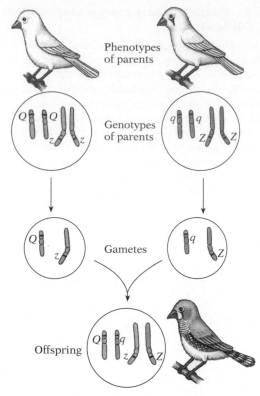

16.8   Complementation test with zebra finches

There are two varieties of albino zebra finch. When these are crossed, the offspring have the slate-gray bodies and colorful markings of normal wild zebra finches. The genetic basis of this phenomenon is shown here. Finches of one strain (B, left) are homozygous for the recessive allele at the Q locus (genotype q/q), and therefore fail to produce functional pigment. Those of the other strain (A, right) are homozygous recessive at the Z locus (genotype z/z), and lack normal color because of flaws in the mechanisms controlling gene activity. The offspring of the cross are heterozygous at both loci, and so can produce the normal amount of pigment. In the example shown here, both parents are homozygous for both genes. If one or both of the parents were heterozygous for one gene (with the genotype Q/q z/z or q/q Z/z), we would expect some of the offspring to be normal and some to be albino.

duced by the action of a specific gene. Most of these enzymes depend on one or more others operating earlier in the chemical chain reaction to provide a substrate with which they can interact, and so in turn provide a substrate suitable for the next enzyme in the series.

Many processes require the cooperation of many genes, but tracking the influence each gene brings to bear in such complicated interactions can be difficult. An individual unable to perform glycolysis at a normal rate, for instance, may be homozygous recessive for any of 12 separate genes. How can we sort out gene functions when genes depend on each other for expressing their true phenotype? We will look briefly at ways of solving these problems, and then at how the problems themselves provide useful clues about the operation of cellular chemistry.

Complementary genes Genes that are mutually dependent are normally detected during crosses between individuals with similar recessive phenotypes. This ***complementation test*** can show whether the trait in question is produced by the same gene in both individuals, or by two separate genes. For example, a normal zebra finch is brightly colored, but there are also two forms of albino finch: one form (strain A), though almost totally white, has a slight brown tinge near its eyes; the other (strain B) is completely white. When these two forms are crossed, their offspring are normal ***wild types*** (the most common naturally occurring form). This result indicates that each form of albinism is produced by a separate gene, at a different locus on the chromosome; the offspring of the cross, being heterozygous in both genes, display the normal phenotype (Fig. 16.8).

Failure in the normal pigmentation system in zebra finches can occur at two known points. The pigment protein can be defective, which is the problem in strain B; or the control mechanisms that turn on pigment production and control the amount made can be defective, which is the problem in strain A. A finch that is homozygous recessive for either element of the pigment system will be an albino. If two white individuals with the *same* defect are crossed, all the progeny will be white, but if two white individuals with different defects are crossed, some or all of the offspring will be normal (Fig. 16.9). This reversion to the normal phenotype in offspring heterozygous at both loci exemplifies the phenomenon known as complementarity, and the two mutually dependent genes are said to ***complement*** each other.

Complementarity can lead to some unusual phenotypic ratios in crosses. Consider the following cross between zebra finches:

P	Q/Q Z/Z	×	q/q z/z
	normal		albino

$\downarrow$

F_1	Q/q Z/z	×	Q/q Z/z
	normal		normal

$\downarrow$

F_2	9 $Q/-$ $Z/-$	3 $Q/-$ z/z	3 q/q $Z/-$	1 q/q z/z
	normal	albino	albino	albino

Were this a dihybrid cross, involving two traits, each at a separate locus, we would have expected four phenotypes in a ratio of 9 : 3 : 3 : 1. Instead, the cross involves two loci but only *one* trait, and hence is digenic but monohybrid. In this example, the 9 : 7 ratio arises because the last three genotype combinations yield the albino phenotype.

Epistasis Complementary genes find expression in the dominant phenotype if and only if the organism is heterozygous or homozygous dominant at both loci. Each gene in some sense has veto power over the other, because their products must cooperate. In *epistasis*, by contrast, these two genes do not have equal "votes": because of the biochemistry of the interaction of their products, only one of the genes can be vetoed by the other. When one gene has the effect of suppressing the phenotypic expression of another gene but not vice versa, the first gene is said to be epistatic to the second. (The Greek root of *epistasis*, logically enough, means "standing upon.")

In guinea pigs, for example, a gene for the production of the skin pigment melanin is epistatic to one for the *deposition* of melanin. The first gene has two alleles: *C*, which causes pigment to be produced, and *c*, which does not; hence a homozygous recessive individual, *c/c*, is albino. The second gene has an allele *B* that causes deposition of much melanin, which gives the guinea pig a black coat, and an allele *b* that causes deposition of only a moderate amount of melanin, producing a brown coat. However, neither *B* nor *b* can cause deposition of melanin if *C* is not present to make the melanin:

Single Pea

Walnut Rose

16.9 Comb types in chickens

P	*C/C B/B* ×	*c/c b/b*	
	black	albino	
	↓		
F₁	*C/c B/b* ×	*C/c B/b*	
	black	black	
	↓		

F₂   9 *C/– B/–*   3 *C/– b/b*   3 *c/c B/–*   1 *c/c b/b*
    black   brown   albino   albino

Instead of an F₂ phenotypic ratio of 9 : 3 : 3 : 1, this monohybrid cross yields a ratio of 9 : 3 : 4.

Notice that epistasis and dominance are quite different. Dominance is the phenotypic expression of one member of a pair of alleles at the expense of the other. Epistasis is the suppression of one gene for a phenotypic effect by another entirely different gene; the epistatic gene prevents natural selection from acting on *either* allele of the second gene. In other words, dominance involves the interaction between alleles, while epistasis refers to the interaction between nonallelic genes.

Collaboration Sometimes two genes interact to produce a novel phenotype. Such collaborative interaction is seen in the control of comb form in chickens (Fig. 16.9). One gene, *R*, produces a rose comb, while its recessive allele, *r*, produces a single comb. Another gene, *P*, produces the "pea" comb, while its recessive, *p*, also produces a single comb. When *R* and *P* occur together, they collaborate to produce the "walnut" comb. When rose-comb Wyandotte chickens are crossed with pea-comb Brahma chickens, four phenotypes result:

P R/R p/p × r/r P/P
         rose   pea

$$\downarrow$$

F$_1$ R/r P/p × R/r P/p
        walnut   walnut

$$\downarrow$$

F$_2$ 9 $R/-$ $P/-$ 3 $R/-$ p/p 3 r/r $P/-$ 1 r/r p/p
    walnut   rose   pea   single

Notice that when collaboration occurs, it is not possible to identify the action of each gene on the basis of its phenotypic effects in the clearcut way Mendel could. The action of selection on any one allele, then, may depend on the alleles present at another locus.

Modifiers Probably no inherited character is controlled exclusively by one gene pair, though for convenience we often analyze crosses as though only a single gene were important for a particular trait. However, even when only one principal gene is involved, its expression is usually influenced to some extent by other genes. An example is eye color in humans.

Human eye color is largely controlled by one gene with two alleles—a dominant allele, *B*, for brown eyes, and a recessive, *b*, for blue eyes. Brown-eyed people (*B/B* or *B/b*) have pigment cells containing melanin in the *front* layer of the iris. Blue-eyed people (*b/b*) lack melanin; the blue is an effect of the black pigment on the *back* of the iris seen faintly through the semiopaque front layer.

Though most eyes can be categorized as brown or blue, eyes can be green or gray (both genetically forms of blue), hazel or black (both forms of brown), as well as shades in between. Obviously, modifier genes are also involved, some affecting the amount of pigment in the iris, some the tone of the pigment (which may be light yellow, dark brown, and so on), some its distribution (uniform over the whole iris, or in scattered spots, or in a ring around the outer edge, and so forth). In fact, in rare cases two blue-eyed people can have a brown-eyed child, because one of them, in whom the lack of pigmentation is a consequence of the action of modifier genes, actually carries the genotype *B/b* instead of *b/b*.

Multiple-gene inheritance Not all characters have a limited number of relatively distinct phenotypes. While pea flowers are either red or white, and eyes blue or brown (with intermediates generated by modifier genes) other traits have no distinct boundaries. Human height, skin pigmentation, and IQ are three examples. What accounts for the absence of distinct phenotypes?

The most common reason is that two or more separate genes can affect the same character in the same way, in an additive fashion. This kind of inheritance is called ***polygenic***. The first clear demon-

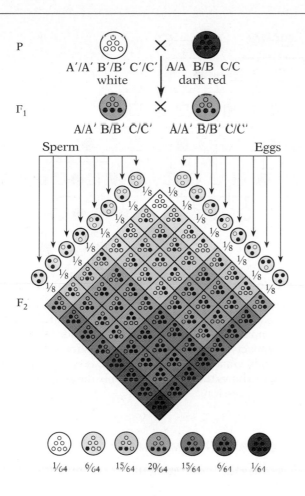

16.10 Nilsson-Ehle's trigenic cross

A line of wheat with pure white kernels and a line with dark red kernels were crossed to produce hybrid medium red F_1 offspring; these were then crossed to produce offspring with the distribution of phenotypes shown here. Nilsson-Ehle concluded from his observations that (1) there are three genes for kernel color in wheat, each with a partially dominant allele for red (*A, B, C*) and a partially dominant allele for white (*A', B', C'*); (2) that the genotype of the dark red line is *A/A B/B C/C*, while that of the white line is *A'/A' B'/B' C'/C'*; and (3) that the F_1 hybrids therefore have the genotype *A/A' B/B' C/C'*. A gradation of phenotypes is seen in the offspring because all the alleles are additive. Eight different kinds of gametes occur because each of the genes for color is on a different chromosome. The existence of the seven distinct phenotypes of F_2 offspring in the ratio $1 : 6 : 15 : 20 : 15 : 6 : 1$ is predicted by this Punnett square of a trigenic cross postulating partial dominance in kernel color.

stration came in 1909, when Herman Nilsson-Ehle showed that the color of wheat kernels, which vary from white through various shades of pink and red to a very dark red, results from the interactions of three genes. Each gene has two alleles, a partially dominant allele for red and a partially dominant allele for white. Dark red kernels are homozygous for the red allele in all three genes, while pure white kernels are homozygous for the white allele in all three genes. All the phenotypes in between result from different heterozygous mixtures of the alleles (Fig. 16.10).

If we graph the probable frequency of colors in F_2, we obtain a jagged approximation of the bell-shaped curve that represents the so-called normal distribution of most continuously varying traits in a population, traits like weight, height, IQ, and the like (Fig. 16.11). The greater the number of phenotypes, the less jagged the distribution should appear. In most cases, however, the genes involved in multiple-gene effects do not all contribute equally to the phenotype, and the effects of modifier genes and of the environment blur the boundaries. As a result, it is usually impossible to determine the number and nature of the genes involved simply by scrutinizing the distribution of phenotypes in F_2. On the other hand, asymmetries in plots of actual phenotypic variation in a population can suggest that selection is operating against certain ranges of the phenotypic spectrum.

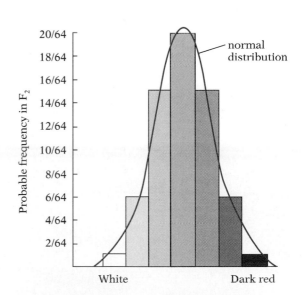

16.11 Frequency distribution of kernel color in wheat

As Nilsson-Ehle showed, a cross of heterozygous wheat kernels results in offspring with seven classes of kernel color, their relative frequencies producing a so-called normal distribution centered around light red.

16.12   Pedigree of syndactyly in seven generations of one family

Squares represent males, and circles represent females. Red indicates syndactyly and gray indicates the normal phenotype. The character appears to be inherited as a simple dominant: there is no correlation with sex, and individuals with syndactyly have a parent who also shows this trait. There is one exception: an individual (black square) who has syndactyly even though neither of his parents shows the trait. Presumably the gene was present in his mother without expression.

■ WHEN GENES ARE ONLY PARTIALLY EXPRESSED

Because genes interact, complementary or epistatic genes, or a combination of modifier genes, may prevent the expression of a dominant allele. And even when an allele is expressed, the effects of other genes may modify the intensity of phenotypic expression. Certain alleles thus have incomplete penetrance and variable expressivity: **penetrance** refers to the percentage of individuals in a population carrying a given allele that actually express the phenotype; **expressivity** denotes the manner in which the phenotype is expressed.

Incomplete penetrance and variable expressivity are illustrated by an allele in humans that causes blue sclera: the whites of the eyes appear bluish. This allele is dominant, but only about 9 out of 10 people who have the allele actually show the phenotype. Thus the penetrance of this gene is about 90% (or 0.9). Among those that do show the phenotype, the expressivity is variable, with the intensity of the bluish coloration ranging from very pale whitish blue to very dark blackish blue.

Figure 16.12 shows a portion of the pedigree of a family in Virginia that for generations has had many members with syndactyly of the ring and little fingers—the two fingers are joined by a web of muscle and skin (Fig. 16.13). This character exhibits variable expressivity: a few individuals have three fingers webbed; most have two fingers fully webbed; some have two fingers partially webbed; and a few have a crooked finger with no webbing.

Syndactyly also exhibits incomplete penetrance. Of the more than 50 individuals in this family known to have had syndactyly, all but one had a parent who also had syndactyly—almost exactly the

16.13   Syndactyly

pattern of a pure dominant. (By contrast, a recessive character often appears in a person neither of whose parents showed the character, because they were heterozygous.) The abnormality in the one man having two parents with normal fingers demonstrates that the allele for syndactyly has incomplete penetrance: the man's mother must have been carrying the allele, but did not exhibit its phenotype because of the allele's incomplete penetrance.

Both examples demonstrate that penetrance and expressivity are aspects of the same phenomenon. Lack of penetrance of the gene for blue sclera produces white color, which is simply one extreme of the expressivity gradient from white through very pale blue to dark blue. Lack of penetrance of the gene for syndactyly is simply one extreme of the expressivity gradient whose other extreme is the full webbing of three fingers.

Penetrance and expressivity are often also affected by the environment. For example, *Drosophila* that are homozygous for the allele for vestigial wings have wings that are only tiny stumps when reared at normal room temperatures (about 20°C), but their wings grow almost as long as normal wings when reared at temperatures as high as 31°C. Himalayan rabbits are normally white with black ears, nose, feet, and tail (Fig. 16.14), but if a patch of fur on a patch on the rabbit's back is plucked and an ice pack is kept on the patch, the new fur that grows there will be black: the allele for black color can express itself only if the temperature is low, which it normally is only at the body extremities.

The same phenomenon underlies the coloring of Siamese cats. An enzyme required for pigment synthesis in this breed retains its normal active conformation only at low temperatures, and denatures—changes to an inactive conformation—when heated to body temperature.

Thus the expression of a gene can depend both on the other genes present (the genetic environment) and on the physical environment (temperature, sunlight, humidity, diet, and so forth). We don't inherit characters. We inherit only genetic potentialities; other factors govern whether the potentialities are realized. All organisms are products of both their inheritance and their environment.

16.14 Effect of temperature on expression of a gene for coat color in the Himalayan rabbit

(A) Normally, only the feet, tail, ears, and nose are black. (B) Fur is plucked from a patch on the back, and an ice pack is applied to the area. (C) The new fur grown under the artificially low temperature is black. Himalayan rabbits are normally homozygous for the gene that controls synthesis of the black pigment, but the enzyme encoded by the gene is active only at low temperatures (below about 33°C).

◾ WHEN GENES HAVE SEVERAL ALLELES

Mendel simplified his analysis by concentrating on just two alleles at each locus. And of course, the maximum number of alleles for each gene that any diploid individual can possess is two, because the organism has only two copies of each gene. However, within a population to which that individual belongs, many other alleles may be present for a given gene, so multiple alleles can play a major role in evolution.

Eye color in *Drosophila* One of the first examples of multiple alleles was discovered in the fruit fly *Drosophila melanogaster*—a species whose genetics have been extensively studied. Though normally red-eyed, fruit flies may have white-, eosin- (a brightly fluorescing red), wine-, apricot-, ivory-, or cherry-colored eyes. Each of these colors is controlled by a different allele of the same gene. The allele for the wild-type eye (red) is dominant over all the rest—that is, the normal red pigment masks the pigment produced by any other allele. When two of the other alleles occur together in a heterozygous fly, however, they produce an intermediate eye-color phenotype.

Blood types A well-known example of multiple alleles in humans is that of the A-B-O blood series, in which four blood types are generally recognized: A, B, AB, and O. The red blood corpuscles (erythrocytes) in type A blood bear antigen A on their surface; in type B, antigen B; in type O, neither A nor B; and in type AB, both A and B.

An antigen is a chemical capable of triggering an immune reaction by which antibodies are produced that bind to and help destroy that particular antigen and the cell that bears it (see Chapter 15). Because the immune system soon becomes insensitive to antigens present from birth, an individual whose red blood corpuscles bear antigen A has no antibodies—called anti-A—to that antigen. Similarly, an individual with antigen B has no anti-B antibodies. The person with type AB blood, therefore, having corpuscles that bear both A and B antigens, will have neither anti-A nor anti-B in the blood plasma; the person with type O blood, on the other hand, having corpuscles that bear neither antigen, will have both anti-A and anti-B in the blood plasma (see Table 16.2).

The presence of these antigens and antibodies in the blood has important implications for transfusions. Because the antibodies present in the plasma of blood of one type tend to react with the antigens on the red blood cells of other blood types and cause clumping, it is always best to obtain a donor who has the same blood type as the patient. Blood of another type may be used, provided that the *plasma of the patient* and the *erythrocytes of the donor* are compatible. The reason is that, unless the transfusion is to be a massive one or is to be made very rapidly, the donor's plasma is sufficiently diluted during transfusion so that little or no agglutination occurs.

This means that type O blood can be given to anyone, because its red blood corpuscles have no antigens and hence are obviously compatible with the plasma of any patient; type O blood is sometimes called the universal donor. But type O patients can receive

Table 16.2 Antigen and antibody content of the blood types of the A-B-O series

BLOOD TYPE	BLOOD CONTAINS	
	CELLULAR ANTIGENS	PLASMA ANTIBODIES
O	None	anti-A and anti-B
A	A	anti-B
B	B	anti-A
AB	A and B	None

transfusions only from type O donors, because their plasma contains both anti-A and anti-B and hence is obviously not compatible with the erythrocytes of any other class of donor. Luckily for those of us with type O blood, it is the most common variety in most parts of the world. Conversely, people with the rare type AB, whose plasma contains no anti-A or anti-B antibodies, are universal recipients, but cannot act as donors for any except type AB patients (Table 16.3).

At first glance, you might suppose that two independent genes are involved in the A-B-O system, one determining whether the A antigen is present and another whether the B antigen is present. Actually the inheritance of the A-B-O groups is for the most part controlled by three alleles of the same gene, I^A, I^B, and i. Both I^A and I^B are dominant over i, but neither I^A nor I^B is dominant over the other. Accordingly the four blood-type phenotypes correspond to the genotypes indicated in Table 16.4. I^A and I^B are alleles that code for different functional proteins, while i codes for a nonfunctional protein. In fact, the protein is an enzyme that modifies the carbohydrate groups on a membrane component of the glycocalyx (Fig. 16.15).

Blood typing is often used as a source of evidence in paternity cases. For example, a man with type O blood could not be the father of a child with type A blood whose mother is type B. The child's true father must be either type A or type AB, because the child must have received its I^A allele from its father; an O man has no such allele. Similarly, a man with type AB blood could not be the father of a type O child, because the child must have received an i allele from each parent, but an AB man has no such allele. Blood-type analysis can only determine who could *not* be the father. Genetic "fingerprinting" techniques, which usually measure the number of copies of each type of repetitive DNA (a pattern that is similar between related individuals, but unique in detail for each

Table 16.3 Transfusion relationships of the A-B-O blood groups

BLOOD GROUP	CAN DONATE BLOOD TO	CAN RECEIVE BLOOD FROM
O	O, A, B, AB	O
A	A, AB	O, A
B	B, AB	O, B
AB	AB	O, A, B, AB

Table 16.4 Genotypes of the A-B-O blood types

BLOOD TYPE	GENOTYPE
O	i/i
A	I^A/I^A or I^A/i
B	I^B/I^B or I^B/i
AB	I^A/I^B

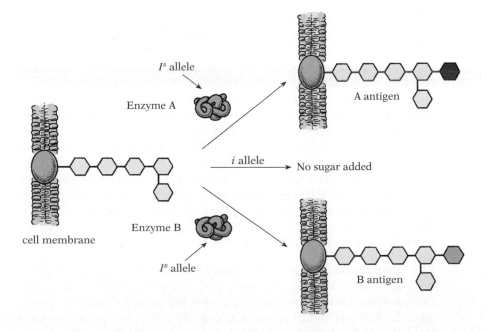

16.15 Molecular basis of the A-B-O blood group antigens

Blood cells produce a carbohydrate chain with five sugar groups that is mounted on a cell-surface lipid or protein. Cells with an I^A allele produce an enzyme that adds a particular sixth sugar (red) to the chain, producing the type A antigen; cells with an I^B allele produce an enzyme that adds a different sixth sugar (blue) to the chain, producing the type B antigen. Cells with both alleles produce both kinds of six-sugar chains, leading to the AB phenotype; cells homozygous for the i allele fail to modify any of the surface-mounted carbohydrates.

Table 16.5 Frequencies of A-B-O blood groups in selected populations

POPULATION	O	A	B	AB
U.S. whites	45%	41%	10%	4%
U.S. blacks	47	28	20	5
African Pygmies	31	30	29	10
African Bushmen	56	34	8	2
Australian aborigines	34	66	0	0
Pure Peruvian Indians	100	0	0	0
Tuamotuans of Polynesia	48	52	0	0

person) can be more reliable, though it is very time consuming to perform.

As indicated in Table 16.5, the frequencies of the various A-B-O blood types vary in populations of different ancestral extraction. Since the most frequent phenotype in most human populations is type O, which corresponds to the homozygous recessive phenotype, the i allele is more common than the I^A or I^B alleles. This illustrates an important point: whether an allele is dominant or recessive does not determine whether it will be common or rare in the population. "Dominant" and "recessive" describe the way the alleles interact when they occur together in a heterozygous individual; they do not indicate which allele determines the more advantageous phenotype.

Rh factors The A and B antigens are not the only surface proteins on red blood corpuscles. Rh factors, named for antigens first identified in rhesus monkeys, are produced by alleles of the Rh gene. Individuals whose two copies of the Rh allele coding for nonfunctional products are said to be Rh-negative (Rh^-). Individuals with at least one functional Rh allele (and there are many, including four common ones) are said to be Rh^+. Another gene for surface proteins on red corpuscles has two alleles, M and N, giving rise to the genotypes M/M, M/N, and N/N. Each of these groups of blood antigens can create immunological problems during transfusions.

◼ HOW DELETERIOUS ALLELES ARISE AND SURVIVE

As discussed in Chapters 8, 9, and 11, a variety of influences can cause mutations in genes. Because cells have elaborate DNA-repair mechanisms, the rate at which any particular gene undergoes mutation is ordinarily extremely low. However, every individual organism has a large number of different genes, and the total number of genes in all the individuals of a species is vast indeed. Hence mutations are constantly occurring within a species.

Every living organism is the product of billions of years of evolution and is a finely tuned, smoothly running, astoundingly intricate mechanism, in which the function of every part in some way influences the function of every other part. By comparison, the most complex computer is simple indeed. If you were to take such a machine and make some random change in its parts, the chances are great that you would make it run worse rather than better: a random change in any delicate and intricate mechanism is far more likely to damage it than to improve it. Thus the vast majority of the mutations that have obvious phenotypic effects are deleterious. Only very rarely is a mutation beneficial. Nevertheless, mutations are a major source of the variation on which natural selection operates. (The other sources include crossing over and sexual recombination, each of which creates new combinations of preexisting genes.)

The protection of heterozygosity When a deleterious allele arises by mutation, natural selection can act against it only if it causes some change in the organism's phenotype. Because dominant deleterious mutations are expressed, they can be eliminated from the population rapidly by natural selection; thus most mutations that persist in a population are recessive to the normal alleles. Deleterious alleles that are not dominant may be retained in the population in a heterozygous condition for a long time. Clearly, organisms that are diploid throughout their lives are much less sensitive to mutation than those that have extended haploid stages. This is probably the main reason diploidy predominates among long-lived organisms.

When a mating occurs between two diploid individuals carrying the same deleterious recessive allele in the heterozygous condition, about one-fourth of the progeny will be homozygous for the deleterious allele, and these homozygous offspring will have the harmful phenotype. Selection (death before reproduction) culls out such organisms, thereby reducing the frequency of the deleterious allele in the population. A recessive allele whose phenotype is fatal is called a *lethal*. The occurrence of lethals can greatly modify the phenotypic ratios obtained in the progeny of some crosses.

In numerous instances alleles harmful or even lethal when homozygous are actually beneficial when heterozygous. The Dexter cattle breed is a good beef producer; however, it is impossible to establish a pure-breeding herd because some of its most desirable characteristics are caused by the heterozygous expression of an allele that is lethal when homozygous.

An example in humans is the allele for *sickle-cell anemia*. In an individual homozygous for the sickle-cell allele, the mutant hemoglobin tends to crystallize under acidic conditions (as when blood levels of CO_2 rise during exercise). When the hemoglobin in a corpuscle crystallizes, the corpuscles become curved like a sickle and bear long filamentous processes (Fig. 16.16). These abnormal corpuscles tend to form clumps and to clog the smaller blood vessels. The resulting impairment of circulation leads to severe pains in the abdomen, back, head, and extremities, and to enlargement of the heart and atrophy of brain cells. In addition, the tendency of the deformed corpuscles to rupture brings about severe anemia. Homozygous victims of sickle-cell anemia usually suffer an early death. Individuals heterozygous for the sickle-cell allele sometimes

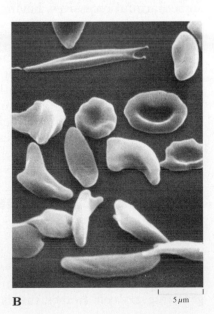

A   5 μm   B   5 μm

16.16 Normal and sickled red blood corpuscles

Normal erythrocytes (A), which are biconcave disks, look dramatically different from sickled cells (B). Some of the sickled cells seen here in this scanning electron micrograph bear the filamentous processes that may cause clogging in the body's smaller blood vessels.

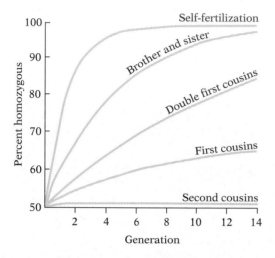

16.17   Percentage of homozygotes in successive generations under different degrees of inbreeding

It is assumed in this graph that the initial condition is two alleles of equal frequency (hence the 50% homozygosity at the start). If a similar graph were drawn for very rare recessive alleles, the initial rises in homozygosity would be much steeper, particularly in the three curves for cousin-to-cousin matings. Thus matings between first or second cousins, though of little effect in the situation graphed here, may greatly increase the percentage of homozygosity of rare, perhaps deleterious recessive alleles. (Double first cousins result when siblings of one family marry siblings of another family—when two brothers marry two sisters, for example. The offspring of such marriages are first cousins through *both* of their parents rather than through just one of them.)

show mild symptoms of the disease, but the condition is usually not serious.

You might suppose that natural selection would operate against the propagation of any allele so obviously harmful and that such an allele would be held at very low frequency in the population. But the allele is surprisingly common in many parts of Africa, being carried by up to 20% of the black population. In fact, individuals heterozygous for this allele have an unusually high resistance to malaria. Since malaria is very common in many parts of Africa, the sickle-cell allele is beneficial when heterozygous. Hence, in Africa there is selection for the allele when it is heterozygous and selection against it when it is homozygous. In other parts of the world, however, malaria is infrequent, and the benefits of the sickle-cell allele are outweighed by its costs. The frequency of the sickle-cell allele in the descendants of African blacks living in the United States is steadily declining, a clear instance of evolution.

Sickle-cell anemia is a dramatic example of an allele that has more than one effect. Such alleles are ***pleiotropic***. Pleiotropy is very common (though its effects may be subtle). A striking example involves the temperature-sensitive coat-color allele of Siamese cats mentioned earlier. This same allele also causes a mysterious misrouting during development of the axons carrying visual information from the eye to the brain. Many Siamese cats are cross-eyed to compensate for their visual miswiring.

The dangers of inbreeding The conditions under which genes cause deleterious phenotypes explain the danger of matings between closely related humans. Everyone carries in heterozygous combination many alleles that would cause harmful effects if present in homozygous combination, including some lethals. Because most of these deleterious alleles originated as rare mutations, however, and are limited to a tiny percentage of the population, the chances are slight that two unrelated individuals will be carrying the same deleterious recessive alleles and produce homozygous offspring that show the harmful phenotype.

The chances that two closely related persons are carrying the same harmful recessives, having received them from common ancestors, are much greater. If they mate, they may have children homozygous for the deleterious traits. In short, close inbreeding increases the percentage of homozygosity (Fig. 16.17). Matings between first cousins cause slight increases in homozygosity; while brother-sister matings and matings between double first cousins (those sharing two sets of grandparents) cause rapid increases. Many species have evolved behavioral or physiological mechanisms that prevent close relatives from mating.

HOW SEX ALTERS INHERITANCE

■ HOW SEX IS DETERMINED

The sex chromosomes We have said that diploid individuals have two of each type of chromosome, and hence two copies of each gene in every cell. In fact, this is not quite correct: in most species with distinct sexes, the chromosomal endowments of males and females are slightly different. In general, one of the two sexes has one pair of nonmatching chromosomes. These are the **sex chromosomes**, which play a fundamental role in determining the sex of the individual; they exhibit their sex-specific effects primarily in the brain, in certain hormone-producing organs, in cells in sexually dimorphic portions of the external and internal anatomy, and in the gonads. All other chromosomes are called **autosomes**.

Let's look first at the chromosomes of *Drosophila* and humans. In each species, one chromosome—the **X chromosome**—bears many genes; the other, which is much smaller—the **Y chromosome**—has only a few genes. Normal females have two X chromosomes, while males have one X and one Y. The diploid number in *Drosophila* is 8 (four pairs); females have three pairs of autosomes and one pair of X chromosomes, while males have three pairs of autosomes and a pair of sex chromosomes consisting of one X and one Y (Fig. 16.18). The diploid number in humans is 46 (23 pairs); females have 22 pairs of autosomes and one pair of X chromosomes, while males have 22 pairs of autosomes plus one X and one Y (see Fig. 12.4, p. 311).

When a female produces egg cells by meiosis, all the eggs receive one of each type of autosome plus one X chromosome. When a male produces sperm cells by meiosis, half the sperm cells receive one of each type of autosome plus one X chromosome, while the other half receive one of each autosome plus one Y chromosome (Fig. 16.18). When fertilization takes place, the chances are approximately equal that the egg will be fertilized by a sperm carrying an X chromosome or a sperm carrying a Y chromosome. If fertilization is by an X-bearing sperm, the resulting zygote will be XX and will develop into a female. If fertilization is by a Y-bearing sperm, the resulting zygote will be XY and will develop into a male. The sex of an individual is normally determined at the moment of fertilization and depends on which of the two types of sperm fertilizes the egg.

This XY-male system, though characteristic of many plants and animals (including all mammals), is not universal. Birds, butterflies and moths, and a few other animals have just the opposite system, where XX is male and XY is female. (To distinguish this from the usual XY system, the symbols Z and W are often substituted: ZZ for male and ZW for female.) In still other species—many reptiles and marine fish, for example—sex is environmentally determined: the sex of alligators depends on the temperature of the eggs during development, while some coral-reef fish reverse sex as adults to take advantage of variations in food and competition. Many species of animals (ranging from insects to mammals) have the ability to manipulate the sex ratio of their offspring to exploit changes in habitat conditions, competition, or individual social status that maximize the reproductive potential of their progeny.

What does the Y chromosome do? Occasionally the members of a homologous pair of chromosomes fail to separate properly in meiosis, so both move to the same pole. The effect of such **nondisjunction** is that one of the daughter cells receives one too many chromosomes while the other cell receives one too few. If a human gamete carrying an extra chromosome is involved in fertilization,

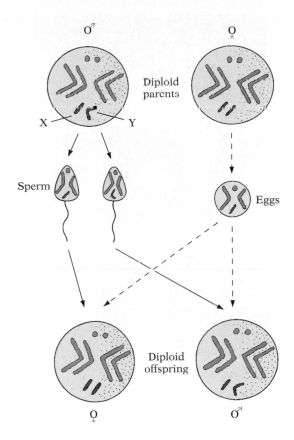

16.18 Chromosomes of male and female *Drosophila melanogaster*

There are three pairs of autosomes and one pair of sex chromosomes. Males (♂) have one X chromosome and one Y chromosome; since these separate at meiosis, half the sperm carry an X and half carry a Y. Since females (♀) have two X chromosomes, all eggs have an X. The sex of the offspring depends on which type of sperm fertilizes the egg.

16.19   Hairy pinna

This trait is thought to be determined by a gene on the Y chromosome.

the zygote produced has 47 chromosomes (22 pairs plus three of one type) instead of 46 (23 pairs).

When the sex chromosomes fail to separate in meiosis, an XXY individual may result. In *Drosophila*, XXY individuals are essentially normal females, which suggests that only the X chromosomes function in sex determination: one produces a male and two a female, regardless of any Y chromosome. And indeed, in *Drosophila*, sex is determined by the ratio of certain gene products produced by the X chromosomes to certain products produced by the autosomes. A low ratio results in a male, while a high ratio produces a female.

This X-based model accounts for sex determination in grasshoppers and other animals that lack the Y chromosome entirely—females have two X chromosomes and males a single unpaired X. Such a system of sex determination is known as the XO system, with "O" denoting the absence of a chromosome.

The X-based system is not universal, however. The human Y chromosome, for example, bears a gene with strong male-determining properties. Its product is a DNA-binding protein that, early in fetal life, binds to numerous sites on the autosomes, initiating the developmental program that will produce a male. In the absence of this gene—even in XY individuals in whom this gene is deleted—a female develops. Genes unique to the Y chromosome are termed **holandric**. The phenotypic traits they control appear, of course, only in males (Fig. 16.19). There are very few genes on the human Y chromosome.

■ THE GENETICS OF SEX-LINKED CHARACTERS

Most genes on the X chromosome are absent from the Y chromosome. Such genes are said to be sex-linked. The inheritance of characteristics controlled by sex-linked genes are completely different from those controlled by autosomal genes: females have two copies of each sex-linked gene, one from each parent, but males have only one copy of each sex-linked gene, and that one copy always comes from the mother. Furthermore, since the male has only one copy of each sex-linked gene, recessive alleles cannot be masked; selection acts directly on deleterious recessives on the X chromosome. Recessive sex-linked phenotypes, such as red-green color blindness, occur much more often in males than in females.

Sex linkage was discovered in 1910 by Thomas Hunt Morgan. It was Morgan who began the systematic use of *Drosophila* in genetic studies. With fruit flies it is possible to perform in a few months experiments that would have taken Mendel years with peas. *Drosophila* can be cultured in large numbers in the laboratory. It produces a new generation in 10–12 days and is subject to many easily detectable genetic variations. Most of the modern knowledge of eucaryotic genetics derives from work on this insect.

The first sex-linked trait discovered by Morgan was white eye color in *Drosophila*. This mutation, like dozens discovered since (including many others affecting eye color) arose spontaneously; it is controlled by a recessive allele *r*. The normal red eye color is controlled by the dominant allele *R*. If homozygous red-eyed females are crossed with white-eyed males, all the F_1 offspring, regardless of sex, have red eyes, since they receive from their mother an X chromosome bearing an allele for red. In addition, the F_1 females receive from their father an X chromosome bearing an allele for white eyes, but the red allele masks its presence. The F_1 males, like the females, receive from their mother an X chromosome bearing an allele for red eyes. Unlike the females, however, they receive no gene for eye color from their father, who contributes a Y chromosome instead of an X. (In writing the genotype of a male for a sex-linked character, the Y is shown to indicate that no second X chromosome is present and hence there is no second copy of the sex-linked gene.) We can summarize this cross as follows (♀ denotes females, ♂ males):

P $\qquad$ R/R $\times$ r/Y
$\qquad\qquad$ red-eyed ♀ $\qquad$ white-eyed ♂

$\qquad\qquad\qquad\qquad\downarrow$

F_1 $\qquad$ R/r $\times$ R/Y
$\qquad\qquad$ red-eyed ♀ $\qquad$ red-eyed ♂

$\qquad\qquad\qquad\qquad\downarrow$

F_2 $\quad$ R/R $\qquad$ r/R $\qquad\qquad$ R/Y $\qquad$ r/Y
$\quad$ red- $\qquad$ red- $\qquad\qquad$ red- $\qquad$ white-
$\quad$ eyed ♀ $\qquad$ eyed ♀ $\qquad\qquad$ eyed ♂ $\qquad$ eyed ♂

Notice that when the F_1 flies of this cross are allowed to mate among themselves, the F_2 flies show the customary 3 : 1 phenotypic ratio of a monohybrid cross where dominance is present. This 3 : 1 ratio is different, however, from the 3 : 1 ratio obtained in a cross involving autosomal genes. In an autosomal cross there is no correlation of phenotype with sex, but in this cross all F_2 individuals showing the recessive phenotype are males. This asymmetry illustrates the potential for selection to operate differentially on the two sexes.

Two well-known examples of recessive sex-linked traits in humans are red-green color blindness and hemophilia (a failure to form blood clots, which can lead to uncontrolled bleeding). Color blindness occurs in about 8% of white males in the United States and in about 4% of black males. It occurs in only about 1% of white females and about 0.8% of black females. More men than women will show such a trait since men need only one copy of the allele to show the phenotype. For a woman to be color-blind, she must have two copies of the allele, and so be homozygous; not only must her father be color-blind, but her mother must be either color-blind or a heterozygous carrier of the allele. Since the allele is not very common in the population, it is not likely that two such people will marry.

Though human females have two copies of each sex chromosome, one of the X chromosomes in each somatic cell condenses into a tiny dark object called a **Barr body** (Fig. 16.20). Most of the genes on this condensed X chromosome are inactive. Hence, a normally functioning female cell contains only one active copy of most sex-linked genes. Why, then, are sex-linked recessive traits expressed in females only when they are homozygous?

The explanation is that it is not always the same X chromosome

5 µm

16.20 Nuclei from epidermal cells of a human female

The arrows indicate the Barr bodies. Since Barr bodies are present in the cells of female fetuses, the sex of an unborn child can be ascertained by examination of the nuclei of cells sloughed off into the fluid of the mother's womb.

16.21 X-chromosome inactivation in cats

One of the most common natural demonstrations of X-chromosome inactivation is the coat color of tortoiseshell and calico cats. A gene for color found on the cat's X chromosome has two common alleles—black and yellow (or orange). Males can have only one allele, so (in the absence of modifier genes) they are either yellow or black, usually with various white markings. Females, however, can have both alleles, and after inactivation some cells will express the black allele and others the yellow allele. Large patches of cells with one allele or the other develop, and the result is a mosaic of yellow, black, and (usually) white patches. Except for rare XXY individuals, all cats with both yellow and black fur are females.

that condenses into a Barr body in the different somatic cells, with no discernible pattern as to which chromosome is active in which cell. For example, consider the distribution of alleles of the sex-linked gene for the enzyme glucose-6-phosphate dehydrogenase; suppose one of the alleles codes for an active version of the enzyme, the other for a defective version. When the red blood corpuscles of women heterozygous for these two alleles are examined, roughly half the cells have normal enzyme activity while the other half have none. Women—and the females of most other mammals—are, in a sense, genetic mosaics for sex-linked traits. For some of these traits, the mosaic pattern finds phenotypic expression; an example is the coat color of tricolor (tortoiseshell and calico) cats (Fig. 16.21). For others, it does not. It seems, for instance, that as long as half the cells are normal in a woman heterozygous for red-green blindness or for hemophilia, she will be phenotypically normal.

The phenomenon of X-chromosome inactivation explains why the sex of mammals depends on the presence or absence of the Y chromosome rather than on the ratio of X chromosomes to autosomes, as in *Drosophila*: because of inactivation each mammalian somatic cell has only one functional X chromosome, and so is either XO (female) or XY (male); both sexes have the same number of X-chromosome genes available for transcription. In *Drosophila*, on the other hand, females have twice the number males have. The resulting difference in the amount of certain gene products determines the sex of the individual. For many other products, various mechanisms have evolved to achieve **dosage compensation**—to equalize the amount resulting from transcription of the two X chromosomes in females with the amount resulting from transcription of the single chromosome in males. On average, genes on the single X chromosome of *Drosophila* males do double duty: they are transcribed twice as often as the corresponding genes on any one female X chromosome.

■ THE EXPRESSION OF SEX-DEPENDENT CHARACTERS

As we have seen, sex-linked genes may control characters not customarily regarded as sexual. Conversely, not all genes commonly associated with sex are sex-linked; many are located on the autosomes. For example, a number of genes that control growth and development of the sexual organs, as well as those that control distribution of body hair, size of breasts, pitch of voice, or other secondary sexual characteristics, are autosomal and are present in individuals of both sexes. That their phenotypic expression is different in the two sexes indicates that they are sex-*limited*, not sex-linked. The sex chromosomes determine what hormones are to be synthesized in each sex, and these hormones influence the activity of the sex-limited autosomal genes secondarily, either inhibiting or stimulating them.

There are other instances of sex-correlated inheritance. One, discussed in Chapter 10, involves the inheritance of mitochondria: in most higher organisms, sperm and pollen only rarely contribute mitochondria to the zygote, so mitochondrial genes are essentially always inherited from the female parent. The other case is imprinting: as described in Chapter 11, progeny can occasionally inherit genes with different patterns of methylation; if the gene in question is permanently inactivated in the gamete of one parent, the phenotype of the zygote is determined by the allele contributed by the gamete from the other parent.

HOW GENES ON THE SAME CHROMOSOME ARE INHERITED

Mendel formulated two generalizations. The first was that each individual carries two copies of every gene, and these copies segregate during gamete formation; this Principle of Segregation is often called Mendel's first law. The second generalization was that when several genes are involved in a cross (as in a dihybrid cross), they sort out into gametes independently of one another; this **Principle of Independent Assortment** is frequently referred to as Mendel's second law. All seven of the traits Mendel reported on from garden peas did indeed assort independently, and were free of the complications of modifiers, partial dominance, sex linkage, multiple alleles, and so on.

In fact, independent assortment describes only the simplest genetic interactions—the behavior of genes on separate chromosomes. Mendel almost certainly knew that some pairs of factors do not assort independently to produce offspring in neat 9 : 3 : 3 : 1 ratios, and that some do not show a simple pattern of recessiveness and dominance. He probably chose to ignore such anomalies. However, these nagging problems—particularly the failure of many crosses to exhibit independent assortment—were well known to geneticists and students of evolution. They were not understood until the discovery and description of chromosomes.

■ HOW GENES COME TO BE LINKED

In 1900, shortly after the rediscovery of Mendel's paper, W. S. Sutton pointed out the striking accord between Mendel's conclusion that hereditary factors (genes) occur in pairs in somatic cells and separate in gametogenesis, and the recent cytological evidence that somatic cells contain two of each kind of chromosome and that these chromosomes segregate in meiosis. Sutton interpreted this agreement as powerful evidence that the chromosomes are the bearers of the genes.

Now that Mendel's ideas could be directly related to observations made with microscopes, scientists could focus on the anomalies. In time it became apparent that traits assort independently only when their respective genes occur on two different chromosomes. Genes that occur on the same chromosome cannot assort independently during meiosis unless separated by crossing over. Genes on the same chromosome are said to be **_linked_**. One of the first examples of linkage was reported in 1906 by R. C. Punnett and William Bateson. They crossed sweet peas that had purple flowers and long pollen with ones that had red flowers and round pollen. All the F_1 plants had purple flowers and long pollen, as expected. (They knew that purple is dominant over red and that long is dominant over round.) The F_2 plants from this cross did not show the expected 9 : 3 : 3 : 1 ratio, however, but a highly anomalous one. Next, Bateson and Punnett crossed the F_1 plants back to homozygous recessive plants (with red flowers and round pollen). These results were equally anomalous. Using the symbols B for purple, b for red, L for long, and l for round, we can summarize Punnett and Bateson's results as follows:

$$BbLl \quad \times \quad bbll$$

purple long   red round

$$\downarrow$$

7 *BbLl*	1 *Bbll*	1 *bbLl*	7 *bbll*
purple	purple	red	red
long	round	long	round

According to the Principle of Independent Assortment, the heterozygous purple long parent should have produced four kinds of gametes (*BL, Bl, bL,* and *bl*) in equal numbers. When united with the *bl* gametes from the homozygous recessive parents, *BL* gametes should have given rise to purple long offspring, *Bl* gametes to purple round, *bL* to red long, and *bl* to red round, and these four phenotypes should have occurred in equal numbers, in a 1 : 1 : 1 : 1 ratio. But the result Bateson and Punnett actually obtained—a ratio of 7 : 1 : 1 : 7—makes it appear that the heterozygous parent produced far more *BL* and *bl* gametes than *Bl* and *bL* gametes.

Only in 1910 did Thomas Hunt Morgan, who had obtained similar results from *Drosophila* crosses, conceive of the answer. He postulated that the anomalous ratios were caused by linkage. Hence we should write the genotypes of the parents in Bateson and Punnett's second cross with slashes as *BL/bl* and *bl/bl*, to show that *B* and *L* are on one chromosome and *b* and *l* on the other. (We would write these genotypes as *B/b   L/l* and *b/b   l/l* if the genes were not linked).

If in Bateson and Punnett's cross the genes for purple and long and the genes for red and round were linked, the *BL/bl* parent in the second cross should have produced only two kinds of gametes, *BL* and *bl*, and the second cross should have yielded offspring of only two phenotypes, purple long and red round, in equal numbers. Yet the cross also yielded some purple round and red long offspring. Morgan suggested that some mechanism occasionally breaks the original linkages between purple and long and between red and round and establishes in a few individuals new linkages between purple and round and between red and long, making possible the production of *Bl* and *bL* gametes. The mechanism of this recombination is, of course, crossing over.

From an evolutionary perspective, crossing over is important because it increases the number of genetic combinations a cross can produce, and therefore the number of phenotypes on which selection can operate. Linkage can also have a major impact. Imagine that there were no crossing over. In such a case, an allele *a* on a chromosome containing a deleterious allele *Z* would be culled by selection operating against *Z*, even though *a* may be innocuous or even beneficial. Crossing over allows *a* to escape from its association with *Z*, but the more tightly linked two genes are—that is, the nearer they are to one another on the chromosome—the less often escape can occur (Fig. 16.22). Two unrelated genes located next to each other may be so tightly linked that selection for or against one is, in practice, selection for or against the other. Thus, it is possible for the frequency of one allele to be strongly affected by selection on an unrelated locus. At the other extreme, genes at opposite ends of a long chromosome in species with high crossover rates will be inherited virtually independently.

■ MAPPING GENES ON CHROMOSOMES

Alfred H. Sturtevant—then an undergraduate working in Morgan's lab—pointed out that if the probability of crossing over is approximately equal at any point along the length of a chromosome, then the greater the distance between two linked genes, the greater the frequency with which they will be separated: the frequency of recombination between any two linked genes will be proportional to the distance between them. Sturtevant realized that the percentage of recombination can serve as a tool for mapping the location of genes on chromosomes. We speak in terms of the percentage of re-

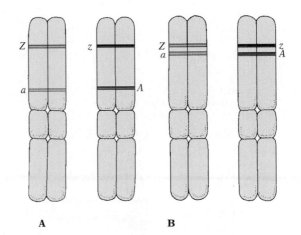

A   **B**

16.22   Effects of linkage on crossing over

When alleles *Z* and *a* are located far apart on a chromosome (A), the odds of a crossover event occurring between the two loci (and thus separating the alleles, leading to a *ZA* or *za* combination on a chromosome) are greater than when the two loci lie close together (B). In the latter case, selection operating on the allele at one locus can strongly affect the allele at neighboring loci.

16.23 Linkage and recombination

After this typical round of crossing over, the three genes originally linked as *ABC* and *abc* are combined in four ways: *abC*, *AbC*, *aBc*, and *ABc*. Notice that the farther apart two genes are, the more likely it is that a crossover event will take place between them. In this example, crossing over between *A* and *C* takes place in every instance, whereas crossing over between *A* and *B* occurs only half as often. Notice also that crossing over between two genes does not always recombine them: because chromatids 2 and 3 have two compensating crossover events, *A* and *C* remain together, as do *a* and *c*, even though segments between the two genes have been interchanged. As a result, recombination frequency is always lower than crossover frequency.

homologous chromosomes

twin chromatids

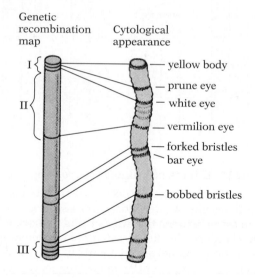

Genetic recombination map

Cytological appearance

— yellow body
— prune eye
— white eye
— vermilion eye
— forked bristles
— bar eye
— bobbed bristles

16.24 Comparison of a genetic-recombination map and cytological appearance of a portion of the X chromosome in *Drosophila melanogaster*

Staining and photographic techniques now exist to localize the genes on chromosomes directly. When these cytological results are compared with crossover frequencies (linkage maps), the effects of variations in crossover rates in different parts of the chromosome become apparent. In this case, crossing over is suppressed within two groups of genes (I and III), and greatly enhanced between the loci for white eyes and vermilion eyes (II). Suppression results from a relative lack of possible crossover sites while enhancement is a consequence of an overabundance of them.

combination rather than the number of crossover events because most chromosomes cross over at more than one place; therefore, two crossover events can cancel each other and so go undetected (Fig. 16.23).

The percentage of recombination gives no information about the absolute *distances* between genes (Fig. 16.24). Crossover events occur primarily at special DNA sequences, and selection can act to increase or decrease the number of such sites in any part of the chromosome if unusual crossover rates between particular genes are adaptive. On the other hand, crossover rates do reveal gene *order*. By convention, one unit of map distance on a chromosome is the distance within which recombination occurs 1% of the time. In Bateson and Punnett's test cross, two out of 16 of the offspring were recombinant products of crossing over. Two is 12.5% of 16; hence the genes controlling flower color and pollen shape in the sweet peas of this cross are located 12.5 map units apart.

Suppose we know that linked genes *B* and *L* are 12.5 map units apart. Further, suppose we find another gene, *A*, linked with these, that crosses over with gene *L* 5% of the time. How do we determine the order of the genes? The order could be *B–A–L*:

or it could be *B–L–A*:

The way to decide between these two alternatives is to determine the frequency of recombination between *A* and *B*. If this frequency is 7.5% (12.5 − 5), then the first alternative is correct; if it is 17.5% (12.5 + 5), then the second alternative is correct. In this way, by determining the frequency of recombination between each gene and at least two other known genes, the arrangement of genes on a chromosome can be mapped. Such maps reveal which traits are so closely linked that selection acts on them as a unit (Fig. 16.24).

■ LINKAGE AND VARIATION

Genetics is critical to understanding evolution because it is the major basis of heritability and variation, two crucial requirements for natural selection. As detailed in Chapter 12, the recombination of chromosomes during gamete formation and fertilization generates enormous diversity (Fig. 16.25). In our species, for instance, even without crossing over, each individual can produce more than 8 million different kinds of gametes, and thus, when two unrelated individuals pair, 70 trillion different zygotes. Crossing over increases these values enormously. The average human gamete, for instance, carries a set of chromosomes that has undergone an average of 30 crossover events during meiosis, so that most chromosomes have recombined internally at least once. The result is that any individual can generate an essentially infinite number of distinct gametes. Recombination and crossing over provide far more variation for natural selection to operate on than does mutation.

THE EFFECTS OF CHROMOSOMAL ALTERATIONS

A number of genetic alterations can have important consequences for selection and the evolution of new species—mutation and transposon movement, for instance, discussed in Chapters 8 through 11. In this section we will look at a few of the other chromosomal events that affect variation, linkage, and the emergence of new species.

■ STRUCTURAL CHANGES

In one form of alteration, called *translocation*, portions of two nonhomologous chromosomes are interchanged, with a consequent modification in linkage groups. Suppose, for example, that a pair of bar-shaped chromosomes in a certain species bear the genes *ABCDEFG* and that a pair of J-shaped chromosomes bear the genes *lmnopqrst*. If the *EFG* end of one bar-shaped chromosome and the *st* end of one J-shaped chromosome were interchanged, the result would be a shorter bar chromosome bearing only the genes *ABCDst* and a longer J-shaped chromosome bearing the genes *lmnopqrEFG*. By changing the linkage relationships of genes, translocations can have important effects on phenotypes and selection.

Sometimes a piece breaks off one chromosome and fuses onto

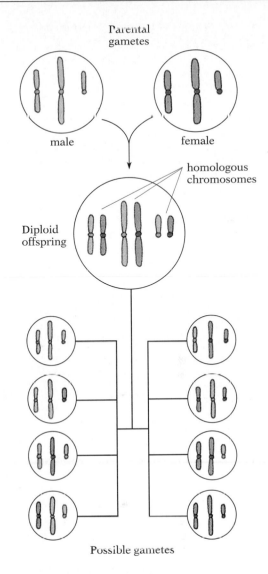

16.25 Gamete diversity in the absence of crossing over

Even in the absence of crossing over, each individual is capable of generating 2^n different kinds of gametes (where n is the number of chromosomes). Hence this hypothetical three-chromosome species produces eight classes of gametes, while our own species, with 23 chromosomes, can generate 2^{23}, or more than 8 million.

the end of the homologous chromosome. Such an alteration is a kind of *duplication*. An example would be loss of the *ABC* portion from one bar chromosome and fusion of this portion onto the homologous chromosome. The chromosome in which the loss occurred would thus bear only the genes *DEFGH*, while the chromosome undergoing the duplication would bear the genes *ABCABCDEFGH*. Translocation and duplication are most often caused by X-ray-induced damage to chromosomes.

Since translocations and duplications do not result in a net loss of genes, they usually have little effect on somatic cells. But these chromosomal rearrangements create problems in meiosis, when homologous chromosomes first synapse and then separate in the formation of gametes: If one chromosome of a pair has lost genes in a duplication event, half the gametes produced will lack these genes. If one chromosome of a pair has undergone a translocation exchange with a nonhomologous chromosome, synapsis with its homologue may be difficult, and meiosis may be blocked altogether. When meiosis does occur, the two nonhomologous chromosomes that have exchanged genes in translocation (losing some and gaining others) will segregate together only half the time. When they do not, half the resulting gametes will have nonhomologous duplications, while the other half will have two chromosomes with genes missing. A gamete that has lost essential genes may be inviable. Any zygote that does result from fusion involving such a gamete will lack the second copy of the genes in question; as a result, deleterious recessive alleles carried by the other gamete may be "exposed" (expressed in the organism's phenotype).

■ CHANGES IN CHROMOSOME NUMBER

As we saw earlier, the separation of chromosomes in cell division does not always proceed normally, and chromosomes that should move to opposite poles of the spindle may move instead to the same pole and become incorporated into the same daughter nucleus. The result of this nondisjunction may be an organism with one or more extra chromosomes. The presence of three chromosomes of one type in an otherwise diploid individual is called *trisomy*.

Occasionally cell division may be so aberrant that all the chromosomes move to the same pole, giving rise to a daughter cell with twice the normal number of chromosomes. If this happens during meiosis, the gamete produced is diploid instead of haploid. If such a gamete unites at fertilization with a normal haploid gamete, a triploid zygote results; if it unites with another diploid gamete, also produced by aberrant meiosis, a tetraploid zygote results. Cells or organisms that have more than two complete sets of chromosomes—that are triploid, tetraploid, hexaploid, and so on—are said to be *polyploid*.

Polyploidy is fairly common in plants. It has sometimes given rise to new species that are adaptively superior to the original diploid species under certain environmental conditions. Polyploidy can be stimulated in the laboratory by treating plants with special chemicals that cause nondisjunction during cell division. This procedure has been used in the production of many of the new strains

EXPLORING FURTHER

TRISOMY IN HUMANS

In humans most trisomies are lethal. Trisomy-18 (Edwards' syndrome) and trisomy-13 (Patau's syndrome), for example, produce physical malformations and mental and developmental retardation so severe that most afflicted infants die within a few weeks after birth. Because trisomies of most other autosomes result in spontaneous abortion, they are not found in live births. Two kinds of trisomy—trisomy-21 (Down's syndrome) and trisomies of the sex chromosomes—are exceptional in that the infants may survive.

Down's syndrome, in which three chromosomes of type 21 occur in the individual's cells, was the first clinical condition ever linked to a chromosomal abnormality. It is associated with a variety of characteristic physical features (broad head, rounded face, perceptible folds in the eyelids, a flattened bridge of the nose, protruding tongue, small irregular teeth, and short stature) and also mental retardation (IQ of about 40). The incidence of Down's syndrome is often related to the age of the mother. It occurs in fewer than one out of 1000 births in women under 20; it is over 7 times more common in women 35–39 years old, more than 20 times more common when the mother is 40–44, and more than 50 times more common when she is 45 or older. A similar association with the age of the mother is seen in Edwards' and Patau's syndromes.

Trisomy of the sex chromosomes can take several forms. In one, called Klinefelter's syndrome, the chromosomal makeup is XXY, and the individuals are males. Most victims show a variety of physical abnormalities and mental retardation; furthermore, they often suffer from thyroid dysfunction, chronic pulmonary distress, and diabetes.

Males with a second type of sex-chromosome trisomy, the XYY syndrome, generally show fewer and less severe abnormalities, though they often have subnormal intelligence. Because the incidence of XYY individuals is often significantly higher in penal institutions than in the general population, some investigators have advanced the controversial suggestion that men with the XYY condition are predisposed to aggressive behavior.

Women with triple-X syndrome (XXX) usually have underdeveloped sexual characteristics and sometimes subnormal intelligence, but their abnormalities are not debilitating.

These trisomic conditions, as well as many other kinds of genetic or chromosomal diseases, can be detected during embryonic development by the process of *amniocentesis*, in which fluid containing sloughed-off epidermal cells from the fetus is withdrawn from the uterus with a long needle inserted through the mother's abdominal wall. The fetal cells are then cultured and examined for abnormalities. A new technique, in which embryonic tissue cells are obtained directly through the maternal vagina, promises earlier and safer detection of these and other chromosomal conditions.

---

of cultivated plants (Fig. 16.26). Polyploidy is rare in animals and has probably not been an important factor in the origin of new animal species. There are, however, at least two exceptions: there exist several species of triploid lizards, which are of special interest because they are asexual. In addition, polyploidy may have played

16.26 Induced polyploidy in alfalfa

Tetraploidy and octoploidy in alfalfa and other commercially valuable crops can be induced experimentally. Tetraploid alfalfa (4×) is the most stress-tolerant, and is the type cultivated by farmers; cultivated alfalfa is a naturally occurring tetraploid. Octoploid alfalfa (8×) grows well in the greenhouse but is sensitive to the stress induced by lack of water in the field. (Diploid alfalfa, 2×, is shown at the upper left.) The polyploids shown here were obtained by a process known as sexual polyploidization, in which gametes with unreduced numbers of chromosomes are produced during meiosis. The union of such gametes results in tetraploid plants; repetition of the process in the next generation yields octoploids. This process is also the principal source of polyploidy of plants in nature.

a role in the evolution of bees and related insects, among which the various species usually have 4, 8, 12, or 16 chromosomes; a similar pattern exists in certain groups of fish and some frogs.

HOW EXPERIMENTAL RESULTS ARE TESTED

■ WHY STATISTICAL ANALYSIS IS NECESSARY

So far we have mentioned Mendelian ratios such as 1 : 2 : 1, 3 : 1, 1 : 1, and 9 : 3 : 3 : 1—all commonly expected results of various types of crosses. Geneticists frequently use the phenotypic ratios obtained in breeding experiments to deduce the underlying genetic phenomena. Evolutionary biologists, on the other hand, look for evidence of evolution—that is, changes in gene frequencies—in the form of departures from the expected Mendelian ratios. For example, when selection is operating against a homozygous dihybrid recessive, the size of the fourth class of the 9 : 3 : 3 : 1 distribution is reduced. However, chance deviations from a predicted distribution are common: although four coin tosses have an expected distribution of two heads and two tails, a 3 : 1 ratio is hardly surprising. However, if the four-toss sequence is repeated 100 times, yielding 300 heads and 100 tails, we would suspect that the coin was biased in some way. To judge whether a phenotypic distribution indicates the operation of selection (or of some other unexpected genetic process), we therefore must consider both the degree of discrepancy from the expected ratio *and* the sample size.

Scientists constantly encounter the same fundamental question—whether the deviations they observe in their experimental results (from a prediction or from another set of results) are significant or not. They cannot rely simply on a subjective impression. To help them decide, they can compute the mathematical probability that any observed deviation in their sample could have occurred by chance alone. The calculated probability is the chance of obtaining the observed results (or results even more extreme) by drawing data at random from the predicted set, or from the other set of results being compared. Statisticians have devised many tests for evaluating experimental or observational data. Though these tests differ in their form and in the sorts of data to which they can validly be applied, all are ways of calculating the probability that any deviations of the observed values might be due to chance alone.

■ THE CHI-SQUARE TEST

One test of statistical significance, the chi-square (χ^2) test, is particularly applicable to many genetic experiments. This test measures whether any deviation from the predicted distribution that occurs in experimental results exceeds the deviation that might occur by chance. It is designed for categorical data—green versus red, for example—rather than continuous data, such as 2.1 cm versus 2.4 cm. The formula for chi-square is

$$\chi^2 = \Sigma\, d^2/e$$

where d is the deviation from the expected value, e is the expected value, and Σ means "the sum of."

Consider two hypothetical crosses in which we expect that the phenotypic ratio should be 1 : 1 in the absence of selection for or against one of the phenotypes. In one cross we get values of 45 and 55 instead of 50 and 50, and in the other we get values of 5 and 15 instead of 10 and 10. We want to know in each case whether the deviation of the observed values from the expected values can reasonably be attributed to chance, or implies that selection is at work.

First we must determine the chi-square value for the two crosses (Table 16.6). In each of these experiments the absolute deviations of the observed values from the expected values are the same: a deviation of 5 in each phenotype. Note, however, that the chi-square values obtained in the two crosses are very different: the one based on a sample of 20 is 5 times as large as the one based on a sample of 100. This illustrates how sensitive the chi-square test is to sample size: the difference in sample size alone has made the great difference in these two chi-square values.

Each of these crosses involved only two classes, in this case two different phenotypes. Hence their chi-square values were calculated on the basis of only two squared deviations. Now suppose we had been analyzing a cross involving three different phenotypes. Then the chi-square would have been calculated on the basis of three squared deviations, and it is only reasonable to expect that the chi-square value obtained would have been higher than one based on only two. It is clear, then, that in evaluating chi-square values we must also take into account the number of classes on which they are based. The number of independent classes in a chi-square test is termed the ***degree of freedom***. The number of independent classes is one fewer than the total number of classes in the cross.

In our crosses involving two phenotypes, there is only one independent class (and so one degree of freedom), while in a cross involving three phenotypes there are two independent classes and two degrees of freedom. A moment's thought will tell you why this is so. In our cross based on a sample of 100, once we know that 45 offspring show the first phenotype, we automatically know that 55 must show the other phenotype. Since we know the total, the number in one class automatically tells us the number in the other class. In other words, the number in the second class is dependent on the number in the first class. Therefore, only the first class is an independent class. The same reasoning applies if we perform a cross involving three different phenotypes, and the total number of observations in our sample is 100; once we know the number showing the first and second phenotypes, we automatically know the number showing the third phenotype, because the number in the third class is dependent on the number in the first two classes.

We now know the chi-square values (1.0 and 5.0) and the degrees

Table 16.6 Chi-square analysis of two crosses

	FIRST PHENOTYPE	SECOND PHENOTYPE
45:55 experiment:		
Observed values	45	55
Expected values (e)	50	50
Deviation (d)	-5	$+5$
Deviation squares (d^2)	225	25
d^2/e	25/50 = 0.5	25/50 = 0.5

$$\chi^2 = \Sigma d^2 e = 0.5 + 0.5 = \mathbf{1.0}$$

	FIRST PHENOTYPE	SECOND PHENOTYPE
5:15 experiment		
Observed values	5	15
Expected values (e)	10	10
Deviation (d)	-5	$+5$
Deviation squared (d^2)	25	25
d^2/e	25/10 = 2.5	25/10 = 2.5

$$\chi^2 = \Sigma d^2 e = 2.5 + 2.5 = \mathbf{5.0}$$

Table 16.7 Probabilities for certain values of chi-square

DEGREES OF FREEDOM	P = 0.20 (1 IN 5)	P = 0.10 (1 IN 10)	P = 0.05 (1 IN 20)	P = 0.01 (1 IN 100)
1	1.64	2.71	3.84	6.64
2	3.22	4.60	5.99	9.21
3	4.64	6.25	7.82	11.34
4	5.99	7.78	9.49	13.28
5	7.29	9.24	11.07	15.09
6	8.56	10.64	12.59	16.81
7	9.80	12.02	14.07	18.48
8	11.03	13.36	15.51	20.09
9	12.24	14.68	16.92	21.67
10	13.44	15.99	18.31	23.21
15	19.31	22.31	25.00	30.58
20	25.04	28.41	31.41	37.57
30	36.25	40.26	43.77	50.89

Source: Based on a larger table in R. A. Fisher, 1946. *Statistical Methods for Research Workers,* 10th ed., Oliver & Boyd, London.

of freedom (one for each experiment) for our two hypothetical crosses. The next step is to consult a table of chi-square values. Table 16.7 gives four different chi-square values for each of a series of different degrees of freedom, and gives the probability (*P*) that a deviation as great as or greater than that represented by each chi-square value would occur simply by chance.

We can now evaluate the results obtained in our hypothetical crosses. In the first cross the deviation from the expected result yielded a chi-square value of 1.0. The cross had one degree of freedom. According to the table, a value as high as or higher than 1.64 has a chance probability of 0.20 (20%); that is, deviation from the expected as great as or greater than that represented by 1.64 will occur about once in five trials by chance alone. Our chi-square value is less than 1.64; hence the deviation in the experiment can be expected to occur by chance even more often than once in five trials. Most biologists agree that deviations having a chance probability as great as or greater than 0.05 (5%, or 1 in 20) are not statistically significant. Thus the deviation in our experiment is not regarded as statistically significant; it is presumed to be a chance deviation, which can be disregarded.

In our second experiment, the chi-square value representing the deviation from the expected results turned out to be 5.0. Again there was one degree of freedom. Looking at the listings in the table for one degree of freedom, we find that the value of 5.0 is greater than 3.84, which has a probability of 0.05 (5%), but less than 6.64, which has a probability of 0.01 (1%). Hence the probability that the deviation in this cross resulted purely from chance is between 1% and 5%, and thus is significant. Some factor other than chance was involved in producing the disagreement between result and prediction.

At this point, a geneticist would begin the search for a reasonable explanation: the original observations are always open to scrutiny. Selection may have acted against one of the phenotypes, so that some of those individuals died, thus leading to fewer representatives of this class than were expected; or the assumptions concerning the genetics involved in this cross may have been wrong. In any event, something interesting is probably occurring.

CHAPTER SUMMARY

HOW SINGLE-GENE TRAITS ARE INHERITED

WHAT MENDEL FOUND Mendel discovered that some traits are dominant, and thus mask other (recessive) traits. Thus the phenotype reflects only the dominant traits; any recessives in the genotype are hidden. Each organism carries two factors for a trait, which can be the same (so the organism is homozygous) or different (heterozygous). In reproduction the factors segregate independently, with one going into each gamete. Factors are now called alleles, and each allele is one form of a gene that occupies a particular locus on a chromosome. (p. 407)

PARTIALLY DOMINANT TRAITS When neither of two alleles is masked by the other, they are said to be partially dominant, and the phenotype is often a blend of the two traits. (p. 412)

HOW MULTIPLE AND MULTIGENIC TRAITS ARE INHERITED

INHERITING TWO TRAITS AT ONCE Traits are inherited completely independently when their loci are on different pairs of chromosomes. The distribution of genotypes can be computed with the Product Law (p. 413)

HOW GENES INTERACT Many genes depend on other genes for their full expression. A complementation test can determine whether a given recessive phenotype in two individuals arises from alleles at a single locus. Some genes can modify the expression of other genes. An epistatic gene determines whether or not one or more other genes will be expressed. Traits controlled by many separate genes often interact to produce a graded range of phenotypes; in such cases the effect of an individual gene is often impossible to discern. (p. 415)

WHEN GENES ARE ONLY PARTIALLY EXPRESSED Not all individuals with a particular genotype will express the associated phenotype; such genes have only partial penetrance. The genetic or physical environment may control the expressivity—the manner in which a genotype is expressed. (p. 420)

WHEN GENES HAVE SEVERAL ALLELES While an individual can only have two alleles of a gene, the population may harbor a large number, yielding a variety of genotypes and phenotypes. Examples are eye color in fruit flies and blood type in humans. (p. 422)

HOW DELETERIOUS ALLELES ARISE AND SURVIVE Recessive deleterious alleles at low frequency are generally hidden in heterozygous individuals. Some deleterious alleles actually confer an advantage in the heterozygous state. Inbreeding vastly increases the probability of generating offspring that are homozygous for a deleterious allele. (p. 424)

HOW SEX ALTERS INHERITANCE

HOW SEX IS DETERMINED Many species have sex-specific chromosomal differences, which result from an absent or much-reduced copy of one member of one chromosomal pair. (p. 426)

THE GENETICS OF SEX-LINKED CHARACTERS When one sex has only one full-length copy of the sex chromosome, it generally expresses any recessive alleles on that chromosome. In many species, the second sex chromosome in the sex with two full-sized sex chromosomes condenses to form a Barr body. (p. 428)

THE EXPRESSION OF SEX-DEPENDENT CHARACTERS Most sex-dependent characters are determined by genes on the autosomes. Some signal that is dependent directly or indirectly on the sex chromosomes determines which sex-dependent genes will be active. (p. 431)

HOW GENES ON THE SAME CHROMOSOME ARE INHERITED

HOW GENES COME TO BE LINKED When genes lie on the same chromosome, they usually do not segregate independently. Their degree of independence depends on the probability of a crossover event occurring between them during meiosis; in some species the crossover rate is so high that genes at opposite ends of long chromosomes are inherited essentially independently. (p. 432)

MAPPING GENES ON CHROMOSOMES The probability of genes on a chromosome becoming unlinked during meiosis can be used to map the approximate distance between them and to establish their order. (p. 433)

LINKAGE AND VARIATION Crossing over enormously increases the number of unique gametes meiosis can produce, and thus increases the range of variation in offspring. (p. 435)

THE EFFECTS OF CHROMOSOMAL ALTERATIONS

STRUCTURAL CHANGES Chromosomes sometimes break and are rejoined incorrectly. Occasionally pieces of chromosomes are copied mistakenly and incorporated into the genome, leading to duplications. Both translocations and duplications make pairing during meiosis difficult. (p. 435)

CHANGES IN CHROMOSOME NUMBER Occasionally chromosomes fail to segregate properly, and a cell winds up with too many or too few copies of a chromosome. Down's syndrome is caused by three copies of chromosome 21. (p. 436)

HOW EXPERIMENTAL RESULTS ARE TESTED

WHY STATISTICAL ANALYSIS IS NECESSARY Experiments either compare two sets of results, or a set of results with a prediction. Statistical techniques determine the probability that the degree of difference observed could have occurred by chance—that is, by drawing the data for one set at random from the other set. (p. 438)

THE CHI-SQUARE TEST The chi-square test is used when the data being collected are categorical. (p. 438)

STUDY QUESTIONS

1 In squash an allele for white color (*W*) is dominant over the allele for yellow color (*w*). Give the genotypic and phenotypic ratios for the results of each of the following crosses:

$$W/W \times w/w$$
$$W/w \times w/w$$
$$W/w \times W/w$$

2 A heterozygous white-fruited squash plant is crossed with a yellow-fruited plant, yielding 200 seeds. Of these, 110 produce white-fruited plants, while only 90 produce yellow-fruited plants. Using the chi-square test, would you conclude that this deviation is the result of chance, or that it probably represents some complicating factor? What if there were 2000 seeds, and 1100 produced white-fruited plants while 900 produced yellow-fruited individuals?

3 In humans, brown eyes are dominant over blue eyes. Suppose a blue-eyed man marries a brown-eyed woman whose father was blue-eyed. What proportion of their children would you predict will have blue eyes?

4 If a brown-eyed man marries a blue-eyed woman and they have 10 children, all brown-eyed, can you be certain that the man is homozygous? If the 11th child has blue eyes, what will that show about the father's genotype?

5 The litter resulting from the mating of two short-tailed cats contains three kittens without tails, two with long tails, and six with short tails. What would be the simplest way of explaining the inheritance of tail length in these cats? Show the genotypes.

6 When Mexican hairless dogs are crossed with normal-haired dogs, about half the pups are hairless and half have hair. When, however, two Mexican hairless dogs are mated, about a third of the pups have hair, about two-thirds are hairless, and some deformed puppies are born dead. Explain these results.

7 In peas an allele for tall plants (*T*) is dominant over the allele for short plants (*t*). An allele of another independent gene produces smooth peas (*S*) and is dominant over the allele for wrinkled peas (*s*). Calculate both phenotypic ratios and genotypic ratios for the results of each of the following crosses:

$$T/t \quad S/s \times T/t \quad S/s$$
$$T/t \quad s/s \times t/t \quad s/s$$
$$t/t \quad S/s \times T/t \quad s/s$$
$$T/T \quad s/s \times t/t \quad S/S$$

8 In some breeds of dogs a dominant allele controls the characteristic of barking while trailing. In these dogs an allele of another independent gene produces erect ears; it is dominant over the allele for drooping ears. Suppose a dog breeder wants to produce a pure-breeding strain of droop-eared barkers, but he knows that the genes for silent trailing and erect ears are present in his kennels. How should he proceed?

9 In Leghorn chickens, colored feathers are produced by a dominant allele, (*C*); white feathers are produced by the recessive allele (*c*). The dominant allele of another independent gene (*I*) inhibits expression of color in birds with genotypes *C/C* or *C/c*. Consequently both *C/–ce I/–* and *c/c –/–* are white. A colored cock is mated with a white hen and produces many offspring, all colored. Give the genotypes of both parents and offspring.

10 If the dominant allele *K* is necessary for hearing, and the dominant allele *M* of another independent gene results in deafness no matter what other genes are present, what percentage of the offspring produced by the cross *k/k M/m × K/k m/m* will be deaf?

11 What fraction of the offspring of parents each with the genotype *K/k L/l M/m* will be *k/k l/l m/m*?

12 Suppose that an allele, *b*, of a sex-linked gene is recessive and lethal. A man marries a woman who is heterozygous for this gene. If this couple had a large number of normal children, what would be the predicted sex ratio of these children?

13 Red-green color blindness is inherited as a sex-linked recessive. If a color-blind woman marries a man who has normal vision, what would be the expected phenotypes of their children with reference to this character?

14 In cats, short hair is dominant over long hair; the gene involved is autosomal. An allele B^1 of another gene, which is sex-linked, produces yellow coat color; the allele B^2 produces black coat color; and the heterozygous combination B^1/B^2 produces tortoiseshell and calico coat color. If a long-haired black male is mated with a tortoiseshell female homozygous for short hair, what kind of kittens will be produced in F_1? If the F_1 cats are allowed to interbreed freely, what are the chances of obtaining a long-haired yellow male?

15 In *Drosophila melanogaster* there is a dominant allele for gray body color and a dominant allele of another gene for normal wings. The recessive alleles of these two genes result in black body color and vestigial wings, respectively. Flies homozygous for gray body and normal wings are crossed with flies that have black bodies and vestigial wings. The F_1 progeny are then crossed, with the following results:

Gray body, normal wings	236
Black body, vestigial wings	253
Gray body, vestigial wings	50
Black body, normal wings	61

Would you say that these two genes are linked? If so, how many units apart are they on the linkage map?

16 The recombination frequency between linked genes *A* and *B* is 40%; between *B* and *C*, 20%; between *C* and *D*, 10%; between *C* and *A*, 20%; between *D* and *B*, 10%. What is the sequence of the genes on the chromosome?

SUGGESTED READING

HOLLIDAY, R., 1989. A different kind of genetic inheritance, *Scientific American* 260 (6). *On genetic "imprinting."*

WHITE, R., AND J. M. LALOUEL, 1988. Chromosome mapping with DNA markers, *Scientific American* 258 (2). *On modern methods for mapping the human genome.*

PART III

EVOLUTIONARY BIOLOGY

◀ **Diversity of species** produced by natural selection is suggested by this colorful array of beetles. The species represented here constitute a minute fraction of the estimated 400,000 extant species in the order Coleoptera, one of the 28 orders that make up the vast class of animals known as Insecta. (Actually, one of the specimens shown is a true bug—a member of the order of insects called Hemiptera.)

CHAPTER 17

VARIATION, SELECTION, AND ADAPTATION

Evolution, by definition, is the change in allelic frequencies in populations across generations. Because an individual's genotype is fixed at conception, an individual cannot evolve; only populations can. At its simplest, evolution can occur through the emigration or immigration of members of a population. In this chapter we will concentrate instead on the roles played by chance and natural selection in evolution. After we look at how these two processes generate changes in allele frequencies, we will examine some especially compelling cases of adaptation.

HOW SELECTION OPERATES

Evolution by natural selection depends on five factors:

1 More offspring are produced than can survive to reproduce (excess progeny).

2 The characteristics of living things differ among individuals of the same species (variability).

3 Many differences are the result of heritable genetic differences (heritability).

4 Some differences affect how well adapted an organism is (differential adaptedness).

5 Some differences in adaptedness are reflected in the number of offspring successfully reared (differential reproduction).

The excess reproductive capacity of organisms is self-evident. Darwin made the point incisively:

> The elephant is reckoned the slowest breeder of all known animals, and I have taken some pains to estimate its probable minimum rate of natural increase; it will be safest to assume that it begins breeding when 30 years old and goes on breeding until 90 years old; if this be so, after a period from 740 to 750 years there would be nearly 19 million elephants descended from this first pair.

After about 1200 years, this hypothetical elephant population would cover the entire land area of the earth, Antarctica included, shoulder to shoulder and head to tail (Fig. 17.1). Clearly, not all offspring survive long enough to reproduce.

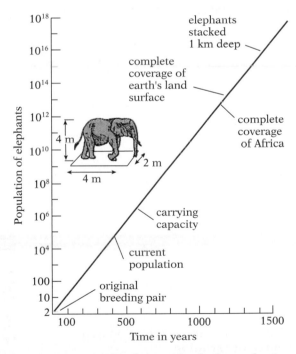

17.1 Unrestrained growth of a hypothetical elephant population

When Darwin's calculations for the slowest reproducing animal known are graphed and extrapolated, his point that not all offspring can survive becomes clear. (Darwin overestimated both the reproductive life span and the age of sexual maturity in elephants, but these minor errors cancel one another out.)

The nature of heritability of variant alleles was described in Chapter 16. As pointed out there, most variants are recessive, and therefore are unexpressed and hidden from selection when present in only one copy. Here we will look at the sources of variation and their consequences, and then consider mechanisms of selection. The second major section of the chapter will explore several types of adaptation.

■ WHAT CAUSES VARIATION?

A population is composed of many individuals. With rare exceptions, no two of these are exactly alike. We are well aware of the uniqueness of individual humans; we know from experience that each has distinctive anatomical and physiological characteristics, as well as distinctive abilities and behavioral traits. We are also fairly well aware of individual variation in dogs, cats, and horses. However, we tend to overlook the similar individual variation in less familiar species, like robins, squirrels, earthworms, sea stars, dandelions, and corn. Yet even though this variation may be less obvious to our unpracticed eye, it exists in all species.

The members of a population, then, share some important features but differ from one another in numerous ways. It follows that, if there is selection for or against certain heritable variants within a population, the makeup of that population may change with time.

Genetic variation Variation in sexual organisms arises from three main sources: crossing over in meiosis, the union of unrelated haploid gametes (sexual recombination; Fig. 17.2), and mutation (Fig. 17.3). The first two processes, discussed in Chapters 12 and 16, do not lead to new alleles but instead to a recombination of existing ones. Although beneficial in the long run, crossing over and sexual recombination are part of the metabolically expensive and geneti-

17.2   Sexual recombination as a source of variation

Offspring of the same parents frequently do not resemble each other or either of their parents. Shown here with their mother are the various kittens of a single litter.

17.3   Spontaneous mutation as a source of variation

(A) Most wisterias, like the one shown here, have lavender flowers. (B) Several decades ago, however, the famous white-flowering Eno wisteria appeared behind a biology building on the Princeton campus. It is almost certainly the result of a spontaneous mutation in a pigment gene of an offspring of one of the many lavender wisterias nearby.

A   B

cally chancy process of sexual reproduction. Since selection operates on individuals, the existence of recombination suggests that sex persists because it creates variation. Mutation, which includes point mutations as well as large-scale exon recombinations, is a potential source of entirely new alleles; we will look at how new alleles arise in more detail in a later section. Taken together, these processes generate new alleles and new combinations of alleles, and they provide the genetic variability on which natural selection can act to produce evolutionary change.

Purely phenotypic variation Some kinds of variation are immune or irrelevant to selection. Natural selection can act on genetic variation only when it is expressed in the phenotype. A completely recessive allele never occurring in the homozygous condition would be totally shielded from the action of natural selection.

Another potential complication is that any phenotypic variation within a population may give rise to reproductive differentials between individuals, whether or not the variation reflects corresponding genetic differences. Thus variations produced by exposure to different environmental conditions during development, or produced by disease or accidents, are subject to natural selection. In general, such selection has no effect on the overall genetic makeup of the population: there is no correlation between an individual's alleles and any chance events that lead to its lowered viability and reproductive potential. However, there are cases in which luck *can* affect evolution. For example, if the population is small and the one carrier of a rare allele is killed by an accident, gene frequencies will change. We will return to the role of chance in small populations in both this chapter and the next.

Variation produced by somatic mutations is not raw material for evolutionary change. It would be possible, for example, for an important mutation to occur in an ectodermal cell of an early animal embryo. All the cells descended from the mutant cell would be of the mutant type. The result might be a major change in the animal's nervous system, but the change could not be passed on to the animal's offspring. The ectodermal cells are not the ones that give rise to gametes. Hence selection that acts on variations produced by somatic mutations cannot result in evolutionary change in sexually reproducing organisms.

The theory of *evolution by natural selection* had a rival during the 19th century in the concept of *evolution by the inheritance of acquired characteristics*—an idea often identified with Jean-Baptiste Lamarck. Lamarck thought that somatic characteristics acquired by an individual during its lifetime could be transmitted to its offspring. Thus the characteristics of each generation would be determined, in part at least, by all the modifications—including those caused by experience, use and disuse of body parts, and accidents—to members of the preceding generations. Evolutionary changes would be the gradual accumulation of such acquired modifications over many generations. The classic example is the evolution of the long neck of the giraffe (Fig. 17.4).

According to the Lamarckian view, ancestral giraffes with short necks tended to stretch their necks as much as they could to reach the tree leaves they fed on. This frequent neck stretching caused

A

B

17.4 An okapi and a giraffe, two related African herbivores

The okapi (A) and giraffe (B) are thought to have had a common ancestor with a relatively short neck. The long-necked giraffe of today can reach food unavailable to shorter individuals.

EXPLORING FURTHER

THE LOGIC OF SEX

At first glance, sex in higher organisms makes no evolutionary sense. Half of a female's reproductive effort goes into sons, who will bear no offspring and so contribute little to her reproductive output. If males could be eliminated, each female could produce twice as many daughters and far more grand-children. This is readily observed in groups of higher plants and animals that have both sexual and asexual species. In nearly every case, however, the asexuals arose relatively recently from sexual precursors.

In addition to reducing reproductive output, sex tampers with success. A constellation of alleles that has worked well enough to allow an organism to reproduce is broken up, and half are assigned arbitrarily to each gamete, rather like taking cards randomly from a royal flush and a full house and recombining them to form a new hand. The novel collection of cards is unlikely to be as good as either "parent" hand. Some asexual species are quite successful: dandelions are, despite their flowers, incapable of sexual reproduction.

The evolution of sex presumably began with haploid asexual species such as bacteria that normally reproduce asexually. *E. coli* can reproduce every 20 min if enough food is available; since it takes 18 min to replicate the chromosome, the contest to generate offspring while conditions are good is a race against the clock.

Diploidy is one step toward sex. A bacterium with two copies of its its chromosome to replicate would be at an enormous reproductive disadvantage, having to collect enough nutrients to synthesize twice as much DNA as its haploid counterpart. The cost of haploidy, however, is the immediate expression of any mutation: since most genetic changes are deleterious, this penalty can be large. This is the most likely explanation of the evolution of diploid dominance in plants and animals.

Diploidy by itself does not entail sexual recombination (dandelions, for example, are diploid), but even without it diploidy confers several advantages. Most obviously, deleterious mutations are usually hidden. In addition, the intact chromosome can be used as a template for the repair of double-stranded damage to the other. Diploids can also accumulate and harbor variants on the "spare" copy that may prove useful when conditions change.

Sexual reproduction adds the features of crossing over and genetic recombination. Since crossing over involves the same enzymes that asexual diploids use for repair and that bacteria use for their rare bouts of conjugation, it seems likely that the first sexual species took advantage of crossing over as a way of correcting errors just prior to gamete formation. But changes bring new opportunities, and crossing over now creates more variation than it repairs.

There are two general hypotheses concerning the role of variation. The first, the **tangled-bank model**, is based on a reference to a diverse, multi-species environment discussed near the end of *The Origin of the Species*. Darwin realized that if conditions were everywhere the same, a few species would come to dominate any habitat. Yet when he surveyed a square meter of his own lawn, he counted more than 20 species of plants. Darwin's description of a tangled bank emphasized the many different microhabitats found in even a small area. According to the tangled-bank model, asexual species are at a disadvantage because they are superior in only one microhabitat, whereas sexual species produce a range of offspring-microhabitat matches. Because of their adaptive narrowness, asexual siblings may even wind up competing primarily with each other, whereas the many unique sexual progeny of one pairing might be different enough to seek out their own most appropriate sites.

A   Alice and the Red Queen

They were unable to outrun their environment, no matter how hard they tried.

The second hypothesis, the **red queen model**, takes its inspiration from the scene in Lewis Carroll's *Through the Looking Glass* in which Alice and the Red Queen are running as quickly as they can and yet make no progress: however fast they move, they cannot outrun their surroundings (Fig. A). The red queen theory suggests that variation is needed to keep up with a rapidly evolving background of predators, prey, and parasites as well as with changing conditions. The emphasis is on variation in time rather than variation in space.

The potential problem from parasites is especially easy to understand: most parasites are strain-specific and can devastate an asexual or inbred population. New strains of barley in Britain, for example, have a useful life of only 3–5 years before fungal parasites adapt enough to ruin crops. Clones of thrips (juice-sucking insects) on long-lived hosts regularly develop specializations for that one plant. This degree of specialization is possible because parasites have a short generation time compared with their hosts; they can go through many rounds of selection before the host reproduces.

One of the predictions of the red queen hypothesis is that crossing over should be most common among the longest-lived species; this would enable the host to compensate for rapid parasite evolution by making its offspring as different as possible. Crossover rates increasing with life span seem to be the pattern at least among mammals. What is hard to understand is why parasites have not wiped out dandelions and other asexual organisms.

With regard to habitat variation, both hypotheses are consistent with pattern seen in many species that can reproduce either sexually or asexually. For instance, the first aphid to find a suitable new plant shoot reproduces asexually, creating a clone of daughters equally well adapted to this local habitat. When conditions deteriorate, as when the shoot becomes overcrowded or suffers from drought, the founder begins producing winged reproductive offspring that leave the shoot, mate with other aphids, and search out a suitable host (Fig. B). Clearly the aphid life cycle can be interpreted as an adaptation to unpredictable spatial *or* temporal variation. There is every reason to believe that the degree and nature of the advantage of sexual reproduction can differ between species.

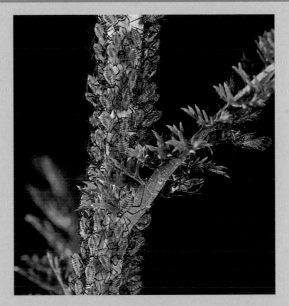

B Sexual (winged) and asexual aphids on a yarrow stem

The common thread, however, is that sex has been selected for because it creates variation. Without the capacity to change faster than mutations alone allow, long-lived species would be in trouble.

their offspring to have slightly longer necks. Since these also stretched their necks, the next generation had still longer necks. As a result of neck stretching to reach higher and higher foliage, each generation had slightly longer necks than the preceding generation.

Natural selection, on the other hand, proposes that ancestral giraffes probably had short necks, but that neck length varied from individual to individual because of their different genotypes. If the supply of food was limited, individuals with longer necks had a better chance of surviving and leaving progeny than those with shorter necks. This does not mean that all individuals with shorter necks perished or that all with longer necks survived to reproduce, but simply that a higher proportion of those with longer necks survived and left offspring. As a result, the proportion of individuals with genes for longer necks increased in the succeeding generation. As the proportion of individuals with somewhat longer necks rose, the increased competition for food higher up on the trees resulted in a selective advantage for those with yet longer necks, and so evolution continued.

■ HOW DO NEW GENES AND ALLELES ARISE?

One of the troubling mysteries of evolution for Darwin and his early supporters was how a complex organ or other structure could arise through selection on small variants of existing traits. How, for example, could the elaborate camera eye of vertebrates evolve? Wouldn't all the intermediate steps, as well as the first, be useless or even maladaptive, altering some important preexisting structure? In fact, new genes, new alleles, new morphological structures generally arise out of previous ones by a combination of well-documented genetic mechanisms.

Point mutations—deletions, substitutions, and additions of one or a few bases—can have two useful effects. First, they can "fine-tune" one allele of an existing gene, safe from negative selection during intermediate steps by virtue of the protection afforded by diploidy. (The "orthodox" version of the gene is on the other copy of the chromosome, performing its job.) Second, when small mutations occur in control regions—the binding sites for transcription factors, for example—they can dramatically alter the specificity, timing, and degree of responsiveness of gene activity in one or a group of alleles on that copy of a chromosome.

Though these are potentially useful sources of variation, the origin of much major evolutionary change is based on two other mechanisms: (a) gene duplication followed by the independent evolution of the spare copies of the gene, and (b) recombinations of preexisting exons.

How new genes arise Three observations suggest a novel molecular basis for evolution:

1 Eucaryotic chromosomes contain introns, exons, seemingly functionless pseudogenes, and repetitive DNA (Chapter 11). More than 90% of eucaryotic DNA is apparently functionless and dormant.

2 The realization that the fetal immune system actively disposes of redundant gene segments (Chapter 15) indicates that the genome is not necessarily as stable as had been thought. The discovery of transposons (Chapter 10) underscores this point.

3 The unique structure and organization of antibody genes, MHC proteins, and T-cell receptors (discussed in Chapter 15) implies that the exons of the variable and constant regions of both the light and heavy chains of antibodies arose from the duplication and rearrangement of some prehistoric gene sequence—probably one of the cell-adhesion molecules. (The cell-adhesion molecules are themselves part of a closely related family believed to be derived by duplication from a still more ancient gene.)

Duplication of genes or exons followed by small-scale evolutionary change in the superfluous copies is more likely to lead to functional genes with novel properties than random changes alone would be. Imagine how rarely we could generate a meaningful sentence by randomly arranging letters and spaces, whereas if we began with a meaningful sentence (a gene for a functional protein) consisting of words and spaces (exons and introns) and changed a few existing letters or words, the odds of ending up with an intelligible sentence with a new meaning would be fairly high. We will look at some documented examples of duplication, and then at the evidence that new genes are created by recombining the exons of old genes.

Evidence for gene duplication The evidence for the role of duplication is particularly clear in the genes for myoglobin and hemoglobin. Myoglobin, the oxygen-storage protein in muscles, consists of a single polypeptide chain, while hemoglobin, which carries oxygen in the blood, has two pairs of chains, for a total of four. The three-dimensional conformations of the α and β chains in hemoglobin are nearly identical, and they also closely resemble the conformation of the single chain in myoglobin; the genes for all are thought to have evolved by duplication from a single ancestral gene (Fig. 17.5). This conclusion is reinforced by the discovery that introns are located in the same places in all these genes. (Myoglobin itself evolved from a bacterial gene whose product is used to capture cellular oxygen that would otherwise destroy a nitrogen-fixing enzyme complex.)

Hemoglobin is synthesized in slightly different forms at different times in an organism's life—during embryonic development and during adult life, for instance—and is thus specialized for differing conditions of pH and oxygen concentration. Again, the genes for these alternative forms of hemoglobin, which in humans lie near one another on chromosome 7, are thought to have originated by duplication and then to have followed independent evolutionary pathways. The same region contains many pseudogenes with sequences very similar to those of the hemoglobin genes. Perhaps the pseudogenes are duplications that never evolved into improved functional gene products.

The same pattern is seen in many groups of enzymes. Elastase, the digestive enzymes trypsin, and chymotrypsin, and the blood-clotting enzyme thrombin all have different functions, but the genes for them have base sequences and intron locations that are nearly identical. It is unlikely that each evolved independently into a near duplicate of the others. The genes for these enzymes probably began separately as duplications of the gene for some primordial enzyme, and then went their separate evolutionary ways.

Gene evolution thus depends in part on gene duplication, followed by changes in base sequence that give rise to functionally different products. Such changes occur slowly, and most mutations result in genes with products of reduced function—or no function at all—rather than of altered specificity. For every new gene that produces a functional protein, there may be tens or hundreds of incomplete or failed "experiments" involving duplications. Unless some process is at work to edit out useless duplications, the chromosomes should be full of nonfunctional base sequences with clear similarities to those of functional genes. And indeed, well over 90% of the mammalian genome does not code for functional products. It may be that the enormous number of pseudogenes in eucaryotic chromosomes is evidence of past duplications that never evolved into functional genes.

Another source of genes is *horizontal transmission*—importation of a gene from another species. This interspecific transfer can occur when a parasite—especially a virus—picks up a host gene, incorporates it into its capsule, and then releases it into its next victim. Lysogenic viruses and retroviruses (discussed in Chapter 10) are especially efficient mediators of horizontal exchange, since they incorporate their genetic material into the host's genome.

Myoglobin

β Hemoglobin

17.5 Myoglobin and the β chain of hemoglobin compared

The similarity in conformation between these peptide chains is evident from this representation. The genes for these two chains are thought to have arisen from a duplication event.

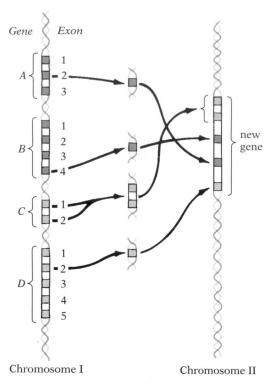

Gene   *Exon*

Chromosome I

Chromosome II

17.6 Hypothetical recombination of duplicate exons to create a new gene

Duplicates of five exons (colored blocks), from four different genes on a single chromosome (designated chromosome I), are imagined to be inserted near one another in a different chromosome (chromosome II). The intron regions (white blocks) flanking each exon that is being inserted provide the signals necessary for the new gene's transcript to be processed correctly. Since many exons code for protein subunits, the newly combined exons are more likely to encode a functional product than would have been the case if they began and ended at random points. If their product is at least partially functional, the new gene will survive in the germ line (the gametes transmitted to offspring) and be subject to improvement through natural selection.

Exon recombination Walter Gilbert and Colin Blake have suggested that individual exons from different genes can be brought together to produce new combinations. This widely entertained proposal makes excellent sense if exons code for parts of the resulting protein called "domains"—distinct subunits that, like building blocks of various shapes, can form new structures when put together in new ways. Careful examination of the genes for dozens of proteins confirms that this is sometimes the case. For instance, introns in some myoglobin genes occur between regions that code for sections joined at a major turn of this highly folded globular protein. Thus each intron defines a boundary of sorts between compact domains, and each myoglobin exon can be thought of as coding for one of these domains, or subunits.

In genes for some other proteins, introns fall at the boundaries between regions coding for sections of α helix and regions coding for sections of the β pleated sheet. Alternatively, the introns sometimes flank regions coding for sections of the protein containing the active site. Because the regions encoded by exons form distinct subunits, a recombination of exons would have a real chance of generating a working enzyme with novel properties. A new gene could evolve by combining, say, exon 2 of gene *A*, exon 4 of gene *B*, exons 1 and 2 of gene *C*, and exon 2 of gene *D* at a new site (Fig. 17.6). Such a recombination could be effected by a simple movement of the exons from their original genes or from duplicates. It might involve movement of the exons themselves, or of copies of them. In either case, insertion of exons (with their flanking introns) within an intron region of the chromosome would improve the chances of generating a functional new gene.

The best evidence for exon shuffling comes from the genes involved in blood clotting. Intron positions in many other genes do not seem to fit neatly with the domain hypothesis. Perhaps, as some suggest, modern exons are the descendants of ancient minigenes that have been strung together to create larger and more efficient proteins; some have taken on domainlike identities, while others have not. Still other researchers believe that many introns are actually dormant transposons (described in Chapter 10) that have inserted themselves at random into modern genes; many exons, therefore, are random slices of genes rather than functional domains.

Despite the controversy over the multiple sources of exons and introns, comparative study of DNA sequences argues strongly for a major role of exon recombination in the evolution of many genes. Gilbert has estimated that all the genes that exist today (50,000 in humans alone) evolved from perhaps as few as 1000 unique exons or minigenes. The processes of duplication and intragenic recombination have almost certainly been efficient mechanisms of major evolutionary innovation. Molecular biologists have recombined existing exons to create hybrid "designer" genes; these hybrid genes can generate working enzymes with novel properties.

■ HOW DO MUTATIONS SPREAD IN POPULATIONS?

Because mutation rates are low, when a novel mutation occurs, it is necessarily rare in the population. The mutation may go extinct (particularly if it is harmful), achieve a stable intermediate frequency (as part of a balanced polymorphism, discussed later in this chapter), or become the most common form of the gene. A rare mutation could spread and become a common allele in two ways: (1) in the absence of any selective advantage, it could nevertheless increase in frequency by chance, or (2) by conferring an advantage on its bearers, it could be favored by natural selection. Let us look at how the frequency of an allele is traced, and then at how chance and selection can drive that frequency up or down.

The gene pool Chapter 16 discusses genetics in terms of the genotype, which is the genetic constitution of an individual. The genetics of populations—the level at which evolution operates—depends on the *gene pool*, which is the sum total of all the genes possessed by all the individuals in the population.

As detailed in Chapter 16, the genome of a diploid individual contains no more than two alleles of any given gene.[1] But the gene pool of a population can contain any number of different allelic forms of a gene. The gene pool is characterized with regard to any given gene by the frequencies of the alleles of that gene in the population. Suppose that gene *A* occurs in only two allelic forms, *A* and *a*, in a particular sexually reproducing population. And suppose that allele *A* constitutes 90% of the total of both alleles, while allele *a* constitutes 10%. The frequencies of *A* and *a* in the gene pool of this population are therefore 0.9 and 0.1. If those frequencies were to change with time, the change would be evolutionary. Hence it is possible to determine what factors cause evolution by determining what produces shifts in allelic frequencies.

For our alleles *A* and *a*, with frequencies of 0.9 and 0.1 in a population, how can we calculate the genotypic frequencies in the next generation? If we assume the population is large and that all genotypes have an equal chance of surviving, this calculation is not hard. If the frequency of *A* in the entire population is 0.9 and the frequency of *a* is 0.1, the alleles carried by sperm and eggs will also appear at these frequencies. Using this information, we can set up a Punnett square analogous to those introduced in Chapter 16:

		Sperm	
		0.9 *A*	0.1 *a*
	0.9 *A*	0.81 *A/A*	0.09 *A/a*
Eggs			
	0.1 *a*	.09 *a/A*	.01 *a/a*

Notice that the only difference between this and a Punnett square for a cross between individuals is that here the sperm and eggs are not those produced by a single male and a single female, but those produced by all the males and females in the population, with the frequency of each type of sperm and egg shown on the horizontal and vertical axes respectively. Filling in the square (by combining

---

[1] There is one obvious exception to this generalization: because the genome contains multiple copies of rRNA genes, it is almost inevitable that each individual will carry more than two alleles of these genes.

the indicated alleles and multiplying their frequencies) tells us that the frequencies of the homozygous dominant genotype (*A/A*) in the next generation of this population will be 0.81, of the heterozygous genotypes (*A/a* and *a/A*) will be 0.18, and of the homozygous recessive genotype (*a/a*) will be 0.01. If the frequencies we have computed change in successive generations, the population has evolved.

Evolution versus equilibrium It is easy to believe that the more frequent allele (*A*) would automatically increase in frequency while the less frequent allele (*a*) would decrease and eventually be lost; thus variation—the raw material for evolution—would inevitably vanish. If there were little variation, evolution would be a minor phenomenon in nature. In fact, however, the rarity of a particular allele in a large population does not doom it to automatic disappearance.

As noted above, the genotypic frequencies in the gene pool of the second generation of our hypothetical population will be 0.81, 0.18, and 0.01. We can use these figures to compute the allelic frequencies of *A* and *a* in this generation. Since the frequency of the *A/A* individuals is 0.81, the frequency of their gametes in the gene pool will be 0.81. All these gametes will contain the *A* allele. Likewise, the frequency of the gametes of the *a/a* individuals will be 0.01, and each gamete will contain an *a* allele. The frequency of the heterozygous (*A/a* and *a/A*) individuals is 0.18, and the frequency of their gametes in the gene pool will be 0.18, but their gametes will be of two types, *A* and *a*, in equal numbers. The frequency of the *A* and *a* alleles in the gametes of the population can be calculated:

Frequency of genotypes	Frequency of *A* gametes	Frequency of *a* gametes
0.81 *A/A*	0.81	0
0.01 *a/a*	0	0.01
0.18 *A/a* + *a/A*	0.09	0.09
	0.9	0.1

The allelic frequencies are 0.9 and 0.1, the same frequencies we started with in the preceding generation.

Since the allelic frequencies are unchanged, the genotypic frequencies in the succeeding generation will again be 0.81, 0.18, and 0.01; the allelic frequencies in turn will be 0.9 and 0.1. In populations large enough to swamp chance effects, variation is retained. Moreover, evolution can occur only when something disturbs this genetic equilibrium. This was first recognized in 1908 by G. H. Hardy and W. Weinberg. According to the ***Hardy-Weinberg Law***, *under certain conditions of stability both phenotypic and allelic frequencies remain constant from generation to generation in sexually reproducing populations.*[2]

---

[2] For this statement to hold true for phenotypes, the initial genotypic frequencies must be in equilibrium. If these frequencies are not in equilibrium, they will change in successive generations until the equilibrium is achieved. For example, if genotypes *A/A* and *a/a* were present in a population but *A/a* was missing (say, because of human intervention), all three genotypes would appear in the next generation, and thereafter the genotypic and phenotypic frequencies would remain constant.

For a population to be in genetic equilibrium, five conditions must be met:

1 The population must be large enough to make it unlikely that chance could significantly alter allelic frequencies.

2 Mutations must not occur, or there must be mutational equilibrium.

3 There must be no immigration or emigration that alters allelic frequencies in the population.

4 Mating must be random with respect to genotype.

5 Reproductive success—the number of offspring and the number of their eventual offspring—must be random with respect to genotype.

The Hardy-Weinberg Law demonstrates that variability and heritability, two bases of natural selection, cannot alone cause evolution. Because the five conditions are *never* completely met, however, evolution does occur. The value of the Hardy-Weinberg Law is that it provides a baseline against which to judge data from actual populations. By defining the criteria for genetic equilibrium, it indicates when a population is not in equilibrium and helps to isolate the causative agents of evolution. The role of the investigator is then to discover the relative contribution of each factor in the evolution of a population.

With regard to the first condition, chance is always a factor: it is unlikely that an allele will appear in the next generation in exactly the same proportion of the individuals that carry it in the current generation; variation arises because gametes carrying any allele tend to contribute to zygotes more or less often than the exact number of times their proportion in the population would predict. This is analogous to the familiar observation that a coin flipped 10 times has only a 25% chance of yielding exactly five heads and five tails. But when the number of flippings is large, the average chance deviation from 50:50 is very small. Similarly, many natural populations are large enough that chance alone is not likely to cause any appreciable alteration in the allelic frequencies in their gene pools. However, in small isolated populations of fewer than 100 breeding-age members, allelic frequencies are highly susceptible to random fluctuations. Moreover, these fluctuations can easily lead to loss of an allele from the gene pool even when that allele is adaptively superior. In the absence of immigration or mutation, such an allele is lost forever. In such populations, in fact, there are relatively few alleles with intermediate frequencies; most alleles are either lost or become fixed as the only allele present. Small populations tend to have a high degree of homozygosity, while large populations tend to be more variable. But since chance changes, called **genetic drift**, are not much influenced by the adaptiveness of alleles, it is an indeterminate evolution, as likely to take one direction as another (Fig. 17.7).

The second condition for genetic equilibrium—either no mutation or mutational equilibrium—is rarely met in populations. Mutations are always occurring. Most genes undergo mutation probably once every 1 million to 100 million replications; the rate of mutation for different genes varies greatly. As for mutational equilibrium, very rarely, if ever, are the mutations of alleles for the

17.7 Possible genetic drift in a cichlid fish

Pseudotropheus zebra, one of hundreds of species of cichlid fish living in the rift lakes of Africa, is divided into numerous isolated populations, many of which have evolved their own distinctive morphology. As there is no known selective force that accounts for this diversity, the varied colors are thought by many researchers to be the result of genetic drift in each small population.

EXPLORING FURTHER

THE HARDY-WEINBERG EQUILIBRIUM

On p. 452 we used a Punnett square to calcualte the frequencies of the genotypes produced by alleles A and a, whose respective frequencies in a hypothetical population were given as 0.9 and 0.1. The same results can be obtained more rapidly algebraically.

Expansion of the binomial expression $(p + q)^2$ where p is the frequency of one allele (in our case, A) and q is the frequency of the other allele (a), yields the formula for the Hardy-Weinberg equilibrium:

$$p^2 + 2pq + q^2 = 1$$

Substituting the allelic frequencies 0.9 and 0.1 for p and q respectively, we obtain

p^2	$+$	$2pq$	$+$	q^2	$=$	1
(0.9)(0.9)	$+$	2(0.9)(0.1)	$+$	(0.1)(0.1)	$=$	1
0.81	$+$	0.18	$+$	0.01	$=$	1

The three terms of the Hardy-Weinberg formula indicate the frequencies of the three genotypes:

p^2	$=$	frequency of A/A	$=$	0.81
$2pq$	$=$	frequency of $A/A + a/A$	$=$	0.18
q^2	$=$	frequency of a/a	$=$	0.01

These are the same results we obtained using the Punnett square.

In this example we have assumed that we know the allelic frequencies and want to compute the corresponding genotypic frequencies. The Hardy-Weinberg formula allows many other sorts of calculations as well. Suppose, for example, we know a certain disease caused by a recessive allele d occurs in 4% of a certain population and we want to find out what percentage are heterozygous carriers of the disease. Since the disease occurs only in homozygous recessive individuals, the frequency of the d/d genotype is 0.04. Letting q^2 stand for the frequency of d/d in the formula, we can write

$$q^2 = 0.04$$

The frequency of allele d, then, is the square root of 0.04:

$$q = \sqrt{0.04} = 0.2$$

If the frequency of allele d is 0.2, the frequency of allele D must be 0.8, because the two frequencies must always add up to 1 (that is, $p + q = 1$). Substituting the frequencies of both alleles in the Hardy-Weinberg formula, we can compute the frequencies of the genotypes:

p^2	$+$	$2pq$	$+$	q^2	$=$	1
(0.8)(0.8)	$+$	2(0.8)(0.2)	$+$	(0.2)(0.2)	$=$	1
0.64	$+$	0.32	$+$	0.04	$=$	1

Since the term $2pq$ stands for the frequency of the heterozygous genotype, which is what we want to know, our answer is that 0.32, or 32%, of the population are heterozygous carriers of the allele d that causes the disease we are studying. Powerful as this method is, however, it is important to remember that these equations apply only to populations in Hardy-Weinberg equilibrium.

This type of reasoning can be used to calculate changes in allelic frequencies when only phenotypes can be measured directly. Suppose in a large population of freely interbreeding plants we find 59% with yellow blossoms (the dominant phenotype) and 41% with white (a recessive phenotype). In the spring of the following year, after a very severe winter, we find 64% with yellow blossoms and 36% with white blossoms. Clearly the plants with the dominant allele survived better, but exactly how much have the allelic frequencies changed?

Since white blossoms indicate a recessive phenotype, the frequency of the genotype y/y was 0.41 initially. Setting $q^2 = 0.41$, we calculate that $q = 0.64$ (approximately), which means that the frequency of allele y was 0.64. The frequency of dominant allele Y was therefore 0.36 ($1 - 0.64 = 0.36$). The next spring the frequency of white blossoms has fallen to 36%, which gives $q^2 = 0.36$ and $q = 0.60$. The frequency of allele y is therefore 0.60, and the frequency of Y must be 0.40. Thus the frequency of allele y has changed from 0.64 to 0.60 and the frequency of Y from 0.36 to 0.40.

Similar procedures, even if considerably more complicated mathematically, can be used for situations involving multiple alleles. Thus the Hardy-Weinberg formula for a triallelic (three-allele) situation requires expansion of the trinomial $(p + q + r)^2$, where r is the frequency of the third allele. A quadriallelic (four-allele) situation uses $(p + q + r + s)^2$.

same character in exact equilibrium: the number of forward mutations per unit time is rarely exactly the same as the number of back mutations.[3] The result of this difference is a ***mutation pressure*** tending to cause a slow shift in the allelic frequencies in the population. More stable alleles increase in frequency while more muta-

---

[3] By convention, the mutation from the more common allele to the less common one is called the forward mutation; the reverse is the back mutation.

ble alleles decrease, unless some other factor offsets the mutation pressure. Eventually, of course, the frequency of the more stable allele will become so high that it will undergo the same number of mutations per unit time as the more mutable allele, despite its lower mutation rate, and equilibrium will be achieved. This requires so much time, however, that other events almost always change allelic frequencies before mutational equilibrium is reached. But even though mutation pressure is almost always present, it is seldom a major factor in producing changes in allelic frequencies in a population in the short run. On the other hand, as we saw earlier, gene duplication, exon recombination, and mutations in gene-control regions do provide much of the basis for long-term genetic change.

According to the third condition for genetic equilibrium, a gene pool cannot have immigrants from other populations that introduce new alleles or different allelic frequencies, and it cannot undergo changes in allelic frequencies by emigration. Most natural populations, however, experience at least a small amount of gene migration, generally called *gene flow*, which tends to upset the Hardy-Weinberg equilibrium and lead to evolution. Such evolution—that is, a change in allelic frequencies—need not, at least in the short run, involve natural selection. On the other hand, there are doubtless populations—those on distant islands or in isolated habitat patches, for instance—that experience no gene flow, and in many instances where flow does occur it is probably so slight as to be negligible. Thus the third condition for genetic equilibrium is sometimes met in nature.

The final two conditions for genetic equilibrium are that mating and reproductive success be random. Among the vast number of factors involved in mating and reproduction are choice of a mate, physical efficiency and frequency of mating, fertility, total number of zygotes produced at each mating, percentage of zygotes that develop successfully, survival of the young to reproductive age, fertility of the young, and even survival of postreproductive adults when their survival affects either the chances of survival or the reproductive efficiency of the young. For mating and reproductive fitness to be random, all these factors must be independent of genotype, so that natural selection cannot operate.

These conditions are probably never met in any real population. An organism's genotype almost always influences each of these factors. To take an obvious case, female guppies do not mate at random but instead choose showy males with large tails (Fig. 17.8). In short, few aspects of reproduction are totally uncorrelated with genotype. Nonrandom reproduction is normal. Aside from mate choice, which is considered part of sexual selection (discussed later in this chapter), nonrandom reproduction is a component of natural selection. Natural selection, then, is almost always operating in populations; there is always *selection pressure* acting to disturb the Hardy-Weinberg equilibrium and cause evolution, even if selection serves merely to limit the frequency of deleterious mutations.

While we will concentrate on natural selection as the most influential mechanism in causing evolution, it is useful to think about Lamarck's long-necked giraffes again in the context of the Hardy-Weinberg factors. Though the Lamarckian concept of inheritance of acquired characteristics is wrong, the Darwinian explanation of

A

B

C

17.8 Female-choice sexual selection in guppies

In most species of vertebrates, including guppies, the female chooses the mate. (A, B) Male guppies exhibit great variety in the size, color, and location of their spots, and in the size and patterning of their tails. These features are largely heritable. Females choose males on the basis of their conspicuousness. (C) Given a choice between males with tails of two different sizes—large and small (bars at left), large and medium (the latter produced by surgical shortening; bars at middle), or medium and small (bars at right)—females prefer to be near males with larger tails. Subsequent findings indicate that they also preferentially mate with them. Females also prefer high display rates and greater coloration (particularly orange spots).

longer necks as a result of natural selection, which would favor them if they offer an adaptive advantage, is by no means the only possibility. There are several other evolutionary scenarios that might conceivably have given rise to the present-day giraffe.

The most plausible alternative to natural selection is that giraffes have long necks as a result of chance. Suppose that the ancestral population had a wide variety of neck lengths, or that a mutation caused a small subset of the population to have unusually long necks. If we assume that long necks are adaptively neutral—neither advantageous nor disadvantageous, or, more likely, that the benefits they confer are balanced by physiological costs—neck length in the population will not change. Now suppose that the population suddenly declined because of some environmental factor like disease or bad weather, so only a few individuals survived the crisis. If, by chance, a disproportionate number of the survivors were long-necked, the trait could become established without the intervention of natural selection (Fig. 17.9).

As this hypothetical example indicates, evolution does not depend exclusively on any one mechanism, whether natural selection, genetic drift, mutation, or migration. It also underscores the potential error of assuming that all the traits of the living things around us are the adaptive result of natural selection. Throughout this chapter and the next we will consider the alternatives to natural selection. Nevertheless, natural selection is apparently the most important factor in the evolution of most populations.

17.9 Natural selection versus chance

In this hypothetical example, a small group of long-necked individuals arises in an otherwise stable population. (For simplicity, each animal in the drawing represents a fraction of the total population.) In evolution by natural selection (A), long-necked individuals become prevalent because they are able to reach more vegetation, and so survive to have proportionately more offspring in succeeding generations. In evolution by chance (B), the frequency of long-necked individuals does not change until a chance catastrophe—fire, flood, heavy predation, or disease, for example—kills most members of the population, or the population size becomes small in some other way. Because only long-necked individuals happen to survive, their offspring multiply to fill the habitat even though their distinctive trait is of no net selective advantage.

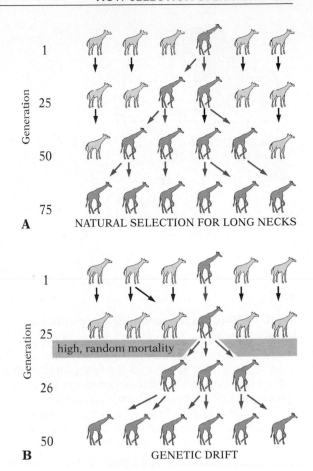

A   NATURAL SELECTION FOR LONG NECKS

B   GENETIC DRIFT

▪ NATURAL SELECTION

How selection changes allelic frequencies Our hypothetical population had alleles A and a, with frequencies 0.9 and 0.1; the genotypic frequencies were 0.81, 0.18, and 0.01. Let's look now at the effects of selection pressure on these alleles.

Suppose that selection acts against the dominant phenotype, and that this negative selection pressure is strong enough to reduce the frequency of A in the present generation from 0.9 to 0.8 before reproduction occurs. (Note that there will be a corresponding increase in the frequency of a from 0.1 to 0.2 among the survivors, since the two frequencies must total 1.) If we set up a Punnett square, we can calculate the genotypic frequencies in the zygotes of the second generation:

	Sperm	
	$0.8A$	$0.2a$
$0.8\ A$	$0.64A/A$	$0.16A/a$
$0.2\ a$	$0.16a/A$	$0.04a/a$

Eggs

The genotypic frequencies of these zygotes are different from those in the parental generation; instead of 0.81, 0.18, and 0.01, the frequencies are 0.64, 0.32, and 0.04. If selection now were to act against the dominant phenotype in this generation, and thereby again reduce the frequency of A, the frequency of A/A will be still lower and that of a/a higher. Eventually the frequency of A/A would fall to a very low level and the frequency of a/a would rise. Natural selection would cause a change from a population in which 99% of the individuals showed the dominant phenotype, to one in which most showed the recessive phenotype. This evolutionary change from the prevalence of one phenotype to the prevalence of another would have occurred without any new mutation, simply as a result of natural selection operating on preexisting mutant alleles.

Selection has been observed to produce radical shifts in allelic frequencies. For example, soon after the discovery of penicillin, *Staphylococcus aureus* (a bacterial species that can cause numernormal medium

medium with penicillin

17.10 Mutation for resistance to penicillin is spontaneous

Bacterial cells were cultured on a normal agar medium, and many colonies developed (upper culture dish). Then a block wrapped with velveteen was pressed against the surface of the culture to pick up cells from each of the colonies. The block was next pressed against the surface of a second culture dish, containing sterile medium to which penicillin had been added; care was taken to align the transfer block and the culture dishes according to markers on the block and dishes (black lines). The cells from most of the colonies on the original dish failed to grow on the penicillin medium, but those of a few colonies (two are shown here) did grow. Had the cells of those two colonies spontaneously become penicillin-resistant before being transferred to the penicillin medium, or did exposure to the penicillin induce a mutation for resistance?

Because the transfer block had been aligned with each dish in the same way, according to the markers, it was possible to tell precisely which original colonies had given rise to the two colonies on the penicillin medium. Cells could therefore be taken from those original colonies, which had never been exposed to penicillin, and tested for resistance. They were found to be resistant. Hence the mutation for resistance must already have arisen; it was not induced by exposure to the drug.

ous infections, including boils and abscesses) developed resistance to the drug. Higher and higher doses of penicillin were necessary to kill the bacteria, and the resistant bacteria became a serious problem in hospitals. The bacterial population evolved in the face of the strong selection exerted by the penicillin. The drug itself does not induce mutations for resistance; it simply selects against susceptible bacteria by killing them (Fig. 17.10). Apparently some alleles determining metabolic pathways that confer resistance to penicillin are already present in low frequency in most populations, having arisen earlier as a result of random mutations. Individuals possessing these alleles are thus **preadapted** to survive the antibiotic treatment, and, since they are the ones that reproduce and perpetuate the population, the next generation shows a marked resistance to penicillin. If such alleles were not already present in a population exposed to penicillin, no cells would survive and the population would be wiped out.

The importance of selection pressure does not mean that new mutations are irrelevant; in fact, continued selection with penicillin usually leads to gradually increased resistance, some of which results from new mutations that enhance resistance. However, mutations beneficial in an environment containing penicillin arise by chance whether or not the drug is administered; they are simply not selected for in the absence of penicillin. They can nevertheless persist if they have no adverse effect.[4]

Evolution in haploid organisms can proceed faster than in conventional biparental diploid organisms. Nevertheless, even very small selection pressures can produce major shifts in gene frequencies in biparental populations over an evolutionarily brief period. J. B. S. Haldane showed that if the individuals carrying a given dominant allele consistently benefit by as little as 0.001 in their capacity to survive (that is, if 1000 *A/A* or *A/a* individuals survive to reproduce for every 999 *a/a* individuals), then the frequency of the dominant allele would increase from 0.00001 to 1.0 in fewer than 24,000 generations. This may sound like a large number of generations, but many plants and animals have at least one generation a year—*Drosophila*, for example, has more than 30. In very few species is the generation time more than 10 years. Hence 24,000 generations often means fewer than 2400 years and rarely more than 240,000 years. These are short time spans when measured on the geological time scale. Moreover, selection pressures in nature are usually much larger than 0.001; hence major changes in allelic frequencies sometimes take less than a century, perhaps less than a decade (and less than a year for microorganisms).

Directional selection of polygenic characters So far, we have discussed idealized situations involving only two clearly distinct phenotypes determined by two alleles of a single gene. In reality, as

[4] There is evidence that in the face of adverse conditions, some species of bacteria can increase the rate at which their genomes mutate, either by neglecting to repair some errors, or by actively creating mistakes (as in the hypermutation process at work on the antibody genes, described in Chapter 15). In the face of declining reproductive potential, such a system would increase the variability in the population. With variability comes the chance of obtaining a favorable mutation that could "rescue" at least one individual, thus allowing it to found a new clone able to thrive despite altered conditions.

17.11 Frequency distribution of number of body segments in a population of the millipede *Narceus annularis*

The pattern of variation in number of segments (shown by the vertical bars) approximates, but does not exactly fit, the bell-shaped normal curve of probability.

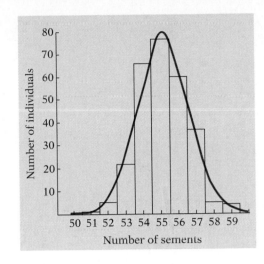

discussed in Chapter 16, the vast majority of characters on which natural selection acts are influenced by many different genes, most of which have multiple alleles in the population; the expression of many characters, moreover, is influenced considerably by environmental conditions. Consequently such characters as height usually vary continuously over a wide range. When graphed, the frequency distribution often approximates the normal, or bell-shaped, curve (Fig. 17.11).

If the environmental conditions change, creating a shift in the selection pressure, we would expect the curve of phenotypic variation to shift as a result of changing allelic frequencies. For example, assume that the conditions under which a certain plant grows best are genetically determined (Fig. 17.12). The first curve (Fig. 17.12A) shows the annual rainfall at which the various plants in a particular population would grow best. The actual rainfall in the area where this population occurs averages 40 cm (arrow 1), though it will vary around this mean from year to year. The population contains a few plants (S) that would grow best if the annual rainfall were about 32 cm and a few (W) that would grow with about 48 cm of rain. Plants that would grow best if the annual rainfall were about 36 cm (T) or 44 cm (V) are fairly common in the population. This phenotypic diversity is maintained because of the variability of the rainfall about the mean, so that in some unusually dry years, for example, the group S plants would have the advantage. In any given year, then, there is selection for an optimum phenotype, but the optimum varies. Since the average rainfall is 40 cm, the U plants are best adapted in the long run and thus are the most common.

Suppose that the average annual rainfall in the area slowly increases until it is 44 cm (arrow 2). Under these new environmental conditions, the V plants (which grow best when the rainfall is about 44 cm) will do better than before; a higher percentage of them can be expected to survive and reproduce, and their frequency should increase. Similarly, the W plants (which grow best with about 48 cm of rain) will now grow better than formerly, and they too should increase in frequency. Conversely, the T plants and

17.12 Evolutionary change of a hypothetical plant population in response to directional selection by changing rainfall

The various phenotypes (S, T, U, etc.) in a hypothetical plant population reflect different genetically determined growth responses to annual rainfall. The four curves (A–D) show the frequency distributions of the phenotypes at different times, under conditions of average annual rainfall (solid black arrows). As described in the text, a systematic change in average rainfall (toward the values shown by dashed arrows) exerts directional selection on the plant population, shifting the curve of relative phenotypic frequencies more and more to the right.

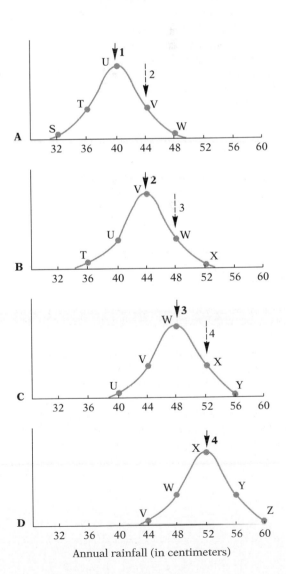

the U plants will not grow as well as formerly, so they will decrease in frequency. Finally, the S plants, only a few of which manage to survive with less than 40 cm of rain, will now be so poorly adapted to the prevalent conditions that none can survive. These changing frequencies, produced by the shift in the selection operating on the population, give rise a new curve (Fig. 17.12B).

If the average annual rainfall continues to increase over the years until it reaches 48 cm (arrow 3), the W and X plants will increase in frequency, the U and V plants will decrease in frequency, and the T plants will disappear (Fig. 17.12C). If the average annual rainfall increases to 52 cm (arrow 4), further shifts in frequencies will occur (Fig. 17.12D).

Changing environmental conditions, then, give rise to **directional selection**, which causes the population to evolve in a particular way. If the population has not been sufficiently variable to respond to the environmental changes, it will be much reduced and might even become extinct.

How selection can create novel phenotypes Notice in Figure 17.12 that directional selection does not simply shift the *peak* of the population curve to the right, as might happen when the variation in the gene pool is small. Instead, it causes the *entire curve—* the extremes as well as the peak—to shift to the right. The shift is eventually so great, in fact, that a class of plants (X) not even present in the original population becomes the largest class. If X, Y, and Z plants are not present initially, how do they arise in the descendant populations? One possibility is that new genes or alleles arise that make their possessors grow better in wetter habitats; such novel genes or alleles would be strongly selected for and would spread rapidly through the population. However, if moisture preference is influenced by many different genes, as is highly likely, new phenotypes such as X, Y, and Z could arise without the necessity of any new genetic variation, simply through the separate increase in frequency of particular alleles already present, which would then be more likely to occur together and produce a new phenotype.

Haldane calculated how long it would take for a new phenotype to be created in this way. He showed that if one particular allele of each of 15 independent genes is present in 1% of the individuals of a population, then all 15 alleles will occur together in only one of 10^{30} individuals. But there has never been a population of higher organisms containing anywhere near 10^{30} individuals. Hence the chance that all 15 alleles would occur together in even one individual in a real population is essentially zero. However, if there is moderate natural selection for each of the 15 alleles, it would take only about 10,000 generations for the frequency of each allele to increase from 1 to 99%. Once each allele is present in 99% of the population, 86% of the individuals in it will have all 15 alleles and hence will show the phenotype that was previously nonexistent in the population. Thus recombination and selection alone can produce new phenotypes by combining old genes in new ways and systematically favoring certain combinations.

This process is illustrated by a long-term experiment performed on corn. Researchers at the University of Illinois systematically se-

HOW WE KNOW:
HOW COMPLEX STRUCTURES EVOLVE

While it is easy enough to imagine directional selection for longer necks leading to the evolution of the giraffe, it is much more difficult to see how a complex structure like the camera eye of vertebrates could be selected for. After all, what good is any part of it—retina, lens, cornea, iris, ciliary body, sclera, choroid, or the suspensory ligament—without the other components already in place? And given the enormous spans of time required for the evolution of even simple structures, what could ever be learned by experimentation?

An increasingly fruitful approach in evolutionary biology involves modeling selection mathematically and then allowing the computer to compress millions of generations into a few minutes of digital activity. Such electronic experiments can determine whether or not a model is plausible and provide a rough idea of the time required for particular changes to occur. Dan Nilsson and Susanne Pelger took this approach in an attempt to see whether camera eyes can evolve from a patch of light-sensitive cells.

They began with a light-detecting patch sandwiched between a transparent covering and a dark supporting layer (Fig. A). In each "generation" they allowed tissue deformations or changes of density (refractive index) of 1% in any direction, and permitted the "progeny" to survive only if the visual resolution of the structure was an improvement over that of the previous generation. Thus every intermediate step had to represent progress.

The evolutionary route that led to a camera eye in this simulation involved a steadily increasing invagination of the bottom layer to create an eyeball over the course of about a thousand generations (Figs. A–E). At this point further improvement came only with a localized increase in the optical density of what had been the protective layer; after another 800 generations of 1% steps, selection led to a lens capable of forming a sharply focused image (Figs. E–H).

Thus, by a series of exceedingly tiny selective events, each required to increase the fitness of the individual involved and constrained to operate on a very small level of morphological variation, the essentials of the most advanced optical system in nature appear spontaneously in fewer than 2000 generations—2000–20,000 years for a typical vertebrate species. So rapid is this process when modeled that the perennial question of how eyes could evolve must now be replaced with an equally puzzling problem: how could eyes have taken so long to evolve?

lected for high oil content in the kernels across 50 generations. There was a steady increase in oil content throughout most of this period (Fig. 17.13). The kernels of the original stock of corn plants averaged about 5% oil; those of the plants in the 50th generation after selection averaged about 15% (higher than any individuals in the first generation), and there was no indication that a maximum had been reached.

17.13   Results of 50 generations of selection for high oil content in corn kernels

This steady change through 50 generations probably resulted mainly from the formation of new genetic combinations through selection rather than from new mutations. In 50 generations the researchers raised between 10,000 and 15,000 plants. The usual rate of mutation per gene in corn is lower than one in 50,000 plants. It is unlikely that a mutation contributing to an increase in oil content occurred in any *particular* gene affecting this phenotype during the experiment. Since the trait is polygenic, the chance of a mutation in one of the relevant genes is greater; if, say, 20 genes were involved, 2 to 10 mutations are likely to have occurred over the brief interval studied, one or more of which could have been favorable. Therefore, though some small part of the increase in oil content may have arisen from mutation, nearly all of the change came about from recombination of existing alleles.

Thus selection—whether natural or artificial—determines the direction of change largely by altering the frequencies of alleles that arose usually many generations before, through duplication, transposition, and random mutation. This establishes new genetic combinations and gene activities that produce new phenotypes. The processes that create new alleles and genes are not usually major *directing* forces in short-term evolution; the principal evolutionary role of genetic changes consists in creating and replenishing the essential store of variability in the gene pool, thus providing the potential on which future selection can act.

Disruptive selection   Sometimes a polygenic character in a population is subject to two (or more) directional selection pressures favoring the two extremes of the distribution. Suppose, for example, that a certain population of birds shows much variation in bill length. Suppose, further, that as conditions change, there are increasingly good feeding opportunities for the birds with the shortest bills and also for those with the longest, but decreasing opportunities for birds with bills of intermediate length. This might happen if the population of plants producing fruits suitable for intermediate-sized bills began to decline, or if a competing species more efficient at harvesting such fruits were to immigrate and become established. The effect of such selection, at least in the short run, would be to divide the population into two distinct

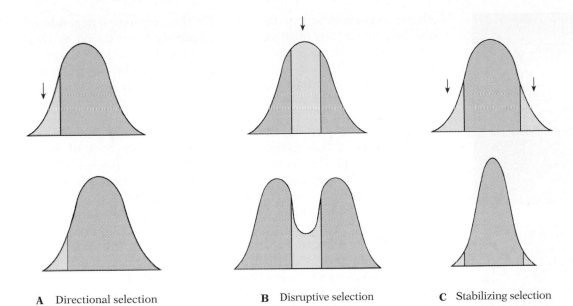

A Directional selection   **B** Disruptive selection   **C** Stabilizing selection

types, one with short bills and one with long bills. The combined action of the opposing directional pressures would thus disrupt the smooth curve of phenotypes in a population (Fig. 17.14B); hence this sort of selection is called ***disruptive selection***.

The most important presumed example of disruptive selection is gamete dimorphism: large eggs and tiny sperm or pollen. If we assume that gametes were originally all the same size, as is still the case in many primitive species, any that were slightly larger would have enhanced the ultimate survival of the zygotes they contributed to by providing more food to fuel initial development. At the same time, organisms that produced small gametes would have gained a reproductive advantage by virtue of making more gametes with the same amount of food, thereby outnumbering the competition from other organisms. In addition, smaller gametes can swim faster because they encounter less resistance, so where speed is important, they would have been at an advantage. If this widely accepted scenario is correct, then all the vast diversity of sex-specific morphology, physiology, and behavior is a consequence of disruptive selection.

Stabilizing selection Most often, when a polygenic character in a population is subject to two or more opposing directional selection pressures operating simultaneously, the pressures select against the two extremes. Plants that are too tall to resist high winds, for example, and those of the same species that are so short that they are shaded by other plants will both be selected against. Similarly, unusually severe winter storms often kill a disproportionate number of the largest and smallest birds in a population. When selection operates against individuals at the two ends of the distribution for a polygenic trait, the process is called ***stabilizing selection*** (Fig. 17.14C).

Costs and benefits: net selection Many characteristics benefit the organisms that possess them in some ways and harm them in

17.14 Directional, stabilizing, and disruptive selection

Each graph indicates the relative abundance of individuals of various heights in a population—in the original condition (above) and later after the specified selection (below). (A) Directional selection acts (arrow) against individuals exhibiting one extreme of a character (here the shortest individuals, represented by the blue area under the upper curve). The eventual result (bottom curve) is that the distribution of heights in the population has shifted to the right, indicating that the population has evolved in the direction of greater height. (B) By contrast, disruptive selection acts against individuals in the middle part of a distribution, thereby favoring both extremes; in our example both the shortest and the tallest individuals would be favored, but individuals of medium height would be selected against. The result is a tendency for the population to split into two contrasting subpopulations. (C) Stabilizing selection acts against both extremes, culling individuals that deviate too far from the mean condition and thus decreasing diversity and preventing evolution away from the standard condition.

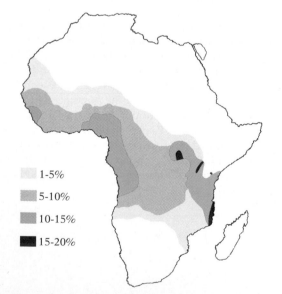

17.15 Male and female guppies
The showy coloration of the male (top) contrasts with the drab gray of the female.

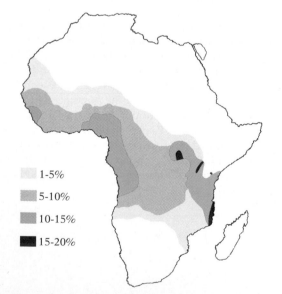

1-5%
5-10%
10-15%
15-20%

17.16 Distribution of sickle-cell anemia in Africa

The various colors indicate the percentage of the population in each area that has the disease.

others. The evolutionary fate of such characteristics depends on whether the various positive selection pressures produced by their advantageous effects outweigh the negative pressures produced by their harmful effects. If the algebraic sum (an addition taking into account plus and minus signs) of all the separate selection pressures is positive, the trait will increase in frequency; if the sum is negative, the trait will decrease in frequency.

Consider, for example, the selection pressures on showiness in male guppies, a fish native to the freshwater streams of South America (Fig. 17.15). As we have already seen, males with larger tails and brighter spots are more likely to mate because females prefer such males. Competition for females should result in very showy males, since males with alleles for large tails and bright spots will leave more offspring; and, indeed, males in large aquarium populations become increasingly showy with succeeding generations. This process, which will be described in more detail shortly, is known as female-choice sexual selection.

No such selection pressure has operated on the females. These gray, nondescript fish closely resemble one another and are much less easily seen by predators than the males, whose bright markings and showy tails make them more subject to predation. This liability causes strong selection against such features; indeed, in laboratory situations, predators capture the showiest males first. As we might expect, males found in the wild are much less conspicuous than their counterparts in predator-free aquaria. The showiness of males, then, exemplifies stabilizing selection—a balance between selection for greater showiness (by the females, who confer reproductive success on the most "attractive" males) and selection for inconspicuousness (by the predators, who can put an end to further mating if a male is too visible).

Just as polygenic traits often have both advantageous and disadvantageous effects, the alleles of a single gene also usually have multiple effects (a condition called pleiotropy), and it is unlikely that all of them will be advantageous. For example, *Drosophila* carrying alleles that disrupt the activity of specific systems involved in communication between nerve cells also have altered patterns of wing veins; those individuals with a muscle defect causing them to hold their wings upright also will not walk toward light, though this is a powerful response in normal fruit flies. *Whether an allele increases or decreases in frequency is determined by whether the sum of the various selection pressures favoring it is greater or smaller than the sum of the selection pressures acting against it.*

Many instances are known in which the effects of a given allele are more advantageous in the heterozygous than in the homozygous condition. As discussed in Chapter 16, for example, in some parts of Africa the allele for sickle-cell anemia occurs in humans much more often than we might expect in view of its lethal effect when homozygous (Fig. 17.16). This is because the allele, when heterozygous, gives the possessor a partial resistance to malaria. The equilibrium frequency of the sickle-cell allele is thus determined by at least four separate selection pressures: (1) the strong selection against the recessive homozygotes, who suffer the full debilitating effects of sickle-cell anemia; (2) the weaker selection against the heterozygotes as a result of their mild anemia; (3) the selection against the dominant homozygotes, who are more susceptible to malaria, and (4) the fairly strong selection favoring the heterozygotes as a result of their resistance to malaria. A similar pattern seems to underlie the cystic fibrosis gene: heterozygotes are less susceptible to cholera.

Balanced polymorphism Polymorphism is the occurrence in a population of two or more distinct forms, or *morphs*, of a genetically determined character. These phenotypes do not grade into one another like those represented by a bell-shaped curve for height, where "short" and "tall" designate the extremes of a continuum. Instead, individuals fall into separate categories known as *discontinuous phenotypes,* and intermediates are rare or absent. For example, Mendel's peas had polymorphic flower color: plants had either red or white flowers but not pink ones. Polymorphism is common in wild populations: several species of snails, for instance, occur in banded and unbanded forms, and the red fox has both red and silver-colored morphs.

When the relative frequencies of the different morphs in a population are stable over time, we speak of *balanced polymorphism*. Sometimes the balance is maintained because the polymorphism itself is advantageous. Thus if a polymorphic species lives in an environment subdivided into many local areas where different conditions prevail, one of the morphs may do better in one area, while others are more likely to succeed in other areas. Sometimes one morph is adaptively superior at one time of year or in one habitat, and another is superior at another time or in another place; hence an individual's descendants may have a better chance of survival if they are polymorphic than if all belong to a single form. Individuals that produce a pure culture of one form—putting all their eggs in one basket—run the risk of having none survive.

There are many instances of balanced polymorphism in which some of the contrasting morphs, instead of being themselves selected for, are the unavoidable by-products of selection for a heterozygous phenotype. As in the case of sickle-cell anemia in Africa, the heterozygotes are sometimes better adapted than the homozygotes—a condition known as *heterozygote superiority* (or sometimes heterosis or overdominance).

Heterozygote superiority favors balanced polymorphism, because it leads to the retention in the population of both alleles of a given gene at frequencies higher than would be predicted from the selection acting on the homozygous phenotypes. Thus, if *A/a* individuals are adaptively superior to both *A/A* and *a/a* individuals, both allele *A* and allele *a* will be retained in fairly high frequency in the population; neither will be eliminated, as might tend to happen if one of the homozygotes were superior. Therefore all three possible genotypes—*A/A*, *A/a*, and *a/a*—will occur frequently in each generation, and if each of these produces a noticeably different phenotype, the population will be polymorphic. The relative frequencies of *A* and *a* and of the three morphs they produce will be determined, as in sickle-cell anemia, by the balance between the several selection pressures acting on the system.

Playing the odds: frequency-dependent selection Balanced polymorphism can also arise when the selection pressure on an individual displaying one morph (whether physiological, anatomical,

17.17 Handedness in scale-eating cichlids
The ratio of the two morphs fluctuates near 1:1; this balanced polymorphism is maintained by frequency-dependent selection: the rarer phenotype enjoys higher success in attacking prey.

or behavioral) depends on the frequency of that morph compared with the alternatives. For example, there are seven species of cichlids in Lake Tanganyika that feed exclusively on the scales they bite off other cichlids. Each species consists of right-mouthed and left-mouthed forms (Fig. 17.17). Right-mouthed cichlids always approach from the rear and attack the left flanks of their victims; left-mouthed cichlids always attack from the right. This behavior, which becomes apparent in the fry while they are still feeding on plankton, is controlled by a simple single-gene, two-allele system in which right-mouthedness is dominant. Field observations show that the proportion of left-mouthed individuals varies from 0.4 to 0.6. The prey species are extremely watchful (only one attack in 500 is successful), and they focus their attention on the flank most often attacked. In any given year, therefore, the scale eaters attacking from the less common side enjoy greater success, driving the phenotypic ratio in the next generation far in the other direction; then the situation is reversed, and the previously less-successful morph does better. Thus the ratio of the two morphs oscillates around 1:1.

■ SEXUAL SELECTION

Darwin distinguished between selection that affects physical survival and selection that operates on traits used exclusively to attract and keep mates. In evolutionary terms, the failure to reproduce is just as fatal as early death, but since selection for characters that enhance an individual's chance of mating are often very different for the two sexes, he considered *sexual selection* to be nearly independent of natural selection.

Contests Darwin identified two basic varieties of sexual selection. The more common form involves contests between members of one sex (usually the males) for access to the other sex. These duels can take the form of dominance fights, which establish a male hierarchy and thus access to reproductively ready females (Fig. 17.18). More often the males fight for the best territories in the habitats fe-

17.18 A dominance ritual in mountain sheep
Male mountain sheep work out a dominance hierarchy through a ritualized duel in which a run on hind legs culminates in a loud collision, followed by a head-turning display. Dominant males obtain unchallenged access to females.

males favor; the strongest males obtain the highest-quality territories and therefore the largest number of matings (or, in monogamous species, the female best able to gain access to the territory). Male-specific features like large size, offensive weapons (including the horns of mountain sheep), and defensive structures (like the lion's mane, which protects his neck from bites inflicted by other males) are each the result of sexual selection.

Choices Darwin proposed that other uniquely male features exist only to attract females. The elaborate tail of the peacock, for instance, is not used to attack males but rather is displayed whenever a peahen is nearby. Experiments in which these dimorphisms have been altered demonstrate that females of some species do indeed select males on the basis of, for example, the length of their tails (Fig. 17.19; see also Fig. 17.8). Since many male dimorphisms exaggerate a species-specific recognition sign, the female may benefit from the increased certainty of mating with a male of the right species, who also carries genes that will enhance the attractiveness (and therefore the reproductive potential) of her sons.

She may also benefit from genes that confer the sort of physiological superiority that have enabled the male to survive despite the burden of his morphological "handicap"—genes that might contribute to the physiological health of her daughters. In other cases the female's attraction to a male's dimorphism seems to have been favored by selection because the presence of bright colors and the like indicate that the male is free from the sorts of parasites that threaten the population, and the progeny would benefit from this immunity. The mating systems of many species include elements of both male-contest and female-choice sexual selection.

17.19 A normal male widow bird in flight
Experimental shortening or lengthening of tails demonstrates that females select males with the longest tails.

ADAPTATION

Every organism is, in a sense, a complex bundle of thousands of *adaptations*. In biology, an adaptation is any genetically controlled characteristic that increases an organism's fitness. *Fitness*, as the term is used in evolutionary biology, is an individual's (or allele's or genotype's) probable genetic contribution to succeeding generations. An adaptation, then, is a characteristic that increases an organism's chances of perpetuating its genes, usually by leaving descendants. Adaptations do not necessarily increase the organism's chances of survival: although it ordinarily enhances prereproductive survival, it need not affect postreproductive survival. In many species it is, in fact, more adaptive for the adults to die soon after they have reproduced.

Adaptations may be structural, physiological, or behavioral. They may be genetically simple or complex. They may involve individual cells or subcellular components, or whole organs or organ systems. They may be highly specific, beneficial only under very limited circumstances, or they may be general and of value under many and varied circumstances.

A population may become adapted to changed environmental conditions very rapidly. W. B. Kemp reported a classic example in

A

B

17.20 Effects of predation on male guppies

When John Endler kept guppies in a predator-free environment, they evolved brighter coloration and larger tails (A), presumably as a result of female-choice sexual selection; when predators were introduced into the same stock for a few generations, males evolved smaller tails and less dramatic colors (B), probably as a result of predator pressure.

1937. The owner of a pasture in southern Maryland had seeded the pasture with a mixture of grasses and legumes. Then he divided the pasture into two parts, allowing one to be heavily grazed by cattle, while protecting the other from the livestock and letting it produce hay. Three years after this division, Kemp obtained specimens of blue grass, orchard grass, and white clover from each part of the pasture and planted them in an experimental garden where all the plants were exposed to the same environmental conditions. He found that the specimens of all three species from the heavily grazed half of the pasture exhibited dwarf, rambling growth, while specimens of the same three species from the ungrazed half grew vigorously upright.

In only three years the two populations of each species had become markedly different in their genetically determined growth pattern. The grazing cattle in the one half of the pasture had cropped most of the upright plants, and only plants low enough to be missed had survived and set seed. There had been intense selection against upright growth in this half of the pasture and correspondingly intense selection for the adaptively superior dwarf, rambling growth pattern. In the half of the pasture where there was no grazing, by contrast, upright growth was adaptively superior, and dwarf plants were unable to compete effectively. Analogous results are observed when animals are subjected to various degrees of predation (Fig. 17.20).

Experimental tests of adaptiveness are usually not as easy to design as straightforward laboratory evaluations of cause and effect. There may be several alternative explanations for why a trait may be adaptive. It is important to remember, furthermore, that not all characters present in an organism need be adaptive in the first place. Some will be incidental pleiotropic effects of genes whose main effects—those that make them adaptive—are quite different. Other traits are essentially historical: we have five fingers because the vertebrate developmental program produces five (or, rarely, six) digits; five may or may not be the optimal number for every kind of amphibian, reptile, bird, and mammal, but selection can only operate on existing variation. As illustrated in Chapter 1, selection has acted to customize digit size rather than number, enlarging or reducing the size of individual digits; the basic five-digit plan remains remarkably consistent. Apparently, some changes are simply not genetically feasible. This sort of evolutionary constraint is often called *phylogenetic inertia*.

Devising tests to discover the selective pressures that have led to particular traits requires ingenuity and persistence. Niko Tinbergen was one of the first scientists to insist on putting his evolutionary theories to the test. When Tinbergen wondered, for instance, why ground-nesting gulls meticulously remove broken eggs from their nests (Fig. 17.21A), he formulated a variety of possible explanations: the damaged eggs might be a source of disease, which would infect the newly hatched young; the jagged edges of the broken shells might endanger the chicks; or the unrelieved white of the exposed interiors of the broken shells might nullify the camouflage provided by the olive-drab exteriors.

Tinbergen solved this problem (and many similar ones) in two steps: first, by making interspecific comparisons and then by con-

A

B

17.21 Eggshell removal by gulls

Ground-nesting gulls remove broken eggshells and other debris from their nests and carry them at least a meter away (A), while cliff-nesting species like the kittiwake do not remove eggshells (B). The nesting habit of the kittiwakes protects them from predation, while ground-nesting gulls must work to keep their nests inconspicuous.

ducting experimental tests. The species-comparison step allowed him to isolate the most likely hypothesis for testing. He observed that kittiwake gulls, which live on cliffs and are therefore virtually immune to predation, do not remove broken eggshells (Fig. 17.21B). Since disease and cuts, if they posed significant threats, ought to be as dangerous for the kittiwakes as for ground-nesting gulls, Tinbergen decided to test the predation hypothesis first. He did this by setting out an array of nests containing normal eggs, with broken eggs placed at varying distances from them. The results were clear-cut: broken eggshells nearby called the attention of predators to an otherwise inconspicuous nest (Table 17.1). In the absence of direct experimental tests or compelling species comparisons, even the most plausible explanations of adaptiveness remain speculative and should serve as working hypotheses to stimulate research.

Let's look at some examples of adaptations, which help clarify the processes of evolution.

Table 17.1 Survival value of eggshell removal[a]

DISTANCE FROM EGG TO EGGSHELL (CM)	PERCENT EGGS TAKEN BY PREDATORS
5	65
15	42
100	32
200	21
No eggshell	22

[a]SOURCE: N. Tinbergen et al., 1963. Egg Shell Removal by the Blackheaded Gull, *Behaviour* 19.

■ ADAPTATIONS FOR POLLINATION

Flowering plants depend on external agents to carry pollen from the male parts in the flowers of one plant to the female parts in the flowers of another plant (Fig. 17.22). The flowers of each species are adapted in shape, structure, color, and odor to the particular pollinating agents on which they depend, and they provide an especially clear illustration of the evolution of adaptedness. Evolving

A

B

C

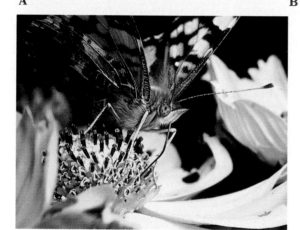
D

17.22 A variety of plant pollinators

(A) A bat at a flower, its face liberally dusted with pollen. (B) A honey possum feeding on nectar from flowers. (C) A hummingbird hovering under nectar-bearing flowers. (D) A butterfly inserting its long proboscis into a flower to obtain nectar. The proboscis is likely to be dusted with pollen, which the butterfly may carry to the next flower it visits.

together, the plants and their pollinators became more finely tuned to each other's peculiarities—a process often termed *coevolution*.

There are indeed striking correspondences between the pollinators and the species they pollinate. Bees are attracted innately to bright colors, UV bull's-eye patterns, and sweet, aromatic, or minty odors. They are active only during the day, and they usually alight on a petal before moving into the part of the flower containing the nectar and pollen. Bee-pollinated flowers have showy, brightly colored petals that are usually blue or yellow (colors bees can see) but seldom red (which bees cannot see at all). In addition, most bee-pollinated flowers have a UV bull's-eye, a sweet, aromatic, or minty fragrance, daytime opening or nectar production, and a special protruding lip or other suitable landing platform. However, these observations do not indicate how these correspondences may have come about—whether the preferences of the pollinators provided the exclusive selective force on flower morphology and odor, or whether early flowers provided the selection pressures leading to the innate preferences of modern bees, or both. A look at other species of pollinators provides some clues.

Hummingbirds, for example, can see red well but blue only poorly; they have a weak sense of smell, and they ordinarily do not land on flowers but hover in front of them while sucking the nectar. Flowers pollinated primarily by hummingbirds are usually red or yellow, are nearly odorless, and lack a landing platform. Since flowers of the same genus can have very different morphologies to suit different pollinators (Fig. 17.23), it is probably the flowers that have done most of the adapting. However, pollinators have probably been adapting to flowers too, though to a lesser extent. Different species of bees, for example, can have very different tongue lengths, suitable for different flower morphologies.

This pattern of coadaptation between pollinators and flowers extends to other nectar-feeding species as well. In contrast to both bees and hummingbirds, moths and bats are generally most active at dusk and during the night. The flowers they pollinate are mostly white, are open only during the late afternoon and night, and often have a heavy fragrance that helps guide the moths and bats to them.

Moths play a role in an interesting adaptation of plants to their pollinators. The flowers produced by scarlet gilia plants near Flagstaff, Arizona, range from red through pink to white. The dark-red flowers are most effective in attracting hummingbirds, but these pollinators emigrate a month after the season begins; the white flowers are most effective in attracting hawk moths, the pollinators available throughout the blooming season. The plants compensate for this shift in relative pollinator abundance by doubling the production of white flowers late in the season and stopping production of red ones (Fig. 17.24).

Unlike bees and moths, the short-tongued flies (which feed primarily on carrion, dung, humus, sap, and blood) are attracted by rank rather than sweet odors, and they rely very little on vision in locating food. The flowers of plants that depend on these flies for pollination are usually dull colored and ill smelling.

A particularly dramatic example of adaptation for pollination is seen in some species of orchids, whose flowers resemble in shape, odor, and color the females of certain species of wasps, bees, or

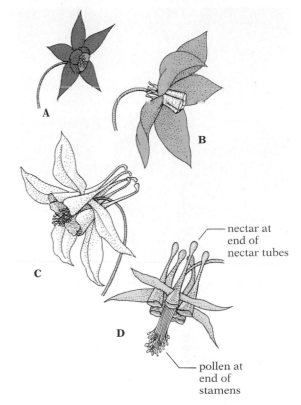

nectar at end of nectar tubes

pollen at end of stamens

17.23 Characters of columbine flowers correlated with their pollinators

(A) *Aquilegia ecalcarata*, pollinated by bees. (B) *A. nivalis*, pollinated by long-tongued bees. (C) *A. vulgaris*, pollinated by long-tongued bumble bees. (D) *A. formosa*, pollinated by hummingbirds. The length and curvature of the nectar tubes of the flowers are correlated with the length and curvature of the bees' tongues and the hummingbirds' bills. The length of the pollen-bearing stamens is suited to the size of the pollinator, and the reduced petal width of *A. formosa* reflects the hummingbird's lack of need for a perch.

A   B

17.24 Pollinator tracking by scarlet gilia (*Ipomopsis aggregata*)

During the early part of the summer, when these plants are pollinated by both hummingbirds and hawk moths, they produce about twice as many red flowers (preferred by hummingbirds) as white flowers (A). Later in the summer, as the hummingbird populations leave the area, scarlet gilia plants shift over to producing pink and especially white flowers, which are more attractive to the sole remaining pollinator, a local species of hawk moth (B).

17.25   An orchid flower that resembles a fly

The flowers of this species (*Ophrys insectifera*) look enough like female flies to attract some male flies to land on them. The males thus become dusted with pollen, which they may carry to other flowers

flies (Fig. 17.25). The male insect is stimulated to attempt to copulate with the flower and becomes covered with pollen in the process. When he attempts to copulate with another flower, some of the pollen from the first flower is deposited on the second.

Flowers pollinated by wind or water rather than animals characteristically lack bright colors, special odors, and nectar. In fact, most of them have no petals, and their sexual parts are freely exposed to the air currents.

The characteristics of flowers are, therefore, not simply pleasing curiosities of nature. They are important adaptations that have evolved in response to fundamental selection pressures.

■ DEFENSIVE ADAPTATIONS

Cryptic appearance   Many animals blend into their surroundings so well as to be nearly undetectable. Frequently their color matches the background almost perfectly (Fig. 17.26). In some cases animals even have the ability to alter the condition of their own pigment cells and change their appearance to harmonize with their background (Figs. 17.27 and 17.28). Often, rather than match the color of the general background, the animals may resemble

17.26   Cryptic coloration of a crab

The sargassum crab, which lives in dense growths of the brown alga *Sargassum* off Bermuda, is the same color as the alga; its rounded body resembles the algal floats.

17.27   Flounders on two different backgrounds

These fish can change color to match the background, whether light-colored (A) or dark (B).

A

B

A   B

17.28 Color change by the frog *Hyla versicolor*

Individuals of this species are able to change color to match either a tree trunk (A) or vegetation (B).

inanimate objects commonly found in their habitat, such as leaves (Fig. 17.29) or twigs (Fig. 17.30). When the shape or color of an animal offers concealment against its background, it is said to have a *cryptic* appearance.

Cryptic appearance is an adaptive characteristic that helps animals escape predation. In one study, F. B. Sumner invested the interaction of Galápagos penguins and mosquito fish (*Gambusia partuelis*). Mosquito fish contract or expand their pigment cells to become lighter or darker, depending on their background. Sumner showed that the penguins catch 70% of the fish that contrast with their background but only 34% that resemble their background. In a similar experiment, F. B. Isely studied predation of grasshoppers of various colors, on differently colored backgrounds, by chickens,

17.29 Leaflike mantis

The mantis (at top in picture) looks strikingly like the green leaves below it. Even the relationship between its thorax and abdomen reflects the way the leaves often occur in pairs.

17.30 A moth that looks like a broken twig

The perfection of the cryptic appearance of this moth (*Phalera bucephala*) is a triumph of evolutionary adaptation.

turkeys, and native birds. He found that 88% of the nonprotected grasshoppers were eaten, but only 40% of the cryptically colored ones.

One of the most extensively studied cases of cryptic coloration is the so-called industrial melanism of moths. Since the mid-1880s many species of moths have become decidedly darker in industrial regions. This is actually a case of polymorphism in which the less frequent of two forms has become the more frequent; the originally predominant light form in these species of moths has given way in industrial areas to the dark (melanic) form. In the Manchester area of England, the first black specimens of the species *Biston betularia* were caught in 1848; by 1895 melanics constituted about 98% of the total population in the area. For such a remarkable shift in frequency to have occurred in so short a time, the melanic form must have had at least a 30% advantage over the light form.

The various species of moths exhibiting the rapid shift to melanism, though unrelated to one another, all habitually rest during the day in an exposed position on tree trunks or rocks, being protected from predation only by their close resemblance to their background. In former years the tree trunks and rocks were light colored and often covered with light-colored lichens. Against this background the light forms of the moths were difficult to see, whereas the melanic forms were quite conspicuous (Fig. 17.31). Under these conditions it seems likely that predators captured melanics far more easily than the cryptically colored light moths. The light forms would thus have been strongly favored, and they would have occurred with much higher frequency than melanics. With the advent of extensive industrialization, tree trunks and rocks were blackened by soot, and the lichens, which are particularly sensitive to such pollution, disappeared. In this altered envi-

A

B

17.31 Cryptic coloration of peppered moths

(A) Light and dark morphs of *Biston betularia* at rest on a tree trunk in unpolluted countryside. (B) Light and dark morphs on a soot-covered tree trunk. Here the light form is easier to see.

ronment the melanic moths resembled the background more closely than did the light moths. Thus selection should now have favored the melanics, which would explain why they increased in frequency.

This scenario was put to a test in the mid-1950s, when H. B. D. Kettlewell released equal numbers of the light and melanic forms of *Biston betularia* onto trees in a rural area in Dorset, England, where the tree trunks were light colored, lichens were abundant, and the wild population of the moth was about 94.6% light colored. Of 190 moths observed to be captured by birds, 164 were melanics and only 26 were light forms. Of hundreds of marked individuals of each form released in another experiment, roughly twice as many light moths as melanic moths were recaptured in traps set up in the woods, an indication that more of the light moths had survived.

Taken alone, however, these experiments did not prove that the factor favoring the light moths over the melanics was their resemblance to their background. Perhaps the birds preferred the melanics because of some difference in flavor. Kettlewell duplicated the experiments under the reverse environmental conditions—in woods near Birmingham, England, where the tree trunks were blackened with soot, lichens were absent, and the wild population of the moth was about 85% melanic. The results of these experiments were the reverse of those in the Dorset experiments. Now birds captured nearly three times as many light moths as melanics, and roughly twice as many melanics were recaptured in traps. Recently, the imposition of pollution controls has led to a shift back toward the light morph.

Warning coloration Whereas some animals have evolved cryptic coloration, others have evolved colors and patterns that contrast boldly with their background and thus render them clearly visible to potential predators. Such warning appearance is called ***aposematic.*** (Fig. 17.32). Nearly all of these animals are in some way disagreeable to predators; they may taste bad, smell bad, sting, or secrete poisonous substances. They are animals that a predator will usually reject after one or two unpleasant encounters. Such animals benefit by being gaudy and conspicuous because predators that have experienced their unpleasant features learn to recognize and avoid them more easily in the future.

The warning is sometimes so effective that, after unpleasant experiences with one or two warning-colored insects, some vertebrate predators simply avoid all flashily colored insects, whether or not they resemble the ones they encountered earlier. G. D. H. Carpenter demonstrated this by offering over 200 different species of insects to an insectivorous monkey. The monkey accepted 83% of the cryptically colored insects but only 16% of those with the warning coloration, even though many of the insects belonged to species the monkey had probably not previously encountered. It is possible, therefore, that avoidance of aposematic insects does not depend entirely on learning, and that some predators have a genetic predisposition to avoid brightly colored prey. They would probably have an adaptive advantage over predators that waste time and energy pursuing inedible prey; hence innate conservatism could evolve.

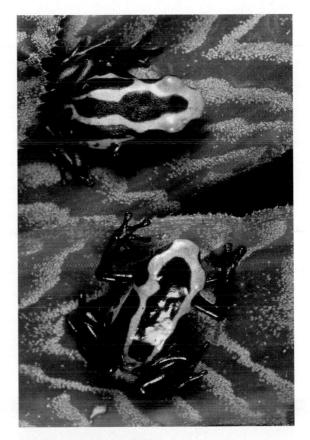

17.32 Aposematic coloration

The bright color of the poison-arrow frog (from which South American Indians obtain poison for their arrow tips) makes it easily recognizable by predators, which carefully avoid it.

Mimicry Species not naturally protected by some unpleasant character of their own may closely resemble (mimic) in appearance and behavior some dangerous or unpalatable aposematic species (called models). Such a resemblance can be adaptive: the mimics may suffer little predation because predators cannot distinguish them from their models, which the predators have learned are unpleasant. This phenomenon is called ***Batesian mimicry*** (Fig. 17.33A,B).

Convincing evidence for the potential effectiveness of this type of mimicry in protecting the mimic species comes from the elegant experiments of Jane van Z. Brower. Brower produced an artificial model-mimic system with starlings as the predators and mealworms, which starlings ordinarily eat voraciously, as prey. She painted a tasteless color band on the mealworms and dipped some in a distasteful solution. She then presented various ratios of noxious (dipped) models and mimics to different groups of birds. After a few unpleasant encounters with the models, the birds learned to recognize and avoid the painted worms, with the result that the mimics also escaped predation, particularly when the percentage of mimics presented to the starlings was 60% or less.

A few species pursue the opposite approach: these "aggressive mimics" provide lures of apparently palatable prey to attract potential victims (Fig. 17.33C).

In addition to Batesian mimicry, which is based on deception—mimicry of a distasteful or dangerous species by individuals of a species that is neither—there is a second kind of mimicry, called

A

B

17.33 Examples of Batesian and aggressive mimicry

(A) The prominent imitation eyes and mouth of the spicebush swallowtail larva give it the appearance of a predator to be avoided. (B) The markings of the harmless syrphid fly (top) resemble those of a stinging bumble bee. (Upon close inspection, flies are readily distinguished from bees because their antennae are very short and they have only a single pair of wings.) (C) The warty frogfish is an aggressive mimic: its body, which resembles an algae-covered rock, is merely cryptic, but it displays a lure that resembles a small fish.

C

Müllerian mimicry, which involves the evolution of a similar appearance by two or more distasteful or dangerous species. In this type of mimicry, individuals of each species act as both model and mimic. The members of each species have some defensive mechanism, but if each species had its own characteristic appearance, the predators would have to learn to avoid each of them separately; the learning process would thus be more demanding, and would involve the death of some individuals of each prey species. Selection favors evolution toward one appearance; the various protected species thus come to constitute a single prey group from the standpoint of the predators, which learn avoidance more easily.

One striking case of Müllerian mimicry involves the monarch butterfly *(Danaus plexippus)* and the unrelated viceroy *(Limenitis archippus)*. These two species look very much alike (Fig. 17.34) and are each distasteful to birds, though in different ways: monarchs sequester poisons from the milkweed plant upon which their caterpillars feed, while viceroys synthesize their own bad-tasting chemical. Until recently, the viceroy was thought to be a Batesian mimic of the monarch because blue jays, the predators used in early tests, are one of the few species of birds that find viceroys only slightly distasteful. Other kinds of birds that taste one member of either species learn to avoid them both.

The selective advantage of Müllerian mimicry may explain the similar markings of many unrelated species of wasps and bees, or of the group of poisonous reptiles known as coral snakes. If avoidance involves any genetic predisposition, then resemblances among the prey animals would facilitate more rapid selection for improved prey-recognition mechanisms in the predators. In fact, there is evidence that some predators recognize coral snakes innately; perhaps only Müllerian mimicry can provide a strong enough selection pressure to cause the evolution of such specialized recognition—a recognition that benefits individuals of both predator and prey species.

A

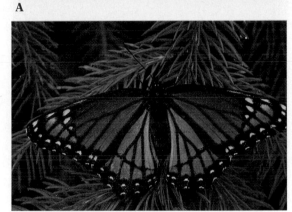

B

17.34 Müllerian mimicry in butterflies

The monarch butterfly (A) and the viceroy (B) which are not closely related, have evolved strikingly similar color patterns.

■ SYMBIOTIC ADAPTATIONS

The term *symbiosis* is used in a variety of ways in the biological literature. Some authors apply it only to cases in which two species live together to their mutual benefit. Etymologically, however, symbiosis simply means "living together," without any implied value judgments. This is the meaning it was given when it was first introduced into biology, and this is the meaning we use here. There are four categories of symbiosis: *commensalism*, a relationship in which one species benefits while the other receives little or no benefit or harm; *mutualism*, in which both species benefit; *parasitism*, in which one species benefits and individuals of the other species are harmed; and *enslavement* in which one species is enslaved by another (Table 17.2).

Commensalism The advantage derived by the commensal species from its association with the host often involves shelter, support, transport, food, or several of these. For example, in tropical forests numerous small plants, called epiphytes, grow on the branches of the larger trees or in forks of their trunks (see Fig. 40.35B, p. 1186). These epiphytes, including species of orchids and bromeliads, are

Table 17.2 Varieties of symbiotic relationships

RELATIONSHIP	SPECIES A	SPECIES B
Commensalism	+	0
Mutualism	+	+
Parasitism	+	−
Enslavement	+	−

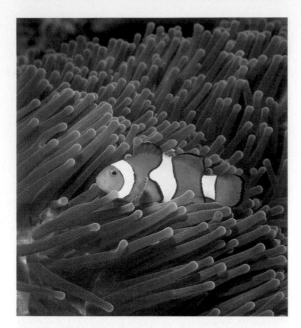

17.35   An example of commensalism

An anemone fish living among the tentacles of a sea anemone from the Palau Islands

often not parasites. They use the host trees as a base of attachment and do not obtain nourishment from them. They apparently do no harm to the host except very rarely, when so many of them are on one tree that they stunt its growth or cause limbs to break. A similar type of commensalism is the use of trees as nesting places by birds.

Sometimes it is difficult to tell what benefit is involved in a commensal relationship. For example, certain species of barnacles occur nowhere except attached to the heads of whales, and other species of barnacles occur nowhere except attached to the barnacles that are attached to whales. Just what advantages either of these groups of barnacles enjoys is not clear.

In some cases of commensalism, however, the benefit is dramatically obvious. For example, certain species of fish live in association with sea anemones, deriving protection and shelter from them and sometimes stealing some of their food (Fig. 17.35). These fish swim freely among the tentacles of the anemones, even though the tentacles quickly paralyze other fish that touch them. The anemones regularly feed on fish, yet the particular species that live as commensals with them sometimes enter the gastrovascular cavity of their host and emerge later with no apparent ill effects. The physiological and behavioral adaptations that make such a commensal relationship possible are extensive.

Another striking example is a small tropical fish that lives in the respiratory tree of a particular species of sea cucumber. The fish emerges at night to feed and then returns to its curious abode by first poking its host's rectal opening with its snout and then quickly turning so that it is drawn tail first through the rectal chamber into the respiratory tree. Still another example is a tiny crab that lives in the mantle cavity of oysters. The crab enters the cavity as a larva and eventually grows too big to escape through the narrow opening between the two valves of the oyster's shell. It is thus a prisoner of its host, but a well-sheltered prisoner. It steals a few particles of food from the oyster but apparently does it no significant harm.

Mutualism Symbiotic relationships beneficial to both species are common. Figure 17.36 illustrates two instances of the widespread phenomenon called cleaning symbiosis, which is patently mutual-

A

17.36   Cleaning symbiosis

(A) A giant seabass being cleaned by a cleaner-fish. (B) Yellow-billed oxpeckers search for parasitic insects on a black rhinoceros. In both cases the symbiosis is mutualistic: the cleaner obtains food, and the host gets rid of parasites that could endanger its health.

B

istic. Other examples of mutualism include the relationship be-tween a termite and the cellulose-digesting microorganisms in its digestive tract, or between a human being and the bacteria in the intestine that synthesize vitamin K. The plants we call lichens are actually formed of an alga or a cyanobacterium and a fungus united in such close symbiosis that they give the appearance of being one plant (see Fig. 23.9, p. 650). The fungus benefits from the photosynthetic activity of its "guest," and in at least some cases the alga or bacterium benefits from the water-retaining properties of the fungal walls.

The division of symbiosis into four subcategories is in many ways arbitrary. In particular, commensalism, mutualism, and para-sitism are all parts of a continuous spectrum of possible interac-tions. The categories are only devices to help us organize what we know about nature and form testable hypotheses. What is impor-tant is to keep in mind how commensalism, mutualism, and para-sitism grade into each other; each case of symbiosis is different from all others and must be studied and analyzed on its own.

Parasitism There is no strict line between parasitism and preda-tion. Mosquitoes and lice both suck mammalian blood, yet we usu-ally call only the latter parasites. Foxes and tapeworms may both attack rabbits, but foxes are called predators and tapeworms are called parasites. The usual distinction is that a predator eats its prey quickly and then goes on its way, while a parasite passes much of its life on or in the body of a living host, from which it de-rives food in a manner harmful to the host. Parasites generally do not kill their hosts; those that eventually do so are called ***para-sitoids*** (Fig. 17.37).

Parasites are customarily divided into two types: external para-sites and internal parasites. The former live on the outer surface of their host, usually either feeding on the hair, feathers, scales, or skin of the host or sucking its blood. Internal parasites may live in the various tubes and ducts of the host's body, particularly the di-gestive tract, respiratory passages, or urinary ducts; or they may bore into and live embedded in tissues such as muscle or liver; or, in the case of viruses and some bacteria and protozoans, they may actually live inside the individual cells of their host.

Internal parasitism is usually marked by more extreme special-izations than external parasitism. The habitats available inside the bodies of living organisms are completely unlike those outside, and the unusual problems the host environments pose have resulted in evolutionary adaptations entirely different from those seen in free-living forms. Internal parasites, for example, have often lost organs that would be essential in a free-living species: tapeworms have no digestive system, but they live in their host's intestine and are bathed by the products of its digestion, which they absorb directly across their body wall.

Because of their frequent evolutionary loss of structures, certain parasites are often said to be degenerate. "Degenerate" implies no value judgment but simply refers to the lack of many structures present in their free-living ancestors. This condition is common in internal parasites, like the tapeworms mentioned above, but also can occur in external parasites. For example, the ant species

17.37   A caterpillar (tomato hornworm) with numerous pupae of a parasitoid wasp attached to its body

The wasp laid her eggs inside the body of the host caterpillar, and the larvae fed on it until they emerged and pupated.

17.38 A parasitic ant

Queens of the "ultimate ant," *Teleutomyrmex*, ride on the host queen and are fed by a host worker (lower right). They dispense their eggs along with those of the host queen, and their offspring, all reproductives, mate in the colony.

Teleutomyrmex (Fig. 17.38) lives in the nests of another species of ant. It is degenerate to the extent that it cannot survive outside its hosts' nest: it has lost many of its glands, its sting, its pigmentation, and its ability to digest anything but liquid fed to it by its hosts. Even its brain is degenerate. From an evolutionary point of view, however, loss of structures that are useless in a new environment is an instance of specialized adaptation.

Specialization, then, does not necessarily mean increased structural complexity; it only means the evolution of characteristics particularly suited to some special situation or way of life. In internal parasites—and cave animals, which frequently lack eyes—the development and maintenance of structures that no longer serve a useful function would require energy that the organism might use to more advantage in some other way. Moreover, some useless structures, such as eyes in internal parasites, deep-sea fish, and cave animals, might well be a handicap in these special environments, because they would be a likely point of infection. Natural selection often favors those individuals in which such useless organs are either relatively small or lacking entirely. Alternatively, the unused structures could have been adaptively neutral and lost through genetic drift.

As parasites evolve, the host species is also evolving, and there is strong selection pressure for the evolution of more effective defenses against the ravages of parasites (Fig. 17.39). This constant interplay between host and parasite is at the heart of the red queen hypothesis. (see "Exploring Further: The Logic of Sex," p. 448). Those individuals of the host species with superior defenses will be better able to survive and reproduce. Correspondingly, those individuals of the parasite species with the best ways of counteracting the defenses will be most likely to prosper. In turn, their counteractions will lead to pressure on the host to evolve still better defenses, against which the parasite may then evolve new means of surviving, and so forth.

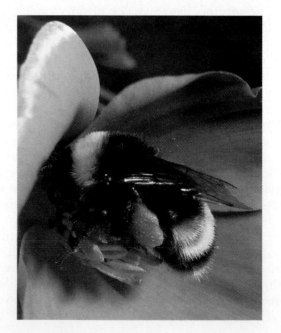

17.39 A bumble bee defense against parasitoids

Up to 70% of foraging bumble bees are parasitized by fly larvae. Infected bees respond by visiting the hive only to deliver their loads of pollen and nectar; they spend the nights in the field on flowers rather than in the colony. This counterploy not only slows parasitoid development by exposing it to cold nighttime temperatures (thus lengthening the useful life of the forager), it also greatly lowers the chance that the parasitoid will chance to emerge and spread within the colony.

Probably most long-established host-parasite relationships are balanced ones. Relationships that result in serious disease in the host tend to be relatively new, or they are relationships in which a new and more virulent form of the parasite has recently arisen or in which the host showing disease symptoms is not the main host of the parasite. Native Americans, for example, suffered severely when exposed to diseases first brought to the Americas by European colonists, even though some of these same diseases caused only mild symptoms in the Europeans, who had been exposed to the disease-causing organisms for many centuries and in whom the host-parasite relationship had nearly reached a balance. Many examples are known in which humans are only occasional hosts for a particular parasite and suffer severe disease symptoms, though the wild animal that is the major host shows few ill effects.

It is worth keeping in mind, however, that benign interactions between the host and a parasite need never evolve regardless of how much time passes. The reason is that the optimal strategy for a parasite depends critically on the life histories of itself and its host, and for some combinations there is no advantage to the parasite in achieving a balance and no way for the host to impose one. In some diseases—for example, rabies (which is fatal to humans and many other mammals) and smallpox (which, untreated, kills about 30% of its victims)—the parasite apparently benefits from a massive attack on the host, which enables it to spread its offspring rapidly. A slower, longer-lasting release of progeny produces lower reproductive success. An evolutionary perspective on host-parasite dynamics—particularly life histories and transmission efficiencies—can lead to optimal designs for disease treatment and prevention, whereas intuitively attractive alternatives that fail to consider these interactions often turn out to be futile.

Enslavement Enslavement occurs when an individual of one species maintains members of another species, controlling their reproduction and using their effort to its own ends. Some cases are clear-cut. Several species of ants enslave workers of other species, either by raiding the host colonies for pupae (which hatch into workers in the slave-making colony), or by killing the host queen and establishing one of their own queens in her place. The algae in some lichens are enslaved in that they cannot escape. Mitochondria and chloroplasts are in the same position, having been enslaved by eucaryotes more than a billion years ago. Other instances shade into parasitism or predation.

CHAPTER SUMMARY

HOW SELECTION OPERATES

Evolution is a change in allelic frequencies. Natural selection operates on heritable variation, favoring better adapted individuals. Chance (genetic drift) can also cause evolution. (p. 444)

WHAT CAUSES VARIATION? Variation arises from crossing over, sexual recombination, and mutation. (p. 445)

HOW DO NEW GENES AND ALLELES ARISE? Alleles can be modified through point mutations. Larger-scale changes arise from gene or exon duplication, exon recombination, and horizontal transfer of genes from other species. (p. 448)

HOW DO MUTATIONS SPREAD IN POPULATIONS? In the absence of immigration and emigration (gene flow), nonrandom mating (sexual selection), natural selection, mutation, and genetic drift, the allelic frequencies in a population (gene pool) will not change from one generation to another. (p. 451)

NATURAL SELECTION Selection operates by increasing the reproductive fitness—the number of viable offspring produced—of individuals carrying well-adapted alleles relative to individuals lacking these alleles. Selection can be directional (favoring one end of a range of variation), stabilizing (favoring a certain value in the continuum of variation), or disruptive (favoring two—or, rarely, more—alternative values). Crossing over and recombination can create novel allelic combinations. Net selection pressure on a trait usually depends on the difference between the trait's costs and benefits. Frequency-dependent selection depends on the relative frequency of a trait in the population: in many cases, rarer traits are favored. (p. 458)

SEXUAL SELECTION Selection favors an ability to increase the quality and quantity of offspring, in part through optimizing the quality and quantity of mates. One sex (usually males) often competes for mates, either through dominance contests, or through courtship displays and sex-specific morphology. (p. 467)

ADAPTATION

Adaptations help an organism survive and reproduce. The most spectacular cases are those that are unique to a particular species and suit it for making its living in a highly specific way. (p. 468)

ADAPTATIONS FOR POLLINATION Plants and their pollinators have coevolved so that each is well suited to the other. (p. 470)

DEFENSIVE ADAPTATIONS Many species are cryptic in their particular habitat, thus avoiding predation. Toxic or dangerous species often have aposematic (warning) colors that may be innately recognized or readily learned by predators. Batesian mimics benefit by having the same warning signals without paying the costs of defense. Müllerian mimics are each toxic but, by resembling each other, "share" the cost of teaching predators to avoid them. (p. 473)

SYMBIOTIC ADAPTATIONS One species in a commensal pair benefits from its association with the other, while the second is not helped or harmed. Mutualistic species each benefit. In a parasitic relationship, one species benefits at the expense of the other. Finally, one species can enslave members of another species. (p. 478)

STUDY QUESTIONS

1 Imagine a relatively large, long-lived animal (say at least 5 kg, living a decade on average) which, because of its habitat and lifestyle, does not benefit from sexual recombination. Derive a list of the rare conditions that would favor cloning in such an animal. (pp. 447–448)

2 Preadaptations can enable a species to expand into a novel niche. Many species—rats, mice, roaches, silverfish, sea gulls, pigeons, and so on appear to have occupied the niches created by humans. What sorts of adaptations, already useful or neutral in their normal habitat, enable these species to exploit new opportunities? Is there a common theme? Is it just a coincidence that several of these species are standard laboratory animals? (pp. 458–459)

3 What sorts of ecological factors might favor male contests over female choice? How might such factors tip the balance toward harems rather than monogamy or bigamy? (pp. 467–468)

4 Some researchers believe that human females are, on average, inherently better at some tasks (verbal expression, making subtle distinctions, and so on) than males, while males are, on average, better at mathematical and geometric problems. (Others, of course, believe the differences that have been observed are the result of conditioning.) Is it possible for natural selection to work in different directions on the two sexes of one species? Cite examples to support your conclusion. If the answer is yes, what is the likely logic of differential specializations? (pp. 463–464, 467–468)

5 What, if anything, keeps parasites from driving their hosts to extinction? (There are several possibilities.) (pp. 480–482)

SUGGESTED READING

AMÁBILE-CUEVAS, C. F., AND M. M. CHICUREL, 1993. Horizontal gene transfer. *American Scientist* 81 (4). *On how gene transfers between different species can fuel evolution.*

DOOLITTLE, R. F., AND P. BORK, 1993. Evolutionary mobile modules in proteins. *Scientific American* 269 (4). *Thoughtful alternatives to Gilbert's model for exon duplication and recombination emphasizing horizontal transmission.*

GOULD, J. L., AND C. G. GOULD, 1989. *Sexual Selection.* Scientific American Books, New York. *Well-illustrated treatment of the evolution of sex and mate choice.*

LAWN, R. M., 1992. Lipoprotein (a) in heart disease. *Scientific American* 266 (6). *Latter half of article details how this protein evolved from exon recombination.*

LEWONTIN, R. C., 1978. Adaptation. *Scientific American* 239 (3). *Points out that most features are compromises between different selection pressures, and that chance plays a role in evolution when more than one solution to a problem is possible.*

LI, W. H., AND D. GRAUR, 1991. *Fundamentals of Molecular Evolution.* Sinauer Associates, Sunderland, Mass. *An excellent account of how gene evolution can occur through duplication and exon recombination.*

RIDLEY, M., 1993. *Evolution.* Blackwell, Oxford. *Excellent text on the subject.*

CHAPTER 18

SPECIATION AND PHYLOGENY

HOW SPECIATION OCCURS

The previous chapter discussed how populations change through time. Eventually such a population may split, giving rise to two or more different descendant populations, perhaps generating new species in the process. This chapter will examine the processes underlying speciation, as well as the ways in which evolutionary history is reconstructed.

■ WHAT ARE SPECIES?

Demes To understand speciation, we need to look at population structure in more detail. With reference to sexually reproducing organisms, we can define a *population* as a group of individuals that interbreed and so share a common gene pool. A *deme*, in contrast, is a small local population, such as all the deer mice or all the red oaks in a certain woodland, or all the perch or all the water striders in a given pond (Fig. 18.1). Though no two individuals in a deme are exactly alike, the members of a deme usually resemble one another more closely than they resemble the members of other demes. There are at least two reasons for this: (1) the individuals in a deme are more closely related, because pairings occur more frequently between members of the same deme than between members of different demes; and (2) the individuals in a deme are exposed to more similar environmental influences and hence to more nearly the same selection pressures.

Demes are not clear-cut permanent units of population. Though the deer mice in one stand of trees are more likely to mate among themselves than with deer mice in nearby woods, there will almost certainly be occasional matings between mice from different woods. Similarly, though the female parts of a particular red oak tree are more likely to receive pollen from another red oak tree in the same stand, they will sometimes receive pollen from a tree in another nearby woods. And woods themselves are not permanent ecological features. They have only a transient existence as separate and distinct ecological units; neighboring stands may fuse after a few years, or a single woods may become divided into two or more separate smaller ones. Such changes in ecological features will produce corresponding changes in the demes of deer mice and oaks. Demes, then, are usually temporary units of population that intergrade with other similar units.

18.1 Aquatic demes

Each of these ponds contains a separate deme of each species of fish and aquatic insect.

487

A

B

C

20 mm

D

20 mm

18.2 Variations in morphology between populations within a species

In some species, populations within a single area show easily observable morphological variations, which may be continuous (such as size) or discrete (distinct alternative morphs). A male of the cichlid species *Cichlasoma minckleyi,* found in the Cuatro Ciénegas basin in Mexico, exhibits the deep-bodied morph (A), while another male of the same species from the same basin exhibits the slender-bodied morph (B); there are no intermediates in this character. An independent variation of the food-grinding structures in the lower pharyngeal jaw of males of this species also exists. Some males, regardless of body form, display the papilliform morph (C), while others display the molariform morph (D). Despite these major differences, individuals of the various morphs readily interbreed and produce fertile offspring; hence they are all considered members of the same species.

Species For centuries it has been recognized that plants and animals seem to fall naturally into many separate and distinct "kinds," or species. This does not mean that all the individuals of any one species are precisely alike—far from it. It does mean that all the members of a single species share certain biologically important attributes and that, as a group, they are genetically separated from other such groups.

A *species,* in the modern view, is a genetically distinctive population, a group of natural demes that share a common gene pool and that are reproductively isolated from all other such groups. A species is the largest unit of population within which effective gene flow occurs or can occur. The key word here is *effective;* we will see later why two species whose members mate but produce infertile hybrids are not classified as a single species.

The modern concept of species says nothing about how different from each other two populations must be to qualify as separate species. Most species can be separated on the basis of fairly obvious anatomical, physiological, or behavioral characters. However, the final criterion for living species is always reproduction—whether or not there is actual or potential gene flow.[1] If there is complete reproductive isolation between two outwardly almost identical populations—that is, if there can be no gene flow between them—then those populations belong to different species despite their similarity. On the other hand, if two populations show striking differences, but there is effective gene flow between them, those populations belong to the same species (Fig. 18.2). Anatomical, physiological, or behavioral characters simply serve as clues toward the identification of reproductively isolated populations; they do not in themselves determine whether a population constitutes a species.

Variation within species Though there is significant variation between individuals in a deme, there is often a superimposed variation correlated with geographic distribution. This leads to

---

[1] Because paleontologists deal almost exclusively with fossils, they must rely to a great extent on morphological criteria in distinguishing between species.

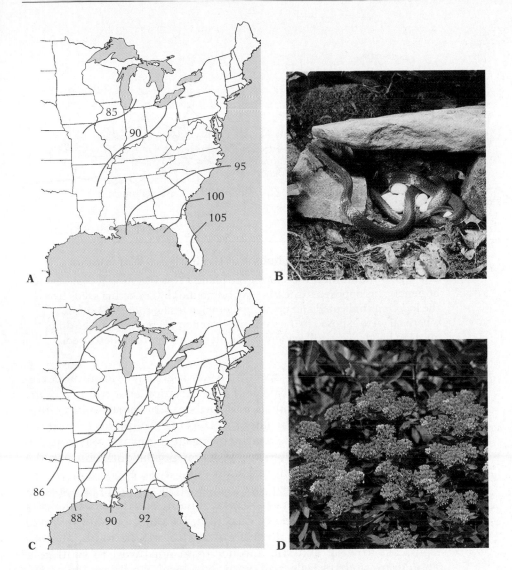

18.3 Clinal variation

(A) Map showing, by means of isophene lines (lines connecting equal values), the geographic variation in the mean number of subcaudal scales of the snake *Coluber constrictor* (the racer; B). Isophene map (C) showing geographic variation in the apical taper of leaves of the milkweed *Asclepias tuberosa* (D). The numbers represent degree of apical taper.

larger-scale differences between demes. Some of this pattern may result from genetic drift, but much of the geographic variation reflects differences in selection pressures operating on the populations, resulting from differences in environmental conditions. Each local population or deme tends to evolve adaptations to the specific environmental conditions in its own small portion of the species range. Such geographic variation is found in the vast majority of animal and plant species.

Environmental conditions often vary geographically in a more or less regular manner. There are changes in temperature with latitude or with altitude on mountain slopes, or changes in rainfall with longitude, as in many parts of the western United States, or changes in topography with latitude or longitude. Such environmental gradients are usually accompanied by gradients of allelic frequency in the species of animals and plants that inhabit the areas involved. Most species show north-south gradients in many characters; east-west gradients (Fig. 18.3A,C) and altitudinal gradients (Fig. 18.4) are not uncommon.

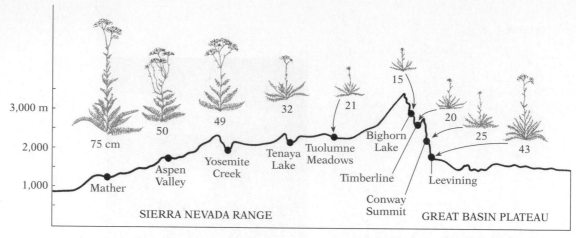

18.4 Altitudinal cline in height of the herb *Achillea lanulosa*

The higher the altitude, the shorter the plants. This variation was shown to be genetic (as opposed to merely environmental) by collecting seeds from the locations indicated and planting them in a test garden at Stanford where all were exposed to the same environmental conditions. The differences in height were still evident in the plants grown from these seeds.

When a character of a species shows a gradual variation correlated with geography, we speak of that variation as forming a ***cline***. For example, many mammals and birds exhibit north-south clines in average body size, being larger in the colder climates toward the poles and smaller in the warmer climates toward the equator. Similarly, many mammalian species show north-south clines in the size of such extremities as tails and ears: these exposed parts are smaller in the more polar demes.[2] A single widespread species often has many characters that vary clinally, but the several clines frequently do not coincide in direction, location, or intensity; one character may show clinal variation from north to south, another from east to west, and still another from northwest to southeast. Though most clines reflect adaptation to local conditions, some appear to be chance patterns.

Sometimes geographically correlated genetic variation is not as gradual as in clines. When an abrupt shift in a genetically determined character occurs in a geographically variable species, the populations involved may be called ***subspecies***. This term is also sometimes applied to more isolated populations—such as those on different islands or in separate mountain ranges or, as in fish, in separate rivers—when the populations are recognizably different genetically but are potentially capable of interbreeding freely. Subspecies, then, are groups of natural populations within a species that differ genetically and that are partly isolated from each other because they have different ranges (Fig. 18.5).

Two subspecies of the same species cannot long occur together geographically, because it is only the limitation on interbreeding imposed by distance that keeps them genetically distinct. If they occur together, they will interbreed and any distinction between

[2] Increase in average body size with increasing cold, called Bergmann's rule, is very common in warm-blooded animals; the tendency toward decrease in the size of the extremities with increasing cold is known as Allen's rule. The adaptive significance of these clines reflects the role of surface-to-volume ratios in heat loss.

A

B

18.5   Two subspecies of Canada goose

These two subspecies of Canada goose have different breeding grounds and different ranges. *Branta canadensis maxima* (A) breeds in the central and south-central United States. *Branta canadensis moffitti* (B) breeds largely in central and western Canada.

them will disappear. Some biologists have argued against recognizing subspecies, mainly because the distinctions are often made arbitrarily on the basis of one morphologically obvious character, while other less obvious characters may form different patterns of variation that are ignored (Fig. 18.6). Nevertheless, assigning names to distinguish separate populations can be a great convenience.

As a result of intraspecific geographic variation, two populations belonging to the same species but occurring in two widely separated localities often show no more resemblance to each other than to populations belonging to other species. Such intraspecific dissimilarity underscores the point made earlier: it is not the degree of morphological resemblance that determines whether or not two populations belong to the same species, but rather whether they are reproductively isolated. There are even instances where two widely separated populations are regarded as belonging to the same species even though the respective individuals, when brought together, are incapable of producing viable offspring. They are members of the same species because the populations are connected by an unbroken chain of intermediate populations that can interbreed.

A   B

C   D

E   F

18.6   Discordant geographic variation in six characters of the snake *Coluber constrictor* in the eastern United States

Since no two of the characters vary together, selection of any one of them as a criterion for recognition of subspecies is largely arbitrary. (A) Red eyes are found in juveniles. (B) Red ventral spots are found on juveniles. (C) The loreal scale is in contact with the first supralabial scale in at least 10% of these specimens. (D) Black adults are found. (E) Dark postocular stripes are found. (F) Full-grown adults have white chins.

A

B

18.7 Squirrels of the Grand Canyon

Two populations of squirrels that live in different ranges of the Grand Canyon in Arizona are morphologically distinct. The Kaibab squirrel (A), which lives on the Kaibab Plateau on the northern rim of the canyon, is darker than the Abert squirrel (B), of the related population that inhabits a range on the southern rim. Whether these two groups are separate species is not known.

■ THE PROCESS OF SPECIATION

New species can arise as an existing species changes, or as it splits into two or more species. We will focus on the latter phenomenon, known as divergent speciation.

The role of geographic isolation Since species are defined in terms of reproductive isolation, the fundamental question of divergent speciation is: How do two populations that initially share a common gene pool come to have completely separate gene pools?

In the majority of cases the initiating factor in speciation is geographic separation. As long as all the populations of a species are in direct or indirect contact, gene flow will normally continue throughout the system, and splitting will not occur. Various populations within the system, however, may diverge in numerous characters and thus give rise to much intraspecific variation. If the continuous system of populations is divided by some geographic barrier to individual dispersal, then the separated populations will no longer be able to exchange genes, and further evolution will be independent (Fig. 18.7).

Isolated populations are said to be *allopatric* (literally "of different groups"); populations that share a habitat, on the other hand, are called *sympatric*. Given sufficient time, the two separate populations may become quite unlike each other as each evolves in its own way. At first, the only reproductive isolation between them will be geographic—isolation by physical separation—and they will still be potentially capable of interbreeding. Eventually, however, they may become genetically so different that there would be no effective gene flow between them even if they should again come into contact. When this point in their divergence has been reached, the two populations constitute separate species. This is not to say that separation inevitably leads to speciation, though that is likely: the sycamores in Europe have been isolated from the population in China for more than 30 million years, and yet they can still interbreed.

There are at least three factors that can make geographically separated populations diverge:

1 The chances are good that the two populations will have different initial gene frequencies. Because most species exhibit geographic variation, it is unlikely that a geographic barrier would divide a variable species into portions that are genetically alike; more likely it would separate populations that are already genetically different, with different alleles adapting them to the local conditions in their geographic regions. In fact, geographic isolation often cuts off the terminal portions of a cline. Alternatively, small numbers of individuals frequently manage to cross existing barriers and found new geographically isolated colonies.

Regardless of how the population becomes divided, if one group is relatively small, its members will carry with them in their own genotypes a relatively small percentage of the total genetic variation present in the gene pool of the parental population. The new colony will have allelic frequencies different from those of the parental population; this is a special form of genetic drift called the *founder effect*. If from the moment of their separation two populations have different genetic potentials, their future evolution may follow different paths. Most cases of geographic isolation do involve small populations that exhibit the founder effect and genetic drift.

2 Separated populations will experience different mutations. Mutations are random; some mutations will occur in one of the populations and not in the other, and vice versa. Since there is no gene flow between the populations, a new mutant allele arising in one of them cannot spread to the other.

3 Isolated populations will almost certainly be exposed to different environmental selection pressures, since they occupy different ranges.

The barriers that cause initial spatial separation leading to speciation can differ from one species to another. A mountain range is a barrier to species that can live only in lowlands, a desert is a barrier to species that require a moist environment, and a valley is a barrier to montane species. On a grander scale, oceans and glaciers have played a role in the speciation of many plants and animals.

Reproductive isolation The initial factor preventing gene flow between two closely related populations is ordinarily an extrinsic one—geography. As two populations diverge, they accumulate differences that will lead, given enough time, to the development of intrinsic isolating mechanisms—biological characteristics involving morphology, physiology, chromosomal compatibility, or behavior that prevent the two populations from occurring together or from interbreeding effectively when (or if) they meet again (Fig. 18.8). There are several kinds of intrinsic isolating mechanisms that can evolve:

1 *Ecogeographic isolation* Two populations, initially separated by some extrinsic barrier, may in time become so specialized for different environmental conditions that even if the original barrier is removed they may never become sympatric, because neither can survive the conditions where the other lives. Two well-known species of the genus *Platanus*—*P. occidentalis* (the buttonwood tree), which occurs in the eastern United States, and *P. orientalis* (the Oriental plane tree), which occurs in the eastern part of the Mediterranean region—can be artificially crossed, and the hybrids are vigorous and fertile. But each species is adapted to the climate in its own native range, and the climates in the two ranges are so different that neither species will long survive in the range of the other. Thus there are genetic differences that under natural conditions prevent gene flow between the two species.

2 *Habitat isolation* When two sympatric populations occupy different habitats within their common range, the individuals of each population will be more likely to encounter and mate with members of their own population than with members of the other population. Their genetically determined preference for different habitats thus helps keep the two gene pools separate.

There are numerous examples of such habitat isolation. *Bufo woodhousei* and *B. americanus* are two closely related toads that

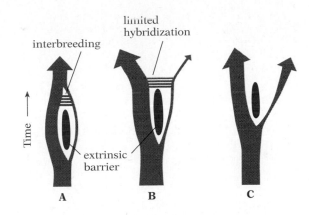

18.8 Model of geographic speciation

An ancestral population is split by an extrinsic (geographic) barrier for a time, and diverges with regard to various traits (only one of which is considered in these hypothetical examples). (A) The barrier breaks down before the two subpopulations have been isolated long enough to have evolved intrinsic reproductive isolating mechanisms, so the populations will interbreed and fuse back together. (B) Two populations are isolated by an extrinsic barrier long enough to have evolved incomplete intrinsic reproductive isolating mechanisms. When the extrinsic barrier breaks down, some hybridization occurs, but the hybrids are not as well adapted as the parental forms. Hence there is a strong selection pressure favoring forms of intrinsic isolation that prevent mating, and the two populations diverge more rapidly until mating between them is no longer possible. This rapid divergence is called character displacement. (C) Two populations are isolated by a geographic barrier so long that by the time the barrier breaks down they are too different to interbreed. In most cases, one population has many fewer members than the other, as indicated by the width of the "branches" where they separate. The smaller population—often quite small—usually diverges more from the common ancestor than does the larger population. This greater divergence results from the founder effect, a greater tendency for genetic drift, and the need to adapt to a smaller and perhaps more specialized habitat.

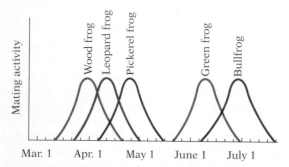

18.9   Mating seasons at Ithaca, New York, for five species of frogs of the genus *Rana*

The period of most active mating is different for each species. Where the mating seasons for two or more species overlap, different breeding sites are used.

18.10   A male fiddler crab (*Uca*)

The animal waves its large cheliped in the air as a courtship display. Critical details of the display differ among the various species of fiddler crabs.

can cross and produce viable offspring. However, in those areas where the ranges of the two toads overlap, *B. woodhousei* elects to breed in the quieter water of streams, while *B. americanus* breeds in shallow rain pools. The dragonfly *Progomphus obscurus* lives in northern Florida, while its close relative *P. alachuensis* lives in southern Florida. When the ranges of the two species overlap in north-central Florida, the two species occupy different habitats, *P. obscurus* being restricted to rivers and streams and *P. alachuensis* to lakes. In California the ranges of *Ceanothus thyrsiflorus* and *C. dentatus*, two species of buckthorn shrubs, overlap broadly, but *C. thyrsiflorus* grows on moist hillsides with good soil, while *C. dentatus* grows on drier, more exposed sites with poor or shallow soil.

3 *Seasonal isolation* If two closely related species are sympatric, but breed during different seasons of the year, interbreeding will be impossible. *Pinus radiata* and *P. muricata*, two species of pine, are sympatric in some parts of California. They are capable of crossing, but rarely do so under natural conditions, because *P. radiata* sheds its pollen early in February while *P. muricata* waits until April. *Reticulitermes hageni* and *R. virginicus*, two closely related species of termites, are sympatric in southern Florida, but the mating flights of the former occur from March through May while those of the latter occur in the fall and winter months. Five species of frogs belonging to the genus *Rana* are sympatric in much of eastern North America, but the period of active mating is different for each species (Fig. 18.9).

4 *Behavioral isolation* Chapter 38 discusses the immense importance of behavior in courtship and mating, particularly with respect to species recognition. In many cases, species have unique courtship patterns, and where two species are sympatric, crosses rarely occur because a courtship between members of different species involves so many wrong responses that it is unlikely to proceed all the way to copulation.

A particularly interesting example of the role of visual displays in species recognition was discovered by Jocelyn Crane. She found 12 species of fiddler crabs of the genus *Uca* actively courting on the same small beach (only about 56 m²) in Panama. Each species had its own characteristic display, which included waving the large claw (cheliped), elevating the body, and moving around the burrow (Fig. 18.10). Crane found that the displays were so distinctive that she could recognize each species from a considerable distance merely by the form of its display; presumably the female crabs do likewise.

Auditory stimuli are important in species recognition among many animals, particularly birds and insects, and help prevent mating between related species. In several instances specialists have noticed that two or three very different calls are produced by what had been considered members of a single species of cricket. On investigation, they found that each call was, in fact, produced by a different species so similar morphologically that no one had previously distinguished among them. They do not hybridize in nature even when they are sympatric, because females do not respond to the sounds made by a male of a different species. ***Sibling*** *species*—species so closely related that humans can hardly distinguish them—may be a fairly common phenomenon.

5 *Mechanical isolation*   If structural differences between two closely related species make it physically impossible for matings between males of one species and females of the other to occur, the two populations will obviously not exchange genes. If, for example, one species of animal is much larger than the other, matings between them may be very difficult, if not impossible. Likewise, if the genital organs do not fit, mating will be prevented.

Mechanical isolation is probably more important in plants than in animals, particularly in plants that depend on insect pollinators. Consider *Salvia apiana* and *S. mellifera*, two closely related species of sages with overlapping ranges in California. They are reproductively isolated by differences in habitat and flowering season and by the behavior of their pollinators. In addition, mechanical features play a role. Whereas *S. mellifera* is pollinated by relatively small bees, the flowers of *S. apiana* can be entered only by very large bees whose weight is sufficient to cause the landing platform (lower lip of the corolla) to unfold and permit free entrance into the flower; a similar isolating mechanism is observed in other plants as well, including Scotch broom (Fig. 18.11).

6 *Gametic isolation*   If individuals of two animal species mate or if the pollen from one plant species gets onto the stigma of another, fertilization still may not take place. Some 68 interspecific combinations are known in tobacco in which no cross-fertilization will occur; the sperm nucleus from the pollen is unable to grow to the egg cell in the ovary. In *Drosophila*, if cross-insemination occurs between *D. virilis* and *D. americana*, the sperm are immobilized by the unsuitable environment in the reproductive tract of the female. In other species of *Drosophila*, interspecific matings cause an antigenic reaction in the female genital tract, which kills the sperm before they reach the eggs.

The mechanisms discussed so far exact little cost from the individual organisms: the machinery of isolation is either imposed from without, or generated internally from existing behavior, physiology, and morphology. As a result, no great loss of fitness is involved in maintaining isolation. The remaining four mechanisms, though effective, exact a significant cost from the individual, and this cost acts as a selection pressure favoring evolution of the more efficient mechanisms of reproductive isolation already mentioned.

7 *Developmental isolation*   Even when cross-fertilization occurs, embryonic development is often irregular and may cease before birth. The eggs of fish can often be fertilized by sperm from a variety of other species, but development usually stops early on. Crosses between sheep and goats produce embryos that die before birth.

8 *Hybrid inviability*   Hybrids are often weak and malformed and frequently die before they reproduce; hence there is no gene flow through them from the gene pool of one parental species to the gene pool of the other. Certain tobacco hybrids, for example, develop tumors in their vegetative parts and die before they flower.

A

B

C

18.11   Pollination of Scotch broom (*Cytisum scoparius*) by a bumble bee

The nectar and pollen are inaccessible to lighter pollinators like honey bees, whose weight is insufficient to trip release of the reproductive structures. (A) An untripped bloom, (B) A bumble bee gathering nectar. (C) A flower after tripping.

9 *Hybrid sterility* Some interspecific crosses produce vigorous but sterile hybrids. The best-known example is the cross between a female horse and a male donkey, which produces a mule. Mules have many characteristics superior to those of both parental species, but they are sterile.

10 *Selective hybrid elimination* The members of two closely related populations may be able to cross and produce fertile offspring. If those offspring and their progeny are as vigorous and well adapted as the parental forms, then the two original populations will not remain distinct if they are sympatric. But if the fertile offspring and their progeny are less well adapted than the parental forms, then they will be eliminated. There will be only limited gene flow between the two parental gene pools through the hybrids. The parental populations are consequently regarded as separate species.

These last four mechanisms isolate species from one another effectively, but their cost is significant: valuable metabolic resources are invested in doomed embryos or in frail or possibly sterile young, and the seasonal nature of many reproductive cycles may offer no second chance to mate and rear young. Mistaken matings waste gametes, whether fertilization takes place or hybrids are viable. Selection, therefore, will strongly favor individuals whose behavior, morphology, or physiology reduces the chance of a mismatch in the first place. If the parental populations are sympatric, there will be strong selection for the evolution of more effective intrinsic isolating mechanisms. Gene combinations that lead to correct mate selection will increase in frequency until eventually all hybridization ceases. The tendency of closely related sympatric species to diverge rapidly in characteristics that reduce the chances of hybridization and/or minimize competition between them is called *character displacement* (see Fig. 18.8). Table 18.1 summarizes the different intrinsic isolating mechanisms.

Situations in which only one of the 10 isolating mechanisms is operative are extremely rare. Ordinarily several contribute to keeping two species apart. For example, closely related sympatric plant species often exhibit habitat and seasonal isolation in addition to some form of hybrid incapacity.

Speciation by polyploidy and chromosomal change Speciation through the divergence of geographically separated populations is not the only way new species can arise. Speciation not involving geographic isolation is called *sympatric speciation*. One important example is speciation by polyploidy, the condition of having more than two sets of chromosomes. This condition can arise so quickly that it is possible for a parent to belong to one species and its offspring to another. (Other forms of sympatric speciation will be discussed later.) Speciation by polyploidy has been common in plants: by some estimates, 70% of the flowering plant species arose this way.

One type of polyploid speciation, called *autopolyploidy*, involves a sudden increase in the number of chromosomes, usually as a result of the nondisjunction of chromosomes during meiosis. An example of this type of polyploidy was discovered by the Dutch

Table 18.1 Intrinsic isolating mechanisms

EFFECT	MECHANISM	INDIVIDUALS AFFECTED
Mating prevented	1. Ecogeographic isolation	Parents: fertilization prevented
	2. Habitat isolation	
	3. Seasonal isolation	
	4. Behavioral isolation	
	5. Mechanical isolation	
Production of hybrid young prevented	6. Gametic isolation	
	7. Developmental isolation	
Perpetuation of hybrids prevented	8. Hybrid inviability	Hybrids: success prevented
	9. Hybrid sterility	
	10. Selective hybrid elimination	

geneticist Hugo De Vries, while he was making extensive studies of the evening primrose, *Oenothera lamarckiana*. This diploid species has 14 chromosomes. During De Vries's studies, a new form suddenly arose (Fig. 18.12). The new form, *Oenothera gigas*, has 28 chromosomes. This tetraploid is reproductively isolated from the parental species because hybrids between *O. lamarckiana* and *O. gigas* are triploid (they received one of each type of chromosome from their *O. lamarckiana* parent and two of each type from their *O. gigas* parent); triploid individuals, because of the highly irregular distribution of their chromosomes at meiosis, are sterile. Hence polyploid populations fulfill all the requirements of the modern definition of species—they are genetically distinct and reproductively isolated—though botanists do not always give each polyploid population a formal species name.

The reproductive isolation of the polyploid daughter species from the ancestral stock sometimes permits adaptive divergence that would not otherwise be possible. For example, new polyploid plant species have become adapted to mineral soils like mine tailings and serpentine outcrops. If the plants were not reproductively isolated, gene flow from the large surrounding population of normal diploids would prevent local adaptation to these special soils. Polyploidy, then, substitutes for geographic isolation.

A second type of polyploid speciation, called ***allopolyploidy***, involves a multiplication (usually a doubling) of the number of chromosomes in a hybrid between two species. The hybrid has one set of chromosomes from each of the two species. Unless these two species are so closely related that they have homologous chromosomes capable of pairing in meiosis, the hybrid will almost certainly fail to produce gametes. However, if the hybrid undergoes chromosome doubling before meiosis, it will have a complete diploid set of chromosomes from each species, and viable gametes are more likely. This type of polyploidy has probably been more important in speciation than autopolyploidy. Allopolyploid individuals are able to breed freely among themselves, but they cannot cross with either of the parental species. Consequently the allopolyploid population is a distinct species.

Allopolyploid plants are only rarely more vigorous than the parental diploid plants, probably because each parental species has a well-balanced mixture of genes, while the allopolyploid combines two sets of gene products, metabolic pathways, control systems, and, especially, developmental instructions. But sometimes allopolyploids, because of this mix of genes, can grow in habitats that neither parental species can colonize. Allopolyploid speciation has played an important role in the perpetuation of some plant groups during periods of widespread environmental change.

Both allopolyploidy and autopolyploidy have been proved to be of great importance in the production of valuable new crop plants. As soon as plant breeders realized that many of our most useful plants—such as oats, wheat, cotton, tobacco, potatoes, bananas, coffee, and sugarcane—are polyploids, they began trying to stimulate polyploidy, and obtained many new varieties. The chemical colchicine readily induces polyploidy. One of the first artificially produced allopolyploids came from a cross made in 1924 between the radish and the cabbage. Unfortunately it had the root of the

18.12 Evening primrose

This species arose in a single generation from an autopolyploidy event in 1904.

A

B

18.13 Speciation through a developmental mutation?

Strepsiptera is an insect order closely related to Diptera, the true flies. The order probably arose through a developmental mutation (a homeotic mutation, discussed in Chapter 14) that reversed two thoracic segments. In flies (A) the club-shaped structures called halteres are located on the segment behind the wings and spin during flight to provide a gyroscopic stability; in strepsipterans (B) the haltere- and wing-bearing segments are reversed.

cabbage and the shoot of the radish. Other crosses have yielded more desirable results (see Fig. 16.26, p. 437).

Chromosome doubling creates an obvious problem: if tetraploids cannot breed with diploids, they are restricted to breeding with each other. Hence, the rare diploid pollen grain must find an equally rare diploid egg cell to produce a fertile tetraploid. Most likely, the resulting plant must then fertilize itself, since other tetraploids will be exceedingly rare. The same is true of allopolyploids: the sole hybrid is extremely unlikely to find a genetically compatible plant, and so will probably have to fertilize itself. The result in both cases will be severe inbreeding of the new species, a strong founder effect, and rapid genetic drift. The new population will persist only if it has a strong competitive advantage in the habitat.

Some researchers believe that chromosomal rearrangements—whether the result of a few major breakage-and-fusion events or of smaller but more numerous transpositions, duplications, and deletions—might give rise to individuals or small populations genetically incompatible with the rest of their species. This would be especially likely if a developmental control gene were to be altered (Fig. 18.13). If the variant organisms are at a competitive advantage, selection would favor further intrinsic isolation, and sympatric speciation might occur as a result. There is some evidence for this mechanism of speciation, at least among the *Drosophila*.

Nonchromosomal sympatric speciation Though most speciation not involving gross chromosomal changes is certainly allopatric (requiring a period of geographic isolation), there is increasing evidence that sympatric speciation can occur without polyploidy or other major chromosomal rearrangements. Reproductive isolation is just as essential to these types of sympatric speciation, but is effected by other means. For example, relatively small changes in habitat preference may produce habitat isolation. Thus a species of clover adapted to soil containing mine tailings is now reproductively isolated from the widespread and closely related species from which it evolved, not by polyploidy but by virtue of having a different flowering season.

Another alternative to geographic isolation is related to host specificity. Treehoppers often mate on the plants on which they feed, so subspecies that are adapted for a particular species of host plant will tend to inbreed. Selection should favor those individuals that breed strictly on the host species; consequently, intrinsic isolating mechanisms might develop that would serve to isolate tree hoppers with adaptations to different host plants (Fig. 18.14).

Yet another mechanism that may contribute to sympatric speciation, particularly in birds, is sexual imprinting (detailed in Chapter 38). While young, the members of one sex of many species memorize features displayed by their parents or siblings. The features may be visual, auditory, or olfactory. Memorization enables these individuals later to identify suitable mates. If the offspring imprint on a parent displaying a mutation in the feature being committed to memory (a brightly colored eye ring, for example, or novel elements in the courtship song), and if the young are later able to locate mates displaying the same mutation, the mutant individuals are likely to pair with each other. The result could be the founding of a population that does not interbreed with the rest of the species. In birds, then, instant reproductive isolation based on imprinting may play an important role in shaping populations.

Imprinting can also create a kind of habitat preference that leads to reproductive isolation. The picture-wing fruit fly, *Rhagoletis pomenella*, courts and deposits its eggs on hawthorn berries; the larvae feed on the fruit, pupate, and emerge to continue the cycle. At some stage in its life, the fly imprints on the odor of hawthorn, and uses that food-based memory to locate a suitable host plant. Jeff Feder has shown that the larvae are well adapted to hawthorn berries, growing and pupating about twice as well as on any other kind of tree fruit. However, there are two specialized species of parasitic wasps that seek out hawthorn berries and insert their eggs into the growing fruit fly larvae, killing about 90% of them.

About 150 years ago, land was cleared and large commercial apple orchards were established in the Hudson Valley of the northeastern United States; hawthorns became less common, and subsequently (probably by mistake) a few *Rhagoletis* began to lay their eggs on apples. Imprinting committed the next generation to apples, so reproductive isolation must have been nearly complete from the outset. Although *Rhagoletis* larvae grow only about half as well on apples, they suffer far less parasitism: not only are the wasps not looking for them on apples, but also the fruits appear earlier (so the larvae probably enter the pupal stage before the height of the season for parasites) and apples are larger, allowing the larvae to burrow so far in that the wasps cannot reach them. The result is that six times as many of the eggs laid on apples produce adult flies as compared with hawthorn. Whether full speciation has yet taken place is unknown, but the potential for this pathway to sympatric speciation is clear.

Adaptive radiation One of the most striking aspects of life is its extreme diversity. A bewildering array of species now occupies this globe. And the fossil record shows that, of the species that have existed at one time or another, those now living represent probably less than 0.1%, all the other species being now extinct. Clearly, then, divergent evolution—the evolutionary splitting of species into many separate descendant species—has been as frequent as

A

B

18.14 Sympatric speciation in treehoppers

Two sympatric populations of the treehopper *Enchenopa binotata* have evolved adaptations to different host plants. The treehopper in A lives on bittersweet, while the one in B lives on butternut. Host specificity may take the place of physical separation (allopatry) in preventing these two populations from interbreeding.

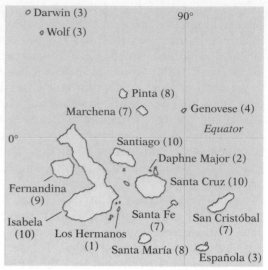

18.15   Galápagos Islands

Left: The islands are located about 950 km off the coast of Ecuador. Cocos Island is about 700 km northeast of the Galápagos. Right: The islands shown in greater detail. The number in parentheses after each name indicates the number of species of Darwin's finches that occur on the island.

extinction. How could opportunities for geographic isolation have been sufficient to lead to most of this speciation? Perhaps the best documented example of rampant divergent evolution—often called ***adaptive radiation***—involves the finches on the Galápagos Islands, which (along with tortoises and mockingbirds on those islands) played a major role in leading Darwin to formulate his theory of evolution by natural selection.

The Galápagos Islands lie astride the equator in the Pacific Ocean roughly 950 km west of the coast of Ecuador (Fig. 18.15). The islands have never been connected to one another. They apparently arose from the ocean floor as volcanoes approximately 7 million years ago. At first they were completely devoid of life, and were therefore open to exploitation by the few species that chanced to reach them from the mainland. The only land vertebrates present on the islands before human beings arrived were at least seven species of reptiles (one or more snakes, a species of huge tortoise, and at least five species of lizards, including two very large iguanas), seven species of mammals (five rats and two bats), and a limited number of birds (including two species of owls, a hawk, a dove, a cuckoo, a warbler, two flycatchers, a martin, four mockingbirds, and Darwin's finches).

The 14 species of Darwin's finches constitute a separate subfamily found nowhere else in the world. Thirteen of them are believed to have evolved on the Galápagos Islands, and one on Cocos Island, from some unknown finch ancestor that colonized the islands from the South or Central American mainland. The descendants of geographically isolated colonizers often undergo so much evolutionary change as to become, in time, very unlike their mainland ances-

18.16 Model of speciation on the Galápagos Islands

An ancestral form colonized the larger of these two hypothetical islands. Later, part of the population dispersed to the smaller island. (1) Eventually the two populations, being isolated from each other, evolved into separate species A and B. (2) Some individuals of A dispersed to B's island. The two species coexisted, but intense competition between them led to rapid divergent evolution. (3) This rapid evolution of the population of A on B's island caused it to become increasingly different from the original species A, until eventually it was sufficiently distinct to be considered a full species, C, in its own right. At the same time, the selection pressure imposed by the small invading population caused the large population of species B to evolve to a small degree as well.

tors. How did the descendants of the original immigrants split into today's 14 species?

The Galápagos group is a cluster of more than 25 separate islands. Finches will not readily fly across wide stretches of water, and they show a strong tendency to remain near their home area. Hence a population on any one of the islands is effectively isolated from the populations on the other islands. The initial colony was probably established on one of the islands where the colonizers, perhaps blown by high winds, chanced to land. Later, stragglers from this colony wandered or were blown to other islands and founded new colonies. Because of the founder effect the allelic frequencies in the new colonies differed from those in the original colonies from the moment they started. In time the colonies on the different islands diverged even more, because of different mutations, different selection pressures, and, in such small populations as some of these must have been, genetic drift. We might expect, therefore, a different species or subspecies on each of the islands. In fact, most of the islands have more than one species of finch, and the larger islands have 10 (Fig. 18.15, right). What has happened?

Let us suppose that form A evolved originally on the island of Santa Cruz and that the closely related form B evolved on Santa María. If form A later spread to Santa María before the two forms were isolated long enough to evolve any but minor differences, they might have interbred freely and merged. However, if A and B were separated long enough to evolve major differences before A invaded Santa María, then A and B might have been intrinsically isolated from each other, having developed into separate species, and they might have been able to coexist on the same island without interbreeding (Fig. 18.16). Selection would have favored individuals that mated only with their own kind, and this selection pressure would have led rapidly to more effective, intrinsic isolating mechanisms. In fact, Darwin's finches readily recognize members of their own species and show little interest in other species.

It would be highly unlikely that A and B could coexist on Santa María indefinitely if they used the same food supply. The ensuing competition would be severe, and the less well adapted species would tend to be eliminated by the other unless it evolved differences that minimized the competition. In short, wherever two or more very closely related species occur together, competition will lead either to the extinction of one or to character displacement—in this instance to the evolution of different feeding specializations.

Character displacement is indeed what we find in Darwin's finches (Fig. 18.17). The 14 species form four groups (genera). One

18.17 Darwin's finches

Darwin's finches fall into four genera: birds 1, 3, 4, 5, 6, and 10 are the tree finches (*Camarhynchus*); birds 7, 8, 11, 12, 13, and 14 are the ground finches (*Geospiza*); bird 2 is the unfinchlike warbler finch; bird 9 is the one species inhabiting Cocos Island.

1. Vegetarian tree finch (*Camarhynchus crassirostris*)

2. Warbler finch (*Certhidea olivacea*)

3. Large insectivorous tree finch (*Camarhynchus psittacula*)

4. Medium insectivorous tree finch (*C. pauper*)

5. Mangrove finch (*C. heliobates*)

6. Small insectivorous tree finch (*C. parvulus*)

7. Large cactus ground finch (*Geospiza conirostris*)

8. Cactus ground finch (*G. scandens*)

9. Cocos finch (*Pinaroloxias inornata*)

10. Woodpecker finch (*C. pallidus*)

11. Large ground finch (*G. magnirostris*)

12. Sharp-beaked ground finch (*G. difficilis*)

13. Medium ground finch (*G. fortis*)

14. Small ground finch (*G. fuliginosa*)

group contains six species that live primarily on the ground; of these, some feed primarily on seeds and others mostly on cactus flowers. Of the species that feed on seeds, some feed on large, some on medium-sized, and some on small seeds. These feeding preferences result from the morphological specialization of the beaks: small beaks are more efficient at handling small seeds, while larger beaks can crack large seeds. From a series of careful beak measurements, David Lack was able to find clear evidence of character displacement. For example, where the small and medium ground finches coexist on the larger islands, their beak sizes are widely separated, with depths averaging about 8.4 and 13.2 mm, respectively. On small islands where only one of the two species exists, the birds of that species tend to have beaks of intermediate size, on the order of 9.7 mm (Fig. 18.18).

The second group of finches contains six species that live primarily in trees. Of these, one is vegetarian and the others eat insects, but the insect eaters differ from one another in the size of their prey and in where they catch them (Fig. 18.19). A third group contains only one species, which has become very unfinchlike and strongly resembles the warblers of the mainland. The fourth group also contains one species, restricted to Cocos Island, which is about 700 km northeast of the Galápagos Islands and about 500 km from Panama. Correlated with the differences in diet among the species are major differences in the size and shape of their beaks. (Noteworthy variations in beak size and shape among birds thought to share a common ancestry can be found in other island

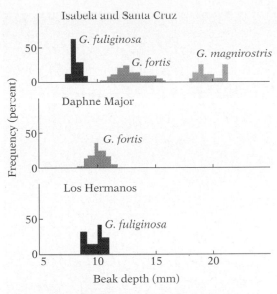

18.18 Beak sizes of ground finches

On large islands like Isabela and Santa Cruz, three species of ground finch coexist. Though the beak sizes of individuals within each species show the sort of variation essential for evolution by natural selection, they fall into three separate ranges, specialized respectively for efficient feeding on small, medium-sized, and large seeds. However, on small islands where only one species exists (either the small or the medium ground finch), the beak sizes fall into an intermediate range regardless of the size specialization found in the same species on the larger islands. This shows that, in the face of competition for food on the larger islands, character displacement has taken place.

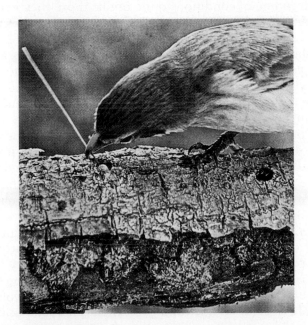

18.19 Tool-using finch

One of the insectivorous tree finches of the Galápagos has evolved a most unusual feeding habit. Sometimes called the woodpecker finch, it chisels into wood after insects, but it lacks the long tongue that a woodpecker uses to probe insects out of a crack. Instead, it pokes into the crack with a cactus spine or twig that it holds in its beak. The mangrove finch does the same thing in a different habitat. These are two of the few known cases of tool use by birds.

18.20 Beak differences in Hawaiian honeycreepers

Differences in beak size and shape are apparent in related species of Hawaiian honeycreepers, which, like Darwin's finches, are thought to have evolved from a common finchlike ancestor.

habitats as well, as shown, for example, in Figure 18.20. These characteristics of the beak are one important means by which the birds recognize other members of their own species; song is another.)

If selection on Santa María favored character displacement between species A and B, the population of species A on Santa María would become less and less like the population of species A on Santa Cruz (Fig. 18.16). Eventually these differences might become so great that the two populations would be intrinsically isolated from each other and would thus be separate species. We might now designate as species C the Santa María population derived from species A. The geographic separation of the two islands would thus have led to the evolution of three species (A, B, and C) from a single original species. The process of island hopping followed by divergence could continue indefinitely, producing many additional species. It was doubtless such a process that led to the 14 species of Darwin's finches.

The sorts of processes seen on islands can account also for radiation in a continental area, where intervening habitats can limit the frequency of movement as much as an expanse of ocean. Adaptive radiation on islands, as in the case of Darwin's finches, is dramatic and lends itself particularly well to analysis, but it does not differ in principle from adaptive radiation under other conditions.

It should be clear, as Darwin himself pointed out, that the rate of evolutionary divergence is not constant. When the first colonizing finches reached the Galápagos Islands, they would have encountered environmental conditions quite unlike those they had left behind in Central or South America. The selection pressures to which they were subjected would probably have been quite different from those in their former home; differences in the resources available, for example, may have led to selection for different morphology, physiology, and behavior. On the other hand, if there was initially little or no competition from other species, the result might have been a temporary relaxation of selection pressures. Only when the new habitat became saturated with finches would intraspecific competition for the available resources have become important. And so, sooner or later, selection pressures must have led to rapid divergence from the ancestral population. Later, as the finches became increasingly well adapted to conditions on the Galápagos, the rate of evolutionary change probably slowed down.

In general, when conditions change radically and organisms have new evolutionary opportunities for which they are at least modestly preadapted, they may undergo an evolutionary burst—a period of rapid adaptational change—which may then be followed by a more stable period, during which any further evolutionary changes are merely a fine-tuning of their characteristics. Such bursts of rapid evolutionary divergence probably characterized the tremendous radiation of amphibians when they moved onto land for the first time, and the explosive radiation of mammals when the demise of the dinosaurs left many niches—that is, ways of surviving—unoccupied. (The concept of niche is detailed in Chapter 39.)

How constant and important is competition? Chapter 17 contrasted the roles played by genetic drift (chance) and natural selection in the evolution of populations. The discussion there emphasized that evolution is a vital, ongoing phenomenon, while drift (potentially very important in small populations) and selection are two contributing—and clearly demonstrated—mechanisms by which it occurs. In our discussion of speciation here we have touched on a similar theme: barriers that contribute to reproductive isolation can arise by chance (primarily genetic drift), by natural selection, or by a combination of the two. We have emphasized the role of competition in the formation of the species: even reproductively isolated populations may not be able to coexist indefinitely if they compete for precisely the same food, since even a slight but systematic superiority of one will tend to lead to the extinction of the other.

G. F. Gause, who first observed this phenomenon in the laboratory in the 1930s, formulated the competitive exclusion principle, often called the **Niche Rule**: no two species occupying the same niche can long coexist (Fig. 18.21). Only the character displacement that natural selection produces, resulting in changes in food preference, habitat choice, and the like, will allow closely related species to coexist. This Darwinian interpretation of how separate species form, with its emphasis on reproductive isolation and character displacement in the face of competition, probably accounts for most speciation.

In the last 20 years, however, the nature and role of competition in speciation have come under critical scrutiny. Competition is no longer thought to be as overwhelmingly important or constant; chance is now seen as a greater force than Darwin thought. There are at least two ways in which chance can take precedence over competition in causing species to diverge. The first we have already discussed: in small populations, genetic drift can be very powerful, driving alleles to extinction before the generally more gradual process of natural selection can produce stability. The other process, however, is more fundamental to our understanding of evolution. Most Darwinian analyses tacitly assume that selection pressures change relatively slowly, and that large continuous populations have ample time to evolve specialized adaptations to their environments. Do conditions really remain sufficiently constant for a species to achieve a stable set of allelic frequencies in the face of all the selection pressures affecting it?

Apparently, many populations are not always at equilibrium. Two things in particular can disrupt the balance: **crises**, which greatly reduce population size, and ecological changes, which alter the competitive "landscape." Both factors are evident in one particularly thorough study by Peter Grant and his colleagues. They tagged hundreds of medium-sized ground finches on one of the smaller Galápagos islands, Daphne Major. In 1977 only 20% of the normal amount of rain fell, and the plants on the island produced many fewer seeds than usual. Relatively dry or wet seasons, a surprisingly early or late hard frost, or especially warm or cold years are familiar conditions that organisms can generally take in stride, but the effects of extreme events on plants and animals can be drastic.

On Daphne Major the consequences of the reduction in food were dramatic for the medium ground finch (*Geospiza fortis*). With only 35% of the usual amount of food available, 387 of the 388

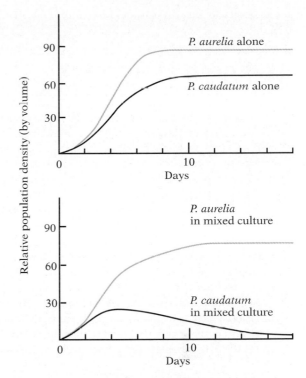

18.21 Competition and extinction

When a few individuals of either species of *Paramecium* are introduced into a tank alone (top), they multiply until they reach a limiting density. When individuals of both species are introduced together (bottom), they multiply independently for the first 3 days, and then begin to compete for resources. *P. aurelia* is in some way more efficient under these conditions, and drives *P. caudatum* to extinction in only 3 weeks.

HOW WE KNOW:
THE BEAK OF THE FINCH

In order to trace evolution as it occurs, researchers must know the distribution of heritable phenotypes in a population, and then monitor any changes as they occur. In the case of the Galapagos finches, this means measuring every size parameter of every bird on the island under study (Fig. A), marking each bird individually, and then recapturing every bird annually both to monitor survivorship and to add each year's crop of youngsters to the inventory. The extraordinary difficulty of such an undertaking probably explains why few researchers have been able to track evolution generation by generation. Let's look at some of the data Peter Grant and his colleagues collected, and how they were thus able to document one bout of evolution.

Two basic requirements for evolution are in place: there is variation in beak parameters and it is heritable: the offspring of parents with large beaks tend to have large beaks themselves (Fig. B). The question Grant sought to answer was whether beak size is subject to selection.

In 1976, before a drought, the beaks of the 741 resident medium ground finches on the island of Daphne Major had a mean depth of about 9.4 mm (Fig. C, blue curve). The range of depths reflects the range of seeds available as food. Thus birds with smaller beaks were better able to harvest smaller seeds, whereas larger or harder seeds were more readily opened by finches with larger beaks.

The drought led to a threefold reduction in seed number. The seeds that remained were 50% larger and harder than normal. Selection strongly favored larger beak size: only 9% of the birds with average beaks survived, whereas 20% of the individuals with beaks only 1 mm deeper lived through the crisis, as did 40% of the finches with beaks 2 mm deeper. The result was a upward shift in the mean beak depth to 10.2 mm (Fig. C, red curve); the only hatchling to survive had an unusually deep beak.

A

B Mean beak depth of parents, (mm) / Beak depth of offspring, (mm)

C Beak depth (mm) / Individuals in each depth class

1976 mean

Population in 1976

1978 mean

Survivors in 1978

Despite the strong selection, the overall shift in mean beak depth was just 0.8 mm. The reason is that there were very few individuals with longer beaks for selection to act on; if the variation in the 1976 population had been greater, the shift in beak depth would doubtless have been more dramatic. (Note, however, the evident role of chance: as Fig. C shows, despite the strong selection, the bird with the longest beak died, while the one with the shortest beak survived.)

The effect of this bout of intensive selection on the next generation of finches was clear: in 1978 the mean depth of hatchling beaks was about 0.8 mm longer. Evolution—a change in allele frequencies—had occurred. By 1978, however, the average seed size and hardness had fallen back to normal levels. In 1978, therefore, many finches, young and old, had beaks that were awkwardly large compared to many of the seeds that were being produced once the rainfall returned to normal levels. This change in the food supply created a less intense directional selection for smaller beaks, and the mean beak depth began to fall; another precisely documented bout of evolution was underway.

nestlings of 1976 died, as did the majority of adults. Furthermore, the adult survivors were not a representative cross section of the former population: roughly 180 of the 500 adult males survived, but only about 30 of the 500 females did so, and the birds that did survive the year of starvation were significantly larger than those that succumbed. Exceptionally intense natural selection in favor of large size—especially larger beaks—had occurred. At the same time, the role of chance was also clear: the individual with the shortest beak in the population nevertheless managed to survive.

Thus one of several rare and unpredictable crises sharply altered the selection pressures for a year, imposed a large element of chance into the struggle to survive, and reduced the population to a level at which genetic drift (and thus chance) became a real possibility. What had seemed to be an adaptive equilibrium was upset by a chance disturbance, and a population with altered characteristics emerged. The finch population had been forced through a period of environmental crisis. When a crisis is so severe as to cause major changes in allelic frequencies in the surviving population, it is called an ***evolutionary bottleneck***.

Is it possible that the role of interspecific and intraspecific competition, leading gradually to character displacement, is dwarfed by chance crises that upset whatever stability exists in the allelic frequencies within a species? Crises can be caused by any environmental factor—pestilence or extreme weather, for example—that severely affects an isolated population, or that serves to isolate one part of a population from the remaining body. Because of their small population size, endangered species are in this position. During such a crisis, as an isolated population passes through the evolutionary bottleneck, one character or another may gain ascendancy. Having been selected by a founder effect, such a character is still at risk if the surviving population is small, because it may decline in frequency or even become extinct through genetic drift. Alternatively, if a character that was not adaptive prior to the crisis proves adaptive thereafter, it may increase in frequency in the population. In the case of the finches, the effects of the drought in 1977 (and another in 1982) were canceled by another freak climatic event: record rainfall in 1983. Unprecedented numbers of seeds were produced, and selection strongly favored smaller beaks.

Though most researchers still believe that competition is the predominant force leading to speciation over time, many studies now

18.22 Gradualism versus punctuated equilibrium

These drawings represent two hypotheses of how different species—here, the okapi and the giraffe—might have evolved from a common ancestor, the "pre-okapi." Forms intermediate between the pre-okapi and the okapi and giraffe are not shown. (A) In gradualism, the conventional interpretation of natural selection, a population is imagined to evolve slowly (as suggested by the gentle leftward movement of the lower part of the arrow, indicating a slow change in one trait) until a reproductive barrier (usually geographic isolation) separates two parts of the population. Generally one of the subgroups is very small, as indicated by the narrowness of the right-hand branch; it may diverge slowly from the larger subgroup through selection, genetic drift, or both. (B) According to the punctuated equilibrium model, there is little or no gradual change in a trait over time. Instead, a crisis or genetic event selects or gives rise by chance to one or more populations with traits very different from those of the ancestral population.

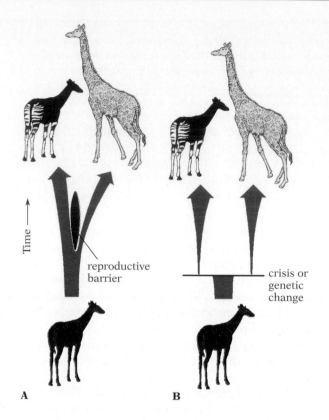

A   B

suggest that rare boom-or-bust events have been important in the evolution of at least certain populations.

Punctuated equilibrium One consequence of the increasing evidence that natural catastrophes can affect the allelic frequencies of populations has been a growing interest in ***punctuated equilibrium***. This much-debated hypothesis concerning the mode and tempo of speciation was originally formulated by Niles Eldredge and Stephen Jay Gould. Based on careful study of certain fossil records, their hypothesis suggests that most allopatric speciation events are the result of crises or major genetic alterations that punctuate long periods of equilibrium, or ***stasis***, in which the morphology of the species remains relatively constant. Gould and Eldredge maintain that the fossil record does not support ***gradualism***—the post-Darwinian view that speciation occurs as a gradual accumulation of morphological and physiological changes (Fig. 18.22). The changes they emphasize are *geologically* instantaneous.

"Instantaneous" speciation, however, is by no means as rapid as it sounds. The time scale of geological layering can be measured in millennia. Gould and Eldredge believe speciation events may have taken thousands of generations, up to 100,000 years, to complete. Perhaps most important, because the population that survives and evolves after a crisis would seem to arise suddenly in the fossil record, punctuated equilibrium provides an explanation of the gaps that exist in the record. The more gradual processes of change we are used to seeing in evolution are, from this point of view, of minor importance in speciation, and serve mostly to fine-tune a species to its environment.

The debate between punctuated equilibrium and the conventional perspective is fueled to a great extent by the many differences in the way the fossil record can be interpreted. In some species, for instance, certain morphological features may change rapidly in response to selection pressures while others may remain relatively constant. A researcher who examines or is only able to discern the characters that remain constant may overlook this *mosaic evolution* and assume that the species exhibits stasis—no change at all. On the other hand, supporters of gradualism often assume that slow morphological or physiological change is occurring even when it is not obvious from the record. Of course, neither physiological nor morphological changes in most soft body parts are preserved in fossils.

Another difference in interpretation concerns the great discontinuities in the record. The conventional view of these gaps is that they are, in most cases, just anomalies, and that the missing pieces in the picture of gradual evolutionary change might, if found, provide valuable information about transitions between the species preceding and following the gaps. The supporter of punctuated equilibrium, by contrast, regards the gaps as the norm, and as proof of the rapid speciation events. However, gaps in the record may frequently result from the kinds of environmental crises already discussed. Consider what happens when a lake, whose sediments have provided an excellent record of its organic inhabitants for hundreds of thousands of years, suddenly dries up as a result of a major climatic change. The formation of the fossil record in this location is terminated just at the time when the inhabitants of the lake are faced with extraordinary selection pressures and potential extinction.

The punctuated equilibrium hypothesis holds that besides crises, punctuational change could be generated by major genetic alterations—polyploidy, hybridization, translocations, and deletions—particularly if they affect developmental-control genes, which can change the expression of entire constellations of other genes. One widely cited example is the idea that vertebrates evolved from the larval form of an aquatic invertebrate (p. 704). A change in developmental control could have blocked transformation into the adult morphology; this would have been followed by an elaboration of the larval phase and the evolution of the first fish. (Another possible example, involving the evolution of arthropods, is discussed in the next section.) Of course, genetic anomalies face the same problems as polyploids: with only siblings to mate with (at best), inbreeding and genetic drift become serious problems and founding a healthy population becomes difficult.

The Cambrian explosion One of the most spectacular cases of rapid evolutionary radiation occurred in the geologic time division called the Cambrian period, about 550 million years ago. The "sudden" appearance of complex animals (older fossils are nearly all the remains of single-celled creatures) is a potentially compelling instance of punctuation. Two sites with well-preserved remains have been discovered, each about the size of a city block. The more extensively studied one is in the Burgess shale deposit in the mountains of western Canada; the other one is near Chengjiang in southern China. Because many of the fossils are similar, the animals probably had a worldwide distribution, and must have evolved over the course of a few million years. Until these discover-

A

B

C

D

E

F

18.23 Some species from the Burgess shale

The Burgess fauna includes annelids (segmented worms) like the polychaete worm shown here (A) and the oldest preserved chordate (B), whose notochord and somite-produced muscle bands are clearly visible. Most of the remaining species, of which only a small fraction are illustrated here (C–F), represent extinct groups of arthropods.

ies, we had no clear picture of the kind of diversity that existed before selection "pruned" the tree of life.

The animals that typify a given place or a particular period are known collectively as a ***fauna***. The Burgess fauna contains a group of polychaetes (a kind of segmented worm, whose physiology will be discussed in Part V) and the oldest known chordate (Fig. 18.23A,B), but mostly it consists of arthropods. These include some trilobites (an extinct group), a few uniramians (the group of invertebrates with mandibles—that is, chewing mouthparts—to which crustaceans and insects belong), and cheliceratans (the group that includes spiders, scorpions, and horseshoe crabs). However, none of these Burgess arthropods look like any modern species. In the same deposits there are at least 20 (and perhaps 30) other classes of arthropods, all now extinct (Fig. 18.23C–F).

The diversity of design is staggering, and poses two evolutionary problems: how did they come into being over such a relatively brief period, and why is it that 80–90% of these creatures have left no modern descendants, while other groups, with no obvious advantages, continue to thrive? We will deal with this second question in the next section. A potential answer to the first, favored by many biologists, is that the explosion in diversity resulted from a biological breakthrough: the invention of a novel developmental strategy. As described in Chapter 14, the embryonic development of animals involves the formation of a blastula, followed by gastrulation. The breakthrough came with the appearance of compartmentalization: the embryo divides along the anterior-posterior axis into a number of nearly identical units, each of which can differentiate to produce specialized organs and appendages. It is likely that the evolution of the first multicellular creatures was followed quickly by the evolution of this more efficient multisegmental plan.

The remarkable diversity that followed would have been a consequence of the absence at that point of the highly redundant and hierarchical system of morphogens that now exist to assure precision and stability during development; as a result, variation would have been generated at an extraordinary rate, while competition in an environment with nothing occupying the many niches available to multicellular organisms must have been relatively lax initially. As the number of different species began to saturate the environment, however, selection would necessarily become more intense, and less efficient designs and less stable developmental programs probably went extinct. This scenario, therefore, postulates a kind of punctuation, followed by more gradual evolution and selective pruning. As we will see, there is another school of thought—an alternative to selection—to explain how many species may go extinct.

Whatever the fate of punctuation theory, the evolution of organisms can be less uniform and gradual than many biologists have supposed. Indeed, Darwin himself rejected simple gradualism, pointing out that both local and global environmental phenomena must redefine and magnify selection pressures enormously and hence alter the direction and rate of evolution. The usual tempo of speciation probably lies somewhere between the gradual-change and the punctuated equilibrium models.

How important is chance? As we saw, in the Galápagos one species of finch underwent enormous evolutionary radiation into a variety of available lifestyles. The first group of birds to reach the islands might just as well have been some other species of finch or some other sort of bird altogether. Chance events led to the creation of the new habitat and the immigration of the first birds. Later avian arrivals failed to thrive because the habitat was already being efficiently exploited by the (now) well-adapted first arrivals.

On a larger scale, there have undeniably been massive bouts of extinction, probably resulting from chance worldwide calamities such as asteroid impacts or other catastrophic events. These random prunings of the tree of life have cleared away thousands of species at a time, leaving the survivors the opportunity to radiate and exploit novel lifestyles in the absence of competition. Perhaps this is what happened to most of the Burgess species. As described in Chapter 24, the enormous radiation of mammals after the Cretaceous-Tertiary boundary probably occurred because a global disaster eliminated the dinosaurs, which had dominated the terrestrial large-animal opportunities. In some sense, therefore, the evolution of humans depended on this accident, and subsequent crises—warmings, ice ages, the continental movement that opened the Rift Valley in Africa (where the earliest human fossils are found), and so on—may also have been critical to the history of our species. We may be where we are now because distant ancestors happened to be in the right place at the right time. If many large-scale evolutionary developments depend on chance, simple luck may often be more important than selection. Had the cards fallen differently, the dominant species of animals today might be something quite different.

■ PROBLEMS IN IDENTIFYING SPECIES

The modern definition of species cannot always be applied without difficulty. Though most biologists accept the major ideas on which the definition rests, and though most of them, if they were to study the same set of natural populations, would probably agree in the great majority of instances about which populations represent full species and which do not, there would be a small percentage of populations on which they could not agree. Let us examine a few such cases.

Asexuals Since the modern definition of species assumes interbreeding, it cannot apply to asexual organisms. Though most so-called asexual organisms actually do have provision for occasional sexual recombination, a few lack sexual mechanisms altogether. Dandelions, for instance, despite their flowers, are totally asexual (Fig. 18.24). Can such truly asexual organisms be said to form species in any sense?

Asexual organisms do seem to form recognizable groups or kinds, even though the members of a group cannot exchange genes. Gaps, or discontinuities in the variation, occur between the various kinds just as they do between sexual species. One possible explanation for the groupings is that each group of asexual organisms evolved from a sexual species. The flowers of dandelions, which now have infertile pollen and diploid eggs that go on to form seeds without fertilization, originated in sexually reproducing ancestors. Since not all the variations that occur over the course of time are likely to be equally well adapted, asexual organisms like dandelions would continue to form recognizable groups: only those individuals whose genotypes produced well-adapted phenotypes would survive in significant numbers. Hence there would be a limited number of types, and all individuals falling within the bounds of one such type would constitute a natural group that could be called a species. In the long run, the inability of asexual organisms to exploit the potential of genetic recombination makes them less able to survive the vicissitudes of environmental change. And, indeed, nearly all macroscopic multicellular asexual species are of recent evolutionary origin. Among short-lived asexuals, however, the combination of mutations and high reproductive rate

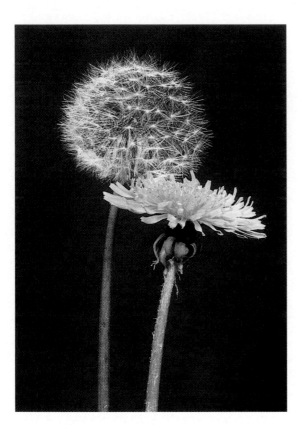

18.24 Flower and seeds of the dandelion, an asexual organism

Despite appearances, the dandelion reproduces asexually at all times. The flower (right) of the dandelion, which was inherited from a sexually reproducing precursor, now has infertile pollen and diploid eggs that go on to form seeds without fertilization.

seem to create sufficient variation for natural selection to operate on, and some of these species are extremely ancient.

Fossil species The modern definition of species can be applied formally only to organisms that coexist, since the criterion of interbreeding cannot be used when comparing an organism with its likely ancestors of a million years earlier. Therefore, paleontologists can compare organisms from different periods in the earth's history only by using morphological criteria and geographic distribution. Two forms can be classified as separate species when they differ to about the same degree as related organisms from the present day that are known to constitute reproductively isolated species. For practical purposes, paleontologists usually regard gaps in the fossil record as breaks between species, even though they are fully aware that no gaps actually occurred in the lineages of the organisms.

Populations versus species Geographically isolated populations can slowly diverge until they have reached the level of full species. The intrinsic reproductive isolation that makes them full species itself usually evolves gradually. Hence there is no precise point at which the diverging populations suddenly reach the level of full species. However, the definition of species makes no provision for such intermediate stages. Consequently, intermediate stages, when they are encountered, pose a problem to any biologist intent on rigid categorization of what is in nature a fluid system.

Allopatric species One of the most obvious and frequently encountered problems in applying the modern definition of species arises when two populations are closely related and completely allopatric. Since they are allopatric, they are obviously not exchanging genes, but lack of actual effective gene flow is not sufficient: there must be no *potential* gene flow for the two populations to be regarded as separate species. How can potential gene flow be determined? One way that immediately comes to mind is to release a large sample of individuals from one population in the range of the other, and then see whether free interbreeding takes place and, if it does, whether the hybrids are as viable as the parents.

The wholesale introduction of foreign plants and animals is seldom desirable or even legal. An alternative would be to bring individuals from the two allopatric populations together in the laboratory to see whether they will interbreed. If the individuals will breed freely with other members of their own population but not with members of the other population, then the two populations are probably intrinsically isolated and are therefore separate species.

Interbreeding between members of different populations in the laboratory, however, does not prove two groups are part of the same species. Instead it demonstrates that certain types of intrinsic isolation do not exist between the populations. It says nothing about other types of intrinsic isolation. For example, behavioral isolation may operate in nature but not in the laboratory. Many species of animals that will have nothing to do with each other in the wild, because of important differences in their behavior, will mate in the laboratory. Lions and tigers, for example, are distinct allopatric species in the wild, and never mate in their small region of range overlap; in the unnatural environment of a zoo, however, they mate and produce living offspring. Clearly, when members of two different allopatric populations mate in the laboratory and produce viable offspring, the question of whether they belong to the same or to different species remains unanswered. The same ambiguity exists for the many organisms that will not breed at all under laboratory conditions: after all, males and females of the *same* species often refuse to mate outside of their natural habitats.

In many cases, then, there is no good test for determining whether two allopatric populations belong to the same or to different species. Although character displacement among sympatric species adds a level of complication, the usual practice in such cases is to determine the extent to which the two populations differ, and then to compare this degree of difference with that seen in related sympatric species. If the differences between the allopatric populations are of the same order of magnitude as (or greater than) those that distinguish sympatric species, the allopatric populations are considered separate species. If the difference is less, the two allopatric populations are regarded as belonging to the same species.

THE CONCEPT OF PHYLOGENY

Evolution implies that many different species have a common ancestor and that all forms of life probably stem from the same remote beginnings. Hence one of the tasks evolution sets for biologists is to discover the relationships among the species alive today and to trace the ancestors from which they descended (Fig. 18.25). Once these relationships are understood, they are summa-

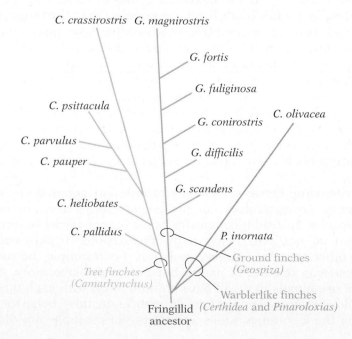

18.25 A phylogenetic tree for Darwin's finches

This tree is based on both the degree and the nature of the morphological differences between species. The distances along branches from any one species to another reflect inferred degrees of relatedness.

rized by grouping species into *taxa* (collections of related organisms, ranging from genera to kingdoms). There is lively disagreement about the best ways to determine ancestry and define groups.

■ DETERMINING PHYLOGENETIC RELATIONSHIPS

When systematists,[3] also known as taxonomists, set out to reconstruct the *phylogeny* (evolutionary history) of a group of species that they think are related, they have before them the species living today and the fossil record. To reconstruct a phylogenetic history as closely as possible, they must make inferences based on observational and experimental data. The difficulty is that what can be measured is *similarity*, whereas the goal is to determine *relatedness*. There are four major approaches to systematics—classical evolutionary taxonomy, phenetics, cladistics, and molecular taxonomy. Each uses different techniques to infer relatedness from similarity.

Classical evolutionary taxonomy Classical systematics depends more than any other approach on experience and judgment. The usual procedure in reconstructing phylogenies by the classical method is to examine as many independent characters of the species in question as possible, and to determine in which characters the species differ and in which they are alike. The assumption is that the differences and resemblances will reflect, at least in part, their true phylogenetic relationships. Ordinarily, as many different types of characters as possible are used in the hope that misleading data from any single character will be detected by a lack of agreement with the data from other characters.

The most easily studied and widely used characters are morphological—external morphology, internal anatomy and histology (tissue types), and the morphology of the chromosomes in cell nuclei. It is particularly helpful when morphological characters of living species can be compared with those of fossil forms. The characters chosen for comparison must be ones that vary within the group being analyzed; especially useful are characters that are unique to the group, and so have a common and relatively recent evolutionary origin. Among Darwin's finches (Fig. 18.17), for example, morphological characters that could also be checked in fossils include beak depth, ratio of beak depth to beak width, angle of the beak relative to the head, linear dimensions of the various bones that fuse to create the skull, relative areas of the skull bones, pattern of muscle insertions (visible as small indentations in bone) in the vocal apparatus, and so on. Characters that are usually lost in fossils but can be measured in live finches include tail length, markings of the young (degree of streaking, for instance), changes in markings with age, and the like. Of these, the wide range of beak sizes, the depressed angle of the beaks, and the relatively short tails set Darwin's finches apart from their cousins on the South American mainland.

---

[3] Systematics, or taxonomy, in the words of G. G. Simpson, is "the scientific study of the kinds and diversity of organisms and of any and all relationships among them."

Characters preserved in fossils are of special importance because the fossil record is the most direct source of evidence about the stages through which ancient forms of life passed. Unfortunately, that record is usually incomplete, and for many groups of organisms there is no suitable record at all. At best, fossils suggest the broad outlines of the evolution of major groups. In some groups, notably the horses, fossils have provided phylogenetic information that could have been obtained from no other source (Fig. 18.26).

Another frequently used source of information is embryology. Morphological characters are often easier to interpret if the manner in which they develop is known. For example, if a particular structure in organism A and a structure of quite different appearance in organism B both develop from the same embryonic primordium, the resemblances and differences between those structures in A and B provide information about the phylogeny of

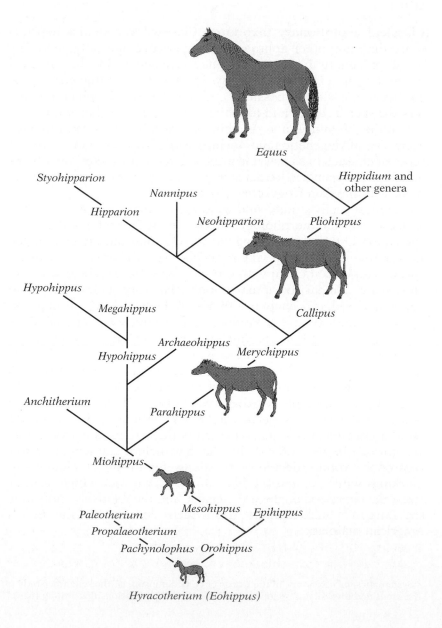

18.26 Presumed evolution of horses

The fairly complete fossil record of horses has enabled paleontologists to work out a reasonable picture of the evolutionary history of the group. The emphasis here is on the direct ancestors of the modern horse; many major branches left no modern descendants. *Hyracotherium* lived in the Eocene epoch about 55 million years ago. It was a small animal, weighing only a few kilograms. It had four toes on each front foot and three on each rear foot. It was a browser, feeding on trees and bushes. *Mesohippus,* which lived during the Oligocene epoch about 35 million years ago, was a bit larger, and its front feet, like the rear feet, had only three toes. *Merychippus,* a grazer, lived during the Miocene about 25 million years ago. It had three toes on each foot, the middle one much larger than the others, which were short and thin and did not reach the ground. *Pliohippus,* of the Pliocene, often had only one toe on each foot, though in some individuals tiny remnants of other toes persisted. *Equus,* the modern horse, is much larger than the ancestors shown here. It has only one toe on each foot.

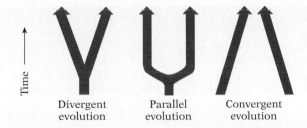

Time →

Divergent evolution   Parallel evolution   Convergent evolution

the organisms. The message would be different if they had developed from entirely separate embryonic structures. Embryological evidence often allows biologists to trace the probable evolutionary changes that have occurred in important structures, and helps them reconstruct the probable chain of evolutionary events that led to the modern forms of life. For example, the brief appearance of pharyngeal gill pouches during the early embryology of mammals, including humans, is thought to indicate that the distant ancestors of land vertebrates were aquatic.

Life histories have also played an important role in classical phylogenetic studies. The stages through which plants pass during their life cycles are particularly important sources of information.

The problem of convergence The classical approach, then, evaluates similarities in a range of characters. However, similarity by itself does not necessarily indicate common descent. A particular similarity might reflect similar adaptations to the same environmental situation. This is common in nature and is a serious source of confusion in phylogenetic studies.

When organisms that are not closely related become more similar in one or more characters because of independent adaptation to similar environmental situations, they are said to have undergone convergent evolution, and the phenomenon is called **convergence** (Fig. 18.27). Whales, which are mammals, have evolved flippers from the legs of their terrestrial ancestors; those flippers superficially resemble the fins of fish, but the resemblances result from convergence; the flippers conceal well-developed finger bones. The "moles" of Australia are not truly moles but marsupials (mammals whose young are born at an early stage of embryonic development and complete their development in a pouch in their mother's abdomen, rather than in a placenta inside the womb). They occupy the same habitat in Australia as do the true moles in other parts of the world and have, as a result, convergently evolved many startling similarities to the true moles. Many Australian marsupials are strikingly convergent with placental mammals of other continents (Fig. 18.28).

As a result, when classical systematists find similarities between two species, they must try to determine whether the similarities are probably **homologous** (inherited from a common ancestor) or merely **analogous** (similar in function and often in superficial structure, but of different evolutionary origins). Thus the wings of robins and those of bluebirds are considered homologous, since both were derived from the wings of a common avian ancestor. But the wings of robins and those of butterflies are only analogous;

18.27 Patterns of evolution

In divergent evolution one stock splits into two, which become less and less like each other as time passes. In parallel evolution two related species evolve in much the same way for a long period of time, probably in response to similar environmental selection pressures. Convergent evolution occurs when two species that are not closely related come to resemble each other more and more as time passes; this is usually the result of living in similar habitats and adopting similar lifestyles; as a result, they experience similar selection pressures.

A

B

C

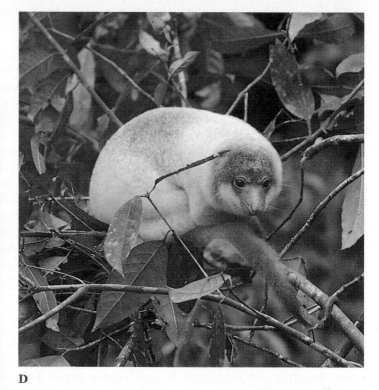

D

18.28   Marsupials that are convergent with placental mammals of other continents

(A) A marsupial mouse. (B) A marsupial glider, convergent with placental flying squirrels. (C) The tiger cat, a marsupial carnivore. (D) A cuscus, a marsupial monkey with a prehensile tail.

they evolved independently and from different ancestral structures.

Two structures can be homologous and analogous at the same time. For example, the wings of birds and of bats are not homologous as wings, for they evolved independently, but they contain homologous bones: both types of wings evolved from the forelimbs of ancient terrestrial vertebrates that were ancestors to both birds and mammals. In short, the wings of birds and bats are analogous as wings and homologous as forelimbs.

Phenetics and molecular techniques The degree of subjectivity evident in classical taxonomy has motivated many systematists to attempt to develop more objective methods. One, which later fell into disfavor, is phenetics. This approach to taxonomy used as many morphological characters as possible, weighted all characters equally, and ignored the issue of analogy versus homology. Since evidence from fossils, embryology, and behavior is difficult to quantify, it was generally not included. The expectation was that if enough characters were compared, the subjective judgements necessary for making relative weightings and identifying cases of analogy would be unnecessary; any errors would be canceled out, or would be swamped by the mass of other data. Cladistic analysis, described in the next section, has replaced simple phenetic analysis by applying an objective rule for excluding many potentially misleading traits.

However, phenetics is not dead: phenetic techniques have been revived now that molecular sequences of genes are available, and "molecular taxonomy" is enjoying enormous popularity. In essence, each nucleotide in a gene (or each amino acid in a protein) is treated as a trait without regard to analogy or homology. The potential number of these "traits" in a genome runs into the millions. Each species' characters (base sequences) are compared with those of each other species, generating a value for the degree of difference for each pairing. A branch diagram is then constructed that places each species at a distance from each other species that corresponds to the calculated difference values (Fig. 18.29, top). Most of the phylogenetic diagrams in Part IV are derived in this way using the sequence of nucleotides in ribosomal RNA.

The analysis is complicated by two problems: First, such a tree is "unrooted"—that is, there is no indication of which species are most like the ancestral form, or the order of branching points in the evolutionary tree. To determine each species' distance from the common ancestor, a suitable outgroup—a species not closely related to those being analyzed—must be added to the calculations to orient the tree correctly (Fig. 18.29, bottom). Second, since errors, uncertainties, and chance effects enter into the measurements and the evolutionary changes they attempt to quantify, the various branch lengths are never entirely consistent from pair to pair. Sometimes these inevitable anomalies are so small that they create no problem: only one branching diagram is consistent with the data, and discordant distances inferred from various pairings can simply be averaged. However, when many species are included and branch points are close together, ambiguities are possible.

In theory, the branch lengths reflect differences that have accu-

18.29 Construction of evolutionary trees from phenetic data on similarity

Species are compared in pairs, and dissimilarity values are obtained (top left). In the simplest case these values are used to construct a branching diagram (top right). In this example, since AB is 4 units greater than AC, the AB side of the BC branch must be 4 units longer than the AC side, which yields this unrooted tree. To locate the point of common ancestry of this group, an outgroup D must be included (bottom left). From the differences in path lengths to each species from the outgroup, the deepest branch among the species being analyzed can be determined (bottom right); in this case, it falls between species A and the branch point between B and C. (The branch point between the ABC group and the outgroup cannot be calculated on the basis of these data; it is placed here arbitrarily as a reminder that the outgroup and the species being analyzed also have a common ancestor.) As the text points out, the measurements usually are slightly inconsistent; you can see how this complicates matters by attempting to recompute this tree after substituting, for instance, a value of 13 for AB.

Sequence of amino acids in cytochrome c for **28** *organisms[a]*

Position		1	5	10	15	20	25	30	35	40	45	50
Mammals	Human, chimpanzee	GDVEKGKKI	FIMKCS	QCHT	VEKGGKHKT	GPNLHGLF	GRKT	GQAPGYSYTA				
	Rhesus monkey	GDVEKGKKI	FIMKCS	QCHT	VEKGGKHKT	GPNLHGLF	GRKT	GQAPGYSYTA				
	Horse	GDVEKGKKI	FVQKCA	QCHT	VEKGGKHKT	GPNLHGLF	GRKT	GQAPGFTYTD				
	Donkey	GDVEKGKKI	FVQKCA	QCHT	VEKGGKHKT	GPNLHGLF	GRKT	GQAPGFSYTD				
	Cow, pig, sheep	GDVEKGKKI	FVQKCA	QCHT	VEKGGKHKT	GPNLHGLF	GRKT	GQAPGFSYTD				
	Dog	GDVEKGKKI	FVQKCA	QCHT	VEKGGKHKT	GPNLHGLF	GRKT	GQAPGFSYTD				
	Rabbit	GDVEKGKKI	FVQKCA	QCHT	VEKGGKHKT	GPNLHGLF	GRKT	GQAVGFSYTD				
	California gray whale	GDVEKGKKI	FVQKCA	QCHT	VEKGGKHKT	GPNLHGLF	GRKT	GQAVGFSYTD				
	Great gray kangaroo	GDVEKGKKI	FVQKCA	QCHT	VEKGGKHKT	GPNLNGI F	GRKT	GQAPGFTYTD				
Other vertebrates	Chicken, turkey	GDIEKGKKI	FVQKCS	QCHT	VEKGGKHKT	GPNLHGLF	GRKT	GQAEGFSYTD				
	Pigeon	GDIEKGKKI	FVQKCS	QCHT	VEKGGKHKT	GPNLHGLF	GRKT	GQAEGFSYTD				
	Pekin duck	GDVEKGKKI	FVQKCS	QCHT	VEKGGKHKT	GPNLHGLF	GRKT	GQAEGFSYTD				
	Snapping turtle	GDVEKGKKI	FVQKCA	QCHT	VEKGGKHKT	GPNLNGL I	GRKT	GQAEGFSYTE				
	Rattlesnake	GDVEKGKKI	FTMKCS	QCHT	VEKGGKHKT	GPNLHGLF	GRKT	GQAVGYSYTA				
	Bullfrog	GDVEKGKKI	FVQKCA	QCHT	CEKGGKHKV	GPNLYGL I	GRKT	GQAAGFSYTD				
	Tuna	GDVAKGKKT	FVQKCA	QCHT	VENGGKHKV	GPNLWGLF	GRKT	GQAEGYSYTD				
	Dogfish shark	GDVEKGKKV	FVQKCA	QCHT	VENGGKHKT	GPNLS GLF	GRKT	GQAQGFSYTD				
Insects[b]	Tobacco hornworm moth	GNADNGKKI	FVQRCA	QCHT	VEAGGKHKV	GPNLHGFF	GRKT	GQAPGFSYSN				
	Fruit fly (*Drosophila*)	GDVEKGKKL	FVQRCA	QCHT	VEAGGKHKV	GPNLHGL I	GRKT	GQAAGFAYTN				
Fungi[b]	Baker's yeast	GSAKKGATL	FKTRCEL	CHT	VEKGGPHKV	GPNLHGI F	GRHS	GQAQGYSYTD				
	Red bread mold	GDSKKGANL	FKTRCAEC	HGEGGNLTQKI	GPALHGLF	GRKT	GS VDGYAYTD					
Plants[b]	Wheat	GNPDAGAKI	FKTKCA	QCHT	VDAGA	GHKQGPNLHGLF	GRQS	GTTAGYSYSA				
	Sunflower	GDPTTGAKI	FKTKCA	QCHT	VEKGA	GHKQGPNLNGLF	GRQS	GTTAGYSYSA				
	Castor bean	GDVKAGEKI	FKTKCA	QCHT	VEKGA	GHKQGPNLNGLF	GRQS	GTTAGYSYSA				
Number of different amino acids		1 3 5 5 4 1 3 3 4 1	4 3 2 1 3	3 1 1 2 3	3 4 2 3 4 2 1 4 1	1 2 1 5 1 3	2 1 1 3	2 1 3 3 6 1 2 3 1 2 4				

[a] Adapted from M. O. Dayhoff, ed., *Atlas of Protein Sequence and Structure* (Washington, D.C.: National Biomedical Research Foundation, 1972), vol. 5; and R. E. Dickerson, The structure and history of an ancient protein, *Sci. Am.*, April 1972, copyright © 1972 by Scientific American, Inc.; all rights reserved.

[b] In cytochrome *c* from insects, fungi, and plants, a few (4–8) amino acids are usually ahead of what is here labeled Position 1; these are omitted from the table.

HOW WE KNOW:
NUCLEIC ACIDS AND PROTEINS AS TAXONOMIC CHARACTERS

A mutation that changes a single base in a gene affects the gene product to varying degrees. At one extreme, it may not affect the gene product at all; if the change is in the third base of a codon, the new codon will often specify the same amino acid as the old one (see Table 9.1, p. 242), and thus be "silent." In most cases, however, the change will result in a codon for either a similar amino acid (for example, one hydrophobic amino acid instead of another, a neutral mutation) or a very different one. Less often, a change creates a termination codon, which will cause translation to end prematurely, or convert a termination codon into a coding sequence so that the transcription continues into the next gene. Base changes that survive are of two major types: silent or neutral mutations that do not significantly alter the activity of the gene product, and mutations that are fixed by selection because they improve the gene product. Distinguishing between silent/neutral and selectively advantageous changes is not easy.

Mutations that do not alter the meaning of the codon are probably nearly neutral, and most of them probably accumu-late randomly with time after two groups of organisms have diverged from a common ancestor. By comparing the extent of single-base changes of this type in the sequences of genes common to a wide range of species—genes for the enzymes of glycolysis, the citric-acid cycle, or the electron-transport chain, for example—we might get some measure of the time elapsed since two groups diverged. As a tool for dating evolutionary events, therefore, neutral single-base changes have great potential.

Another approach is to compare the amino acid sequences of the gene products. Consider, for example, cytochrome *c*, an essential component of the respiratory chain in mitochondria. The complete amino acid sequence of this enzyme has been worked out for a variety of organisms. The table above shows the sequence for some of the species so far examined, with the various functional groupings of amino acids indicated by a color code.

Perhaps the most obvious feature of this table is that cytochrome *c* is remarkably similar in all the species, even

```
        55      60      65      70      75      80      85      90      95      100   104

ANKNKGI IWGEDTLMEYLENPKKYIPGTKMIFVGIKKKEERADLIAYLKKATNE
ANKNKGI TWGEDTLMEYLENPKKYIPGTKMIFVGIKKKEERADLIAYLKKATNE
ANKNKGI TWKEETLMEYLENPKKYIPGTKMIFAGIKKKTEREDLIAYLKKATNE
ANKNKGI TWKEETLMEYLENPKKYIPGTKMIFAGIKKKTEREDLIAYLKKATNE
ANKNKGI TWGEETLMEYLENPKKYIPGTKMIFAGIKKKGEREDLIAYLKKATNE
ANKNKGI TWGEETLMEYLENPKKYIPGTKMIFAGIKKTGERADLIAYLKKATKE
ANKNKGI TWGEDTLMEYLENPKKYIPGTKMIFAGIKKKDERADLIAYLKKATNE
ANKNKGI IWGEDTLMEYLENPKKYIPGTKMIFAGIKKKGERADLIAYLKKATNE

ANKNKGI TWGEDTLMEYLENPKKYIPGTKMIFAGIKKKSERVDLIAYLKDATSK
ANKNKGI TWGEDTLMEYLENPKKYIPGTKMIFAGIKKKAERADLIAYLKQATAK
ANKNKGI TWGEDTLMEYLENPKKYIPGTKMIFAGIKKKSERADLIAYLKDATAK
ANKNKGI TWGEETLMEYLENPKKYIPGTKMIFAGIKKKAERADLIAYLKDATSK
ANKNKGI IWGDDTLMEYLENPKKYIPGTKMIFTGLSKKKERTNLIAYLKEKTAA
ANKNKGI TWGEDTLMEYLENPKKYIPGTKMIFAGIKKKGERQDLIAYLKSACSK
ANKSKGI VWNNDTLMEYLENPKKYIPGTKMIFAGIKKKGERQDLVAYLKSATS-
ANKSKGI TWQQETLRIYLENPKKYIPGTKMIFAGLKKKSERQDLIAYLKKTAAS

ANKAKGI TWQDDTLFEYLENPKKYIPGTKMVFAGLKKANERADLIAYLKQATK-
ANKAKGI TWQDDTLFEYLENPKKYIPGTKMIFAGLKKPNERGDLIAYLKSATK-

ANIKKNVLWDENNMSEYLTNPKKYIPGTKMAFGGLKKEKDRNDLITYLKKACE-
ANKQKGI TWDENTLFEYLENPKKYIPGTKMAFGGLKKDKDRNDIITFMKEATA-

ANKNKAVEWEENTLYDYLLNPKKYIPGTKMVFPGLKKPQDRADLIAYLKKATSS
ANKNMAVIWEENTLYDYLLNPKKYIPGTKMVFPGLKKPQERADLIAYLKTSTA-
ANKNMAVQWGENTLYDYLLNPKKYIPGTKMVFPGLKKPQDRADLIAYLKEATA-

1 1 2 5 2 3 2 6 1 6 4 3 2 2 5 3 1 1 3 1 1 1 1 1 1 1 1 1 1 1 1 1 1 3 1 5 1 2 2 1 6 9 2 1 7 2 2 2 2 2 2 1 6 4 3 5 4
```

Symbol	Amino acid
□	NONPOLAR
G	glycine
A	Alanine
V	Valine
L	Leucine
I	Isoleucine
M	Methionine
F	Phenylalanine
W	Tryptophan
P	Proline
□	POLAR
S	Serine
T	Threonine
C	Cysteine
Y	Tyrosine
N	Asparagine
Q	Glutamine
□	ACIDIC
D	Aspartic acid
E	Glutamic acid
□	BASIC
K	Lysine
R	Arginine
H	Histidine

G is printed in color, because glycine, despite its technically nonpolar R group, behaves like a polar amino acid.

though some of them have probably not had a common ancestor for more than 1 billion years. For example, all the cytochromes have the same amino acid sequence from positions 70 through 80. In fact, cytochrome *c* is an evolutionarily conservative protein; its amino acid sequence has changed at a considerably slower average rate (about 20 million years for a 1% change) than, for example, that of hemoglobin (5.8 million years) or fibrin (only 1 million years). The minimal change in cytochrome c suggests that only minor alterations can be tolerated if the enzyme is to continue functioning properly. Even at points along the chain where there are differences, the amino acids are often functionally similar (with one polar amino acid substituted for another, one nonpolar amino acid for another, and so on). Some of the alterations may be neutral, having essentially no effect on the activity of the gene product, while others may represent minor but adaptive species-specific modifications of the protein. Some, however, almost certainly indicate major changes in the gene product. Among these are substitution of a polar for a nonpolar amino acid, and alterations involving proline (which induces turns in polypeptide chains) or cysteine (which forms strong covalent bonds with other cysteines).

If we compare various species in the table with one another, we find that the number of differences in amino acids usually agrees reasonably well with the presumed evolutionary distances among the species. Thus the mammals differ less among themselves than any of them differ from the fish. Humans and chimpanzees do not differ at all; both differ by one amino acid from the rhesus monkey, by an average of 10.4 amino acids from the other mammals, 14.5 from the reptiles, 18 from the amphibian, and 22.5 from the fish. This is an accurate reflection of the generally accepted evolutionary age of vertebrate classes, with fish predating amphibia, which are more ancient than reptiles, which in turn gave rise to mammals.

Because cytochrome c is evolutionarily so conservative, its value as a taxonomic character is limited to studies of the relationships among evolutionarily distant organisms. More rapidly changing proteins, such as fibrin, may prove useful in cases of closely related species. In any comparison, reliable conclusions depend on using several different proteins; because of unusual selection pressures or an unusually large or small number of mutations in the gene, the rate of change in one particular protein may be too great or too small to be representative.

Species and their character traits:

1. A B C D
2. A B C D
3. A B C D
4. A B C D
5. A B C D
6. A   C D

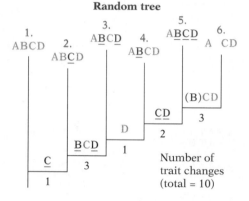

Random tree

Number of trait changes (total = 10)

Parsimonious tree

(total number of changes = 5)

18.30 An example of cladistic analysis

Six species in a group of organisms with a set of traits A–D are compared with one another to construct a phylogenetic tree. Each trait can exist in two forms, indicated by blue versus red letters. Cladistic analysis treats the absence of B in one species as a possible case of secondary loss; character A is ignored since it is common to all members. Cladistic analysis tries many different arrangements of the species to find the tree with a minimum number of changes. Two of the possible 954 trees are shown; one, selected at random, has 10 changes; the other, much simpler one, has five. The tree with the minimum number is used to infer the primitive state (ABCD); thus the loss of B is taken to be secondary.

mulated as a result of drift and selection. If only gradual drift were at work, the total distance from the deepest branch point to each present-day species would be the same, reflecting the time that has passed since the last common ancestor. Deviations—that is, longer paths—imply unusual degrees of selective pressure, and suggest something about the tempo and mode of evolution in the group.

Cladistics Cladists seek to avoid confusing analogy with homology by focusing on ***shared derived characters***—traits that are common to most but not all of the several species in question and thus probably represent relatively recent rather than ancient adaptation.[4] Hence, in comparing two species of bats, the traits shared by mammals in general would not be considered. The cladistic approach is helpful in providing a rule for ignoring the absence (secondary loss) of traits—the hind limbs of aquatic mammals, for example—that causes many species to stand out as obvious exceptions to the general taxonomic patterns of their group. Cladists weight each of the traits they consider equally, but by ignoring features presumed to be shared at the point at which the speciation event in question occurred, they can be more selective than pheneticists: they have a criterion, or rule, for choosing the traits to be included.

While phenetic approaches focus on degrees of difference—the evolutionary "distance" that translates into branch lengths—cladists concentrate on discrete differences, or character states. They use the number of trait differences between each pair of species under consideration to create the simplest, or most parsimonious, branching diagram (Fig. 18.30). There are some drawbacks to cladistics. For one thing, evolution need not occur by the simplest route. For another, cladistic analysis usually ignores fossils, embryology, and behavior. The result is sometimes controversial. For example, cladistic analysis places the crocodiles with the birds, and places mammals close to snakes and turtles (Fig. 18.31). In addition, like phenetics and classical taxonomy, cladistic analysis can be fooled by convergences, such as those exhibited by cer-

---

[4] Cladistics takes its name from the term *clade*, which refers to any monophyletic group—that is, a group with a single common ancestor. Examples are a genus, a class, a kingdom (all described at the end of this chapter), or all living things.

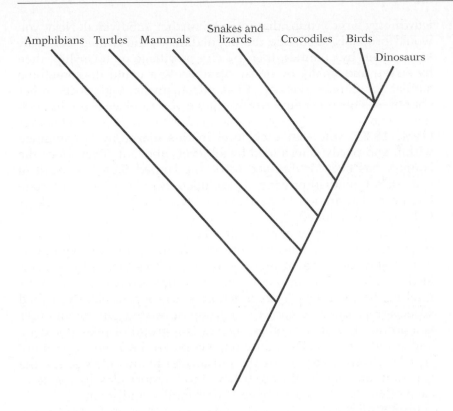

18.31 Cladistic analysis of the phylogeny of higher vertebrates

The major feature of this reconstruction is that reptiles do not belong to a single class. Depending on how a cladist chooses to group species, mammals, snakes, and turtles could form one class, with birds and crocodiles in the other; alternatively, there could be five different classes of reptiles, with birds and mammals in their own classes.

tain Australian marsupials. For analyzing the phylogeny of closely related genera or species, however, cladistics is now widely accepted. (Fig. 18.25, for instance, is based on cladistic techniques.) And cladistic analysis of molecular data is generally considered at least as reliable as phenetic calculations.

■ HOW ORGANISMS ARE CLASSIFIED

Over 1 million species of animals and over 325,000 species of plants are known (and many more are yet to be discovered). To deal with this vast array of organic diversity, we need a system by which species can be classified in a logical and meaningful manner. Many different kinds of classification are possible. We could, for example, classify flowering plants according to their color: all white-flowered species in one group, all red-flowered species in a second group, all yellow-flowered species in a third group, and so on. This sort of system has its usefulness in field guides, but the information such categories convey fails to set apart fundamentally different organisms. The classification system used in biology today attempts to encode the evolutionary history of the organisms; it is thus often a means of conveying information about many of their characteristics to those familiar with the various taxonomic groups (taxa) to which the organisms are assigned.

The classification hierarchy Suppose you had to classify all the people on earth on the basis of where they live. You would probably begin by dividing the world population into groups based on country. This subdivision separates inhabitants of the United States from the inhabitants of France or Argentina, but it still

Table 18.2 Classification hierarchies

BIOLOGICAL	POSTAL
Kingdom	Country
Phylum/division	State/province
Class	City
Order	Street
Family	Number
Genus	Last name
Species	First name

leaves very large groups that must be further subdivided. Next, you would probably subdivide the population of the United States by states, then by counties, then by city or village or township, then by street, and finally by house number. You could do something similar for Canada, Mexico, England, Australia, and all the other countries. This procedure would enable you to place every individual in an orderly system of hierarchically arranged categories (Table 18.2). Note that each level in this hierarchy is contained within and partly determined by all levels above it. Thus, once the country has been determined to be the United States, a Mexican state or a Canadian province is excluded. Similarly, once the state has been determined to be Pennsylvania, a city in New York or California is excluded.

The same principles apply to the classification of living things on the basis of phylogenetic relationships. A hierarchy of categories is used (Table 18.2). Each category (taxon) in this hierarchy is a collective unit containing one or more groups from the next-lower level in the hierarchy. Thus a genus is a group of closely related species (Fig. 18.32), a family is a group of related genera, an order is a group of related families, a class is a group of related orders, and so on. The species in any one genus are believed to be more closely related to each other than to species in any other genus, the genera in any one family are believed to be more closely related to each other than to genera in any other family, and so on.

Ideally, all the organisms in a taxon share a common ancestor—a progenitor not in the ancestry of any other taxa at the same level of the classification hierarchy. Pure taxa of this sort, which include the common ancestor and all its descendants, are **homophyletic** (Fig. 18.32A–C). Sometimes species are grouped into two taxa on the basis of similarities and differences in characters, thus separating species with a common ancestor and mixing species with different lineages; the classification of birds and mammals as taxa distinct from reptiles (see Fig. 18.31) is a clear example of this. When some but not all descendants of a common ancestor are grouped into a taxon, separating other descendants on the basis of dissimilarity, the classification is **paraphyletic** (Fig. 18.32D). When species are grouped into a taxon on the basis of similarities in such a way that their common ancestor is excluded, the taxon is **polyphyletic** (Fig. 18.32E). Two of the kingdoms discussed in Part IV (Archaebacteria and Archezoa) are probably paraphyletic, while another (Protista) is polyphyletic.

Table 18.3 gives the classification of six species. The table shows us immediately that the six species are not closely related, but that a human and a wolf are more closely related to each other than either is to the herring gull. It also shows us that the mammals and the bird are more closely related to each other than to the house fly, which is in a different phylum, or to moss or red oaks, which are in a different kingdom.

The current system dates from the work of Carolus Linnaeus (1707–1778). His system used kingdoms, classes, orders, genera, and species; the phylum and family categories were added later. The rationale on which Linnaeus based his system was very different from the phylogenetic one employed today. He worked a century before Darwin; he had no conception of evolution, thinking of each species as an immutable entity. He simply grouped organisms

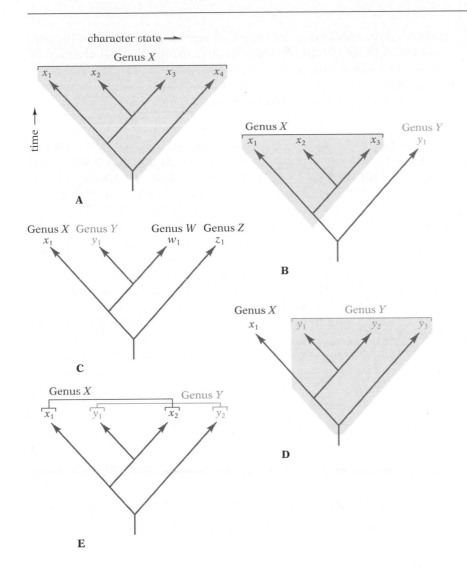

18.32 Alternative generic groupings for four related species

Biologists try to group species in a way that will indicate their phylogenetic relationships. Thus a genus is a group of related species. But how closely related? There is no absolute answer to this question. Some biologists (the "lumpers") like large genera containing many species (A); others (the "splitters") prefer small, compact genera containing only species that are very closely related (C). In the drawing, five alternative ways of grouping four related species are shown.

The first recognizes only one genus, the second recognizes two, and the third recognizes four. All are homophyletic: genera include all species sharing common descent. The fourth classification is paraphyletic: three very similar species, y_1, y_2, and y_3, are placed in the same genera, but species x_1, though closely related to y_1 and y_2, is so different that it is placed in a separate genera. The fifth classification is polyphyletic because it excludes the common ancestor of y_1 and y_2 from genus Y.

Table 18.3 Classification of six species

CATEGORY	HAIRCAP MOSS	RED OAK	HOUSE FLY	HERRING GULL	WOLF	HUMAN
Kingdom	Plantae	Plantae	Animalia	Animalia	Animalia	Animalia
Phylum or Division	Bryophyta	Tracheophyta	Arthropoda	Chordata	Chordata	Chordata
Class	Musci	Angiospermae	Insecta	Aves	Mammalia	Mammalia
Order	Bryales	Fagales	Diptera	Charadriiformes	Carnivora	Primata
Family	Polytrichaceae	Fagaceae	Muscidae	Laridae	Canidae	Hominidae
Genus[a]	*Polytrichum*	*Quercus*	*Musca*	*Larus*	*Canis*	*Homo*
Species[a]	*commune*	*rubra*	*domestica*	*argentatus*	*lupus*	*sapiens*

[a]The name of a particular species consists of the genus name and the species designation, both customarily in italics, as shown in this table.

according to similarities, primarily morphological. His results were similar to those obtained today because morphological characters, as products of evolution, usually reflect evolutionary relationships.

An outline of a modern classification of living things is given in the Appendix and is discussed in some detail in Part IV.

Naming species The modern system of naming species also dates from Linnaeus. Linnaeus gave each species a name consisting of two words: the name of the genus to which the species belongs and a designation for that particular species. Thus one species of carnation is *Dianthus caryophyllus*. Other species in the genus *Dianthus* have the same first word in their names, but each has its own specific designation (*Dianthus prolifer, Dianthus barbatus, Dianthus deltoides*). No two species can have the same name.[5] The names are always Latin (or Latinized), and the genus name is capitalized while the species name is not.[6] Both names are customarily printed in italics. The correct name for any species, according to the present rules, is usually the oldest validly proposed name.[7]

The same Latin scientific names are used throughout the world. This uniformity of usage ensures that each scientist will know exactly which species another scientist is discussing. Thus the plant *Bidens frondosa* can be unambiguously designated even though it is commonly known, just in English, as beggar-tick, sticktight, bur marigold, devil's bootjack, pitchfork weed, and rayless marigold.

[5] More precisely, no two species of plants can have the same name, and no two species of animals can have the same name.

[6] This rule always holds for zoological names, but specific botanical names are sometimes capitalized when they are derived from the name of a person or from other proper nouns.

[7] For purposes of priority, botanical naming dates from the publication of *Species Plantarum* by Linnaeus in 1753, and zoological naming dates from the publication of the 10th edition of his *Systema Naturae* in 1758.

CHAPTER SUMMARY

HOW SPECIATION OCCURS

WHAT ARE SPECIES? A species is a group of organisms that are capable of breeding among themselves and are reproductively isolated from other populations. Species generally exist as groups of local demes that may be somewhat different genetically from one another; systematic differences that vary with habitat are called clines. (p. 487)

THE PROCESS OF SPECIATION The reproductive isolation of two populations required for speciation can occur quickly through chromosomal changes, or slowly as habitat or geographic isolation keeps the populations separate while reproductive differences evolve. Many speciation events involve the isolation of small founder populations in which genetic drift is likely to play a major role. Reproductive isolation can involve genetic or habitat separation, or behavioral isolation based on mating habitat, times, or displays, or genetic or developmental incompatibilities. Speciation often involves adaptive radiation, as populations of a single species become specialized for particular ways of making a living. Evolution can be gradual when selection pressure is mild, or relatively rapid when selection is intense, drift is important, or major genetic or developmetal changes occur. (p. 492)

PROBLEMS IN IDENTIFYING SPECIES Asexuals form distinct groups even though they do not interbreed. Whether allopatric populations are distinct species often cannot be established for certain since they have no opportunity to interbreed. (p. 512)

THE CONCEPT OF PHYLOGENY

DETERMINING PHYLOGENETIC RELATIONSHIPS Phylogeny uses similarity to infer relatedness. A major problem lies in distinguishing evolutionary convergence from true kinship—analogous versus homologous characters. The classical approach depends on subjective weighting of different kinds of similarity. Molecular taxonomy relies on the large numbers of bases in genetic sequences to swamp out chance convergence. Cladistics focuses on identifying shared derived characters to exclude convergence. (p. 515)

HOW ORGANISMS ARE CLASSIFIED Taxonomists use a hierarchical system of kingdoms, phyla or divisions, classes, orders, families, genera, and species; the system reflects evolutionary relatedness rather than mere similarity. (p. 523)

STUDY QUESTIONS

1 One problem with molecular techniques is the issue of whether a given difference in base sequence between two species is neutral or adaptive. After all, there can be subtle differences created by the substitution of one nonpolar amino acid for another, and even third-codon changes can have an effect since they use different tRNAs, which may have different abundances, specificities, and activities. How might one go about attempting to resolve this issue for one specific gene product—an rRNA or a particular enzyme, for example? Feel free to imagine plausible advances in molecular techniques. (pp. 519–522)

2 The argument is sometimes made (almost always in the context of humans) that the concept of subspecies is useless because the variation within a subspecies for most characters exceeds the difference between the two subspecies. Is this line of reasoning correct? Does it work for species within a genus? Could it be used to prove that there are no "important" differences between males and females of a species? (pp. 490–491, 512–514)

3 How could you use one of the taxonomic techniques described in this chapter to determine which if any species in a genus were likely to have been produced through an evolutionary bottleneck? (pp. 507, 515–523)

4 For what sorts of species (including factors like size, generation time, reproductive strategy, habitat, and niche) is chance more likely to have been important than competition? What kinds of patterns would you expect to find through molecular-taxonomic analysis in such cases? (pp. 504–509, 519–522)

SUGGESTED READING

GRANT, P. R., 1991. Natural selection and Darwin's finches. *Scientific American* 265(4). *On the effects of a season of drought on the finch population in the Galápagos. A well-written and modern analysis of speciation.*

LEVINTON, J. S., 1992. The big bang of animal evolution. *Scientific American* 267 (5). *On the apparent explosion of diversity at the beginning of the Cambrian.*

LI, W. H., AND D. GRAUR, 1991. *Fundamentals of Molecular Evolution.* Sinauer, Sunderland, Mass. *Excellent treatment of how molecular techniques can be applied to phylogeny.*

RIDLEY, M., 1993. *Evolution.* Blackwell, Oxford. *Excellent text on the subject.*

SCHOENER, T. W., 1982. The controversy over interspecific competition, *American Scientist* 70, 586–595.

SIBLEY, C. G., AND J. E. AHLQUIST, 1986. Reconstructing bird phylogeny by comparing DNAs, *Scientific American* 254 (2). *On the DNA hybridization technique.*

SIMPSON, G. G., 1983. *Fossils and the History of Life.* Scientific American Books, New York. *Clearly written and well-illustrated exposition of evolution.*

STEBBINS, G. L., AND F. J. AYALA, 1985. The evolution of Darwinism, *Scientific American* 253 (1). *A modern summary of evolutionary theory, with particular attention to the claims for punctuated equilibrium.*

WILSON, A. C., 1985. The molecular basis of evolution, *Scientific American* 253 (4). *On methods for tracing evolution through similarities in nucleotide or amino acid sequences.*

APPENDIX:
A CLASSIFICATION OF LIVING THINGS

The classification given here is one of many in current use. Some other systems recognize more or fewer kingdoms and phyla (or divisions), and combine or divide classes in a variety of other ways. Chapters 19–25 discuss certain of the points at issue between advocates of different systems.

Botanists have traditionally used the term *division* for the major groups that zoologists have called phyla. A recent change in the botanical code allows the use of phyla for divisions, as we do here.

Most classes within a phylum are listed here, but where there is only one class, it is not named. For some classes—Insecta and Mammalia, for example—orders are given too. Except for a few extinct groups of particular evolutionary importance, like Placodermi, only groups with living representatives are included. A few of the better-known genera are mentioned as examples in most of the taxa.

Whenever possible, a (very) rough estimate of the number of living species is provided for higher taxa.

KINGDOM EUBACTERIA   (3000)*

PHYLUM PROTEOBACTERIA Purple bacteria
   CLASS PURPLE SULFUR BACTERIA. *Pseudomonas, Escherichia*
   CLASS RHODOPSEUDOMONAS. *Rhodopseudomonas, Rhizobium*
   CLASS PURPLE NONSULFUR. *Sphaerotilus, Alcaligenes*
   CLASS DESULFOVIBRIO. *Desulfovibrio*
   CLASS RICKETTSIAE. *Rickettsia, Coxiella*
   CLASS MYXOBACTERIA. *Myxococcus, Chondromycos*

PHYLUM PROCHLOROPHYTA. *Chloroxybacteria, Prochloron*

PHYLUM CYANOBACTERIA. *Oscillatoria, Nostoc, Gloeocapsa, Microcystis*

PHYLUM GREEN SULFUR BACTERIA. *Chlorobium*

PHYLUM SPIROCHETES. *Treponema, Spirochaeta*

PHYLUM GRAM-POSITIVE BACTERIA
   CLASS CLOSTRIDIA. *Bacillus, Clostridium, Mycoplasma, Staphylococcus, Streptococcus*
   CLASS ACTINOMYCES. *Micrococcus, Actinomyces, Streptomyces*
   CLASS MYCOPLASMA. *Mycoplasma, Acholeplasma*

KINGDOM ARCHAEBACTERIA   (200)

PHYLUM METHANOGENS. *Methanobacterium*

PHYLUM HALOPHILES. *Halobacterium*

PHYLUM THERMOACIDOPHILES. *Thermoplasma*

PHYLUM SULFUR REDUCERS.

KINGDOM ARCHEZOA   (400)†

PHYLUM METAMONADA. *Giardia, Spironucleus*

PHYLUM MICROSPORIDIA. *Vairimorpha, Nosema*

PHYLUM ARCHAEAMOEBA (or PELOBIONTA). *Pelomyxa, Mastigamoeba*

KINGDOM PROTISTA   (27,000)†

KINETOPLASTID SUBKINGDOM

TRICHOMONADS

PHYLUM PARABASALA. Trichomonads and hypermastigids. *Trichomonas, Trichonympha, Calonympha*

EUGLENOIDS

PHYLUM EUGLENOZOA (600)
   SUBPHYLUM KINETOPLASTIDA. Trypanosomes and bodonids. *Trypanosoma, Bodo*
   SUBPHYLUM EUGLENIDA. Euglenids. *Euglena, Eutreptia, Phacus, Colacium*

---

*There is no generally accepted classification for bacteria at the higher taxon level. One recent classification divides the bacteria into 17 distinct phyla, some without formal names. Another important classification assigns them to 19 "parts," most without formal names. This one, admittedly incomplete, is based for the most part on similarities in the sequences of ribosomal RNAs, as described in G. E. Fox et al., 1980, *Science* 209:457–463; and M. L. Sogin et al., 1989, *Science* 243:75–78.

†There are many alternative classification systems for protists and algae. The underlying uncertainty is reflected here by our use of colloquial groupings. The scheme used here, based primarily on rRNAs, is derived largely from Sogin et al. (1989); T Cavalier-Smith, 1989, *Nature* 339:100–101; and reviews for this edition by D. J. Patterson and T. Cavalier-Smith.

SARCODINE PROTISTS**

PHYLUM HETEROLOBOSEA (or PERCOLOZOA).
Amoeboflagellates. *Percolomonas, Acrasis*

PHYLUM RIZOPODA. Amoebae with filose pseudopodia.
Euglypha

PHYLUM AMOEBOZOA. Lobose amoebae. *Amoeba, Arcella, Entamoeba*

PHYLUM GRANULORETICULOSEA (or RETICULOSA).
Foraminiferans. *Globigerina, Textularia*

SUBKINGDOM ACTINOPODA

PHYLUM HELIOZOA. Heliozoans. *Actinosphaerium, Actinophrys*

PHYLUM POLYCYSTINEA (or RADIOZOA.) Radiolarians.
Aulacantha, Acanthometron

SUBKINGDOM AVEOLATA

PHYLUM DINOFLAGELLATA. Dinoflagellates (1000).
Gonyaulax, Gymnodinium, Ceratium, Oxyrrhis, Noctiluca

PHYLUM APICOMPLEXA. Spore-forming parasitic protists.
Plasmodium, Eimeria

PHYLUM CILIOPHORA. Ciliates (9000). *Paramecium, Stentor, Vorticella, Spirostomum*

AEROBIC ZOOFLAGELATE SUBKINGDOM

PHYLUM CHOANOFLAGELLATA (or CHOANOZOA).
Choanoflagellates. *Daphanoeca, Monosiga*

PHYLUM OPALOZOA

CLASS OPALINEA. **Opalinids.** *Opalina, Zelleriella*

CLASS PLASMODIOPHOREA. **Plasmodiophorids.** *Plasmodiophora, Woronina*

SUBKINGDOM MYCETOZOA: Slime Molds

PHYLUM MYCETOZOA. Slime molds (500)

CLASS MYXOGASTREA. **True slime molds.** *Physarum, Hemitrichia, Stemonitis*

CLASS DICTYOSTELEA. **Cellular slime molds.** *Dictyostelium, Polysphondylium*

CLASS PROTOSTELEA. **Protostelids.** *Protostelium*

KINGDOM CHROMISTA†

SUBKINGDOM CRYPTISTA

PHYLUM CRYPTISTA

CLASS CRYPTOMONADEA. **Cryptomonads.** *Cryptomonas, Chroomonas, Chilomonas, Hemiselmis*

CLASS GONIOMONADEA. *Goniomones*

SUBKINGDOM HETEROKONTA: The Chrysophyta

PHYLUM SAGENISTA

CLASS BICOSECIA. **Biocosoecids.** *Bicosoeca, Cafeteria*

CLASS LABYRINTHULEA. **Net slime molds.** *Labyrinthula*

PHYLUM PSEUDOFUNGI

CLASS OOMYCETES. **Water molds, white rusts, downy mildews (400).** *Saprolegnia, Phytophthora, Albugo*

CLASS HYPHOCHYTRIOMYCETES. **Hyphochytrids (25).** *Rhizidiomyces*

PHYLUM OCHRISTA

CLASS PHAEOPHYCEAE. **Brown algae (1500).** *Sargassum, Ectocarpus, Fucus, Laminaria*

CLASS CHRYSOPHYCEAE. **Golden-brown algae (650).** *Chrysamoeba, Chromulina, Synura, Mallomonas*

CLASS XANTHOPHYCEAE. **Yellow-green algae (360).** *Halosphaera, Tribonema*

CLASS EUSTIGMATOPHYCEAE. **Eustigmatophytes.** *Pleurochloris, Vischeria*

CLASS RAPHIDIOPHYCEAE. *Gonyostomum*

CLASS BACILLARIOPHYCEAE (OR DIATOMEAE). **Diatoms (10,000).** *Pinnularia, Pleurosigma, Navicula, Melosira*

SUBKINGDOM HAPTOPHYTA

PHYLUM HAPTOPHYTA. Haptophytes and coccolithophores.
Isochrysis, Prymnesium, Phaeocystis, Coccolithus, Emiliania

** This classification, though useful, has no phylogenetic validity.

† There are many alternative classification systems for protists and algae. The underlying uncertainty is reflected here by our use of colloquial groupings. The scheme used here, based primarily on rRNAs, is derived largely from Sogin et al. (1989); T Cavalier-Smith, 1989, *Nature* 339:100–101; and reviews for this edition by D. J. Patterson and T. Cavalier-Smith.

KINGDOM PLANTAE

SUBKINGDOM BILIPHYTA[†]

PHYLUM RHODOPHYTA. Red algae (4000). *Nemalion, Polysiphonia, Dasya, Chondrus, Batrachospermum*

PHYLUM GLAUCOPHYTA. Glaucophytes (13)

SUBKINGDOM VIRIDIPLANTAE[†]

SECTION CHLOROPHYTA

PHYLUM CHLOROPHYTA. Green algae (7000).
Chlamydomonas, Volvox, Ulothrix, Oedogonium, Ulva

SECTION STREPTOPHYTA

SUBSECTION CHAROPHYTA

PHYLUM CHAROPHYTA. Stoneworts (300). *Chara, Nitella, Tolypella, Spirogyra*

SUBSECTION EMBRYOPHYTA

PHYLUM BRYOPHYTA (23,600)

CLASS HEPATICAE. Liverworts. *Marchantia, Conocephalum, Riccia, Porella*

CLASS ANTHOCEROTAE. Hornworts. *Anthoceros*

CLASS MUSCI. Mosses. *Polytrichum, Sphagnum, Mnium*

PHYLUM TRACHEOPHYTA. Vascular plants

SUBPHYLUM PSILOPSIDA. *Psilotum, Tmesipteris*

SUBPHYLUM LYCOPSIDA. Club mosses (1500). *Lycopodium, Phylloglossum, Selaginella, Isoetes, Stylites*

SUBPHYLUM SPHENOPSIDA. Horsetails (25). *Equisetum*

SUBPHYLUM PTEROPSIDA. Ferns (10,000). *Polypodium, Osmunda, Dryopteris, Botrychium, Pteridium*

SUBPHYLUM SPERMOPSIDA. Seed plants

CLASS PTERIDOSPERMAE. Seed ferns. No living representatives

CLASS CYCADAE. Cycads (100). *Zamia*

CLASS GINKGOAE (1). *Gingko*

CLASS CONIFERAE. Conifers (500). *Pinus, Tsuga, Taxus, Sequoia*

CLASS GNETEAE (70). *Gnetum, Ephedra, Welwitschia*

CLASS ANGIOSPERMAE. Flowering plants

SUBCLASS DICOTYLEDONEAE. Dicots (225,000). *Magnolia, Quercus, Acer, Pisum, Taraxacum, Rosa, Chrysanthemum, Aster, Primula, Ligustrum, Ranunculus*

SUBCLASS MONOCOTYLEDONEAE. Monocots (50,000). *Lilium, Tulipa, Poa, Elymus, Triticum, Zea, Ophyrys, Yucca, Sabal*

† There are many alternative classification systems for protists and algae. The underlying uncertainty is reflected here by our use of colloquial groupings. The scheme used here, based primarily on rRNAs, is derived largely from Sogin et al. (1989); T Cavalier-Smith, 1989, *Nature* 339:100–101; and reviews for this edition by D. J. Patterson and T. Cavalier-Smith.

KINGDOM FUNGI

PHYLUM CHYTRIDIOMYCOTA. Chytrids (1000). *Olpidium, Rhizophydium, Diplophylctis, Cladochytrium*

PHYLUM ZYGOMYCOTA. Conjugation fungi (250)

CLASS ZYGOMYCETES. *Rhizopus, Mucor, Phycomyces, Choanephora, Entomophthora*

CLASS TRICHOMYCETES. *Stachylina*

PHYLUM ASCOMYCOTA. Sac fungi (12,000)

CLASS HEMIASCOMYCETES. Yeasts and their relatives. *Saccharomyces, Schizosaccharomyces, Endomyces, Eremascus, Taphrina*

CLASS PLECTOMYCETES. Powdery mildews, fruit molds, etc. *Erysiphe, Podosphaera, Aspergillus, Penicillium, Ceratocystis*

CLASS PYRENOMYCETES. *Sordaria, Neurospora, Chaetomium, Xylaria, Hypoxylon*

CLASS DISCOMYCETES. *Sclerotinia, Trichoscyphella, Rhytisma, Xanthoria, Pyronema*

CLASS LABOULBENIOMYCETES. *Herpomyces, Laboulbenia*

CLASS LOCULOASCOMYCETES. *Cochliobolus, Pyrenophora, Leptosphaeria, Pleospora*

PHYLUM BASIDIOMYCOTA. Club fungi (15,000)

CLASS HETEROBASIDIOMYCETES. Rusts and smuts. *Ustilago, Urocystis, Puccinia, Phragmidium, Melampsora*

CLASS HOMOBASIDIOMYCETES. Toadstools, bracket fungi, mushrooms, puffballs, stinkhorns, etc. *Coprinus, Marasmius, Amanita, Agaricus, Lycoperdon, Phallus*

KINGDOM ANIMALIA[††]

SUBKINGDOM PARAZOA

PHYLUM PORIFERA. Sponges (5000)

CLASS CALCAREA. Calcareous (Chalky) sponges. *Scypha, Leucosolenia, Sycon, Grantia*

CLASS HEXACTINELLIDA. Glass sponges. *Euplectella, Hyalonema, Monoraphis*

CLASS DEMOSPONGIAE. *Spongilla, Euspongia, Axinella*

CLASS SCLEROSPONGIAE. Coralline sponges. *Ceratoporella, Stromatospongia*

SUBKINGDOM EUMETAZOA

SECTION RADIATA

PHYLUM CNIDARIA (or COELENTERATA)

CLASS HYDROZOA. Hydrozoans (3700). *Hydra, Obelia, Gonionemus, Physalia*

CLASS CUBOZOA. Sea wasps (20). *Tripedalia*

CLASS SCYPHOZOA. Jellyfish (200). *Aurelia, Pelagia, Cyanea*

CLASS ANTHOZOA. Sea anemones and corals (6100). *Metridium, Pennatula, Gorgonia, Astrangia*

†† The classification of Animalia has been adapted from *Synopsis and Classification of Living Organisms,* 1982, ed. S. P. Parker, McGraw-Hill, New York.

PHYLUM PLACOZOA@ (1). *Trichoplax*

PHYLUM CTENOPHORA. Comb jellies (90). *Pleurobrachia, Haeckelia*

SECTION MYXOZOA

PHYLUM MYXOZOA. Myxosporians (800). *Myxobolus, Myxidium, Ceratomyxa*

SECTION PROTOSTOMIA

PHYLUM PLATYHELMINTHES. Flatworms (10,000)

CLASS TURBELLARIA. **Free-living flatworms.** *Planaria, Dugesia, Leptoplana*

CLASS TREMATODA. **Flukes.** *Fasciola, Schistosoma, Prosthogonimus*

CLASS CESTODA. **Tapeworms.** *Taenia, Dipylidium, Mesocestoides*

PHYLUM GNATHOSTOMULIDA (100). *Gnathostomula, Haplognathia*

PHYLUM NEMERTEA (or RHYNCHOCOELA). Proboscis or ribbon worms (650)

CLASS ANOPLA. *Tubulanus, Cerebratulus*

CLASS ENOPLA. *Amphiporus, Prostoma, Malacobdella*

PHYLUM ACANTHOCEPHALA. Spiny-headed worms (500). *Echinorhynchus, Gigantorhynchus*

PHYLUM MESOZOA (50)

CLASS RHOMBOZOA. *Dicyema, Pseudicyema, Conocyema*

CLASS ORTHONECTIDA. *Rhopalura*

PHYLUM ROTIFERA.∞ Rotifers (1700). *Asplanchna, Hydatina, Rotaria*

PHYLUM GASTROTRICHA∞ (2000). *Chaetonotus, Macrodasys*

PHYLUM KINORHYNCHA∞ **(or ECHINODERA)** (100). *Echinoderes, Semnoderes*

PHYLUM NEMATA.∞ Roundworms or nematodes (12,000). *Ascaris, Trichinella, Necator, Enterobius, Ancylostoma, Heterodera*

PHYLUM NEMATOMORPHA.∞ Horsehair worms (230). *Gordius, Paragordius, Nectonema*

PHYLUM CHAETOGNATHA.# Arrow worms (70). *Sagitta, Spadella*

PHYLUM ENTOPROCTA (150). *Urnatella, Loxosoma, Pedicellina*

PHYLUM LORICIFERA (3, newly discovered and probably very common)

PHYLUM PRIAPULIDA (8). *Priapulus, Halicryptus*

PHYLUM BRYOZOA† **(or ECTOPROCTA).** Bryozoans, moss animals (4000)

CLASS GYMNOLAEMATA. *Paludicella, Bugula*

CLASS PHYLACTOLAEMATA. *Plumatella, Pectinatella*

CLASS STENOLAEMATA

PHYLUM PHORONIDA† (10). *Phoronis, Phoronopsis*

PHYLUM BRACHIOPODA.† Lamp shells (300)

CLASS INARTICULATA. *Lingula, Glottidia, Discina*

CLASS ARTICULATA. *Magellania, Neothyris, Terebratula*

PHYLUM MOLLUSCA. Molluscs

CLASS CAUDOFOVEATA (70). *Chaetoderma*

CLASS SOLENOGASTRES. **Solenogasters** (180). *Neomenia, Proneomenia*

CLASS POLYPLACOPHORA. **Chitons** (600). *Chaetopleura, Ischnochiton, Lepidochiton, Amicula*

CLASS MONOPLACOPHORA (8). *Neopilina*

CLASS GASTROPODA. **Snails and their allies (univalve molluscs)** (25,000). *Helix, Busycon, Crepidula, Haliotis, Littorina, Doris, Limax*

CLASS SCAPHOPODA. **Tusk shells** (350). *Dentalium, Cadulus*

CLASS BIVALVIA. **Bivalve molluscs** (7500). *Mytilus, Ostrea, Pecten, Mercenaria, Teredo, Tagelus, Unio, Anodonta*

CLASS CEPHALOPODA. **Squids, octopuses, etc.** (600). *Loligo, Octopus, Nautilus*

PHYLUM POGONOPHORA. Beard worms (100). *Siboglinum, Lamellisabella, Oligobrachia, Polybrachia*

PHYLUM SIPUNCULA (300). *Sipunculus, Phascolosoma, Dendrostomum*

PHYLUM ECHIURA (140)

CLASS ECHIUROINEA. *Echiurus, Ikedella*

CLASS XENOPNEUSTA. *Urechis*

CLASS HETEROMYOTA. *Crisia, Tubulipora*

PHYLUM ANNELIDA. Segmented worms

CLASS POLYCHAETA (INCLUDING ARCHIANNELIDA). **Sandworms, tubeworms, etc.** (8000). *Nereis, Chaetopterus, Aphrodite, Diopatra, Arenicola, Hydroides, Sabella*

CLASS OLIGOCHAETA. **Earthworms and many freshwater annelids** (3100). *Tubifex, Enchytraeus, Lumbricus, Dendrobaena*

CLASS HIRUDINOIDEA. **Leeches** (500). *Trachelobdella, Hirudo, Macrobdella, Haemadipsa*

PHYLUM ONYCHOPHORA (65). Peripatus. *Peripatopsis*

PHYLUM TARDIGRADA. Water bears (300). *Echiniscus, Macrobiotus*

PHYLUM ARTHROPODA (at least 2 million)

SUBPHYLUM TRILOBITA. No living representatives

SUBPHYLUM CHELICERATA

CLASS EURYPTERIDA. No living representatives

CLASS MEROSTOMATA. Horseshoe crabs (4). *Limulus*

@ Formerly considered a separate subkingdom.

∞ Formerly considered a class of Phylum Aschelminthes.

Formerly considered a phylum in the section Deuterostoma.

† Bryozoa, Phoronida, and Brachiopoda are often referred to as the lophophorate phyla.

CLASS ARACHNIDA. Spiders, ticks, mites, scorpions, whipscorpions, daddy longlegs, etc. (55,000; at least 500,000 undiscovered species of mites are thought to exist). *Archaearanea, Latrodectus, Argiope, Centruroides, Chelifer, Mastigoproctus, Phalangium, Ixodes*

CLASS PYCNOGONIDA. Sea spiders (1000). *Nymphon, Ascorhynchus*

SUBPHYLUM CRUSTACEA (26,000). *Homarus, Cancer, Daphnia, Artemia, Cyclops, Balanus, Porcellio*

SUBPHYLUM UNIRAMIA

CLASS CHILOPODA. Centipedes (2500). *Scolopendra, Lithobius, Scutigera*

CLASS DIPLOPODA. Millipedes (10,000; another 50,000 species thought to exist). *Narceus, Apheloria, Polydesmus, Julus, Glomeris*

CLASS PAUROPODA (500). *Pauropus*

CLASS SYMPHYLA (160). *Scutigerella*

CLASS INSECTA. Insects (900,000; another 1 million species are thought to exist)

ORDER THYSANURA. **Bristletails, silverfish, firebrats.** *Machilis, Lepisma, Thermobia*

ORDER EPHEMERIDA. **Mayflies.** *Hexagenia, Callibaetis, Ephemerella*

ORDER ODONATA. **Dragonflies, damselflies.** *Archilestes, Lestes, Aeshna, Gomphus*

ORDER ORTHOPTERA. **Grasshoppers, crickets, etc.** *Schistocerca, Romalea, Nemobius, Megaphasma*

ORDER PHASMATOPTERA. **Walking sticks.** *Phyllium*

ORDER BLATTARIA. **Cockroaches.** *Blatta, Periplaneta*

ORDER MANTODEA. **Mantids.** *Mantis*

ORDER GRYLLOBLATTARIA. *Grylloblatta*

ORDER ISOPTERA. **Termites.** *Reticulitermes, Kalotermes, Zootermopsis, Nasutitermes*

ORDER DERMAPTERA. **Earwigs.** *Labia, Forficula, Prolabia*

ORDER EMBIIDINA (OR EMBIARIA OR EMBIOPTERA). *Oligotorna, Anisembia, Gynembia*

ORDER PLECOPTERA. **Stoneflies.** *Isoperla, Taeniopteryx, Capnia, Perla*

ORDER ZORAPTERA. *Zorotypus*

ORDER PSOCOPTERA. **Book lice.** *Ectopsocus, Liposcelis, Trogium*

ORDER MALLOPHAGA. **Chewing lice.** *Cuclotogaster, Menacanthus, Menopon, Trichodectes*

ORDER ANOPLURA. **Sucking lice.** *Pediculus, Phthirius, Haematopinus*

ORDER THYSANOPTERA. **Thrips.** *Heliothrips, Frankliniella, Hercothrips*

ORDER HEMIPTERA. **True bugs.** *Belostoma, Lygaeus, Notonecta, Cimex, Lygus, Oncopeltus*

ORDER HOMOPTERA. **Cicadas, aphids, leafhoppers, scale insects, etc.** *Magicicada, Circulifer, Psylla, Aphis, Saissetia*

ORDER NEUROPTERA. **Dobsonflies, alderflies, lacewings, mantispids, snakeflies, etc.** *Corydalus, Hemerobius, Chrysopa, Mantispa, Agulla*

ORDER COLEOPTERA. **Beetles, weevils.** *Copris, Phyllophaga, Harpalus, Scolytus, Melanotus, Cicindela, Dermestes, Photinus, Coccinella, Tenebrio, Anthonomus, Conotrachelus*

ORDER HYMENOPTERA. **Wasps, bees, ants, sawflies.** *Cimbex, Vespa, Glypta, Scolia, Bembix, Formica Bombus, Apis*

ORDER STREPSIPTERA. **Endoparasites**

ORDER MECOPTERA. **Scorpionflies.** *Panorpa, Boreus, Bittacus*

ORDER SIPHONAPTERA. **Fleas.** *Pulex, Nosopsyllus, Xenopsylla, Ctenocephalides*

ORDER DIPTERA. **True flies, mosquitoes.** *Aedes, Asilus, Sarcophaga, Anthomyia, Musca, Chironomus, Tabanus, Tipula, Drosophila*

ORDER TRICHOPTERA. **Caddisflies.** *Limnephilus, Rhyacophilia, Hydropsyche*

ORDER LEPIDOPTERA. **Moths, butterflies.** *Tinea, Pyrausta, Malacosoma, Sphinx, Samia, Bombyx, Heliothis, Papilio, Lycaena*

SECTION DEUTEROSTOMIA

PHYLUM ECHINODERMATA

SUBPHYLUM CRINOZOA

CLASS CRINOIDEA. **Crinoids, sea lilies (630).** *Antedon, Ptilocrinus, Comactinia*

SUBPHYLUM ASTEROZOA

CLASS STELLEROIDEA. **Sea stars, brittle stars (2600).** *Asterias, Ctenodiscus, Luidia, Oreaster, Asteronyx, Amphioplus, Ophiothrix, Ophioderma, Ophiura*

SUBPHYLUM ECHINOZOA

CLASS ECHINOIDEA. **Sea urchins, sand dollars, heart urchins (860).** *Cidaris, Arbacia, Strongylocentrotus, Echinanthus, Echinarachnius, Moira*

CLASS HOLOTHUROIDEA. **Sea cucumbers (900).** *Cucumaria, Thyone, Caudina, Synapa*

PHYLUM HEMICHORDATA (90)

CLASS ENTEROPNEUSTA. **Acorn worms.** *Saccoglossus, Balanoglossus, Glossobalanus*

CLASS PTEROBRANCHIA. *Rhabdopleura, Cephalodiscus*

CLASS PLANCTOSPHAEROIDEA

PHYLUM CHORDATA. Chordates

SUBPHYLUM TUNICATA (or UROCHORDATA). Tunicates (2000)

CLASS ASCIDIACEA. **Ascidians or sea squirts.** *Ciona, Clavelina, Molgula, Perophora*

CLASS THALIACEA. *Pyrosoma, Salpa, Doliolum*

CLASS APPENDICULARIA. *Appendicularia, Oikopleura, Fritillaria*

SUBPHYLUM CEPHALOCHORDATA. Lancelets, amphioxus (30). *Branchiostoma, Asymmetron*

SUBPHYLUM VERTEBRATA. Vertebrates

CLASS AGNATHA. **Jawless fish (50).** *Cephalaspis,[*] Pteraspis,[*] Petromyzon, Entosphenus, Myxine, Eptatretus*

CLASS ACANTHODII. **No living representatives**

CLASS PLACODERMI. **No living representatives**

CLASS CHONDRICHTHYES. **Cartilaginous fish, including sharks and rays (800).** *Squalus, Hyporion, Raja, Chimaera*

CLASS OSTEICHTHYES. **Bony fish (18,000)**

SUBCLASS SARCOPTERYGII

ORDER CERATODIFORMES. **Australian lungfish.** *Neoceratodus*

ORDER LEPIDOSIRENIFORMES. **Lungfish.** *Protopterus, Lepidosiren*

SUBCLASS ACTINOPTERYGII. **Ray-finned fish.** *Amia, Cyprinus, Gadus, Perca, Salmo*

CLASS AMPHIBIA (3100)

ORDER ANURA. **Frogs and toads.** *Rana, Hyla, Bufo*

ORDER CAUDATA (or URODELA). *Salamanders, Necturus, Triturus,*

[*] Extinct.

Plethodon, Ambystoma

ORDER GYMNOPHIONA(or APODA). *Ichthyophis, Typhlonectes*

Class Reptilia (6500)

ORDER TESTUDINES. **Turtles.** *Chelydra, Kinosternon, Clemmys, Terrapene*

ORDER RHYNCHOCEPHALIA. *Tuatara, Sphenodon*

ORDER CROCODYLIA. **Crocodiles and alligators.** *Crocodylus, Alligator*

ORDER LEPIDOSAURIA. **Snakes and lizards.** *Iguana, Anolis, Sceloporus, Phrynosoma, Natrix, Elaphe, Coluber, Thamnophis, Crotalus*

Class Aves. Birds (8600). *Anas, Larus, Columba, Gallus, Turdus, Dendroica, Sturnus, Passer, Melospiza*

Class Mammalia. Mammals (4100)

SUBCLASS PROTOTHERIA

ORDER MONOTREMATA. **Egg-laying mammals.** *Ornithorhynchus, Tachyglossus*

SUBCLASS THERIA. Marsupial and placental mammals

ORDER METATHERIA (or MARSUPIALIA). **Marsupials.** *Didelphis, Sarcophilus, Notoryctes, Macropus*

ORDER INSECTIVORA. **Insectivores (moles, shrews, etc.).** *Scalopus, Sorex, Erinaceus*

ORDER DERMOPTERA. **Flying lemurs.** *Galeopithecus*

ORDER CHIROPTERA.** *Bats, Myotis, Eptesicus, Desmodus*

** Some authorities propose dividing the echo-locating bats and fruit-eating bats into separate orders.

ORDER PRIMATA. **Lemurs, monkeys, apes, humans.** *Lemur, Tarsius, Cebus, Macacus, Cynocephalus, Pongo, Pan, Homo*

ORDER EDENTATA. **Sloths, anteaters, armadillos.** *Bradypus, Myrmecophagus, Dasypus*

ORDER PHOLIDOTA. **Pangolin.** *Manis*

ORDER LAGOMORPHA. **Rabbits, hares, pikas.** *Ochotona, Lepus, Sylvilagus, Oryctolagus*

ORDER RODENTIA. **Rodents.** *Sciurus, Marmota, Dipodomys, Microtus, Peromyscus, Rattus, Mus, Erethizon, Castor*

ORDER ODONTOCETA. **Toothed whales, dolphins, porpoises.** *Delphinus, Phocaena, Monodon*

ORDER MYSTICETA. **Baleen whales.** *Balaena*

ORDER CARNIVORA. **Carnivores.** *Canis, Procyon, Ursus, Mustela, Mephitis, Felis, Hyaena, Eumetopias*

ORDER TUBULIDENTATA. **Aardvark.** *Orycteropus*

ORDER PROBOSCIDEA. **Elephants.** *Elephas, Loxodonta*

ORDER HYRACOIEDEA. **Hyraxes, conies.** *Procavia*

ORDER SIRENIA. **Manatees.** *Trichechus, Halicore*

ORDER PERISSODACTYLA. **Odd-toed ungulates.** *Equus, Tapirella, Tapirus, Rhinoceros*

ORDER ARTIODACTYLA. **Even-toed ungulates.** *Pecari, Sus, Hippopotamus, Camelus, Cervus, Odocoileus, Giraffa, Bison, Ovis, Bos*

CREDITS

1.1 Uniphoto Picture Agency, London. 1.3 Reunion des Musees Nationaux, Paris 1.5 National Portrait Gallery, London. 1.6 Rijksmuseum, Amsterdam. 1.7 Photo No. #326662. Courtesy Department Library Services, American Museum of Natural History. 1.8 (a) Sovfoto, New York. (b) F. Gohier, Photo Researchers, Inc. 1.9 Cop. Bibliotheque centrale M. N. H. N. Paris. 1.11 The New York Public Library, Astor, Tilden, and Lenox Foundations. 1.12 Tim Davis, Photo Researchers, Inc. 1.13 ©Lennart Nilsson, *Behold Man*, Little, Brown, and Company. 1.14 Modified from The Illustrated Origin of Species, by Charles Darwin, abridged and introduced by Richard E. Leakey, 1979; courtesy of Hill and Wang, a division of Farrar, Straus & Giroux, Inc. 1.15 Pages 36–37 of Darwin notebook "B," by permission of the Syndics of Cambridge University. 1.16 National Portrait Gallery, London. 1.17 (center) Kenneth W. Fink, Bruce Coleman, Inc. (all others) Louise B. Van der Meid. 1.18 Clockwise from left: (archaebacterium) H. W. Jannasch and C. O. Wirsen, *Bio Science*, vol. 29, 1979; copyright ©1979 by the American Institute of Biological Science; (archaezoan) Courtesy E. W. Daniels; (protist) © M. I. Walker, Science Source, Photo Researchers, Inc.; (fly agaric) © G. R. Roberts; (great egret) © M. P. Kahl, 1972, Photo Researchers, Inc.; (dahlias) Gene Ahrens, Bruce Coleman, Inc.; (kelp) © Jeff Rotman; (a true bacterium) CNRI, Photo Science Library, Photo Researchers, Inc. 1.19 Courtesy of Cold Spring Harbor Laboratory.

PART I: THE CHEMICAL AND CELLULAR BASIS OF LIFE

Powerhouse of the Cell: Bill Longcore/Photo Researchers, Inc.

2.1 NASA 2.3 Science Photo Library/Photo Researchers, Inc. 2.4 R. M. Feenstra and J. A. Stroscio, IBM Research 2.10 © David Newman/Visuals Unlimited 2.11a © Dwight Kuhn 2.21 Photo by Ann F. Purcell 2.26 Herman Eisenbeiss/Photo Researchers, Inc. 2.27 © Dwight Kuhn, 1986. 2.29b © Zefa Sauer, The Stock Market (#K8949) 2.30 Spenser Swanger/Tom Stack & Associates.

3.12 a Dwight Kuhn © 1980, b Courtesy W. Cheng, International Paper Company 3.14 Reproduced from R. G. Kessel and Randy H. Kardon, *Tissues and Organs: A Text-Atlas of Scanning Electron Microscopy*. W. H. Freeman, 1979. 3.24 Adapted by permission from *The Structure and Action of Proteins* by Richard E. Dickerson and Irving Geis, W. A. Benjamin, Inc., Menlo Park, Calif., Publisher; copyright © by Dickerson and Geis. 3.26 Adapted by permission from *The Structure and Action of Proteins* by Richard E. Dickerson and Irving Geis, W. A. Benjamin, Inc. Menlo Park, Calif., Publisher; copyright © by Dickerson and Geis. 3.27 Adapted by permission from *The Structure and Action of Proteins* by Richard E. Dickerson and Irving Geis, W. A. Benjamin, Inc. Menlo Park, Calif., Publisher; copyright © by Dickerson and Geis. 3.29 Courtesy John Kendrew, Medical Research Center of Molecular Biology 3.30 Adapted by permission from *The Structure and Action of Proteins* by Richard E. Dickerson and Irving Geis, W. A. Benjamin, Inc., Menlo Park, Calif., Publisher; copyright © by Dickerson and Geis. 3.38 Adapted from Bruce Alberts et al, *Molecular Biology of the Cell*, 3e, 1994, by permission of Garland Publishing, New York. 3.39 Jack Barrie/Bruce Coleman, Inc. 3.49 From Lehninger, *Biochemistry*, 2e, 1975. Worth Publishers, New York. 3.52 © K. Talaro/VU.

4.1a Science Photo Library, Photo Researchers, Inc. 4.3 Courtesy J. D. Pickett-Heaps, University of Melbourne. 4.4d J. J. Cardamore and B. A. Phillips, University of Pittsburgh. 4.5 Reprinted with permission from M. Amrein, *Science*, 240:515. Copyright ©1988 American Association for the Advancement of Science. 4.6 Reprinted with permission from Dr. E. Henderson, from his article in *Science*, 257·1944 Copyright ©1992, American Association for the Advancement of Science. 4.11 © David M. Phillips/Visuals Unlimited. 4.13 The Liposome Company. 4.14 Courtesy J. David Robertson, Duke University. 4.16 Courtesy Daniel Branton. 4.25 Photo by Dorothy F. Bainton, M. D. 4.26 © R. L. Roberts 4.27a J. Ross, J. Olmstead, and J. Rosenbaum, *Tissue and Cell*, vol. 7, 1975. 4.28 M. M. Perry and A. B. Gilbert, *J. Cell. Sci.*, vol. 39, 1979 by copyright permission of The Rockefeller University Press. 4.29a N. Hirokawa and J. Heuser, *Cell*, vol. 30, 395–406. © 1982 by M. I. T., b courtesy of Dr. Richard G. W. Anderson, University of Texas Health Center, Dallas, Texas. 4.31b V. Herzog, H. Sies, and F. Miller, *J. Cell. Biol.*, vol. 70, 1976, by copyright permission of the Rockefeller University Press. 4.32 © Biophoto Associates/Science Source, Photo Researchers, Inc. 4.33 Courtesy R. D. Preston and E. Frei, University of Leeds. 4.34b H. Latta, W. Johnson, and T. Stanley, *J. Ultrastruct. Res.*, vol. 51, 1975.

5.1 Biophoto Associates/Science Source, Photo Researchers, Inc. 5.2 Dr. J. Nickerson, G. Krockmalinic, et al, *Science*, 259:1257, 1993. Reproduced by permission of the American Association for the Advancement of Science and the authors. 5.3 Photo courtesy of the Biology Department, Brookhaven National Laboratory. 5.4 Courtesy of Dr. Barbara Hamkalo and Dr. J. B. Rattner. 5.6 W. G. Whaley, H. H. Mollenhauer, and J. H. Leech, *Am. J. Bot.*, vol. 47, 1960. 5.7 Courtesy Daniel Branton, Harvard University. 5.8 Courtesy K. R. Porter, University of Colorado. 5.10 Micrograph courtesy D. S. Friend, University of California, San Francisco. 5.12 Courtesy D. S. Friend, University of California, San Francisco. 5.14 Courtesy D. S. Friend, University of California, San Francisco. 5.16 Micrograph by S. E. Frederick and E. H. Newcomb, J. Cell. Biol., vol. 43, 1969. 5.17 Courtesy D. S. Friend, University of California, San Francisco. 5.18top Micrograph by W. P. Wergin, courtesy E. H. Newcomb, University of Wisconsin. **bottom** M. C. Ledbetter, Photo Researchers, Inc. 5.19 Courtesy Gordon Leedale, Biophoto Associates, Leeds, England. 5.22 Courtesy Elias Laxarides, California Institute of Technology. 5.23 Dr. Susumu Ito, Department of Neurobiology, Harvard Medical School. 5.25c © Boehringer Ingelheim International GmbH, photo Lennart Nilsson. d Photograph by C. Lin, courtesy of Paul Forscher, Yale University. 5.26 Dr. John Hartwig et al., *Sci. Am.*, September 1994, used by permission of authors and W. H. Freeman & Co., Inc. 5.27a H. Kim, L. I. Binder, J. L. Rosenbaum; The Journal of Cell Biology, vol. 80, 266–267, 1979 by copyright permission of the Rockefeller University Press. b Courtesy of Don Fawcett, Harvard Medical School. 5.30 Photo from A. Ashkin, K. Schutze, J. M. Dziedic, U. Euteneurt, and M. Schliwa, *Nature*, vol. 348; 1990; pp. 346–48. 5.32 U. Aebi, University of Basel, Switzerland. 5.33 Micrograph from J. Heuser and S. R. Salpeter, *J. Cell. Biol.* vol. 82, 1979 by copyright permission of the Rockefeller University Press. 5.34 M. McGill, D. P. Highfield, T. M. Monahan, and B. R. Brinkley, *J. Ultrastruct. Res.*, vol. 578, 1976. 5.35 Courtesy R. W. Linck, Harvard Medical School, and D. T. Woodrum. 5.36 Photo provided by Dr. E. R. Dirksen, University of California, Los Angeles. 5.37 Micrograph by K. Roberts, John Innes Institute, Norwich, England; from B. Alberts et al, Molecular Biology of the Cell, Garland Press, New York, 1983. 5.39 C. J. Brokaw, California Institute of Technology, from *Science*, vol. 178, 1972, copyright © by the American Association for the Advancement of Science. 5.41 Micrograph produced by J. W. Heuser, et al, *J. Cell Biol.*, vol. 95, p. 800, 1982. Reproduced by copyright permission of the Rockefeller University Press. 5.40 (mitochondrion) Courtesy K. R. Porter, University of Colorado; (lysosome) Courtesy A. B. Novikoff, Albert Einstein College of Medicine; (glycocalyx) Courtesy A. Ryter, Institut Pasteur, Paris; (Golgi apparatus) D. W. Fawcett/Visuals Unlimited; (centrioles) M. McGill, D. P. Highfield, T. M. Monahan, and B. R. Brinkley, *J. Ultrastruct. Res.*, vol. 57, 1976; (all other micrographs) Courtesy N. B. Gilula, Baylor College of Medicine. 5.41 (plasmodesma) Courtesy W. G. Whaley, et al., *J. Biophys. Biochem. Cytol.*, (now *J. Cell Biol.*) vol. 5, 1959; by copyright permission of the Rockefeller University Press; (nucleus, chloroplast, leucoplast, endoplasmic reticulum, mitochondrion, and Golgi apparatus) Courtesy A. Ryter, Institut Pasteur, Paris. 5.43 Courtesy J. Griffith, School of Medicine, University of North Carolina, Chapel Hill.

copyright © 1984 by Scientific American, Inc., all rights reserved. **14.22** Modified from J. E. Sulston and H. R. Horvitz, *Dev. Biol.*, vol. 56, 1977, copyright © 1977 by Academic Press.

15.3 Photo from R. G. Kessel and R. H. Kardon, *Tissues and Organs: A Text-Atlas of Scanning Electron Microscopy*, W. H. Freeman and Company, San Francisco, California, 1970. **15.4** Photo courtesy Peter Marks and Frederick Maxfield, from P. Marks and F. R. Maxfield, *J. Cell. Biol.*, 110:43-52 (1990). Reprinted by permission of the Rockefeller University Press. **15.5b** Computer graphic image by Arthur J. Olsen, © The Scripps Research Institute, 1986. **15.17** Reproduced from D. Lawson, C. Fewtrell, B. Gomperts, and M. Raff, *Journal of Experimental Medicine*, 142:391-402, 1975. Used by copyright permission of Rockefeller University Press. **15.19** E. Gueho/Photo Researchers, Inc. *Exploring Further*, p. 402, photo courtesy Stephen T. Brentano.

16.5 Jane Burton/Bruce Coleman, Inc. **16.9** Courtesy M. C. Latham, Cornell University. **16.10** Modified from Francisco H. Alaya and John A. Kiger, Jr., *Modern Genetics*, Benjamin-Cummings, Menlo Park, California, 1980. **16.13** Courtesy S. B. Moore. **16.14** Modified from A. M. Winchester, *Genetics*, 5th ed., Houghton Mifflin, Boston, 1977. **16.16** Courtesy Marion I. Barnhart, Wayne State University Medical School, Detroit, Michigan. **16.19** K. R. Dronamraju, in E. J. Gardner, *Principles of Genetics*, copyright ©1975 by John Wiley & Sons, Inc.; reprinted with their permission. **16.20** Courtesy M. L. Barr, *Can. Cancer Conf.*, vol. 2, Academic Press, New York, 1957. **16.21** H. Chaumeton/Nature. **16.24** Modified from *Biological Science: An Ecological Apprach*, 4th ed., Houghton Mifflin, Boston, 1982. Used by permission. **16.26** E. J. Bingham, University of Wisconsin, Madison.

PART III: EVOLUTIONARY BIOLOGY

Diversity of Species: Photo courtesy Bob Natalini

17.2 F. Gohier/Photo Researchers, Inc. **17.3** Courtesy J. L. Gould **17.4a** George Holton/Photo Researchers, Inc.; **b** Masud Quraishy/Bruce Coleman, Inc. *Exploring Further*, Fig. 2. © Kim Taylor/Bruce Coleman, Inc. **17.5** Adapted by permission from *The Structure and Action of Proteins* by Richard E. Dickerson and Irving Geis, W. A. Benjamin, Inc., Menlo Park, California, Publisher; copyright © 1969 by Dickerson and Geis. **17.7** A. J. Ribbink, *S. Afr. J. Zool.*, vol. 18, 1985. **17.8** Hans Reinhard/Bruce Coleman, Inc. **17.13** Data from C. M. Woodworth et al., *Agron, J.*, vol. 44, 1952. **17.15** Hans Reinhard/Bruce Coleman, Inc. **17.17** Photo courtesy Dr. Michio Hori, Department of Zoology, Kyoto University, Japan. **17.18** Animals Animals/ © Stouffer Prod. **17.19** John Wightman/Ardea London Ltd. **17.20** Photographs by John A. Endler, University of California at Santa Barbara. **17.21a** Courtesy Colin G. Beer, Rutgers University; **b** Courtesy John Sparks, BBC (Natural History). **17.20a** Courtesy D. J. Howell, Purdue University; **b** Courtesy M. Morcombe; **c** Bob & Clara Calhoun/Bruce Coleman, Inc.; **d** Courtesy E. S. Ross. **17.24** Courtesy Ken Paige, University of Illinois. **17.25** Derek Washington/Bruce Coleman, Inc. **17.26** Oxford Scientific Films/Bruce Coleman, Inc. **17.27a** Courtesy Jeffrey L. Rotman; **b** Jane Burton/Bruce Coleman, Inc. **17.28a** David Overcash/Bruce Coleman, Inc.; **b** John Shaw/Bruce Coleman, Inc. **17.29** Peter Ward/Bruce Coleman, Inc. **17.30** Jane Burton/Bruce Coleman, Inc. **17.31** Animals, Animals/© Breck Kent. **17.32** Art Wolfe/Tony Stone Images, New York. **17.33a** Courtesy James L. Castner, University of Florida; **b** Courtesy E. S. Ross; **c** Charles Angelo/Photo Researchers, Inc. **17.34a** D. Overcash; **b** J. Shaw/Bruce Coleman, Inc. **17.35** Courtesy Douglas Faulkner. **17.36a** © Steinhardt Aquarium/Tom McHugh/Photo Researchers, Inc.; **b** Joe McDonald/Bruce Coleman, Inc. **17.37** R. P. Carr/Bruce Coleman, Inc. **17.38** Photograph courtesy Department Library Services, American Museum of Natural History, from H. Kutter, *Neujahrsblatt herausgegeben von der Naturforschenden Gesellschaft in Zurich*, vol. 171, 1969. **17. 39** Courtesy E. S. Ross.

18.1 Photo by W. R. Eisenbohr, Courtesy Cies Sexton, U. S. Department of the Interior Geological Survey. **18.2** Irv Kornfield and Jeffrey N. Taylor, *Proc. Biol. Soc. Washington*, vol. 96, 1983. **18.3a** Modified from W. Auffenburg, *Tulane Stud. Zool. Bot.*, vol. 2, 1955; **b,d** Grant Heilman Photography; **c** Modified from R. E. Woodson, *Ann. Mo. Bot. Gard.*, vol. 34, 1947. **18.4** Adapted from Jens Clausen, David Keck, and William Hiesey, *Experimental Studies on the Nature of Species, III* (Carnegie Institution publication 581), 1948. **18.5a** Ed Degingger and **b** Josephy VanWormer/Bruce Coleman, Inc. **18.6** Redrawn from W. Auffenberg, *Tulane Stud. Zool. Bot.*, vol. 2, 1955. **18.7** Pat and Tom Leeson/Photo Researchers, Inc. **18.9** Modified from B. Wallace and A. M. Srb, *Adaptation*, copyright © 1964 by Prentice-Hall, Inc., Englewood Cliffs, New Jersey. **18.10** J. Shaw/Bruce Coleman, Inc. **18.11** Courtesy E. S. Ross. **18.12** Phil Deginger/Bruce Coleman, Inc. **18.13** Courtesy Michael F. Whiting, from *Insect Homeotic Transformation*, p. 696, *Nature*, vol. 368, 21 April 1994 by Dr. Michael F. Whiting and Dr. Ward C. Wheeler, American Museum of Natural History. **18.14** Courtesy Thomas K. Wood, Department of Entomology, University of Delaware, Newark, Delaware. **18.15** Modified from D. Lack, *Darwin's Finches*, Cambridge, The University Press, 1947. **18.17** Drawing courtesy Sophie Webb **18.18** Modified from D. Lack, *Darwin's Finches*, Cambridge, The University Press, 1947. **18.19** Photograph by Irenaus Eibl-Eibesfeldt, Max Planck Institute for Behavioral Physiology. **18.20** Courtesy H. Douglas Pratt. **18.21** Modified from G. F. Gause, *Science*, vol. 79, 1934; copyright © 1934 by A. A. A. S. *How We Know* p. 506 Photos courtesy Peter Grant, Princeton University. **18.23** Illustrations by Marianne Collins are reproduced from *Wonderful Life: The Burgess Shale and the Nature of History*, by Stephen Jay Gould, with permission of W. W. Norton & Company, Inc.; copyright © 1989 Stephen Jay Gould. **18.24** Jane Burton/Bruce Coleman, Inc. **18.25** Redrawn from D. Lack, *Darwin's Finches*, Cambridge, The University Press, 1947. **18.26** Modified from G. G. Simpson, *Horses*, copyright © 1951 by Oxford University Press, Inc., renewed 1979 by G. G. Simpson. Used by permission. **18.28a** Chicago Zoological Park; photograph by Tom McHugh/Photo Researchers, Inc.; **b,c** Michael Morcombe; **d** Jack Fields/Photo Researchers, Inc.

PART IV: THE GENESIS AND DIVERSITY OF ORGANISMS

Major Milestones: Photo courtesy William E. Ferguson

19.1 Courtesy NASA **19.2** Courtesy David W. Deamer, University of California, Santa Cruz **19.3** Modified from R. E. Dickerson, *Sci. Am.*, September 1978; copyright © 1978 by Scientific American, Inc.; all rights reserved. **19.4** Courtesy Sigurgeir Jonasson. **19.6** Courtesy Sidney W. Fox, University of Miami, and Steven Brooke Studios, Coral Gables, Florida. **19.9** Courtesy J. W. Schopf, University of California, Los Angeles. **19.10** G. R. Roberts. **19.11a** Courtesy R. E. Lee, University of the Witwatersrand, South Africa; **b** MMJP Plant Research Lab. **19.12** D. A. Stetler and W. M. Laetsch, *Am. J. Bot.*, vol. 56, 1969, reprinted with permission from the Botanical Society of America. **19.13** W. J. Larsen, *J. Cell Biol.*, vol. 47, 1970, by copyright permission of the Rockefeller University Press. **19.14** Courtesy I. B. Dawid, National Intitutes of Health, Bethesda, Maryland, and D. R. Wolstenholme, University of Utah. **19.15** *Joenia annectens* adapted from drawing by Robert Golder, courtesy of Lynn Margulis and Karlene V. Schwartz, University of Massachusetts. **19.17a** Science Photo Library/Photo Researchers, Inc. **b** Biophoto Associates/Photo Researchers, Inc. **c** Courtesy of E. B. Daniels, from K. W. Jeon, ed. *The Biology of the Amoeba*, 1973, Academic Press, Orlando.

20.1a Courtesy M. Wurtz, University of Basel; **b** Courtesy M. Gomersall, McGill University; **c** Courtesy R. C. Williams, University of California, Berkeley. **20.3** Photo from K. Corbett, *Virology*, vol. 22, 1964; reprinted with permission of Academic Press, Inc., New York. **20.4** Photo courtesy T. O. Diener U.S.D.A. **20.5** From T. O. Diener, *Am. Sci.*, vol. 71, 1983. **20.6** Drawing adapted from one by Fred E. Cohen, which appeared on p. 33 of the January 1995 *Sci. Am.*, in the article by Stanley B. Prusiner, reprinted with permission of the author. **20.8b** Courtesy Esther R. Angert/Indiana University. Originally appeared on p. 1629 of *Science*, vol. 256, 19 June 1992. Reprinted with permission of A. A. A. S. **20.9a** Courtesy David Scharf/Peter Arnold, Inc.; **b** Center for Disease Control, Atlanta, Georgia. **20.10** Turtox/Cambosco, Macmillan Science Co., Inc. **20.11** Courtesy Z. Skobe, Forsyth Dental Center/BPS. **20.12** Courtesy M.

Gomersall, McGill University. **20.13** S. Kimoto and J. C. Russ, *Am. Sci.*, vol. 57, 1969. **20.15** G. B. Chapman, *J. Bacteriol.*, vol. 71, 1956. **20.16** Photograph by Ginny Fonte, from H. C. Berg, *Sci. Am.*, August 1975; copyright © 1975 by Scientific American, Inc.; all rights reserved. **20.17** From R. C. Johnson, M. P. Walsh, B. Ely, and L. Shapiro, *J. Bacteriol.*, vol. 138, 1979. **20.18** Courtesy W. Burgdorfer, S. F. Hayes, and D. Corwin, Rocky Mountain Laboratories. **20.19** Courtesy David Chase, Veterans Hospital, Sepulveda, California. **20.20** Murti/Phototake, New York. **20.21** H. Reichenbach, Gesellschaft fur Biotechnologische Forschung, mbH. **20.22** Elliot Scientific Corp. **20.23** Elliot Scientific Corp. **20.24a** Science Source/Photo Researchers, Inc.; **b,c** © Thomas E. Adams/Visuals Unlimited. **20.25** Courtesy Malcolm Brown, Jr., University of Texas. **20.26** © S. Thompson/Visuals Unlimited. **20.27** This EM was made by Zell A. McGee in the course of research funded by the National Institute of Allergy and Infectious Diseases and was provided to the publisher without charge in the interest of science. **20.28** Photograph by Walther Stoeckenius, courtesy Carl Woese, University of Illinois. **20.29a** Courtesy of H. W. Jannasch, Woods Hole Oceanographic Institution; **b** From W. J. Jones, J. A. Leigh, F. Mayer, C. R. Woese, and R. S. Wolfe, *Arch. Microbiol.*, vol. 136, 1983, copyright © 1983 by Springer-Verlag, New York.

21.1 Photograph courtesy E. W. Daniels, art modified from K. W. Jeon, ed., *The Biology of the Ameoba*, 1973, Academic Press, Orlando. **21.2** From J. M. Jensen and S. R. Wellings, *J. Protozool.*, vol. 19, p. 297–305, 1971. **21.4** Courtesy E. V. Grave. **21.5** Photo courtesy Ed Reschke. **21.6** Photo courtesy E. V. Grave. **21.7** Courtesy E. V. Grave. **21.8** Edward Degginger/Bruce Coleman, Inc. **21.8** Courtesy E. V. Grave. **21.10** Courtesy E. V. Grave. **21.12a** Courtesy E. V. Grave; **b** Biophoto Associates/Photo Researchers, Inc.; **c** Courtesy Roman Vishniac; **d** Courtesy A. Fleury, Service d'Imagerie Cellulaire, Orsay, France. Originally appeared on the cover of *Science*, 9 December 1994. **21.13** Courtesy Thomas Eisner, Cornell University. **21.14** Manfred Kage/Peter Arnold, Inc. **21.16** Courtesy E. V. Grave. **21.19** Ray Simons/Photo Researchers, Inc. **21.22a,b** Courtesy K. B. Raper, *Proc. Am. Philos. Soc.*, vol. 104, 1960.; **c–f** Carolina Biological Supply Company.

22.2 B. S. C. Leadbeter, University of Birmingham. **22.4** Turtox/Cambosco, Macmillan Science Co., Inc. **22.5** Edward Degginger/Bruce Coleman, Inc. **22.6** H. Chaumeton/Nature. **22.7** Edward Degginger/Bruce Coleman, Inc. **22.8** Courtesy Sally Faulkner. **22.10b** Robert Carr/Bruce Coleman, Inc. **22.13b** Photo courtesy Susumu Honjo, Woods Hole Oceanographic Institution, Woods Hole, Massachusetts. **22.14** H. Chaumeton/Nature. **22.17** F. Sauer/Nature. **22.18** Courtesy Ed Reschke. **22.19** Courtesy Roman Vishniac. **22.20** Courtesy Roman Vishniac. **22.22** Courtesy Roman Vishniac. **22.27** Edward Degginger/Bruce Coleman, Inc. **22.30** Modified from H. J. Fuller and O. Tippo, *College Botany*, Holt, Reinhart & Winston, Inc., New York, 1954. **22.32** Jane Burton/Bruce Coleman, Inc. **22.33a–c** From G. Shih and R. Kessel, *Living Images: Biological Microstructures Revealed by Scanning Electron Microscopy*. © 1982 by Science Books International; **d** Stephen Dalton/Photo Researchers, Inc. **22.34** Adrian Davies/Bruce Coleman, Inc. **22.36** Adrian Davies/Bruce Coleman, Inc. **22.38** Courtesy Edward Degginger. **22.40** Carnegie Museum of Natural History, Pittsburgh, Pennsylvania. **22.41** Courtesy Field Museum of Natural History, Chicago. **22.42** Courtesy E. S. Ross. **22.43** John Shaw/Bruce Coleman, Inc. **22.44** Courtesy G. R. Roberts. **22.45** Stephen Parker/Photo Researchers, Inc. **22.49** Ed Reschke/Peter Arnold, Inc. **22.50** Courtesy Thomas Esiner, Cornell University. **22.51** Photo courtesy Nels R. Lersten, Iowa State University. **22.52a** Courtesy Victor B. Eichler; **b** Courtesy Thomas Eisner, Cornell University. **22.55** Redrawn from H. N. Andrews, *Science*, vol. 142, 1963; copyright © 1963 by A.A.A.S. **22.56** Courtesy E. S. Ross. **22.57** Courtesy Roman Vishniac. **22.58** From W. Krommenhoek, J. Sebus, and G. J. van Esch, *Biological Structures*, copyright © 1979 by L. C. G. Malmberg B. V., The Netherlands. **22.59a,b** Courtesy Thomas Eisner, Cornell University; **c** Courtesy E. S. Ross. **22.60** Modified from H. J. Fuller and O. Tippo, *College Botany*, Holt, Reinhart & Winston, Inc., New York, 1954. **22.64** M. I. Walker/Photo Researchers, Inc. **22.66** James H.

Robinson/Photo Researchers, Inc. **22.68a** Jane Burton and **b** R. P. Carr/Bruce Coleman, Inc. **22.69** Modified from H. J. Fuller and O. Tippo, *College Botany*, Holt, Reinhart & Winston, Inc., New York, 1954.

23.1 Science Photo Library/Photo Researchers, Inc. **23.3** Photograph by W. H. Amos/Bruce Coleman, Inc. **23.5** M. P. L. Fogden/Bruce Coleman, Inc. **23.6** From L. W. Sharp, *Fundamentals of Cytology*, McGraw-Hill Book Co., New York, copyright © 1943; used by permission. **23.9a** Jane Burton and **b** L. West/Bruce Coleman; **c,d** V. Ahmadjian and J. B. Jacobs, *Nature*, vol. 289, 1981 ©1981, Macmillan Journals Ltd. **23.10a** Masana Izawa and **b** Satoshi Kuribyashi/Nature Production. **23.11** Photograph by B. A. Roy, reprinted by permission of *Nature*, vol. 362, page 57. **23.12** From L. W. Sharp, *Fundamentals of Cytology*, McGraw-Hill Book Co., New York, copyright © 1943; used by permission.

24.1 Courtesy Jeff Rotman. **24.4** Courtesy Oxford Scientific Films. **24.5** H. Chaumeton/Nature. **24.8** Carolina Biological Supply Company. **24.11** R. N. Mariscal/Bruce Coleman, Inc. **24.12** Courtesy Howard Hall. **24.15** After K. G. Grell, University of Tubingen (1974). **24.16** Courtesy Dr. A. Ruthmann, from J. Rassat and A. Ruthmann, *A. Zoomorphology* vol. 93, p. 59–72, Springer-Verlag, New York. **24.18** After K. G. Grell, University of Tubingen (1974). **24.20** H. Chaumeton/Nature. **24.24** CNRI/Science Photo Library/Photo Researchers, Inc. **24.26a** Courtesy Ed Reschke; **b** H. Chaumeton/Nature. **24.27** H. Chaumeton/Nature. **24.32** Adapted from *Life* by W. K. Purves and G. H. Orians, Sinauer Associates, Sunderland, Massachusetts; copyright © 1983. **24.33** Courtesy T. E. Adams. **24.35** H. Chaumeton/Nature. **24.36** H. Chaumeton/Nature. **24.38** Fred Bavendam/Peter Arnold, Inc. **24.39** Oxford Scientific Films. **24.41** H. Chaumeton/Nature. **24.42top** H. Chaumeton/Nature; **bottom** Rod Borland/Bruce Coleman, Inc. **24.43** Based in part on drawings by Louise G. Kingsbury. **24.44** F. Sauer/Nature. **24.45** H. Chaumeton/Nature. **24.46** Modified from W. Stempell, *Zoologie im Grundriss*, Borntraeger, 1926. **24.47** H. Chaumeton/Nature. **24.49** Photography by Joseph Pawlik, University of North Carolina at Wilmington. **24.50** Edward Degginger/Bruce Coleman, Inc. **24.51** Oxford Scientific Films. **24.52** Courtesy E. S. Ross. **24.53** Jane Burton/Bruce Coleman, Inc. **24.55** Courtesy Ed Reschke. **24.56** Scott Camazine/Photo Researchers, Inc. **24.57** Edward Degginger/Bruce Coleman, Inc. **24.58** H. Chaumeton/Nature. **24.59a** H. Chaumeton/Nature; **b** Hans Reinhard/Bruce Coleman, Inc. **24.60a** H. Chaumeton/Nature; **b** Robert Gossington/Bruce Coleman, Inc.; **c** Gilbert S. Grant/Photo Researchers, Inc.; **d** H. Chaumeton/Nature. **24.61** Oxford Scientific Films. **24.62** Courtesy G. R. Roberts. **24.63** A. Cosmos Blank, National Audobon Society/Photo Researchers, Inc. **24.65** Modified from T. I. Storer and R. L. Usinger, *General Zoology*, McGraw-Hill Book Co., New York, copyright © 1957; used by permission. **24.66** Courtesy Stephen Dalton/NHPA. **24.70** Modified from L. H. Hyman, *The Invertebrates*, McGraw-Hill Book Co., New York, copyright © 1955; used by permission. **24.71a** Robert Dunne and **b** Charlie Ott/Photo Researchers, Inc. **24.72** Andrew J. Martinez/Photo Researchers, Inc. **24.73** A. Kerstitch, Sea of Cortez Enterprises. **24.74b** Courtesy Smithsonian Institution, photo 62419. **24.75** H. Chaumeton/Nature.

25.1 Photograph by Bill Wood/Bruce Coleman, Ltd., London. **25.2** Turtox/Cambosco, Macmillan Science Co., Inc. **25.3** Photograph by H. Chaumeton/Nature. **25.4** Courtesy Ed Reschke. **25.6** Courtesy Ed Reschke. **25.7** Courtesy Ed Reschke. **25.8a** D. Claugher, courtesy of the Trustees, The British Museum (Natural History); **b** Courtesy Jerome Gross, Massachusetts General Hospital. **25.10a** Ed Reschke/Peter Arnold, Inc.; **b** Manfred P. Kage/Peter Arnold, Inc. **25.12** Photograph courtesy Ed Reschke. **25.14** Courtesy Heather Angel, Biofotos. **25.15** Chip Clark, National Museum of Natural History, Washington. **25.16** Modified from A. S. Romer, *The Vertebrate Body*, W. B. Saunders, New York, 1949. **25.17** Courtesy Howard Hall. **25.18** Courtesy the Royal Society, London. **25.19** Painting photograph by Chip Clark, courtesy National Museum of Natural History, Washington. **25.20a** S. C. Bisserot and **b** Hans Reinhard/Bruce Coleman, Inc. **25.21** Courtesy E. S. Ross. **25.22** Hans Pfletschinger/

Wisconsin. **32.35** Courtesy Sylvan Wittwer. **32.37** Redrawn from *Biologie: Ein Lehrbuch*, edited by G. Czihak et al., 2nd ed., Springer-Verlag, 1978. **32.40** Photograph by Stephen Gladfelter, Standford University, from research by P. W. Oeller, Lu M.-W., L. P. Taylor, D. A. Pike, and A. Theologis at the Plant Gene Expression Center, University of California, Berkeley. **32.45** After W. A. Jensen and F. B. Salisbury, *Botany: An Ecological Approach*; copyright © 1972 by Wadsworth Publishing Co., Inc., Belmont, California; used by permission of the publisher. **32.46** Photo by M. F. Nuttall, DuPont Experimental Station, courtesy Ilya Raskin, Rutgers State University.

33.1 Plate I from *Insect Hormones* by V. B. Wigglesworth, published in 1970 by Oliver & Boyd, Edinburgh, and W. H. Freeman, New York. **33.2** Modified from H. A. Schneiderman and L. I. Gilbert, *Science*, vol. 143, 1964; copyright © 1964 by A.A.A.S. ***How We Know***, p. 939. Photo from Michael Gadomski/Bruce Coleman, Inc. **33.4** Painting by Frank H. Netter, M.D.; reprinted with permission from *The CIBA Collection of Medical Illustrations*, copyright © 1965 by CIBA Pharmaceutical Company, division of CIBA-GEIGY Corporation; all rights reserved. **33.6** Carolina Biological Supply Company. **33.11** © John P. Kay/Peter Arnold, Inc. **33.13** From A. J. Carlson et al., *The Machinery of the Body*, University of Chicago Press, 1961. **33.16** Modified from H. Curtis, *Biology*, 4th ed.; copyright © 1983 by Worth Publishers, Inc. **33.26** Modified from an original model of a morphine molecule, courtesy Maitland Jones, Jr.

34.1 Hans Pfletschinger/Peter Arnold, Inc. **34.2** Michael Fogden/Bruce Coleman, Inc. **34.3** Ken M. Highfill/Photo Researchers, Inc. **34.7a** Courtesy D. M. Phillips, Population Council, New York; **b** David M. Phillips/Photo Researchers, Inc. **34.11a,c** Courtesy Thomas Eisner, Cornell University; **b** C. Edelman-La Villette/Photo Researchers, Inc.

35.2 Courtesy S. L Palay, Harvard Medical School. **35.4a** Courtesy H. deF. Webster, as printed in W. Bloom and D. W. Fawcett, *A Textbook of Histology*, 10th ed., W. B. Saunders Co., 1975; **b** R. L. Roberts, R. G. Kessel and H. N. Tung, *Freeze Fracture Images of Cells and Tissues*, Oxford University Press, 1991. **35.6a** H. Chaumeton/Nature; **b** adapted from E. R. Kandel, *Cellular Basis of Behavior*, W. H. Freeman, San Francisco, 1976. **35.7** Modified from E. R. Kandel, *Cellular Basis of Behavior*, W. H. Freeman, San Francisco, 1976. **35.10** Modified from T. H. Bullock and G. A. Horridge, *Structure and Function of the Nervous System of Invertebrates*, W. H. Freeman, San Francisco, 1965. **35.11** Modified from L. H. Hyman, *The Invertebrates*, vol. 2, McGraw-Hill Book Co., New York, copyright © 1951, and from Ralph Buschbaum, *Animals Without Backbones*, University of Chicago Press, copyright © 1948 by the University of Chicago, used by permission of the publisher. **35.12** Modified from R. Goldschmidt, *Z. Wiss. Zool., Abt. A.*, vol. 92, 1909. **35.13** Photograph courtesy Thomas Eisner, Cornell University. **35.17** Courtesy A. L. Hodgkin, *J. Physiol. (London)*, vol. 131, 1956. **35.19** Adapted from W. A. Catterall, *Science*, vol. 242, 1988; copyright © 1988 by A.A.A.S. **35.24** E. R. Lewis et al., *Science*, vol. 165, 1969; copyright © 1969 by A.A.A.S. **35.25** R. L. Roberts, R. G. Kessel and H. N. Tung, *Freeze Fracture Images of Cells and Tissues*, Oxford University Press, 1991. **35.30** Photograph courtesy Ed Reschke. **35.31** Courtesy J. E. Heuser, Washington University Medical Center.

36.1 Redrawn from B. Katz, *J. Physiol. (London)*, vol. 111, 1950. **36.3** Graph modified from S. Skoglund, *Acta Physiol. Scand., Suppl.* 124, vol. 36, 1956. **36.6** Courtesy Ed Reschke. **36.7** After Murray, 1973. **36.8** D. Claugher, by courtesy of the Trustees, The British National Museum of Natural History. **36.10** J. L. Lepore/Photo Researchers, Inc. **36.13** J. A. L. Cooke, Oxford Scientific Films. **36.14** Modified from R. E. Snodgrass, *Principles of Insect Morphology*, McGraw-Hill Book Co., New York, copyright © 1934; used with permission of McGraw-Hill Book Co. **36.15** Courtesy J. L. Gould and C. G. Gould. **36.16** Douglas Faulkner/Photo Researchers, Inc. **36.19** Photograph by L. Nilsson; from L. Nilsson, *Behold Man*, English translation © 1974, Albert Bonniers Forlag, Stockholm; Little Brown and Co. (Canada) Ltd. **36.20** Omikron/Photo Researchers, Inc. **36.24** After A. F. MacNichol. **36.26a** Courtesy Thomas Eisner, Cornell Univer-

sity; **b** courtesy E. S. Ross. **36.28** Courtesy A. J. Hudspeth and R. A. Jacobs from A. J. Hudspeth, *Nature*, 341:397-404 (1989). **36.29** Photos from Stuart Ira Fox, *Human Physiology*, 3rd ed., fig. 13.26, page 355, copyright © 1990 William C. Brown Communications, Inc. **36.30** Modified from G. von Bekesy, *Sump. Soc Exp. Biol.*, vol. 16, 1962; G. von Bekesy, *Experiments in Hearing*, McGraw-Hill Book Co., New York, copyright © 1960, used with the permission of McGraw-Hill Book Co.; M. S. Gordon, G. Bartholomew, A. D. Grinnell, C. B. Jorgensen, and F. N. White, *Animal Function: Principles and Adaptations*, Macmillan, New York, 1968, copyright © 1968 by M. S. Gordon. **36.31** L. Nilsson; from L. Nilsson, *Behold Man*, English translation 1974, Albert Bonniers Forlag, Stockholm; and Little, Brown and Co., (Canada) Ltd. **36.33a** William E. Ferguson; **b** S. L. Craig/Bruce Coleman, Inc.; **c** Edward Degginger/Bruce Coleman, Inc. **36.35** H. Vali, courtesy R. J. P. Williams, Oxford University. **36.37** Modified from W. M. Cornsweet, *Visual Perception*, Academic Press, New York, 1970; and S. Coren et al., *Sensation and Perception*, Academic Press, New York, 1978. **36.40** Modified from A. S. Romer and T. S. Parsons, *The Vertebrate Body*, 5th ed., copyright © 1977 by W. B. Saunders Company, reprinted by permission of CBS College Publishing. **36.41** Modified from A. S. Romer and T. S. Parsons, *The Vertebrate Body*, 5th ed., copyright © 1977 by W. B. Saunders Company, reprinted by permission of CBS College Publishing; and G. G. Simpson, C. S. Pittendrigh, and L. H. Tiffany, *Life: An Introduction to Biology*, copyright © 1957 by Harcourt Brace Jovanovich, Inc., used by permission of the publishers. **36.45** Modified from W. Penfield and T. Rasmussen, *The Cerebral Cortex of Man*, Macmillan, New York, 1950. **36.50** Based on data from John Allman, California Institute of Technology. **36.51** Institute of Medicine, National Academy of Sciences Press.

37.1 G. R. Roberts. **37.3** Photograph by D. Claugher, by courtesy of the Trustees, The British Museum of Natural History; drawing modified from J. Gray and H. W. Lissman from *J. Exp. Biol.*, vol. 15, 1938, by permission of Company of Biologists, Ltd. **37.4** H. Chaumeton/Nature. **37.7** © Biophoto Association, Photo Researchers, Inc. **37.9** Courtesy Ed Reschke. **37.14** Courtesy H. E. Huxley, Cambridge University. **37.15** Courtesy H. E. Huxley, Cambridge University. **37.18** Courtesy J. E. Heuser, Washington University Medical Center.

38.1 Ian Wyllie/Survival Anglia. London. **38.2a** Courtesy Paul Trotschel; **b** Stephen Dalton/Photo Researchers, Inc. **38.3** Courtesy John Sparks, BBC, Natural History. **38.4** Courtesy Thomas Eisner, Cornell University. **38.5** Modified from N. Tinbergen, *The Study of Instinct*, Oxford, The University Press, 1951. **38.6** Modified from N. Tinbergen, *The Study of Instinct*, Oxford, The University Press, 1951. **38.7** Courtesy E. R. Willis, Illinois State University. **38.8** Modified from N. Tinbergen and A. C. Perdeck, *Behaviour*, vol. 3, 1950; published by Oxford University Press. **38.9** Courtesy John Sparks, BBC, Natural History. **38.10** Modified from J.-P. Ewert, *Sci. Am.*, March 1974; copyright © 1974 by Scientific American; all rights reserved. **38.11** Adapted from K. Lorenz and N. Tinbergen, *Z. Tierpsychol.*, vol. 2, 1938. **38.12a,b,d,e** from E. H. Hess, *Sci. Am.*, July 1956; copyright © 1956 by Scientific American, Inc.,; all rights reserved; **c** Wallace Kirkland, copyright © 1954, Time, Inc. **38.13** Reprinted with permission of the author and publishers from D. G. Freedman, *Human Infancy: An Evolutionary Perspective*, Lawrence Erlbaum Associates, Hillsdale, New Jersey, 1975. **38.14** Adapted from N. E. Collias and E. C. Collias, *Auk*, vol. 79, 1962. **38.15** After Drees. **38.16** Courtesy R. Silver, Barnard College. **38.17** From K. Lorenz, *Symp. Soc. Exp. Biol.*, vol. 4, 1950. **38.18** Based on J. E. Lloyd, *Misc. Publ. Mus. Zool. Univ. Mich.*, vol. 130, 1966; and A. D. Carlson and J. Copeland, *Am. Sci.*, vol. 66, 1978, reprinted by permission of American Scientist, Journal of Sigma Xi, the Scientific Research Society. **38.19** D. R. Bentley, *Science*, vol. 174, 1971; copyright © 1971 by A.A.A.S. **38.20** From L. P. Brower, J. V. Z. Brower, and F. P. Cranston, *Zoologica*, vol. 50, 1965; used by permission of the New York Zoological Society. **38.21** Courtesy R. Thornhill, University of New Mexico. **38.22** Redrawn from K. von Frisch, *Bees: Their Vision, Chemical Senses, and Language*, copyright © 1950 by Cornell University; used by permission of Cornell University Press; photograph courtesy Kenneth Lorenzen, University of California, Davis. **38.23** Based on data from

R. Buch, *Z. Vergl. Physiol.*, vol. 40, 1957. **38.24** From K. Lorenz, *Zool. Anz.*, vol. 17, 1953. **38.25** TASS from Sovfoto. **38.26** From W. R. Miles, *J. Comp. Psychol.*, vol. 10, 1930; copyright © 1930 by the Williams and Wilkins Co., Baltimore, Maryland. **38.27a** Eric Hosking/ Bruce Coleman, Inc.; **b** Nina Leen, *Life*, copyright © 1964 by Time, Inc. **38.28** From E. H. Hess, *Science*, vol. 130, 1959; copyright © 1959 by A.A.A.S. **38.30** Warren Gurst and Genny Gurst, Tom Stack and Associates. **38.31** Adapted by permission from *The Honey Bee* by James L. Gould and Carol Grant Gould, W. H. Freeman, New York, copyright © 1988. **38.32** Adapted by permission from E. Gwinner and W. Wiltschko, *J. Comp. Physiol.*, vol. 125, 1978; copyright © 1978 by Springer-Verlag, New York. **38.33a** From B. Elsner, *Animal Migration, Navigation, and Homing*, edited by K. Schmidt-Koenig and W. T. Keeton, Springer-Verlag, New York, 1978; **b** from M. Michener and C. Walcott, *J. Exp. Biol.*, vol. 47, 1967. **38.34**, left, Steve Johnson, courtesy Charles Walcott, Cornell University; right, from K. Schmidt-Koenig and C. Walcott, *Animal Behav.*, vol. 26, 1978. **38.36** Modified from N. Tinbergen and A. C. Perdeck, *Behaviour*, vol. 3, 1950. **38.38** Photograph by Russ Charig, courtesy Charles Walcott, Cornell University.

PART VI: ECOLOGY

Threatened reservoirs of diversity: © Art Wolfe/Tony Stone Images.

39.3 Photo by G. W. Barlow, from *Anim. Beh.*, vol. 22, 1974. **39.4** Francois Gohier/Photo Researchers, Inc. **39.6** Modified from T. Carlson, *Biochem. Z.*, vol. 57, 1913. **39.7** Modified from J. Davidson, *Trans. R. Soc. South Aust.*, vol. 62, 1938. **39.9** Modified from E. P. Odum, *Fundamentals of Ecology*, W. B. Saunders, 1959 after Deevey. **39.10** Office of Population Research, Washington, D. C. **39.12** Courtesy E. S. Ross. **39.13** Courtesy L. David Mech, USFWS Minnesota Research Group, St. Paul, MN. **39.15** Photograph by Hans Reinhard/Bruce Coleman, Inc.; drawing based on H. N. Southern, *J. Zool.*, vol. 162, 1970. **39.16** Based on H. N. Southern, *J. Zool.*, vol. 162, 1970. **39.17** From John L. Harper, *Population Biology of Plants*, Academic Press, 1977, p. 153, copyright © 1977 Academic Press, Orlando, Florida. **39.18** Redrawn after J. M. Scriber and G. Slankey, *Am. Rev. Entomology*, vol. 26, 1981. **39.19** Adapted from *Principles of Animal Ecology* by W. C. Allee, A. E. Emerson, O. Park, T. Park, and K. P. Schmidt, W. B. Saunders, Philadelphia; copyright © 1949. **39.20** Drawing adapted from R. H. MacArthur, *Ecology*, vol. 39, 1958, copyright © 1959 by the Ecological Society of America. Photos: **a** Philip Boyer and **b** R. Austing/Photo Researchers, Inc.; **c** Edgar T. Jones/Bruce Coleman, Inc. **39.21** G. Torlofi/F. A. O. Photo. **39.22** Carol Hughes/Bruce Coleman, Inc. **39.23** Norman Meyers/Bruce Coleman, Inc. **39.24** Redrawn by permission from E. O. Wilson, *Sociobiology*, Harvard University PRess, Cambridge, Massachusetts, 1975. **39.25** Hans Pfletschinger/Peter Arnold, Inc. **39.26** Adapted from R. E. Kenward, *J. Anim. Ecol.*, vol. 47, 1978. **39.27** Courtesy E. D. Estes, San Diego State University. **39.28** Coutesy L. T. Nash, Arizona State University. **39.30** George D. Lepp, Bio-Tec Images. **39.31** Roberto Bunge, Ardea London Ltd. **39.32** Redrawn with permission of Macmillan Publishing Co., Inc. from David R. Pilbeam, *The Ascent of Man: An Introduction to Human Evolution*; copyright 1972 by David R. Pilbeam. **39.33a** J. Tanaka and **b** M. Shostak/Anthro-Photo. **39.34** I. De Vore/Anthro-Photo. **39.35** Modified from J. H. Connell, *Ecology*, vol. 142, 1961; copyright © 1961 by the Ecological Society of America. **39.36** Photographs reproduced with permission of the Department of Lands, Queensland, Australia. **39.37** Redrawn after R. M. May, *Sci. Am.*, September, 1978; copyright 1978 by Scientific American, Inc., all rights reserved. **39.39** Courtesy E. J. Kormondy, *Smithsonian Magazine*, vol. 1, 1970. **39.40** Jeff Foote/Bruce Coleman, Inc. **39.41** S.

Hurwitz/Bruce Coleman, Inc. **39.42** Based on data in E. P. Odum, *Fundamentals of Ecology*, W. B. Saunders, 1959. **39.43** Redrawn from R. M. May, *Sci. Am.*, September 1978; copyright 1978 by Scientific American, Inc.; all rights reserved. **39.44** Redrawn from R. H. Whittaker, *Communities and Ecosystems*, 2nd ed., Macmillan, New York, 1975; after B. Holt and G. M. Woodwell. **39.45** Redrawn from R. H. Whittaker, *Communities and Ecosystems*, 2nd ed., Macmillan, New York, 1975; after B. Holt and G. M. Woodwell. **39.46** Gene Ahrens/Bruce Coleman, Inc.

40.2a C. J. Tucker, J. R. G. Towshend, and T. E. Goff, *Science*, vol. 227, 1985. Redrawn with permission of the A.A.A.S.; **b** from Fig. 9-5 in R. E. Ricklefs, *Ecology*, 3rd ed., W. H. Freeman, 1990, p. 144. Based on data in H. H. Clayton and F. L. Clayton, 1947, World Weather Records 1931–1940, *Smithsonian Misc. Coll.* 105, 1–646. **40.3** Photographs courtesy E. S. Ross. **40.4** From fig. 12-3a and 12-4a, pp. 243–244 in Helmut Leith, Modeling the Primary Productivity of the World, in H. Leith and R. H. Whittaker, *Primary Productivity in the Biosphere*, Springer-Verlag, 1975. **40.5** From fig. 12.12c, in Helmut Leith, Modeling the Primary Productivity of the World, in Helmut Leith and R. H. Whittaker, *Primary Productivity in the Biosphere*, Springer-Verlag, 1975. **40.10** Modified from E. P. Odum, *Fundamentals of Ecology*, W. B. Saunders, 1959. **40.12** Photos courtesy James Hornbeck, U. S. Forest Service, Durham, New Hampshire. **40.13** From Shukla et al., *Science*, vol. 247, pp. 1322–25, fig. 3b, 1990. Copyright © 1990 A.A.A.S. **40.15** Redrawn after T. E. Graedel and P. J. Crutzen, *Sci. Am.*, September, 1989; copyright © 1989 by Scientific American, Inc.; all rights reserved. **40.16** Redrawn after S. H. Schneider, *Sci. Am.*, September, 1989; copyright © 1989 by Scientific American, Inc.; all rights reserved. **40.18** Photo by Will Owens, from Will Plants Profit From High CO_2? *Science*, vol. 268, 5 May 1995, p. 655. Copyright © 1995 by A.A.A.S. **40.19** Modified from M. H. Pelczar et al., *Microbiology*, 4th ed., McGraw-Hill Book Co., New York, copyright © 1977; used by permission. **40.20** Dr. Jeremy Burgess/Photo Researchers, Inc. **40.22** Courtesy D. W. Schindler, *Science*, col. 184, 1974; copyright © 1974 by A.A.A.S. **40.24** Courtesy Peter L. Ames. **40.25** Redrawn after K. P. Bowman, *Science*, vol. 239, 1988; copyright © 1988 by A.A.A.S. ***How We Know***, p. 1178 David Hall/Photo Researchers, Inc. **40.26** Larry Ditto/Bruce Coleman, Inc. **40.27** U.S.D.A. Forest Service, Washington, D.C. **40.28** Redrawn from G. E. Likens et al., *Ecol. Monogr.*, vol. 40, 1970; copyright © 1970 by the Ecological Society of America. **40.29** Courtesy National Archives. **40.30** Oxford Scientific Films/Animals, Animals, Earth Scenes. **40.31** After Whittaker (1970) Reprinted with permission from Macmillan Science Co., Inc., from *Communities and Ecosystems* by Robert Whittaker. Copyright © Robert Whittaker. **40.32** Photo by Tom Bean/DRK Photo, Sedona, Arizona. **40.33a** Paolo Koch/ Photo Researchers, Inc.; **b** Wolfgang Bayer/Bruce Coleman, Inc. **40.34** Zig Leszczynski, Animals, Animals, Earth Scenes. **40.35a** Ferrero/ Nature; **b** N. De Vore III/Bruce Coleman, Inc. **40.36** David Cayless/ Oxford Scientific Films. **40.37a** Ronald Toms/Oxford Scientific Films; **b,c** courtesy E. S. Ross. **40.39** John Elk III/Bruce Coleman, Inc. **40.41** After Schwarzbach, 1950. **40.42a** Courtesy Howard Hall; **b** courtesy James D. Jordan; **c** courtesy E. H. Newcomb and T. D. Pugh, University of Wisconsin/BPS. **40.44** Redrawn from John Napier, *The Roots of Mankind*, copyright © 1970, Smithsonian Institution Press, Washington, D. C.; used by permission. **40.46** Ferrero/Nature. **40.50a** Redrawn from R. H. MacArthur and E. O. Wilson, *The Theory of Island Biogeography*, copyright 1967 by Princeton University Press; used by permission; **b** redrawn by C. J. Krebs, *Ecology*, 2nd ed., Harper & Row, New York, 1978; after Preston. **40.53a** C. C. Reijnvaan from W. M. Doctors van Leeuwen, *Ann. Jard. Bot. Buitenzorg*, 1936; **b** courtesy Stephen Self, University of Texas, Arlington.

GLOSSARY

The Glossary gives brief definitions of the most important terms that recur in the text, excluding taxonomic designations. For fuller definitions, consult the index, where italicized page numbers refer you to explanations of key terms in context.

Of the basic units of measurement, some are tabulated on p. A16, others have their own alphabetical entries.

Interspersed alphabetically with the vocabulary are the main prefixes and combining forms used in biology. You will notice that, while they are generally of Greek or Latin origin, many of them have acquired a new meaning in biology (examples: *blasto-, -cyte, caryo-, -plasm*). Familiarity with these forms will make it easier for you to learn and remember the numerous terms in which they are incorporated.

TABLE 1   *Standard prefixes of the metric system*

kilo- (k)	1.000	10^3
deci- (d)	0.1	10^{-1}
centi- (c)	0.01	10^{-2}
milli- (m)	0.001	10^{-3}
micro- (μ)	0.000001	10^{-6}
nano- (n)	0.000000001	10^{-9}

TABLE 2   *Common units of length, weight, and liquid capacity*

kilometer (km)	1,000 m	0.62137 mile
meter (m)		39.37 inches
centimeter (cm)	0.01 m	0.39 inch
millimeter (mm)	0.001 m	0.039 inch
micrometer* (μm)	10^{-6} m	
nanometer (nm)	10^{-9} m	
angstrom† (Å)	10^{-10} m	
kilogram (kg)	1,000 g	2.2 pounds
gram (g)		0.035 ounce
milligram (mg)	0.001 g	
microgram (μg)	10^{-6} g	
liter (l)	1,000 cm^3	1.057 quarts
milliliter (ml)	0.001 l	

*Formerly called micron.
†No longer used; nanometer used instead

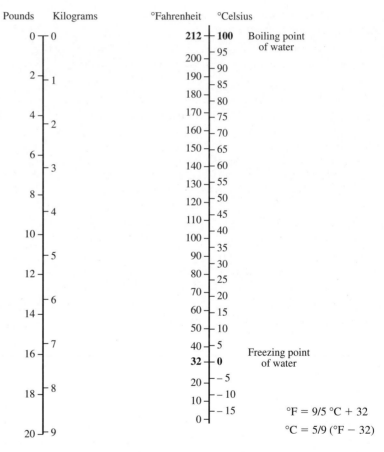

$$°F = 9/5 \ °C + 32$$
$$°C = 5/9 \ (°F - 32)$$

a- Without, lacking.

ab- Away from, off.

abdomen [L belly] In mammals, the portion of the trunk posterior to the thorax, containing most of the viscera except heart and lungs. In other animals, the posterior portion of the body.

absolute zero The temperature (−273°C) at which all thermal agitation ceases. The lowest possible temperature.

acellular Not constructed on a cellular basis.

acid [L *acidus* sour] A substance that increases the concentration of hydrogen ions when dissolved in water.

acoelomate A body plan in which there is no cavity between the digestive tract and the body wall.

ACTH *See* adrenocorticotropic hormone.

action potential *See* potential.

active site In an enzyme, the portion of the molecule that reacts with a substrate molecule.

active transport Movement of a substance across a membrane by a process requiring expenditure of energy by the cell.

ad- Next to, at, toward.

adaptation In evolution, any genetically controlled characteristic that increases an organism's fitness, usually by helping the organism to survive and reproduce in the environment it inhabits. In neurobiology, the process that results in a short-lasting decline in responsiveness of a sensory neuron after repeated firing; *cf.* habituation, sensitization.

adenosine diphosphate (ADP) A doubly phosphorylated organic compound that can be further phosphorylated to form ATP.

adenosine monophosphate (AMP) A singly phosphorylated organic compound that can be further phosphorylated to form ADP.

adenosine triphosphate (ATP) A triply phosphorylated organic compound that functions as "energy currency" for organisms.

adipose [L *adeps* fat] Fatty.

ADP *See* adenosine diphosphate.

adrenal [L *renes* kidneys] An endocrine gland of vertebrates located near the kidney.

adrenalin A hormone produced by the adrenal medulla that stimulates "fight-or-flight" reactions.

adrenocorticotropic hormone (ACTH) A hormone produced by the pituitary that stimulates the adrenal cortex.

adsorb [L *sorbēre* to suck up] Hold on a surface.

advanced New, unlike the ancestral condition.

aerobic [L *aer* air] With oxygen.

alcohol Any of a class of organic compounds in which one or more —OH groups are attached to a carbon backbone.

alkaline Having a pH of more than 7. *See* base.

all-, allo- [Gk *allos* other] Other, different.

allele Any of several alternative gene forms at a given chromosomal locus.

allopatric [L *patria* homeland] Having different ranges.

allosteric Of an enzyme: one that can exist in two or more conformations. *Allosteric control:* control of the activity of an allosteric enzyme by determination of the particular conformation it will assume.

altruism The willingness of an individual to sacrifice its fitness for the benefit of another. *Reciprocal altruism:* the performance of a favor by one individual in the expectation of a favor in return, as when two animals groom each other.

alveolus [L little hollow] A small cavity, especially one of the microscopic cavities that are the functional units of lungs.

amino acid An organic acid carrying an amino group (—NH$_2$); the building-block compound of proteins.

amnion [Gk caul] An extraembryonic membrane that forms a fluid-filled sac containing the embryo in reptiles, birds, and mammals.

amoeboid [Gk *amoibē* change] Amoebalike in the tendency to change shape by protoplasmic flow.

AMP *See* adenosine monophosphate.

amylase [L *amylum* starch] A starch-digesting enzyme.

an- Without.

anabolism [Gk *ana-* upward; *metabole* change] The biosynthetic building-up aspects of metabolism.

anaerobic [L *aer* air] Without oxygen.

analogous Of characters in different organisms: similar in function and often in superficial structure but of different evolutionary origins.

anemia A condition in which the blood has lower than normal amounts of hemoglobin or red blood corpuscles.

angio-, -angium [Gk *angeion* vessel] Container, receptacle.

anisogamous Reproducing by the fusion of gametes that differ only in size, as opposed to gametes that are produced by oogamous species. Gametes of oogamous species, such as egg cells and sperm, are highly differentiated.

anterior Toward the front end.

antheridium [Gk *anthos* flower] Male reproductive organ of a plant; produces sperm cells.

antibody A protein, produced by the B lymphocytes of the immune system, that binds to a particular antigen.

antigen A substance, usually a protein or polysaccharide, that activates an organism's immune system.

anus [L ring] Opening at the posterior end of the digestive tract, through which indigestible wastes are expelled.

aorta The main artery of the systemic circulation.

apical At, toward, or near the apex, or tip, of a structure such as a plant shoot.

apo- Away from.

apoplast The network cell walls and intercellular spaces within a plant body; permits extensive extracellular movement of water within the plant.

aposematic [Gk *sēma* sign] Serving as a warning, with reference particularly to colors and structures that signal possession of defensive devices.

arch- [Gk *archein* to begin] Primitive, original.

archegonium [Gk *archegonos* the first of a race] Female reproductive organ of a higher plant; produces egg cells.

archenteron [Gk *enteron* intestine] The cavity in an early embryo that becomes the digestive cavity.

arteriole A small artery.

artery A blood vessel that carries blood away from the heart.

articulation A joint between bones. Articulating surfaces are those formed between bones and joints.

artifact A by-product of scientific manipulation rather than an inherent part of the thing observed.

ascus [Gk *askos* bag] The elongate spore sac of a fungus of the Ascomycota group.

asexual Without sex.

atmosphere (atm) (unit of pressure) The normal pressure of air at sea level: 101,325 newtons per square meter (approx. 14.7 pounds per square inch).

atom [Gk *atomos* indivisible] The smallest unit of an element, not divisible by ordinary chemical means.

atomic mass unit (amu) *See* dalton.

atomic weight The average weight of an atom of an element relative to ^{12}C, an isotope of carbon with six neutrons in the nucleus. The atomic weight of ^{12}C has arbitrarily been fixed as 12.

ATP *See* adenosine triphosphate.

auto- Self, same.

autonomic nervous system A portion of the vertebrate nervous system, comprising motor neurons that innervate internal organs and are not normally under direct voluntary control.

autosome [Gk *sōma* body] Any chromosome other than a sex chromosome.

autotrophic [Gk *trophē* food] Capable of manufacturing organic nutrients from inorganic raw materials.

auxin [Gk *auxein* to grow] Any of a class of plant hormones that promote cell elongation.

axon [Gk *axōn* axis] A fiber of a nerve cell that conducts impulses away from the cell body and can release transmitter substance.

bacteriophage [Gk *phagein* to eat] A virus that attacks bacteria, *abbrev.* phage.

basal At, near, or toward the base (the point of attachment) of a structure such as a limb.

basal body A structure, identical to the centriole, found at the base of cilia and eucaryotic flagella; consists of nine triplet microtubules arranged in a circle.

base (or alkali) A substance that increases the concentration of hydroxyl ions when dissolved in water. It has a pH higher than 7.

basidium The spore-bearing structure of Basidiomycota (club fungi).

bi- Two.

bilateral symmetry The property of having two similar sides, with definite upper and lower surfaces and anterior and posterior ends.

binary fission Reproduction by the division of a cell into two essentially equal parts by a nonmitotic process.

bio- [Gk *bios* life] Life, living.

biogenesis [Gk *genesis* source] Origin of living organisms from other living organisms.

biological magnification Increasing concentration of relatively stable chemicals as they are passed up a food chain from initial consumers to top predators.

biomass The total weight of all the organisms, or of a designated group of organisms, in a given area.

biome A large climatic region with characteristic sorts of plants and animals.

biotic Pertaining to life.

blasto- [Gk *blastos* bud] Embryo.

blastocoel [Gk *koilos* hollow] The cavity of a blastula.

blastopore [Gk *poros* passage] The opening from the cavity of the archenteron to the exterior in a gastrula.

blastula An early embryonic stage in animals, preceding the delimitation of the three principal tissue layers; frequently spherical and hollow.

B lymphocyte *See* lymphocyte.

buffer A substance that binds H^+ ions when their concentration rises and releases them when their concentration falls, thereby minimizing fluctuations in the pH of a solution.

C_3 plants Plants in which the Calvin cycle is the only pathway of CO_2 fixation. One of the products of photosynthesis is a three-carbon intermediary (C_3) of the Calvin cycle, from which the plants derive their name.

C_4 plants Also called **Kranz plants.** Plants in which one of the main early products of photosynthesis is a four-carbon compound (C_4). Kranz plants can carry out photosynthesis under conditions that are inhospitable to other plants.

caecum [L *caecus* blind] A blind diverticulum of the digestive tract.

calorie [L *calor* heat] The quantity of energy, in the form of heat, required to raise the temperature of one gram of pure water one degree from 14.5 to 15.5°C. The nutritionists' Calorie (capitalized) is 1,000 calories, or one kilocalorie.

cambium [L *cambiare* to exchange] The principal lateral meristem of vascular plants; gives rise to most secondary tissue.

cAMP *See* cyclic adenosine monophosphate.

capillarity [L *capillus* hair] The tendency of aqueous liquids to rise in narrow tubes with hydrophilic surfaces.

capillary [L *capillus*] A tiny blood vessel with walls one cell thick, across which exchange of materials between blood and the tissues takes place; receives blood from arteries and carries it to veins. Also, a similar vessel of the lymphatic system.

carbohydrate Any of a class of organic compounds composed of carbon, hydrogen, and oxygen in a ratio of about two hydrogens and one oxygen for each carbon; examples are sugar, starch, cellulose.

carbon fixation The process by which CO_2 is incorporated into organic compounds, primarily glucose; energy usually comes from the ATP and $NADP_{re}$ generated by photophosphorylation, and the metabolic pathway utilizing this energy is usually the Calvin cycle.

carboxyl group The —COOH group characteristic of organic acids.

cardiac [Gk *kardia* heart] Pertaining to the heart.

carnivore [L *carnis* of flesh; *vorare* to devour] An organism that feeds on animals.

carotenoid [L *carota* carrot] Any of a group of red, orange, and yellow accessory pigments of plants, found in plastids.

carrying capacity The maximum population that a given environment can support indefinitely.

cartilage A specialized type of dense fibrous connective tissue with a rubbery intercellular matrix.

caryo- [Gk *karyon* kernel] Nucleus.

Casparian strip A waterproof thickening in the radial and end walls of endodermal cells of plants.

cata- Down.

catabolism [Gk *katabolē* a throwing down] The degradational breaking-down aspects of metabolism, by which living things extract energy from food.

catalysis [Gk *katalyein* to dissolve] Acceleration of a chemical reaction by a substance that is not itself permanently changed by the reaction.

catalyst A substance that produces catalysis.

cation A positively charged ion.

caudal [L *cauda* tail] Pertaining to the tail.

cell cycle The cycle of cellular events from one mitosis through the next. Four stages are recognized, of which the last—distribution of genetic material to the two daughter nuclei—is mitosis proper.

cell sap *See* sap.

cellulose [L *cellula* cell] A complex polysaccharide that is a major constituent of most plant cell walls.

centi- [L *centum* hundred] One hundredth.

central nervous system A portion of the nervous system that contains interneurons and exerts some control over the rest of the nervous system. In vertebrates, the brain and the spinal cord.

centri- [L *centrum* center] Center.

centrifugation [L *fugere* to flee] The spinning of a mixture at very high speeds to separate substances of different densities.

centriole A cylindrical cytoplasmic organelle located just outside the nucleus of animal cells and the cells of some lower plants; associated with the spindle during mitosis and meiosis.

centromere [Gk *meros* part] A special region on a chromosome from which kinetochore microtubules radiate during mitosis or meiosis.

cephalization [Gk *kephalē* head] Localization of neural coordinating centers and sense organs at the anterior end of the body.

cerebellum [L small brain] A part of the hindbrain of vertebrates that controls muscular coordination.

cerebrum [L brain] Part of the forebrain of vertebrates, the chief coordination center of the nervous system.

channel *See* membrane channel.

character Any structure, functional attribute, behavioral trait, or other characteristic of an organism.

character displacement The rapid divergent evolution in sympatric species of characters that minimize competition and/or hybridization between them.

chemiosmotic gradient The combined electrostatic and osmotic concentration gradient generated by the electron-transport chains of mitochondria and chloroplasts; the energy in this gradient is used, for the most part, to synthesize ATP.

chemosynthesis Autotrophic synthesis of organic materials, energy for which is derived from inorganic molecules.

chitin [Gk *chitōn* tunic] Polysaccharide that forms part of the hard exoskeleton of insects, crustaceans, and other invertebrates; also occurs in the cell walls of fungi.

chlorophyll [Gk *chlōros* greenish yellow; *phyllon* leaf] The green pigment of plants necessary for photosynthesis.

chloroplast A plastid containing chlorophyll.

chrom-, -chrome [Gk *chrōma* color] Colored; pigment.

chromatid A single chromosomal strand.

chromatin The mixture of DNA and protein (mostly histones in the form of nucleosome cores) that comprises eucaryotic nuclear chromosomes.

chromatography Process of separating substances by adsorption on media for which they have different affinities.

chromosome [Gk *sōma* body] A filamentous structure in the cell nucleus (or nucleoid), mitochondria, and chloroplasts, along which the genes are located.

cilium [L eyelid] A short hairlike locomotory organelle on the surface of a cell (*pl* cilia).

cisterna [L cistern] A cavity, sac, or other enclosed space serving as a reservoir.

classical conditioning *See* conditioning.

cleavage Division of a zygote or of the cells of an early embryo.

climax (ecological) A relatively stable stage reached in some ecological successions.

cline [Gk *klinein* to lean] Gradual variation, correlated with geography, in a character of a species.

cloaca [L sewer] Common chamber that receives materials from the digestive, excretory, and reproductive systems.

clone [Gk *klōn* twig] A group of cells or organisms derived asexually from a single ancestor and hence genetically identical.

co- With, together.

codon The unit of genetic coding, three nucleotides long, specifying an amino acid or an instruction to terminate translation.

coel-, -coel [Gk *koilos* hollow] Hollow, cavity; chamber.

coelom A body cavity surrounded by mesoderm.

coenocytic [Gk *koinos* common] Having more than one nucleus in a single mass of cytoplasm.

coenzyme A nonproteinaceous organic molecule that plays an accessory role, but a necessary one, in the catalytic action of an enzyme.

coevolution Two or more organisms evolving, each in response to the other.

coleoptile [Gk *koleon* sheath, *ptilon* feather] A sheath around the young shoot of grasses.

collagen A fibrous protein; the most abundant protein in mammals.

collenchyma [Gk *kolla* glue] A supportive tissue in plants in which the cells usually have thickenings at the angles of the walls.

colloid [Gk *kolla*] A stable suspension of particles that, though larger than in a true solution, do not settle out.

colon The large intestine.

com- Together.

commensalism [L *mensa* table] A symbiosis in which one party is benefited and the other party receives neither benefit nor harm.

community In ecology, a unit composed of all the populations living in a given area.

competition In ecology, utilization by two or more individuals, or by two or more populations, of the same limited resource; an interaction in which both parties are harmed.

condensation reaction A reaction joining two compounds with resultant formation of water.

conditioning Associative learning. *Classical conditioning:* the association of a novel stimulus with an innately recognized stimulus. *Operant conditioning:* learning of a novel behavior as a result of reward or punishment; trial-and-error learning.

conformation (of a protein) [L *conformatio* symmetrical forming] The three-dimensional pattern according to which the polypeptide chains of a protein coil (secondary structure), fold (tertiary structure), and—if there is more than one chain—fit together (quarternary structure).

conjugation [L *jugare* to join, marry] Process of genetic recombination between two organisms (e.g., bacteria, algae) through a cytoplasmic bridge between them.

connective tissue A type of animal tissue whose cells are embedded in an extensive intercellular matrix; connects, supports, or surrounds other tissues and organs.

contractile vacuole An excretory and/or osmoregulatory vacuole in some cells, which, by contracting, ejects fluids from the cell.

cooperativity The phenomenon of enhanced reactivity of the remaining binding sites of a protein as a result of the binding of substrate at one site.

cork [L *cortex* bark] A waterproof tissue, derived from the cork cambium, that forms at the outer surfaces of the older stems and roots of woody plants; the outer bark or periderm.

corpus luteum [L yellow body] A yellowish structure in the ovary, formed from the follicle after ovulation, that secretes estrogen and progesterone (*pl.* corpora lutea).

cortex [L bark] In plants, tissue between the epidermis and the vascular cylinder of stems and roots. In animals, the outer barklike tissue of some organs, as *cerebral cortex, adrenal cortex,* etc.

cotyledon [Gk *kotyle* cup] A "seed leaf," a food-digesting and -storing part of a plant embryo.

countercurrent exchange A strategy in which two streams move past each other in opposite directions, facilitating the exchange of substances between them across a membrane. The gills in gas exchange and kidneys in the production of concentrated urine are two sites of countercurrent exchange.

covalent bond A chemical bond resulting from the sharing of a pair of electrons.

Crassulacean acid metabolism (CAM) A variation of photosynthesis found in some plants that grow in hot, dry environments. Plants such as succulents avoid water loss by closing their stomata during the day and opening them at night.

crossing over Exchange of parts between two homologous chromosomes.

cross section *See* section.

cryptic [Gk *kryptos* hidden] Concealing.

cuticle [L *cutis* skin] A waxy layer on the outer surface of leaves, insects, etc.

cyclic adenosine monophosphate (cyclic AMP or cAMP) Compound, synthesized in living cells from ATP, that functions as an intracellular mediator of hormonal action; also plays a part in neural transmission and some other kinds of cellular control systems.

cyst [Gk *kystis* bladder, bag] (1) A saclike abnormal growth. (2) Capsule that certain organisms secrete around themselves and that protects them during resting stages.

-cyte, cyto- [Gk *kytos* container] Cell.

cytochrome Any of a group of iron-containing enzymes important in electron transport during respiration or photophosphorylation.

cytokinesis [Gk *kinēsis* motion] Division of the cytoplasm of a cell.

cytoplasm All of a cell except the nucleus.

cytosol The relatively fluid, less structured part of the cytoplasm of a cell, excluding organelles and membranous structures.

dalton A unit of mass equal to one twelfth the atomic weight of ^{12}C, or 1.66024×10^{-24} gram. Formerly called atomic mass unit (amu).

deamination Removal of an amino group.

deciduous [L *decidere* to fall off] Shedding leaves each year.

dehydration reaction A condensation reaction.

deme [Gk *dēmos* population] A local unit of population of any one species.

dendr-, dendro- [Gk *dendron* tree] Tree; branching.

dendrite A short unsheathed fiber of a nerve cell—often spiny, usually branched and tapering—that receives many synapses and carries excitation and inhibition toward the cell body.

deoxyribonucleic acid (DNA) A nucleic acid found in most viruses, all bacteria, chloroplasts, mitochondria, and the nuclei of eucaryotic cells, characterized by the presence of a deoxyribose sugar in each nucleotide; the genetic material of all organisms except the RNA viruses.

-derm [Gk *derma* skin] Skin, covering; tissue layer.

di- Two.

dicot A member of a subclass of the angiosperms, or flowering plants, characterized by the presence of two cotyledons in the embryo, a netlike system of veins in the leaves, and flower petals in fours or fives; *cf.* monocot. *Herbaceous dicot:* a perennial whose aboveground parts die annually. *Woody dicot:* a perennial whose above ground parts—trunk and branches—remain alive and grow annually.

differentiation The process of developmental change from an immature to a mature form, especially in a cell.

diffusion The movement of dissolved or suspended particles from place to place as a result of their heat energy (thermal agitation).

digestion Hydrolysis of complex nutrient compounds into their building-block units.

diploid [Gk *diploos* double] Having two of each type of chromosome.

disaccharide A double sugar, one composed of two simple sugars.

distal [L *distare* to stand apart] Situated away from some reference point (usually the main part of the body).

diverticulum [L *devertere* to turn aside] A blind sac branching off a cavity or canal.

DNA *See* deoxyribonucleic acid.

dominant (1) Of an allele: exerting its full phenotypic effect despite the presence of another allele of the same gene, whose phenotypic expression it blocks or masks. *Dominant phenotype, dominant character:* one caused by a dominant allele. (2) Of an individual: occupying a high position in the social hierarchy.

dormancy [L *dormire* to sleep] The state of being inactive, quiescent. In plants, particularly seeds and buds, a period in which growth is arrested until environmental conditions become more favorable.

dorsal [L *dorsum* back] Pertaining to the back.

drive *See* motivation.

duodenum [From a Latin phrase meaning 12 (*duodecin*) finger's-breadths long] The first portion of the small intestine of vertebrates, into which ducts from the pancreas and gallbladder empty.

ecosystem [Gk *oikos* habitation] The sum of physical features and organisms occurring in a given area.

ecto- Outside, external.

ectoderm The outermost tissue layer of an animal embryo. Also, tissue derived from the embryonic ectoderm.

ectothermic *See* poikilothermic.

effector The part of an organism that produces a response, e.g., muscle, cilium, flagellum.

egg An egg cell or female gamete. Also a structure in which embryonic development takes place, especially in birds and reptiles; consists of an egg cell, various membranes, and often a shell.

electrochemical gradient Combined electrostatic and osmotic-concentration gradient, such as the chemiosmotic gradient of mitochondria and chloroplasts.

electron A negatively charged primary subatomic particle.

electronegativity The formal measure of an atom's attraction for free electrons. Atoms with few electron vacancies in their outer shell tend to be more electronegative than those with more. In covalent bonds, the shared electrons are, on average, nearer the more electronegative atom; this asymmetry, in part, gives rise to the polarity of certain molecules.

electronic charge unit The charge of one electron, or 1.6021×10^{-19} coulomb.

electron-transport chain A series of enzymes found in the inner membrane of mitochondria and (with somewhat different components) in the thylakoid membrane of chloroplasts. The chain accepts high-energy electrons and uses their energy to create a chemiosmotic gradient across the membrane in which it is located.

electrostatic force The attraction (also called *electrostatic attraction*) between particles with opposite charges, as between a proton and an electron, or between H⁺ and OH⁻; and the repulsion between particles with like charges, as between two H⁺ ions.

electrostatic gradient The free-energy gradient created by a difference in charge between two points, generally the two sides of a membrane.

elimination (or defecation) The release of unabsorbed wastes from the digestive tract. *Cf.* excretion.

embryo A plant or animal in an early stage of development; generally still contained within the seed, egg, or uterus.

emulsion [L *emulsus* milked out] Suspension, usually as fine droplets, of one liquid in another.

-enchyma [Gk *parenchein* to pour in beside] Tissue.

end-, endo- Within, inside; requiring.

endergonic [Gk *ergon* work] Energy-absorbing; endothermic.

endocrine [Gk *krinein* to separate] Pertaining to ductless glands that produce hormones.

endocytosis The process by which the cell membrane forms an invagination which becomes a vesicle, trapping extracellular material that is then transported within the cell; in general, the invagination is triggered by the binding of membrane receptors to specific substances used by the cell.

endoderm The innermost tissue layer of an animal embryo.

endodermis A plant tissue, especially prominent in roots, that surrounds the vascular cylinder; all endodermal cells have Casparian strips.

endonuclease An enzyme that breaks bonds within nucleic acids, as opposed to an exonuclease, which can digest only a terminal group. *Restriction endonuclease:* an enzyme that breaks bonds only within a specific sequence of bases.

endoplasmic reticulum [L *reticulum* network] A system of membrane-bounded channels in the cytoplasm.

endoskeleton An internal skeleton.

endosperm [Gk *sperma* seed] A nutritive material in seeds.

endosymbiotic hypothesis Hypothesis that certain eucaryotic organelles—in particular mitochondria and chloroplasts—originated as free-living procaryotes that took up mutalistic residence in the ancestors of modern eucaryotes.

endothermic In thermodynamics, energy-absorbing (endergonic). In physiology, warm-blooded (homeothermic).

entropy Measure of the disorder of a system.

enzyme [Gk *zymē* leaven] A compound, usually a protein, that acts as a catalyst.

epi- Upon, outer.

epicotyl A portion of the axis of a plant embryo above the point of attachment of the cotyledons; forms most of the shoot.

epidermis [Gk *derma* skin] The outermost portion of the skin or body wall of an animal.

episome [Gk *sōma* body] Genetic element at times free in the cytoplasm, at other times integrated into a chromosome.

epithelium An animal tissue that forms the covering or lining of all free body surfaces, both external and internal.

equilibrium constant The ratio of products of a reaction to the reactants after the reaction has been allowed to proceed until there is no further change in these concentrations.

erythrocyte [Gk *erythros* red] A red blood corpuscle, i.e., a blood corpuscle containing hemoglobin.

esophagus [Gk *phagein* to eat] An interior part of the digestive tract; in mammals it leads from the pharynx to the stomach.

essential fatty acid A fatty acid an organism needs but cannot synthesize, and so must obtain preformed (or in a precursor form) from its diet.

estrogen [L *oestrus* frenzy] Any of a group of vertebrate female sex hormones.

estrous cycles [L *oestrus*] In female mammals, the higher primates excepted, a recurrent series of physiological and behavioral changes connected with reproduction.

estuary That portion of a river that is close enough to the sea to be influenced by marine tides.

eu- [Gk *eus* good] Most typical, true.

eucaryotic cell A cell containing a distinct membrane-bounded nucleus, characteristic of all organisms except bacteria.

evaginated [L *vagina* sheath] Folded or protruded outward.

eversible [L *evertere* to turn out] Capable of being turned inside out.

evolution [L *evolutio* unrolling] Change in the genetic makeup of a population with time.

ex-, exo- Out of, outside; producing.

excretion Release of metabolic wastes and excess water. *Cf.* elimination.

exergonic [Gk *ergon* work] Energy-releasing; exothermic.

exocytosis The process by which an intracellular vesicle fuses with the cell membrane, expelling its contents into its surroundings.

exon A part of a primary transcript (and the corresponding part of a gene) that is ultimately either translated (in the case of mRNA) or utilized in a final product, such as tRNA.

exoskeleton An external skeleton.

extrinsic External to, not a basic part of; as in *extrinsic isolating mechanism.*

fauna The animals of a given area or period.

feature detector A circuit in the nervous system that responds to a specific type of feature, such as a vertically moving spot or a particular auditory time delay.

feces [L *faeces* dregs] Indigestible wastes discharged from the digestive tract.

feedback The process by which a control mechanism is regulated through the very effects it brings about. *Positive feedback:* the process by which a small effect is amplified, as when a depolarization triggers an acton potential. *Negative feedback* (or feedback inhibition): the process by which a control mechanism is activated to restore conditions to their original state.

fermentation Anaerobic production of alcohol, lactic acid, or similar compounds from carbohydrates via the glycolytic pathway.

fertilization Fusion of nuclei of egg and sperm.

fetus [L *fetus* pregnant] An embryo in its later development, still in the egg or uterus.

fitness The probable genetic contribution of an individual (or allele or genotype) to succeeding generations. *Inclusive fitness:* the sum of an individual's personal fitness plus the fitness of that individual's relatives devalued in proportion to their genetic distance from the individual.

fixation (1) Conversion of a substance into a biologically more usable form, as the conversion of CO_2 into carbohydrate by photosynthetic plants or the incorporation of N_2 into more complex molecules by nitrogen-fixing bacteria. (2) Process of treating living tissue for microscopic examination.

flagellum [L whip] A long hairlike locomotory organelle on the surface of a cell.

flora The plants of a given area or period.

follicle [L *follis* bag] A jacket of cells around an egg cell in an ovary.

follicle-stimulating hormone (FSH) A gonadotropic hormone of the anterior pituitary that stimulates growth of follicles in the ovaries of females and function of the seminiferous tubules in males.

food chain Sequence of organisms, including producers, consumers, and decomposers, through which energy and materials may move in a community.

foot-candle Unit of illumination; the illumination of a surface produced by one standard candle at a distance of one foot; *cf.* lambert.

founder effect The difference between the gene pool of a population as a whole and that of a newly isolated population of the same species.

free energy Usable energy in a chemical system; energy available for producing change.

fruit A mature ovary or cluster of ovaries (sometimes with additional floral structures associated with the ovary).

fruiting body A spore-bearing structure (e.g., the above-ground portion of a mushroom).

FSH *See* follicle-stimulating hormone.

gamete [Gk *gamete(s)* wife, husband] A sexual reproductive cell that must usually fuse with another such cell before development begins; an egg or sperm.

gametophyte [Gk *phyton* plant] A multicellular haploid plant that can produce gametes.

ganglion [Gk tumor] A structure containing a group of cell bodies of neurons (*pl.* ganglia).

gastr-, gastro- [Gk *gastēr* belly] Stomach; ventral; resembling the stomach.

gastrovascular cavity An often branched digestive cavity, with only one opening to the outside, that conveys nutrients throughout the body; found only in animals without circulatory systems.

gastrula A two-layered, later three-layered, animal embryonic stage.

gastrulation The process by which a bastula develops into a gastrula, usually by an involution of cells.

gated channel A membrane channel that can open or close in response to a signal, generally a change in the electrostatic gradient or the binding of a hormone, transmitter, or other molecular signal.

gel Colloid in which the suspended particles form a relatively orderly arrangement; *cf.* sol.

-gen; -geny [Gk *genos* birth, race] Producing; production, generation.

gene [Gk *genos*] The unit of inheritance; usually a portion of a DNA molecule that codes for some product such as a protein, tRNA, or rRNA.

gene amplification Any of the strategies that give rise to multiple copies of certain genes, thus facilitating the rapid synthesis of a product (such as rRNA for ribosomes) for which the demand is great.

gene flow The movement of genes from one part of a population to another, or from one population to another, via gametes.

gene pool Thes sum total of all the genes of all the individuals in a population.

gene regulation Any of the strategies by which the rate of expression of a gene can be regulated, as by controlling the rate of transcription.

generator potential *See* potential.

genetic drift Change in the gene pool as a result of chance and not as a result of selection, mutation, or migration.

genome The cell's total complement of DNA: in eucaryotes, the nuclear and organelle chromosomes; in procaryotes, the major chromosome, episomes, and plasmids. In viruses and viroids, the total complement of DNA or RNA.

genotype The particular combination of genes present in the cells of an individual.

germ cell A sexual reproductive cell; an egg or sperm.

gibberellin A plant hormone—one of its effects is stem elongation in some dwarf plants.

gill An evaginated area of the body wall of an animal, specialized for gas exchange.

gizzard A chamber of an animal's digestive tract specialized for grinding food.

glucose [Gk *glykys* sweet] A six-carbon sugar; plays a central role in cellular metabolism.

glycocalyx The layer of protein and carbohydrates just outside the plasma membrane of an animal cell; in general, the proteins are anchored in the membrane, and the carbohydrates are bound to the proteins.

glycogen [Gk *glykys*] A polysaccharide that serves as the principal storage form of carbohydrate in animals.

glycolysis [Gk *glykys*] Anaerobic catabolism of carbohydrates to pyruvic acid.

Golgi apparatus Membranous subcellular structure that plays a role in storage and modification particularly of secretory products.

gonadotropic Stimulatory to the gonads.

gonadotropin A hormone stimulatory to the gonads, a gonadotropic hormone.

gonads [Gk *gonos* seed] The testes or ovaries.

gram molecule *See* mole.

granum [L grain] A stacklike grouping of photosynthetic membranes in a chloroplast (*pl.* grana).

guard cell A specialized epidermal cell that regulates the size of stoma of a leaf.

habit [L *habitus* disposition] In biology, the characteristic form or mode of growth of an organism.

habitat [L it lives] The kind of place where a given organism normally lives.

habituation The process that results in a long-lasting decline in the receptiveness of interneurons (primarily) to the input from sensory neurons or other interneurons; *cf.* sensitization, adaptation.

haploid [Gk *haploos* single] Having only one of each type of chromosome.

hem-, hemat-, hemo- [Gk *haima* blood] Blood.

hematopoiesis [Gk *poiēsis* making] The formation of blood.

hemoglobin A red iron-containing pigment in the blood that functions in oxygen transport.

hepatic [Gk *hēpar* liver] Pertaining to the liver.

herbaceous [L *herbaceus* grassy] Having a stem that remains soft and succulent; not woody.

herbaceous dicot *See* dicot.

herbivore [L *herba* grass; *vorare* to devour] An animal that eats plants.

Hertz A unit of frequency (as of sound waves) equal to one cycle per second.

hetero- [Gk *heteros* other] Other, different.

heterogamy [Gk *gamos* marriage] The condition of producing gametes of two or more different types.

heterotrophic [Gk *trophē* food] Incapable of manufacturing organic compounds from inorganic raw materials, therefore requiring organic nutrients from the environment.

heterozygous [Gk *zygōtos* yoked] Having two different alleles of a given gene.

Hg [L *hydrargyrum* mercury] The symbol for mercury. Pressure is often expressed in *mm Hg*—the pressure exerted by a column of mercury whose height is measured in millimeters (at $0°$ C, 1 mm Hg = 133.3 newtons per square meter).

hilum Region where blood vessels, nerves, ducts, enter an organ.

hist- [Gk *histos* web] Tissue.

histology The structure and arrangement of the tissues of organisms; the study of these.

histone One of a class of basic proteins serving as structural elements of eucaryotic chromosomes.

homeo-, homo- [Gk *homoios* like] Like, similar.

homeostasis The tendency in an organism toward maintenance of physiological and psychological stability.

homeothermic [Gk *thermē* heat] Capable of self-regulation of body temperature; warm-blooded, endothermic.

home range An area within which an animal tends to confine all or nearly all its activities for a long period of time.

homologous Of chromosomes: bearing genes for the same characters. Of characters in different organisms: inherited from a common ancestor.

homozygous [Gk *zygōtos* yoked] Having two copies of the same allele of a given gene.

hormone [Gk *horman* to set in motion] A control chemical secreted in one part of the body that affects other parts of the body.

hybrid In evolutionary biology, a cross between two species. In genetics, a cross between two genetic types.

hydr-, hydro- [Gk *hydor* water] Water; fluid; hydrogen.

hydration Formation of a sphere of water around an electrically charged particle.

hydrocarbon Any compound made of only carbon and hydrogen.

hydrogen bond A weak chemical bond formed when two polar molecules, at least one of which usually consists of a hydrogen bonded to a more electronegative atom (usually oxygen or nitrogen), are attracted electrostatically.

hydrolysis [Gk *lysis* loosing] Breaking apart of a molecule by addition of water.

hydrophilic Readily entering into solution by forming hydrogen bonds with water or other polar molecules.

hydrophobic Incapable of entering into solution by molecules that are neither ionic nor polar, and therefore cannot dissolve in water.

hydrostatic [Gk *statikos* causing to stand] Pertaining to the pressure and equilibrium of fluids.

hydroxyl ion The $OH^{-2'}$ ion.

hyper- Over, overmuch; more.

hypertonic Of a solution (or colloidal suspension): tending to gain water from some reference solution (or colloidal suspension) separated from it by a selectively permeable membrane—usually because it has a higher osmotic concentration than the reference solution.

hypertrophy [Gk *trophē* food] Abnormal enlargement, excessive growth.

hypha [Gk *hyphē* web] A fungal filament.

hypo- Under, lower; less.

hypocotyl The portion of the axis of a plant embryo below the point of attachment of the cotyledons; forms the base of the shoot and the root.

hypothalamus [Gk *thalamos* inner chamber] Part of the posterior portion of the vertebrate forebrain, containing important centers of the autonomic nervous system and centers of emotion.

hypotonic Of a solution (or colloidal suspension): tending to lose water to some reference solution (or colloidal suspension) separated from it by a selectively permeable membrane—usually because it has a lower osmotic concentration than the reference solution.

imprinting A kind of associative learning in which an animal rapidly learns during a particular critical period to recognize an object, individual, or location in the absence of overt reward; distinguished from most other associative learning in that it is retained indefinitely, being difficult or impossible to reverse.

independent assortment The Principle of Independent Assortment is frequently referred to as Mendel's second law. Genes found on different chromosomes, so-called unlinked genes, assort independently in meiosis unless they are recombined by crossing over.

inducer In embryology, a substance that stimulates differentiation of cells or development of a particular structure. In genetics, a substance that activates particular genes.

inorganic compound A chemical compound not based on carbon.

in situ [L in place] In its natural or original position.

instinct Heritable, genetically specified neural circuitry that guides and directs behavior.

insulin [L *insula* island] A hormone produced by the β islet cells in the pancreas that helps regulate carbohydrate metabolism, especially conversion of glucose into glycogen.

integument [L *integere* to cover] A coat, skin, shell, rind, or other protective surface structure.

inter- Between (e.g., *interspecific,* between two or more different species).

interneuron A neuron that receives input from and synapses on other neurons, as distinguished from a sensory neuron (which receives sensory information) and a motor neuron (which synapses on a muscle).

intra- Within (e.g., *intraspecific,* within a single species).

intrinsic Inherent in, a basic part of; as in *intrinsic isolating mechanism.*

intron A part of a primary transcript (and the corresponding part of a gene) that lies between exons, and is removed before the RNA becomes functional.

invaginated [L *vagina* sheath] Folded or protruded inward.

invertebrate [L *vertebra* joint] Lacking a backbone, hence an animal without bones.

in vitro [L in glass] Not in the living organism, in the laboratory.

in vivo [L in the living] In the living organism.

ion An electrically charged atom.

ionic bond A chemical bond formed by the electrostatic attraction between two oppositely charged ions.

iso- equal, uniform

isogamy [Gk *gamos* marriage] The condition of producing gametes of only one type, with no distinction existing between male and female.

isolating mechanism An obstacle to interbreeding, either extrinsic, such as a geographic barrier, or intrinsic, such as structural or behavioral incompatibility.

isotonic Of a solution (or colloidal suspension): tending neither to gain nor to lose water when separated from some reference solution (or colloidal suspension) by a selectively permeable membrane—usually because it has the same osmotic concentration as the reference solution.

isotope [Gk *topos* place] An atom differing from another atom of the same element in the number of neutrons in its nucleus.

kilo- A thousand.

kin-, kino- [Gk *kinema* motion] Motion, action.

kinase An enzyme that catalyzes the phosphorylation of a substrate by ATP.

lactic acid A three-carbon organic acid produced in animals and some microorganisms by fermentation.

lambert In metric system, unit of brightness of a light source; approximately equivalent to 929 foot-candles.

lamella [L thin plate] A thin platelike structure; a fairly straight intracellular membrane.

larva [L ghost, mask] Immature form of some animals that undergo radical transformation to attain the adult form.

lateral Pertaining to the side.

lateral inhibition Process by which adjacent sensory cells or their targets interact to inhibit one another when excited, the result being an exaggerated contrast; neural basis of feature-detector circuits.

lenticel [L *lenticella* small lentil] A porous region in the periderm of a woody stem through which gases can move.

leukocyte [Gk *leukos* white] A white blood cell; *cf.* lymphocyte, macrophage.

LH *See* luteinizing hormone.

ligament [L *ligare* to bind] A type of connective tissue linking two bones in a joint.

ligase An enzyme that catalyzes the bonding between adjacent nucleotides in DNA and RNA.

lignin [L *lignum* wood] An organic compound in wood that makes cellulose harder and more brittle.

linkage The presence of two or more genes on the same chromosome, which, in the absence of crossing over, causes the characters they control to be inherited together.

lip- [Gk *lipos* fat] Fat or fatlike.

lipase A fat-digesting enzyme.

lipid Any of a variety of compounds insoluble in water but soluble in ethers and alcohols; includes fats, oils, waxes, phospholipids, and steroids.

locus [L place] In genetics, a particular location on a chromosome, hence often used synonymously with gene (*pl.* loci).

lumen [L light, opening] The space or cavity within a tube or sac (*pl.* lumina).

lung An internal chamber specialized for gas exchange in an animal.

luteinizing hormone (LH) A gonadotropic hormone of the pituitary that stimulates conversion of a follicle into corpus luteum and secretion of progesterone by the corpus luteum; also stimulates secretion of sex hormone by the testes.

lymph [L *lympha* water] A fluid derived from tissue fluid and transported in special lymph vessels to the blood.

lymphocyte A white blood cell that responds to the pres-

ence of a foreign antigen. *B lymphocyte:* a cell that upon stimulation by an antigen secretes antibodies. *T lymphocyte:* a cell that attacks infected cells and modulates the activity of B lymphohocytes.

-lysis, lyso- [Gk *lysis* loosing] Loosening, decomposition.

lysogenic Of bacteria: carrying bacteriophage capable of lysing, i.e., destroying, other bacterial cells.

lysosome A subcellular organelle that stores digestive enzymes.

macro- Large.

macrophage A phagocytic white blood cell that ingests material—particularly viruses, bacteria, and clumped toxins—bound by circulating antibodies.

Malpighian tubule An excretory diverticulum of the digestive tract in insects and some other anthropods.

mast cell Cells that are specialized for the secretion of histamine and other local chemical mediators as part of the immune response.

matrix [L *mater* mother] A mass in which something is embedded, e.g., the intercellular substance of a tissue.

medulla [L marrow, innermost part] (1) The inner portion of an organ, e.g., *adrenal medulla,* (2) The *medulla oblongata,* a portion of the vertebrate hindbrain that connects with the spinal cord.

medusa [*after* Medusa, mythological monster with snaky locks] The free-swimming stage in the life cycle of a coelenterate.

mega- Large.

megaspore A spore that will germinate into a female plant.

meiosis [Gk *meiōsis* diminution] A process of nuclear division in which the number of chromosomes is reduced by half.

membrane A structure, formed mainly by a double layer of phospholipids, which surrounds cells and organelles.

membrane channel A pore in a membrane through which certain molecules may pass.

membrane pump A permease that uses energy, usually from ATP, to move substances across the membrane against their osmotic-concentration or electrostatic gradients.

meristematic tissue [Gk *meristos* divisible] A plant tissue that functions primarily in production of new cells by mitosis.

meso- Middle.

mesoderm The middle tissue layer of an animal embryo.

mesophyll [Gk *phyllon* leaf] The parenchymatous middle tissue layers of a leaf.

meta- Posterior, later; change in.

metabolism [Gk *metabolē* change] The sum of the chemical reactions within a cell (or a whole organism), including the energy-releasing breakdown of molecules (catabolism) and the synthesis of complex molecules and new protoplasm (anabolism).

metamorphosis [Gk *morphē* form] Transformation of an immature animal into an adult. More generally, change in the form of an organ or structure.

micro- Small. Male. In units of measurement, one millionth.

microfilament A long, thin structure, usually formed from the protein actin; when associated with myosin filaments, as in muscles, microfilaments are involved in movement.

microorganism A microscopic organism, especially a bacterium, virus, or protozoan.

microspore A spore that will germinate into a male plant.

microtubule A long, hollow structure formed from the protein tubulin; found in cilia, eucaryotic flagella, basal bodies/centrioles, and the cytoplasm.

middle lamella A layer of substance deposited between the walls of adjacent plant cells.

milli- One thousandth.

mineral In biology, any naturally occurring inorganic substance, excluding water.

mitochondrion [Gk *mitos* thread; *chondrion* small grain] Subcellular organelle in which aerobic respiration takes place.

mitosis [Gk *mitos*] Process of nuclear division in which complex movements of chromosomes along a spindle result in two new nuclei with the same number of chromosomes as the original nucleus.

modulator A control chemical that stabilizes an allosteric enzyme in one of its alternative conformations.

mold Any of many fungi that produce a cottony or furry growth.

mole The amount of a substance that has a weight in grams numerically equal to the molecular weight of the substance. One mole of a substance contains 6.023×10^{23} molecules of that substance; hence one mole of a substance will always contain the same number of molecules as a mole of any other substance.

molecular weight The weight of a molecule calculated as the sum of the atomic weights of its constituent atoms.

molecule A chemical unit consisting of two or more atoms bonded together.

mono- One.

monocot A member of a subclass of angiosperms, or flowering plants, characterized by the presence of a single cotyledon in the embryo, parallel veins in the leaves, and flower petals in threes; *cf.* dicot.

-morph, morpho- [Gk *morphē* form] Form, structure.

morphogenesis The establishment of shape and pattern in an organism.

morphology The form and structure of organisms or parts of organisms; the study of these.

motivation The internal state of an animal that is the immediate cause of its behavior; drive.

motor neuron A neuron, leading away from the central nervous system, that synapses on and controls an effector.

motor program A coordinated, relatively stereotyped series of muscle movements performed as a unit, either innate (as the movements of swallowing) or learned (as in speech); also, the neural circuitry underlying such behavior; *cf.* reflex.

mouthparts Structures or appendages near the mouth used in manipulating food.

mucosa Any membrane that secretes mucus (a slimy protective substance), e.g., the membrane lining the stomach and intestine.

muscle [L *musculus* small mouse, muscle] A contractile tissue of animals.

mutation [L *mutatio* change] Any relatively stable heritable change in the genetic material.

mutualism A symbiosis in which both parties benefit.

mycelium [Gk *mykēs* fungus] A mass of hyphae forming the body of a fungus.

myo- [Gk *mys* mouse, muscle] Muscle.

NAD *See* nicotinamide adenine dinucleotide.

NADP *See* nicotinamide adenine dinucleotide phosphate.

nano- [L *nanus* dwarf] One billionth.

natural selection Differential reproduction in nature, leading to an increase in the frequency of some genes or gene combinations and to a decrease in the frequency of others.

navigation The initiation and/or maintenance of movement toward a goal.

negative feedback *See* feedback.

nematocyst [Gk *nēma* thread; *kystis* bag] A specialized stinging cell in coelenterates; contains a hairlike structure that can be ejected.

neo- New.

neocortex Portion of the cerebral cortex in mammals, of relatively recent evolutionary origin; often greatly expanded in the higher primates and dominant over other parts of the brain.

nephr- [Gk *nephros* kidney] Kidney.

nephridium An excretory organ consisting of an open bulb and a tubule leading to the exterior; found in many invertebrates, such as segmented worms.

nephron The functional unit of a vertebrate kidney, consisting of Bowman's capsule, convoluted tubule, and loop of Henle.

nerve [L *nervus* sinew, nerve] A bundle of neuron fibers (axons).

nerve net A nervous system without any central control, as in coelenterates.

neuron [Gk nerve, sinew] A nerve cell.

neutron An electrically neutral subatomic particle with approximately the same mass as a proton.

niche The functional role and position of an organism in the ecosystem; the way an organism makes its living, including, for an animal, not only what it eats, but when, where, and how it obtains food, where it lives, etc.

nicotinamide adenine dinucleotide (NAD) An organic compound that functions as an electron acceptor, e.g., in respiration.

nicotinamide adenine dinucleotide phosphate (NADP) An organic compound that functions as an electron acceptor, e.g., in biosynthesis.

nitrogen fixation Incorporation of nitrogen from the atmosphere into substances more generally usable by organisms.

node (of plant) [L *nodus* knot] Point on a stem where a leaf or bud is (or was) attached.

nonhomologous Of chromosomes: two chromosomes that do not share the same genes and thus do not pair during meiosis.

notochord [Gk *nōtos* back; *chordē* string] In the lower chordates and in the embryos of the higher vertebrates, a flexible supportive rod running longitudinally through the back just ventral to the nerve cord.

nucleic acid Any of several organic acids that are polymers of nucleotides and function in transmission of hereditary traits, in protein synthesis, and in control of cellular activities.

nucleoid A region, not bounded by a membrane, where the chromosome is located in a procaryotic cell.

nucleolus A dense body within the nucleus, usually attached to one of the chromosomes; consists of multiple copies of the genes for certain kinds of rRNA.

nucleosome A complex consisting of several histone proteins, which together form a "spool," and chromosomal DNA, which is wrapped around the spool.

nucleotide A chemical entity consisting of a five-carbon sugar with a phosphate group and a purine or pyrimidine attached; building-block unit of nucleic acids.

nucleus (of cell) [L kernel] A large membrane-bounded organelle containing the chromosomes.

nutrient [L *nutrire* to nourish] A food substance usable in metabolism as a source of energy or of building material.

nymph [Gk *nymphē* bride, nymph] Immature stage of insect that undergoes gradual metamorphosis.

olfaction [L *olfacere* to smell] The sense of smell.

omnivorous [L *omnis* all; *vorare* to devour] Eating a variety of foods, including both plants and animals.

oncogene A gene that causes one of the biochemical changes that lead to cancer.

ontogeny [Gk *ōn* being] The course of development of an individual organism.

oo- [Gk *ōion* egg] Egg.

oogamy A type of heterogamy in which the female gametes are large nonmotile egg cells.

oogonium Unjacketed female reproductive organ of a thallophyte plant.

operant conditioning *See* conditioning.

operator A region on the DNA to which a control substance can bind, thereby altering the rate of transcription.

oral [L *oris* of the mouth] Relating to the mouth.

organ [Gk *organon* tool] A body part usually composed of several tissues grouped together into a structural and functional unit.

organelle A well-defined subcellular structure.

organic compound A chemical compound containing carbon.

organism An individual living thing.

orientation The act of turning or moving in relation to some external feature, such as a source of light.

osmol Measure of osmotic concentration; the total number of moles of osmotically active particles per liter of solvent.

osmoregulation Regulation of the osmotic concentration of body fluids in such a manner as to keep them relatively constant despite changes in the external medium.

osmosis [Gk *ōsmos* thrust] Movement of a solvent (usually water in biology) through a selectively permeable membrane.

osmotic potential The free energy of water molecules in a solution or colloid under conditions of constant temperature and pressure; since this free energy decreases as the proportion of osmotically active particles rises, a measure of the tendency of the solution or colloid to lose water.

osmotic pressure The pressure that must be exerted on a solution or colloid to keep it in equilibrium with pure water when it is separated from the water by a selectively permeable membrane; hence a measure of the tendency of the solution or colloid to take in water.

ov-, ovi- [L *ovum* egg] Egg.

ovary Female reproductive organ in which egg cells are produced.

ovulation Release of an egg from the ovary.

ovule A plant structure, composed of an integument, sporangium, and megagametophyte, that develops into a seed after fertilization.

ovum A mature egg cell (*pl.* ova).

oxidation Energy-releasing process involving removal of electrons from a substance; in biological systems, generally by the removal of hydrogen (or sometimes the addition of oxygen).

pancreas In vertebrates, a large glandular organ located near the stomach that secretes digestive enzymes into the duodenum and also produces hormones.

papilla [L nipple] A small nipplelike protuberance.

para- Alongside of.

parapodium [Gk *podion* little foot] One of the paired segmentally arranged lateral flaplike protuberances of polychaete worms.

parasitism [Gk *parasitos* eating with another] A symbiosis in which one party benefits at the expense of the other.

parasympathetic nervous system One of the two parts of the autonomic nervous system.

parathyroids Small endocrine glands of vertebrates located near the thyroid.

parenchyma A plant tissue composed of thin-walled, loosely packed, relatively unspecialized cells.

parthenogenesis [Gk *parthenos* virgin] Production of offspring without fertilization.

pathogen [Gk *pathos* suffering] A disease-causing organism.

pectin A complex polysaccharide that cross-links the cellulose fibrils in a plant cell wall and is a major constituent of the middle lamella.

pellicle [L *pellis* skin] A thin skin or membrane.

pepsin [Gk *pepsis* digestion] A protein-digesting enzyme of the stomach.

peptide bond A covalent bond between two amino acids resulting from a condensation reaction between the amino group of one acid and the acidic group of the other.

perennial A plant that lives for several years, as compared to annuals and biennials, which live for one and two years respectively.

peri- Surrounding.

pericycle A layer of cells inside the endodermis but outside the phloem of roots and stems.

periderm The corky outer bark of older stems and roots.

peristalsis [Gk *stalsis* contraction] Alternating waves of contraction and relaxation passing along a tubular structure such as the digestive tract.

permeable [L *permeare* to go through] Of a membrane: permitting other substances to pass through.

permease [L *permeare*] A protein that allows molecules to move across a membrane; *cf.* gated channel, membrane channel, membrane pump.

petiole [L *pediculus* small foot] The stalk of a leaf.

PGAL *See* phosphoglyceraldehyde.

pH Symbol for the logarithm of the reciprocal of the hydrogen ion concentration; hence a measure of acidity. A pH of 7 is neutral; lower values are acidic, higher values alkaline (basic).

phage *See* bacteriophage.

phagocytosis [Gk *phagein* to eat] The active engulfing of particles by a cell.

pharynx Part of the digestive tract between the oral cavity and the esophagus; in vertebrates, also part of the respiratory passage.

phenotype [Gk *phainein* to show] The physical manifestation of a genetic trait.

pheromone [Gk *pherein* to carry + hormone] A substance that, secreted by one organism, influences the behavior or physiology of other organisms of the same species when they sense its odor.

phloem [Gk *phloios* bark] A plant vascular tissue that transports organic materials; the inner bark.

-phore [Gk *pherin* to carry] Carrier.

phosphoglyceraldehyde (PGAL) A three-carbon phosphorylated carbohydrate, important in both photosynthesis and glycolysis.

phospholipid A compound composed of glycerol, fatty acids, a phosphate group, and often a nitrogenous group.

phosphorylation Addition of a phosphate group.

photo- [Gk *phōs* light] Light.

photon A discrete unit of radiant energy.

photoperiodism A response by an organism to the duration and timing of the light and dark conditions.

photophosphorylation The process by which energy from light is used to convert ADP into ATP.

photosynthesis Autotrophic synthesis of organic materials in which the source of energy is light; *cf.* photophosphorylation.

-phyll [Gk *phyllon* leaf] Leaf.

phylogeny [Gk *phylē* tribe] Evolutionary history of an organism.

physiology [Gk *physis* nature] The life processes and functions of organisms; the study of these.

-phyte, phyto- [Gk *phyton* plant] Plant.

phytochrome A protein pigment of plants sensitive to red and farred light.

pinocytosis [Gk *pinein* to drink] The active engulfing by cells of liquid or of very small particles.

pistil The female reproductive organ of a flower, composed of one or more megasporophylls.

pith A tissue (usually parenchyma) located in the center of a stem (rarely a root), internal to the xylem.

pituitary An endocrine gland located near the brain of vertebrates; known as the master gland because it secretes hormones that regulate the action of other endocrine glands.

placenta [Gk *plax* flat surface] An organ in mammals, made up of fetal and maternal components, that aids in exchange of materials between the fetus and the mother.

plasm-, plasmo-, -plasm [Gk *plasma* something formed or molded] Formed material; plasma; cytoplasm.

plasma Blood minus the cells and platelets.

plasma membrane The outer membrane of a cell.

plasmid A small circular piece of DNA free in the cytoplasm of a bacterial or yeast cell and replicated independently of the cell's chromosome.

plasmodesma [Gk *desma* bond] A connection between adjacent plant cells through tiny openings in the cell walls (*pl.* plasmodesmata).

plasmolysis Shrinkage of a plant cell away from its wall when in a hypertonic medium.

plastid Relatively large organelle in plant cells that functions in photosynthesis and/or nutrient storage.

pleiotropic [Gk *pleiōn* more] Of a gene: having more than one phenotypic effect.

poikilothermic [Gk *poikilos* various; *thermē* heat] Incapable of precise self-regulation of body temperature, dependent on environmental temperature; cold-blooded, ectothermic.

polar molecule A molecule with oppositely charged sections; the charges, which are far weaker than the charges on ions, a rise from differences in electronegativity between the constituent atoms.

pollen grain [L *pollen* flour dust] A microgametophyte of a seed plant.

poly- Many.

polycistronic Pertaining to the transcription of two or more adjacent cistrons (structural genes) into a single messenger RNA molecule.

polymer [Gk *meros* part] A large molecule consisting of a chain of small molecules bonded together by condensation reactions or similar reactions.

polymerase An enzyme complex that catalyzes the polymerization of nucleotides; examples are DNA polymerase, which is involved in replication, and RNA polymerase, which is involved in transcription.

polymorphism [Gk *morphē* form] The simultaneous occurrence of several discontinuous phenotypes in a population.

polyp [Gk *polypous* many-footed] The sedentary stage in the life cycle of a Cnidarian.

polypeptide chain A chain of amino acids linked together by peptide bonds.

polyploid Having more than two complete sets of chromosomes.

polysaccharide Any carbohydrate that is a polymer of simple sugars.

population In ecology, group of individuals belonging to the same species.

portal system [L *porta* gate] A blood circuit in which two beds of capillaries are connected by a vein (e.g., *hepatic portal system*).

positive feedback *See* feedback.

posterior Toward the hind end.

potential Short for *potential difference:* the difference in electrical charge between two points. *Resting p.:* a relatively steady potential difference across a cell membrane, particularly of a nonfiring nerve cell or a relaxed muscle cell. *Action p.:* a sharp change in the potential difference across the membrane of a nerve or muscle cell that is propagated along the cell; in nerves, identified with the nerve impulse. *Generator p.:* a change in the potential difference across the membrane of a sensory cell that, if it reaches a threshold level, may trigger an action potential along the associated neural pathway.

preadapted A structure that evolved for one function but can at least minimally perform another function when placed in a new environment.

predation [L *praedatio* plundering] The feeding of free-living organisms on other organisms.

presumptive Describing the developmental fate of a tissue that is not yet differentiated. Presumptive neural tissue, for example, is destined to become part of the nervous system once it has differentiated.

primary transcript Newly synthesized RNA—generally mRNA—before the introns are moved.

primitive [L *primus* first] Old, like the ancestral condition.

primordium [L *primus; ordiri* to begin] Rudiment, earliest stage of development.

pro- Before.

proboscis [Gk *boskein* to feed] A long snout; an elephant's trunk. In invertebrates, an elongate, sometimes eversible process originating in or near the mouth that often serves in feeding.

procaryotic cell A type of cell that lacks a membrane-bounded nucleus; found only in bacteria.

progesterone [L *gestare* to carry out] One of the principal female sex hormones of vertebrates.

promoter The region of DNA to which the transcription complex binds.

prot-, proto- First, primary.

protease A protein-digesting enzyme.

protein A long polypeptide chain.

proteolytic Protein-digesting.

proton A positively charged primary subatomic particle.

proto-oncogene A gene that can, after certain sorts of mutation or translocation, or after mutation or translocation in associated control regions, become an oncogene and cause one of the changes leading to cancer.

protoplasm Living substance, the material of cells.

provirus Viral nucleic acid integrated into the genetic material of a host cell.

proximal Near some reference point (often the main part of the body).

pseudo- False; temporary.

pseudocoelom A functional body cavity not entirely enclosed by mesoderm.

pseudogene An untranscribed region of the DNA that closely resembles a gene.

pseudopod, pseudopodium [L *podium* foot] A transitory cytoplasmic protrusion of an amoeba or an amoeboid cell.

pulmonary [L *pulmones* lungs] Relating to the lungs.

purine Any of several double-ringed nitrogenous bases important in nucleotides.

pyloric [Gk *pylōros* gatekeeper] Referring to the junction between the stomach and the intestine.

pyrimidine Any of several single-ringed nitrogenous bases important in nucleotides.

pyruvic acid A three-carbon compound produced by glycolysis.

race A subspecies.

radial symmetry A type of symmetry in which the body parts are arranged regularly around a central line (in animals, running through the oral-anal axis) rather than on the two sides of a plane.

radiation As an evolutionary phenomenon, divergence of members of a single lineage into different niches or adaptive zones.

receptor In cell biology, a region, often the exposed part of a membrane protein, that binds a substance but does not catalyze a reaction in the chemical it binds; the membrane protein frequently has another region that, as a result of the binding, undergoes an allosteric change and so becomes catalytically active.

recessive Of an allele: not expressing its phenotype in the presence of another allele of the same gene, therefore expressing it only in homozygous individuals. *Recessive character, recessive phenotype:* one caused by a recessive allele.

reciprocal altruism *See* altruistic behavior.

recombination In genetics, a novel arrangement of alleles resulting from sexual reproduction and from crossing over (or, in procaryotes and eucaryotic organelles, from conjugation). In gene evolution, a novel arrangement of exons resulting from a variety of processes that duplicate and transport segments of the chromosomes within the genome; these processes include transposition, unequal crossing over, and chromosomal breakage and fusion.

rectum [L *rectus* straight] The terminal portion of the intestine.

redox reaction [*from reduction-oxidation*] A reaction involving reduction and oxidation, which inevitably occur together; *cf.* reduction, oxidation.

reduction Energy-storing process involving addition of electrons to a substance; in biological systems, generally by the addition of hydrogen (or sometimes the removal of oxygen).

reflex [L *reflexus* bent back] An automatic act consisting, in its pure form, of a single simple response to a single stimulus, as when a tap on the knee elicits a knee jerk. Distinguished from a motor program, which involves a coordinated response of several muscles.

reflex arc A functional unit of the nervous system, involving the entire pathway from receptor cell to effector.

reinforcement (psychological) Reward for a particular behavior.

releaser *See* sign stimulus.

renal [L *renes* kidneys] Pertaining to the kidney.

respiration [L *respiratio* breathing out] (1) The release of energy by oxidation of fuel molecules. (2) The taking of O_2 and release of CO_2; breathing.

resting potential *See* potential.

restriction endonuclease *See* endonuclease.

reticulum [L little net] A network.

retina The tissue in the rear of the eye that contains the sensory cells of vision.

retrovirus An RNA virus that, by means of a special enzyme (reverse transcriptase), makes a DNA copy of its genome which is then incorporated into the host's genome.

rhizoid [Gk *rhiza* root] Rootlike structure.

ribonucleic acid (RNA) Nucleic acid characterized by the presence of a ribose sugar in each nucleotide. The primary classes of RNA are mRNA (messenger RNA, which carries the instructions specifying the order of amino acids in new proteins from the genes to the ribosomes where protein synthesis takes place), rRNA (ribosomal RNA, which is incorporated into ribosomes), and tRNA (transfer RNA, which carries amino acids to the ribosomes as part of protein synthesis).

ribosome A small cytoplasmic organelle that functions in protein synthesis.

RNA *See* ribonucleic acid.

salt Any of a class of generally ionic compounds that may be formed by reaction of an acid and a base, e.g., table salt, NaCl.

sap Water and dissolved materials moving in the xylem; less commonly, solutions moving in the phloem. *Cell sap:* the fluid content of a plant-cell vacuole.

saprophyte [Gk *sapros* rotten] A heterotrophoic plant or bacterium that lives on dead organic material.

sarcomere [Gk *sarx* flesh; *meros* part] The region of a skeletal-muscle myofibril extending from one Z line to the next; the functional unit of skeletal-muscle contraction.

sclerenchyma [Gk *sclēros* hard] A plant supportive tissue composed of cells with thick secondary walls.

section *Cross or transverse s.:* section at right angles to the longest axis. *Longitudinal s.:* section parallel to the longest axis. *Radial s.:* longitudinal section along a radius.

Sagittal s.: vertical longitudinal section along the midline of a bilaterally symmetrical animal.

seed A plant reproductive entity consisting of an embryo and stored food enclosed in a protective coat.

segmentation The subdivision of an organism into more or less equivalent serially arranged units.

selection pressure In a population, the force for genetic change resulting from natural selection.

semipermeable Permeable only to solvent (usually water); less strictly; selectively permeable, i.e., permeable to some substances but not to others.

sensitization The process by which an unexpected stimulus alerts an animal, reducing or eliminating any preexisting habituation; *cf.* adaptation, habituation.

sensory neuron A neuron, leading toward the central nervous system, that receives input from a receptor cell or is itself responsive to sensory stimulation.

septum [L barrier] A partition or wall (*pl.* septa).

sessile [L *sessilis* of sitting, low] Of animals, sedentary. Of plants, without a stalk.

sex-linked Of genes: located on the X chromosome.

sexual dimorphism Morphological differences between the two sexes of a species, as in the size of tails of peacocks as compared to peahens.

sexual selection Selection for morphology or behavior directly related to attracting or winning mates. *Male-contest sexual selection:* selection for morphology or behavior that enables a male to win fights or contests for access to females, gaining a high position in a dominance hierarchy, for example, or possession of a territory. *Female-choice sexual selection:* selection for morphology or behavior that enables a male to attract females directly.

shoot A stem with its leaves, flowers, etc.

sieve element A conductile cell of the phloem.

sign stimulus (or releaser) A simple cue that orients or triggers specific innate behavior.

sinus [L curve, hollow] (1) A channel for the passage of blood lacking the characteristics of a true blood vessel. (2) A hollow within bone or another tissue (e.g. the air-filled sinuses of some of the facial bones).

sol Colloid in which the suspended particles are dispersed at random; *cf.* gel.

solute Substance dissolved in another (the solvent).

solution [L *solutio* loosening] A homogeneous molecular mixture of two or more substances.

solvent Medium in which one or more substances (the solute) are dissolved.

-soma, somat-, -some [Gk *soma* body] Body, entity.

somatic Pertaining to the body; to all cells except the germ cells; to the body wall. *Somatic nervous system:* a portion of the nervous system that is at least potentially under control of the will; *cf.* autonomic nervous system.

specialized Adapted to a special, usually rather narrow, function or way of life.

speciation The process of formation of new species.

species [L kind] The largest unit of population within which effective gene flow occurs or could occur.

sperm [Gk *sperma* seed] A male gamete.

sphincter [Gk *sphinktēr* band] A ring-shaped muscle that can close a tubular structure by contracting.

spindle A microtubular structure with which the chromosomes are associated in mitosis and meiosis.

sporangium A plant structure that produces spores.

spore [Gk *spora* seed] An asexual reproductive cell, often a resting stage adapted to resist unfavorable environmental conditions.

sporophyll [Gk *phyllon* leaf] A modified leaf that bears spores.

sporophyte [Gk *phyton* plant] A diploid plant that produces spores.

stamen [L thread] A male sexual part of a flower; a microsporophyll of a flowering plant.

starch A glucose polymer, the principal polysaccharide storage product of vascular plants.

stele [Gk *stēlē* upright slab] The vascular cylinder in the center of a root or stem, bounded externally by the endodermis.

stereo- [Gk *stereos* solid] Solid; three-dimensional.

steroid Any of a number of complex, often biologically important compounds (e.g., some hormones and vitamins), composed of four interlocking rings of carbon atoms.

stimulus Any environmental factor that is detected by a receptor.

stoma [Gk mouth] An opening, regulated by guard cells, in the epidermis of a leaf or other plant part (*pl.* stomata).

stroma [Gk *stroma* bed, mattress] The ground substance within such organelles as chloroplasts and mitochondria.

subspecies A genetically distinct geographic subunit of a species.

substrate (1) The base on which an organism lives, e.g., soil. (2) In chemical reactions, a substance acted upon, as by an enzyme.

succession In ecology, progressive change in the plant and animal life of an area.

sucrose A double sugar composed of a unit of glucose and a unit of fructose; table sugar.

suspension A heterogeneous mixture in which the particles of one substance are kept dispersed by agitation.

sym-, syn- Together.

symbiosis [Gk *bios* life] The living together of two organisms in an intimate relationship.

sympathetic nervous system One of the two parts of the autonomic nervous system.

sympatric [L *patria* homeland] Having the same range.

symplast In a plant, the system constituted by the cytoplasm of cells interconnected by plasmodesmata.

synapse [Gk *haptein* to fasten] A juncture between two neurons.

synapsis The pairing of homologous chromosomes during meiosis.

synergistic [Gk *ergon* work] Acting together with another substance or organ to achieve or enhance a given effect.

systemic circulation The part of the circulatory system supplying body parts other than the gas-exchange surfaces.

-tactic Referring to a taxis.

taxis A simple continuously oriented movement in animals (e.g., phototaxis, geotaxis) (*pl.* taxes).

taxonomy [Gk *taxis* arrangement] The classification of organisms on the basis of their evolutionary relationships.

tendon [L *tendere* to stretch] A type of connective tissue attaching muscle to bone.

territory A particular area defended by an individual against intrusion by other individuals, particularly of the same species.

testis Primary male sex organ in which sperm are produced (*pl.* testes).

thalamus [Gk *thalamos* inner chamber] Part of the rear portion of the vertebrate forebrain, a center for integration of sensory impulses.

thallus [Gk *thallos* young shoot] A plant body exhibiting relatively little tissue differentiation and lacking true roots, stems, and leaves.

thorax [Gk *thōrax* breastplate] In mammals, the part of the trunk anterior to the diaphragm, which partitions it from the abdomen. In insects, the body region between the head and the abdomen, bearing the walking legs and wings.

thymus [Gk *thymos* warty excrescence] Glandular organ that plays an important role in the development of immunologic capabilities in vertebrates.

thyroid [Gk *thyreoeidēs* shield-shaped] An endocrine gland of vertebrates located in the neck region.

thyroxin A hormone, produced by the thyroid, that stimulates a speedup of metabolism.

tissue [L *texere* to weave] An aggregate of cells, usually similar in both structure and function, that are bound together by intercellular material.

T lymphocyte *See* lymphocyte.

toxin A proteinaceous substance produced by one organism that is poisonous to another.

trachea In vertebrates, the part of the respiratory system running from the pharynx into the thorax; the "windpipe." In land arthropods, an air duct running from an opening in the body wall to the tissues.

tracheid An elongate thick-walled tapering conductile cell of the xylem.

trans- Across; beyond.

transcription In genetics, the synthesis of RNA from a DNA template.

transduction [L *ducere* to lead] In genetics, the transfer of genetic material from one host cell to another by a virus. In neurobiology, the translation of a stimulus like light or sound into an electrical change in a receptor cell.

transformation The incorporation by bacteria of fragments of DNA released into the medium from dead cells.

translation In genetics, the synthesis of a polypeptide from an mRNA template.

translocation In botany, the movement of organic materials from one place to another within the plant body, primarily through the phloem. In genetics, the exchange of parts between nonhomologous chromosomes.

transpiration Release of water vapor from the aerial parts of a plant, primarily through the stomata.

transposition The movement of DNA from one position in the genome to another. *Transposon:* a mobile segment of DNA, usually encoding the enzymes necessary to effect its own movement.

-trophic [Gk *trophē* food] Nourishing; stimulatory.

tropic hormone A hormone produced by one endocrine gland that stimulates another endocrine gland.

tropism [Gk *tropos* turn] A turning response to a stimulus, primarily by differential growth patterns in plants.

turgid [L *turgidus* swollen] Swollen with fluid.

turgor pressure [L *turgēre* to be swollen] The pressure exerted by the contents of a cell against the cell membrane or cell wall.

tympanic membrane [Gk *tympanon* drum] A membrane of the ear that picks up vibrations from the air and transmits them to other parts of the ear; the eardrum.

urea The nitrogenous waste product of mammals and some other vertebrates, formed in the liver by combination of ammonia and carbon dioxide.

ureter The duct carrying urine from the kidney to the bladder in higher vertebrates.

urethra The duct leading from the bladder to the exterior in higher vertebrates.

uric acid An insoluble nitrogenous waste product of most land arthropods, reptiles, and birds.

uterus In mammals, the chamber of the female reproductive tract in which the embryo undergoes much of its development; the womb.

vaccine [L *vacca* cow] Drug containing an antigen, administered to induce active immunity in the patient.

vacuole [L *vacuus* empty] A membrane-bounded vesicle or chamber in a cell.

valence A measure of the bonding capacity of an atom, which is determined by the number of electrons in the outer shell.

vascular tissue [L *vasculum* small vessel] Tissue concerned with internal transport, such as xylem and phloem in plants and blood and lymph in animals.

vaso- [L *vas* vessel] Blood vessel.

vector [L *vectus* carried] Transmitter of pathogens.

vegetative Of plant cells and organs: not specialized for reproduction. Of reproduction: asexual. Of bodily functions: involuntary.

vein [L *vena* blood vessel] A blood vessel that transports blood toward the heart.

vena cava [L hollow vein] One of the two large veins that return blood to the heart from the systemic circulation of vertebrates.

ventral [L *venter* belly] Pertaining to the belly or underparts.

vessel element A highly specialized cell of the xylem, with thick secondary walls and extensively perforated end walls.

villus [L shaggy hair] A highly vascularized fingerlike process from the intestinal lining or from the surface of some other structure (e.g., a chorionic villus of the placenta) (*pl.* villi).

virus [L slime, poison] A submicroscopic noncellular, obligatorily parasitic entity, composed of a protein shell and a nucleic acid core, that exhibits some properties normally associated with living organisms, including the ability to mutate and to evolve.

viscera [L] The internal organs, especially those of the great central body cavity.

vitamin [L *vita* life] An organic compound, necessary in small quantities, that a given organism cannot synthesize for itself and must obtain prefabricated in the diet.

woody dicot *See* dicot.

X chromosome The female sex chromosome.

xylem [Gk *xylon* wood] A vascular tissue that transports water and dissolved minerals upward through the plant body.

Y chromosome The male sex chromosome.

yolk stored food material in an egg.

zoo- [Gk *zōion* animal] Animal, motile.

zoospore A ciliated or flagellated plant spore.

zygote [Gk *zygōtos* yoked] A fertilized egg cell.

zymogen [Gk *zymē* leaven] An inactive precursor of an enzyme.

Index

Page numbers in **boldface** refer to illustrations; those in *italics* identify definitions or main treatment of subjects mentioned in several parts of the book.